Environmental Radioactivity and Emergency Preparedness

Series in Medical Physics and Biomedical Engineering

Series Editors: John G Webster, E Russell Ritenour, Slavik Tabakov, and Kwan-Hoong Ng

Other recent books in the series:

Environmental Radioactivity and Emergency Preparedness
Mats Isaksson and Christopher L. Rääf

Graphics Processing Unit-Based High Performance Computing in Radiation Therapy
Xun Jia and Steve B. Jiang (Eds)

Statistical Computing in Nuclear Imaging
Arkadiusz Sitek

The Physiological Measurement Handbook
John G Webster (Ed)

Radiosensitizers and Radiochemotherapy in the Treatment of Cancer
Shirley Lehnert

Diagnostic Endoscopy
Haishan Zeng (Ed)

Medical Equipment Management
Keith Willson, Keith Ison, and Slavik Tabakov

Targeted Muscle Reinnervation: A Neural Interface for Artificial Limbs
Todd A Kuiken; Aimee E Schultz Feuser; Ann K Barlow (Eds)

Quantifying Morphology and Physiology of the Human Body Using MRI
L Tugan Muftuler (Ed)

Monte Carlo Calculations in Nuclear Medicine, Second Edition: Applications in Diagnostic Imaging
Michael Ljungberg, Sven-Erik Strand, and Michael A King (Eds)

Vibrational Spectroscopy for Tissue Analysis
Ihtesham ur Rehman, Zanyar Movasaghi, and Shazza Rehman

Webb's Physics of Medical Imaging, Second Edition
M A Flower (Ed)

Correction Techniques in Emission Tomography
Mohammad Dawood, Xiaoyi Jiang, and Klaus Schäfers (Eds)

Series in Medical Physics and Biomedical Engineering

Environmental Radioactivity and Emergency Preparedness

Mats Isaksson

*Department of Radiation Physics, Institute of Clinical Sciences,
The Sahlgrenska Academy, University of Gothenburg, Sweden*

Christopher L. Rääf

*Medical Physics, Department of Translational Medicine,
Lund University, Sweden*

CRC Press
Taylor & Francis Group
Boca Raton London New York

CRC Press is an imprint of the
Taylor & Francis Group, an **informa** business

CRC Press
Taylor & Francis Group
6000 Broken Sound Parkway NW, Suite 300
Boca Raton, FL 33487-2742

First issued in paperback 2020

© 2017 by Taylor & Francis Group, LLC
CRC Press is an imprint of Taylor & Francis Group, an Informa business

No claim to original U.S. Government works

ISBN 13: 978-0-367-57402-4 (pbk)
ISBN 13: 978-1-4822-4464-9 (hbk)

Visit the Taylor & Francis Web site at
http://www.taylorandfrancis.com

and the CRC Press Web site at
http://www.crcpress.com

To our unbelievably patient families.

Contents

CHAPTER 4 ▪ Radiometry

List of Figures

List of Tables

Foreword

Nuclear and other radiological accidents are now an inevitable fact of life. Nuclear power plants provide a visible reminder of this fact, and judging from past experience a relatively serious nuclear accident happens about once every fifteen years. Some of us may feel that the advantages of a well-operated and properly regulated and supervised nuclear energy production outweigh the risks; others may wish to reject nuclear power entirely in order to minimise the risk of nuclear accidents.

While exaggerations, omissions of scientifically evidenced facts, and sometimes extreme claims can be heard from both camps, in principle both of these standpoints are intellectually defensible. So as to avoid nuclear accidents, some countries have never opted to use nuclear power, and some countries are winding down their existing nuclear power programmes. However, globally the use of nuclear energy is increasing and is likely to continue to increase at least for several decades. Any major nuclear accident will have world-wide consequences – not necessarily including any measurable radiogenic health effects, but at any rate certainly attracting much public attention. Incidentally, the actual process of abolishing nuclear power production has its own risks, due to competence depletion and loss of key staff from an industry with no perceived future and possibly including the risk of increasing proliferation of fissile material.

And while major nuclear accidents are rare, local radiological accidents are reported almost weekly to the International Atomic Energy Agency and other organisations keeping track of incidents – and probably many cases remain unrecorded. Furthermore, there is the threat of malevolent uses of ionising radiation.

Thus all countries, whether 'nuclear' or not, need to be properly prepared for nuclear and radiological emergencies, and to maintain a reasonable level of expertise on ionising radiation, its detection and measurement in the environment, its biological and medical effects, the remediation of contaminated land and objects, and the handling and treatment of persons accidentally exposed to such radiation.

Such expertise is not always present; not even in some countries with a long nuclear and/or radiological tradition. And it does not help that some self-proclaimed 'experts' travel around touting their own agendas, purportedly to share their wisdom and 'assist' but more likely causing increased confusion among the public and even the decision-makers.

This is where Isaksson's and Rääf's book can help. Obviously in line with the title, it discusses radioactive material in the environment, and emergency preparedness planning. But it is so much more! It describes carefully all sorts of sources of radiation, from nuclear weapons to the naturally occurring radionuclides in our own bodies. It provides a fairly detailed overview of radiation biology and radiation dosimetry, including the sometimes quite esoteric computations involved in assessing radiation doses. Moving on to the transfer of radiation and radionuclides from these sources, the book covers exposure pathways and environmental modelling, and of course it covers radiometry and the sampling required. There is an overview of radiation safety, ranging from risk communication and radiological protection to the various aspects of the nuclear fuel cycle. A final chapter on emergency preparedness draws on the authors' own experience of national emergency arrangements as well as the vast body of information and advice provided by international organisations.

All of this is provided with a surprising amount of precise detail. No relevant equation is forgotten, no technical characteristic is omitted – and yet, in view of the scope of the book this is quite a concise presentation. In spite of this it makes relatively easy reading. A layman permitting themselves to skim through some of the mathematics will be able to achieve quite a good understanding of the general topic and the major complications; a professional will appreciate the handy overview of all of the pertinent topics and the carefully selected references leading advanced experts on to the most erudite sources for homing in om particular topics. Thus I believe and hope that you will find this book helpful – it certainly helps me to come to grips with concepts that I have found difficult.

Dr Jack Valentin
Fellow of the Institute of Physics, Honorary Fellow of the Society for Radiological Protection

Preface

Like a bird on a wire, we performed some sort of balancing act when deciding what to include in this book. Ten years ago, we launched our first course in emergency preparedness for medical physicists in Sweden. This course, called *Emergency Preparedness and Radiation Protection in Radiological and Nuclear Emergencies*, was an introduction to the Swedish emergency preparedness organization and also included a basic exercise in radiation measurements with the equipment used in the emergency preparedness. The course aims to give the radiation protection expert a deeper understanding of radiation safety in emergency situations, and an overview of the roles of the first responders, hospitals, universities and other authorities during a radiological or nuclear emergency situation. The medical physicist is a very useful resource in these situations, provided that she or he is well prepared by offering a suitable education and that the role of the medical physicist is recognized by the national authorities.

This course was followed by three other courses: *Radiation Protection and Environmental Impact of the Nuclear Fuel Cycle, Detectors and Measurement Methods in Radiation Protection and Emergency Preparedness* and *Radiation Protection in Medical Emergency Preparedness*. After completing this course package, the medical physicist will have a good understanding of possible source terms related to nuclear fuel production, as well as the instrumentation and methods used in a radiological emergency, including mobile measurements. The fourth course will prepare the medical physicist for assignments related to the treatment and care of contaminated patients at the emergency ward, for example, performing measurements and providing radiological advice, both to staff and affected patients. During the preparation of these courses for continuous profesional development (CPD) of medical physicists, we also prepared university courses at the advanced level, for inclusion in a master program. These courses were built from the CPD courses and extended to meet the demands for courses at the advanced level at the university.

We have now been giving these courses for four years and even during the planning stage we realized the need for a textbook that could cover the topics in the courses. Certainly, there are many good textbooks available, but we had the impression that the more advanced topics are discussed in technical reports that would require substantial knowledge, while texts dealing with the more basic topics did not go into enough detail for our needs. In preparing the text, we have assumed that the reader has some basic knowledge in physics and mathematics, but not necessarily in radiation physics and radiation protection. We have included some sections about interaction mechanisms and dosimetry that may help some readers to follow the discussions. It is our intention that this book will give the reader the background needed to understand the rationale of emergency preparedness in nuclear and radiological emergencies, as well as an ability to contribute to emergency planning and other tasks.

Each chapter ends with exercises, which are intended as a help for the reader to synthesize the various topics in the text. While some of them have definite answers, some more open problems aimed as discussion topics. For some of the exercises, the solution may depend on local conditions prevailing in the reader's country or region.

This book would not have been possible to write without substantial help from some of our colleagues. However, we take full responsibility for any mistakes regarding the choice of content and other kinds of errors. We especially want to acknowledge Robert Finck, Elis Holm, Sören Mattsson and Christer Samuelsson for valuable suggestions about the contents and for scrutinizing the manuscript for relevance; Carl-Erik Magnusson for enlightening discussions about atmospheric physics and the Coriolis effect in particular; and Juan Mantero Cabrera and Jenny Nilsson for radiochemical discussions and atmospheric dispersion calculations, respectively. Other colleagues consulted are Xiaolin Hou, Johan Kastlander, Valery Ramzaev, Per Roos and Andrzej Wojcik. Last, but not least, we thank Mattias Jönsson for invaluable help with, for example, preparation of many figures. There are many more colleagues who have inspired us and contributed to the text by an excellent phrasing of a topic or by making us aware of a problem poorly dealt with in other texts. From both of us to all of you: many thanks. We also wish to thank Helen Sheppard and Struan Gray for thoroughly correcting our sometimes inadequate English and for valuable suggestions that significantly enhanced the readability of the text. We also wish to thank the Swedish Radiation Safety Authority for financial support of this book project and by contributing with funding for development of education in radiation safety. During the whole process, we have also felt great support from our editors Francesca McGowan and Emily Wells, who have led these two, sometimes quite unstructured, university teachers towards the distant goal of realizing this book.

Gothenburg and Malmö, 30 March 2016
Mats Isaksson & Christopher L. Rääf

Authors

Mats Isaksson
Department of Radiation Physics, Institute
of Clinical Sciences, Sahlgrenska
Academy, University of Gothenburg
Gothenburg, Sweden

Christopher L Rääf
Medical Radiation Physics, Department of
Translational Medicine, Lund University
Malmö, Sweden

Author Bios

Mats Isaksson (b. 1961) is professor in radiation physics at the University of Gothenburg in western Sweden. Following his undergraduate education in physics at Lund University, he earned a Master's degree in Radiation Physics from the University of Gothenburg and a PhD in Nuclear Physics from Lund University 1997. Later, he also became a licensed Medical Physicist.

For more than 30 years, he has been working with research concerning environmental radiology, and since more than 20 years with internal dosimetry, whole-body counting and modelling. He is engaged in radiological and nuclear emergency preparedness in Sweden, both on a national and on a regional level and has developed several courses in emergency preparedness. He has also been cooperating with the IAEA in the development of educational material and courses. For many years, he has been program manager for the medical physicist education at the University of Gothenburg and was in 2015 awarded the title "excellent teacher".

He is former chairman of the Swedish Society of Radioecology and secretary of the National Committee of Radiation Protection Research at the Swedish Royal Academy of Sciences.

Christopher L. Rääf (b. 1968) is an associate professor in radiation physics at Lund University in South of Sweden. Following his undergraduate education in physics at Uppsala University, followed by two year studies in medical radiation physics at Lund University, he earned a Master's degree in Medical Radiation Physics in 1996. He continued with PhD studies focussing on the metabolism and ecological behaviour of radioactive caesium and obtained a PhD diploma in 2000. He then for a period practiced as a clinical radiation physics at the Department of Radiation Physics in Malmö, specialising in teaching radiation protection for radiology staff at various levels.

For more than 20 years, he has been working with research concerning environmental radiology and emergency preparedness, including topics such as the biokinetics of radioactive elements in humans, modelling of radioecological transfer, and rapid assessment of radiation exposures in radiological and nuclear emergencies. He has been engaged in radiological and nuclear emergency preparedness in Sweden, both on a national and on a regional level, including developing both practical exercises and theoretical courses in radiation protection and emergency preparedness. He has, together with professor Mats Isaksson from University of Gothenburg, been program manager for a number of courses in applied radiation protection, which have been run regularly since 2012. In the period 2010 to 2011 he was deputy research director at the Department of Medical Radiation Physics, Malmö at Lund University. He has supervised a number of PhD students and has since 2010 been co-ordinating the research in environmental radioactivity at Lund University.

Sources of Radiation

CONTENTS

IN THIS CHAPTER, we give an overview of the sources of radionuclides found in our environment. Some, such as ^{238}U and ^{40}K occur naturally, while others, such as ^{137}Cs and ^{60}Co are man-made, and result from industrial or military activities.

The first artificial nuclear reaction was performed by Ernest Rutherford in 1917, when he bombarded stable ^{14}N with α-particles, each reaction creating an ^{17}O nucleus and a proton. However, no radionuclides were produced because ^{17}O is stable. With the discovery of the neutron by James Chadwick in 1932, a number of radionuclides of non-natural origin began to be created. Using neutrons, Enrico Fermi and others attempted to produce elements heavier than uranium using nuclear reactions, but in many of the experiments the radionuclides produced were lighter than uranium, for example, barium. This was explained by Lise Meitner and Otto Frisch in 1938, who showed that bombarding uranium with neutrons caused the uranium nucleus to split into two parts. This process, called *nuclear fission*, was found to release large amounts of energy, and attempts were made to construct a device for controlled nuclear fission on a larger scale. These attempts proved successful in 1942, when Fermi was able to produce a self-sustaining chain reaction, heralding the start of the nuclear era.

1.1 NATURALLY OCCURRING RADIATION

The naturally occurring radionuclides can be divided into two groups, depending on their origin. *Cosmogenic* radionuclides are produced in nuclear reactions between cosmic radiation and either the constituents of the atmosphere or the surface of the earth, while *primordial* radionuclides have existed since the earth was formed, approximately 4.6 billion ($4.6 \cdot 10^9$) years ago.

1.1.1 Cosmic Radiation

There are many kinds of radiation in outer space, including the microwave background radiation originating from the Big Bang, γ-rays and x-rays. However, there are three main sources of radiation which affect living beings on earth: *galactic cosmic radiation, solar cosmic radiation* and *radiation from the Van Allen belts* which surround the earth (UNSCEAR 2008). Radiation from these sources is usually referred to as *primary cosmic radiation*, while their interaction with the atmosphere gives rise to *secondary cosmic radiation*.

The origin of galactic cosmic radiation is still not fully understood, but it has been suggested that supernovas could impart sufficient energy to the particles making up the galactic radiation field. These include protons (85.5%), α-particles ($\sim$12%), and electrons (2%) (UNSCEAR 2008). The nuclei of various elements, even those as heavy as uranium, are also found in the galactic cosmic radiation ($\sim$1%). The energy of these cosmic particles ranges from 100 MeV to over 10^{17} MeV! Most of this radiation emanates from within our own galaxy, the Milky Way. The energy spectrum of the galactic cosmic radiation is described by two different power functions, depending on the energy of the particles. For energies below 10^{12} MeV, the particles have an energy distribution that is described by

$$\Phi_E \propto E^{-2.7}, \tag{1.1}$$

where Φ_E is the spectral energy fluence (number of particles per unit area per unit energy) and E is given in eV. For higher energies the relation is instead given by

$$\Phi_E \propto E^{-3}. \tag{1.2}$$

The cosmic radiation arising from the sun (solar cosmic radiation) consists mainly of protons ($\sim$ 99%) with energies below 100 MeV, which are emitted by solar flares. The fluence varies with solar activity, which follows roughly an 11-year cycle. The intensity of the galactic cosmic radiation is affected by the highly ionized plasma of the solar wind, and thus it also varies with solar activity, being at a maximum when solar activity is at its minimum. To a large extent, the earth is shielded from cosmic radiation by its magnetic field, and also by the atmosphere. The dose rate resulting from galactic cosmic radiation can thus be quite high at high altitudes, but only the particles with the highest energies contribute to the dose rate at the surface of the earth.

The Van Allen belts consist of charged particles, mainly protons but also electrons, captured by the earth's magnetic field. The energies of the protons can reach several hundred MeV, while the electrons have much lower energies, on the order of a few MeV. There is an internal Van Allen belt whose mid-point is about 3000 km from the earth's surface, while the external belt is at an altitude of about 22 000 km.

When high-energy particles in the primary cosmic radiation enter the atmosphere, they interact with atoms and molecules to produce several kinds of charged and uncharged particles. This *secondary cosmic radiation* consists of particles such as protons, neutrons and pions. However, at lower altitudes, neutrons dominate the particle fluence due to their longer range.

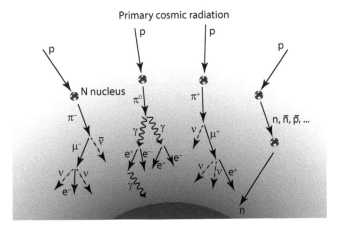

FIGURE 1.1 Protons (p) from the primary cosmic radiation enter the atmosphere and collide with nuclei, such as nitrogen (N), producing secondary particles including pions (π), neutrons (n), antineutrons ($\bar{n}$) and antiprotons ($\bar{p}$). These secondary particles may then decay into muons (μ), neutrinos (ν) and electrons (e), and cause nuclear reactions in other atoms in the atmosphere. The interaction between electrons from the secondary cosmic radiation and molecules in the atmosphere causes the Northern and Southern Lights (*Aurora Borealis* and *Aurora Australis*).

The unstable particles, for example, positively charged pions, decay into muons, electrons, neutrinos and photons. The muons, which have a low probability of interacting with the atoms in the atmosphere, reach the earth with energies between 1 and 20 GeV, before they decay. Muons are thus the largest source of the absorbed dose at the surface of the earth (UNSCEAR 2008). Uncharged pions produce a cascade of photons and electrons since they decay into photons with high energies. These photons form electrons and positrons through pair production, which then produce new photons by annihilation, and the cycle repeats. Figure 1.1 depicts a simplified series of events resulting from the interaction of high-energy protons with nitrogen nuclei in the upper atmosphere. In this way, a proton with a very high energy can produce a so-called *cosmic ray shower* consisting of photons, muons and electrons, which covers an area on the earth's surface of several square kilometres.

Another effect of the earth's magnetic field is to modify the intensity of the primary cosmic radiation in the upper atmosphere. The shape of the earth's magnetic field may be envisaged in two dimensions as two circular lobes extending from the South Geomagnetic Pole (which is a magnetic north pole, situated close to the Geographic South Pole) to the North Geomagnetic Pole (a magnetic south pole, in the Canadian Arctic). This causes the *geomagnetic latitude effect*, which reduces the intensity of cosmic radiation, and thus the dose rate, at lower latitudes. The geomagnetic latitude effect will lead to variations in dose depending on the latitude, but this is quite small and the dose rate from directly ionizing (charged) particles and photons at the equator is only about 10% lower than at the poles. The earth's magnetic field has a more pronounced effect on the neutron fluence at sea level, since neutrons are produced by protons in the primary cosmic radiation. It is estimated that the fluence rate at sea level at the equator is only 20% of the fluence rates at the poles (UNSCEAR 2008).

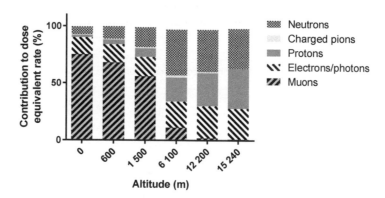

FIGURE 1.2 Changes in the proportions of different particles in the cosmic radiation with altitude. Data from O'Brien (1996).

By taking into account how populations vary with latitude, UNSCEAR has adopted an average dose rate in air at sea level of 31 nGy h^{-1} (UNSCEAR 2008). The muon fluence at sea level is about $6 \cdot 10^4$ m^{-2} s^{-1} and muons account for about 80% of the absorbed dose to air from directly ionizing particles in the secondary cosmic radiation. The average annual effective dose to the world's population, *outdoors at sea level*, is 0.27 mSv. Electrons account for the remaining 20%, but their relative contribution increases with altitude, as can be seen in Figure 1.2. One-third of the world's population lives at altitudes higher than sea level, and this increases the average dose from charged particles and photons by a factor of 1.25. However, since people spend a large part of their time indoors, the effective dose from the ionizing component of cosmic radiation is estimated to be 0.28 mSv (UNSCEAR 2008).

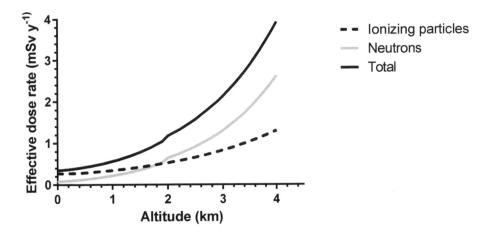

FIGURE 1.3 Effective dose rate from ionizing (charged) particles and neutrons as a function of altitude. The neutron contribution is described by two equations, one for altitudes from zero to 2 km and the other for altitudes above 2 km, hence the slight discontinuities in the curves. Calculated from equations given by Bouville and Lowder (1988) and UNSCEAR (2008).

The neutron contribution increases from 8% at sea level to about 35% at an altitude of 4 km (UNSCEAR 2008). This must be taken into account for airline personnel who spend most of their working hours at altitudes of 10–12 km, where the dose rate can be 4–8 μSv h^{-1}. At even higher altitudes, for instance during space missions, protons and atomic nuclei must also be taken into account. As a rule of thumb, the dose rate from cosmic radiation doubles for every km increase in altitude, as illustrated in Figure 1.3.

Table 1.1 presents the annual effective dose rates for a number of high-altitude locations, some of which are very large metropolitan areas, together with the effective dose rate at sea level, and the mean effective dose, which takes into account shielding in the indoor environment. There are, however, slight discrepancies between the dose estimates in the table and those presented above. For example, the UNSCEAR report from 2008 assumes a value of 0.28 mSv y^{-1} as the mean for the world's population, excluding the contribution from neutrons. The value given in Table 1.1, 0.38 mSv y^{-1}, consists of 0.30 mSv y^{-1} from the directly ionizing component, and 0.08 mSv y^{-1} from neutrons. However, in the UNSCEAR report from 2000, the contributions from the two types of radiation were 0.28 mSv y^{-1} and 0.10 mSv y^{-1}, respectively (UNSCEAR 2000).

In order to calculate the effective dose from neutrons, the shape of the neutron spectrum must also be taken into account, since the cross sections for neutron interactions are strongly dependent on the neutron energy. The average population-weighted annual effective dose from neutrons, outdoors, at sea level, has been estimated to be 0.048 mSv (UNSCEAR 2008). Since most people spend part of their time indoors, the mean dose will be lower, and this can be accounted for using an *occupancy factor*, defined as the fraction of time spent indoors. UNSCEAR uses an occupancy factor of 0.8 and a *shielding factor* of 0.8 (the

TABLE 1.1   Effective dose rate, E (mSv y^{-1}), from directly and indirectly ionizing particles at some high-altitude locations and at sea level, together with the population-weighted mean for the world.

Location	Altitude (m)	E (mSv y^{-1})
La Paz (Bolivia)	3900	2.02
Lhasa (Tibet/China)	3600	1.71
Quito (Ecuador)	2840	1.13
Mexico City (Mexico)	2240	0.82
Nairobi (Kenya)	1660	0.58
Denver (USA)	1610	0.57
Tehran (Iran)	1180	0.44
Sea level	0	0.27
Mean		0.38

Source: Data from UNSCEAR (1993).

ratio of the dose indoors to that outdoors). When including the altitude dependence of the neutron fluence and the population distribution, the mean is assumed to be 0.1 mSv y^{-1}.

In summary, the average effective dose rate to the world's population from cosmic radiation varies between 0.3 and 2 mSv y^{-1}, depending on latitude and altitude. The population-weighted average effective dose rate to the world's population is estimated to be 0.38 mSv y^{-1} (UNSCEAR 2008).

1.1.2   Cosmogenic Radionuclides

Cosmogenic radionuclides are produced mainly by interactions between cosmic radiation and atomic nuclei in the atmosphere. The radionuclides produced in this way are subsequently transported to the earth's surface by precipitation and gravitational settling. The most abundant elements in the atmosphere are nitrogen, oxygen and argon, and so these are the most probable targets for interactions. Although cosmogenic radionuclides can be produced on earth, the attenuation of cosmic radiation in soil and water generally permits only very small amounts to be created. However, due to the high abundance of ^{35}Cl in the earth's crust, the cosmogenic radionuclide ^{36}Cl, which is formed by neutron capture of ^{35}Cl, is an exception. A list of cosmogenic radionuclides is given in Table 1.2.

Because cosmogenic radionuclides all belong to the light elements, they decay through isobaric decay (beta decay and electron capture). Their half-lives vary between minutes and millions of years, which affects the inventory of cosmogenic radionuclides on earth. Table 1.3 gives the global inventory and yearly production rate of some of the most abundant cosmogenic radionuclides (UNSCEAR 2000).

Cosmogenic radionuclides decay to stable isotopes and do not give rise to any of the decay chains we will encounter later. It should also be noted that they are isotopes of rather light elements, with atomic numbers (Z) ranging from 1 (H) to 36 (Kr). Since the atmosphere is composed mainly of light elements, heavier elements cannot be produced by the interaction of cosmic radiation and atomic nuclei in the atmosphere.

Some of the radionuclides listed in Table 1.2 are of concern in terms of their radiation dose to humans as they are involved in human metabolism (UNSCEAR 2008). These radionuclides will be briefly discussed below.

TABLE 1.2 Cosmogenic radionuclides, half-lives and type of decay. The half-lives are given in years (y), days (d), hours (h) or minutes (m). Electron capture is denoted ε, and negative and positive beta decay β^- and β^+, respectively.

Nuclide	Half-life		Decay	Nuclide	Half-life		Decay
^{3}H	12.3	d	β^-	^{32}P	14.3	d	β^-
^{7}Be	53.3	d	ε	^{33}P	25.3	d	β^-
^{10}Be	1.51	My	β^-	^{35}S	87.5	d	β^-
^{11}C	20.4	m	β^+, ε	^{38}S	2.84	h	β^-
^{14}C	5730	y	β^-	^{34m}Cl	32.0	m	β^+
^{18}F	109.8	m	β^+	^{36}Cl	0.3	Ma	β^-, ε
^{22}Na	2.60	y	β^+	^{38}Cl	37.2	m	β^-
^{24}Na	15.0	h	β^-	^{39}Cl	55.6	m	β^-
^{28}Mg	20.9	h	β^-	^{37}Ar	35.0	d	ε
^{26}Al	0.74	My	β^+	^{39}Ar	269	y	β^-
^{31}Si	157.3	m	β^-	^{53}Mn	3.74	My	ε
^{32}Si	172	y	β^-	^{81}Kr	0.23	My	ε

1.1.2.1 Tritium

Tritium (^{3}H) is produced when cosmic radiation, in the form of fast neutrons, interacts with nitrogen and oxygen in the atmosphere according to the reaction:

$$^{14}N + n \rightarrow ^{12}C + ^3H$$

The inventory of naturally produced tritium has been estimated to be 1275 PBq (1 PBq $= 10^{12}$ Bq), with a production rate of 72 PBq y^{-1} (UNSCEAR 2000). However, tritium is also produced in nuclear weapons testing and in nuclear reactors, a factor we will return to later. The effective dose from tritium to the world's population is mainly due to the consumption of water, and the average yearly effective dose has been estimated to be 0.01μSv (UNSCEAR 2008). Since tritium occurs almost only in water or water vapour, it will be distributed according to the global circulation of water, and it has been estimated that the concentration of tritium in the oceans is about 100 Bq m^{-3} and in freshwater, 400 Bq m^{-3} (UNSCEAR 2008). The global circulation of tritium has been modelled with a seven-compartment model, as has been described by UNSCEAR (UNSCEAR 2000).

1.1.2.2 Beryllium-7

The production of ^{7}Be in the atmosphere is mainly due to *spallation* reactions between galactic cosmic rays (neutrons and protons) and nitrogen and oxygen nuclei. Spallation is a process in which a number of nucleons are emitted from the atomic nucleus after a collision with a high-energy particle. The ^{7}Be atoms then become attached to aerosols and are transported through the atmosphere. Since other pollutants are also attached to aerosols, ^{7}Be can be used as a tracer of atmospheric pollutants. The activity concentration of ^{7}Be in outdoor air is about 3 mBq m^{-3}, which will give an average yearly effective dose of 0.01 μSv worldwide (UNSCEAR 2008). The main transport route is via precipitation and, because of its rather short half-life (53.3 d), the main intake route is via fresh vegetables. Since this radionuclide is easily washed out of the atmosphere by rain, the concentration at ground level varies considerably depending on the weather.

This radionuclide has also been used to study atmospheric phenomena. For example, Doering and Saey (2014) measured the activity concentration in surface air at several locations

TABLE 1.3 Annual production rate and global inventory of some common cosmogenic radionuclides.

Radionuclide	Production rate (PBq y^{-1})	Global inventory (PBq)
^{3}H	72	1275
^{7}Be	1960	413
^{10}Be	0.000 064	230
^{14}C	1.54	12 750
^{22}Na	0.12	0.44
^{26}Al	0.000 001	0.71
^{32}Si	0.000 87	0.82
^{32}P	73	4.1
^{33}P	35	3.5
^{35}S	21	7.1
^{36}Cl	0.000 013	5.6
^{37}Ar	31	4.2
^{39}Ar	0.074	28.6
^{81}Kr	0.000 000 017	0.005

Source: Data from UNSCEAR (2000).

in Australia. The observed variations could be explained by a meteorological phenomenon, the migration of the so-called Hadley cell, as is discussed in Section 3.2.1.

1.1.2.3 Carbon-14

This radioactive carbon isotope is produced in the atmosphere by slow neutrons through the reaction ^{14}N+n→^{14}C+p. Estimates of the annual production have changed over the years, from 1 PBq y^{-1} (UNSCEAR 1982) to 1.54 PBq y^{-1} (UNSCEAR 2000); see Table 1.3. However, the 2008 UNSCEAR report gives the value 1.4 PBq y^{-1} (UNSCEAR 2008). The resulting global inventory (PBq) can be calculated by multiplying the annual production rate (PBq y^{-1}) by 1.44 times the half-life in years (as is explained below). Assuming an annual production rate of 1.54 PBq y^{-1} and a half-life of 5730 y, yields a global inventory of 12 706 PBq. Taking rounding errors into account, this is in accordance with the value of 12 750 PBq given in Table 1.3. The atmospheric inventory of ^{14}C has been estimated to be 140 PBq (UNSCEAR 2008).

The relation used above follows from the activation equation, stating that

$$N = \frac{P}{\lambda}(1 - e^{-\lambda t}),\tag{1.3}$$

where N is the number of radionuclides produced as a result of production and simultaneous decay, P is the production rate, the number of radionuclides produced per unit time, λ is the decay constant, and t is time. At equilibrium, when t approaches infinity, the equation reduces to

$$N = \frac{P}{\lambda}.\tag{1.4}$$

Since the decay constant is given by the natural logarithm of 2, divided by the half-life, $T_{1/2}$, $1/\lambda$ is approximately equal to $T_{1/2}/0.693 = 1.44 \cdot T_{1/2}$. The activity, A, is directly proportional to the number of radionuclides, so Eqn. 1.4 can be rewritten as

$$A = \frac{R}{\lambda} = R \cdot 1.44 \cdot T_{1/2}, \qquad (1.5)$$

where R is the annual production rate (PBq y^{-1}) and the half-life is given in the unit y.

The specific activity (based on the total mass of carbon in living matter) has been estimated to be about 230 Bq kg^{-1}, of which naturally produced ^{14}C (see below) accounts for 222 Bq kg^{-1}. The amount of carbon in the human body is about 18.5% by mass, which gives 12 kg C for a body mass of 65 kg. Using the value for the specific activity given above (230 Bq kg^{-1}), the activity of ^{14}C is 2770 Bq. On average, the human body contains about 2700 Bq, which will lead to an effective dose of 12 μSv y^{-1} (UNSCEAR 2008). A compartment model for the global circulation of ^{14}C is described in the UNSCEAR report from 2000 (UNSCEAR 2000).

The ^{14}C produced by cosmic radiation will become incorporated into atmospheric CO_2 and all living organisms that assimilate carbon therefore contain ^{14}C at the same concentration as in the ambient atmosphere. When the organism dies, assimilation of CO_2 stops and the concentration of ^{14}C will decrease due to radioactive decay. A method called carbon dating is used in archaeological studies to determine the age of organic samples by comparing the concentration of ^{14}C in the sample with the present concentration in the atmosphere.

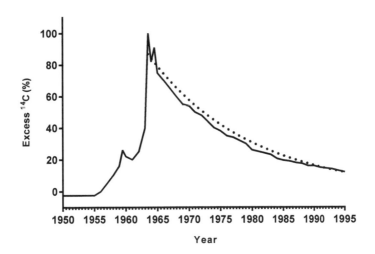

FIGURE 1.4   The decay-corrected excess of ^{14}C (in relation to the international standard activity) in atmospheric CO_2 increased due to nuclear weapons testing, giving rise to a peak in 1963, after which atmospheric testing of nuclear weapons was banned. The figure shown here is a schematic view of the bomb peak and data from actual measurements can be found in e.g. Levin and Hessheimer (2000). The dotted line depicts a decrease corresponding to a "half-life" of 11 years.

Like ^{3}H, ^{14}C is also produced in nuclear weapons testing and in nuclear reactors, and it has been estimated that atmospheric nuclear weapons tests during the 1950s and 1960s introduced 350 PBq of ^{14}C into the atmosphere (UNSCEAR 2008). This doubled the previous atmospheric concentration of ^{14}C and gave rise to a peak in atmospheric ^{14}C in 1963 (Buchholz 2007). This so-called *bomb peak* is schematically shown in Figure 1.4 as the decay-corrected excess of ^{14}C, above the international standard activity, in atmospheric CO_2. The retention time in the atmosphere is, however, rather short due to mixing with the carbon present in the marine and terrestrial environments, and the peak has decreased at a rate corresponding to a "half-life" of approximately 11 years. Since the variation in atmospheric ^{14}C was rapid during the period after the bomb peak, it can be used to determine, for example, when an organism or a cell was formed. This method can thus be used for dating objects much younger than archaeological samples. However, due to the long half-life of ^{14}C, the number of decaying nuclei per unit time will be low and hence the activity in small samples is generally very low. Therefore, methods such as *accelerator-based mass spectrometry* must be used, which allow the carbon atoms to be counted almost individually.

1.1.2.4 Sodium-22

^{22}Na is produced in the atmosphere by spallation reactions of galactic cosmic rays with argon nuclei. The annual effective dose from ^{22}Na has been reported to be 0.15 μSv UNSCEAR (2008).

1.1.3 Primordial Radionuclides

This group of naturally occurring radionuclides takes its name from the Latin *primordialis*, meaning original. Some of them were created before the earth was formed and were a part of the material that condensed to form our planet. Consequently, they have half-lives comparable to the age of the earth, or longer. Others have been produced on earth by the decay of long-lived radionuclides of stellar origin. Primordial radionuclides can thus be divided into *singly occurring*[1] radionuclides and members of a *decay chain*.

The singly occurring primordial radionuclides decay to stable daughter products and these radionuclides generally have half-lives on the order of 10^{10} years or more, as can be seen from Table 1.4. Most of the radionuclides in this table are found on earth at minor concentrations, and the main contributor to the radiation dose to humans is ^{40}K (UNSCEAR 2008).

Some of the primordial radionuclides decay to radioactive daughters, and if their decay products are also radioactive, this will give rise to decay chains. Although the radionuclide at the top of each chain is primordial, the daughters are produced on earth, and have half-lives ranging from about 10^{10} years to tenths of microseconds. When the earth was young, the neptunium series was present. However, the half-life of ^{237}Np is 2.14·10^6 years and the half-lives of the other members of this series are so short that they are no longer found on earth in significant amounts. The only remaining radionuclide from this series is ^{209}Bi, which decays to ^{205}Tl by alpha decay, with a half-life of 2·10^{19} years (Beeman et al. 2012). There are presently three decay chains: the *uranium series*, the *thorium series* and the *actinium series*; all three of which end in a stable isotope of lead. The three series and their radionuclides are presented in Tables 1.5–1.7 and Figure 1.5.

The uranium series starts with the uranium isotope ^{238}U, which decays to ^{234}Th through alpha decay. When the chain reaches ^{218}Po, two types of decay are possible: negative beta decay leads to the formation of ^{218}At, while alpha decay leads to ^{214}Pb. The thin lines in Figure 1.5 indicate that beta decay is less probable than alpha decay. This kind of branched decay is also seen for the decay of ^{218}At, ^{214}Bi and ^{210}Bi. In the thorium series, a branched decay is seen for ^{212}Bi. Branched decays take place in the actinium series for ^{227}Ac, ^{215}Po and ^{211}Bi. The actinium series is named after ^{227}Ac, which was the first radionuclide in the decay series to be discovered.

---

[1]Some of the singly occurring primordial radionuclides actually appear in decay series, for example, ^{152}Gd→^{148}Sm→^{144}Nd→^{140}Ce. These decay series are called short decay series.

Sources of Radiation ■ 11

TABLE 1.4 Singly occurring primordial radionuclides of natural origin. Electron capture is denoted ε, alpha decay α, and negative beta decay β^-. The abundance is given as the relative mass of the element found on earth. For example, ^{40}K constitutes 0.012% of the potassium present on earth. In addition to the radionuclides listed below, others with half-lives $>10^{17}$ years have been found, including ^{50}V, ^{76}Ge, ^{82}Se, ^{96}Zr, ^{100}Mo, ^{128}Te, ^{130}Te, ^{150}Nd and ^{209}Bi.

Radionuclide	Half-life (y)	Decay	Abundance (%)
^{40}K	$1.25\cdot10^9$	β^-,ε	0.0117
^{87}Rb	$4.93\cdot10^{10}$	β^-	27.83
^{113}Cd	$9.0\cdot10^{15}$	β^-	12.22
^{115}In	$4.41\cdot10^{14}$	β^-	95.72
^{123}Te	$1.3\cdot10^{13}$	ε	0.905
^{138}La	$1.04\cdot10^{11}$	β^-,ε	0.092
^{142}Ce	$5\cdot10^{16}$	$2\beta^-$	11.11
^{144}Nd	$2.3\cdot10^{15}$	α	23.80
^{147}Sm	$1.07\cdot10^{11}$	α	15.0
^{148}Sm	$7\cdot10^{15}$	α	11.3
^{149}Sm	$4\cdot10^{14}$	α	13.82
^{152}Gd	$1.1\cdot10^{14}$	α	0.20
^{174}Hf	$2.0\cdot10^{15}$	α	0.162
^{176}Lu	$3.8\cdot10^{10}$	β^-	2.59
^{180m}Ta	$1.2\cdot10^{15}$	β^-,ε	0.012
^{186}Os	$2.0\cdot10^{15}$	α	1.58
^{187}Re	$4.3\cdot10^{10}$	β^-	62.60
^{190}Pt	$6.5\cdot10^{11}$	α	0.012
^{192}Pt	$1\cdot10^{15}$	α	0.78

Source: Data from Atwood (2010) and Bé et al. (2014).

Since the half-life of the radionuclide at the top of each chain is so long, the other radionuclides in the chain will be produced continuously. This means, for example, that if thorium is present in the bedrock, all the other radionuclides in this series will also be present. However, transport processes can change the proportions of the elements at a given site. An example of this is the rare gas radon, which easily migrates through fissures and cracks in the bedrock. Atoms of a radionuclide in a series may thus be transported far away from the place where they were formed by decay of their precursor. It follows that, at a given site, the activity of the radionuclides in a decay chain will generally not be equal. However, due to short half-lives, some radionuclides will not be transported far before they decay, and *secular equilibrium* may prevail for parts of a decay series (at secular equilibrium, the activity of the daughter equals the activity of the mother; see Section 1.1.4). These sub-series are indicated by grey shading in Figure 1.5.

One interesting feature of all these series is that the mass number of each element in a series can be expressed as $4n$ (thorium series), $4n+2$ (uranium series) or $4n+3$ (actinium series). The series represented by $4n+1$ is the extinct neptunium series, which started with ^{237}Np.

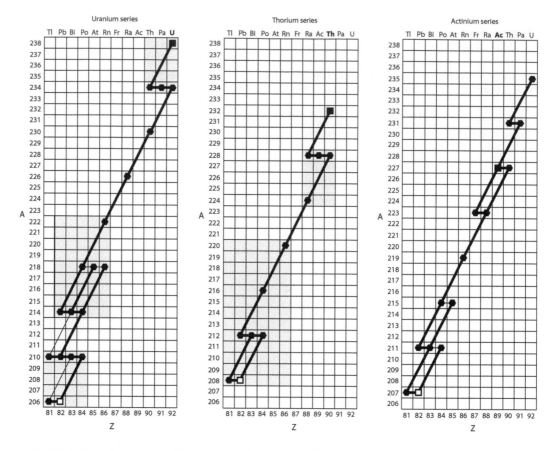

FIGURE 1.5 The three naturally occurring decay series. Sub-series, in which the radionuclides have similar properties regarding transport in the environment, are indicated by grey shading for the uranium and thorium series. The first and the last nuclide in each sub-series tend to have the same activity. Each decay chain ends with a stable isotope of lead, indicated by an open symbol.

1.1.3.1 Potassium

The relative abundance of ^{40}K is 0.012%, giving an activity concentration in naturally occurring K of 30.6 kBq kg^{-1}. According to UNSCEAR, the median value and population-averaged mean value of the concentration of ^{40}K in soil are 400 Bq kg^{-1} and 420 Bq kg^{-1}, respectively (UNSCEAR 2000). Since potassium is an essential element in the human body, ^{40}K will contribute to the background radiation dose. It is generally assumed that the potassium concentration is 2 g per kg body mass, which gives an activity of 61 Bq per kg body mass. The activity of an individual weighing 60 kg will thus be approximately 3700 Bq.

1.1.3.2 Uranium

The relative natural abundances (expressed as a percentage of the total mass of uranium) of the three uranium isotopes, ^{238}U, ^{235}U and ^{234}U are 99.27%, 0.72% and 0.0055%, respectively (Atwood 2010). Due to the short half-lives of the intermediate radionuclides in the uranium series, ^{238}U and ^{234}U reach secular equilibrium (see Fig. 1.5), and the activity

TABLE 1.5 The uranium series. (Some radionuclides may decay by other modes, although these have low probability.)

Radionuclide	Half-life	Decay
^{238}U	$4.5 \cdot 10^9$ y	α
^{234}Th	24.1 d	β^-
^{234m}Pa	1.17 m	β^-
^{234}U	$2.5 \cdot 10^5$ y	α
^{230}Th	$7.5 \cdot 10^5$ y	α
^{226}Ra	1600 y	α
^{222}Rn	3.82 d	α
^{218}Po	3.10 m	α
^{214}Pb	26.8 m	β^-
^{214}Bi	19.9 m	β^-
^{214}Po	$1.64 \cdot 10^{-4}$ s	α
^{210}Pb	22.3 y	β^-
^{210}Bi	5.01 d	β^-
^{210}Po	138.4 d	α
^{206}Pb	Stable	—

per unit mass (the *specific activity*) in natural uranium is therefore the same for the two isotopes ^{238}U and ^{234}U, i.e., about 12 MBq kg^{-1}. The activity concentration of ^{235}U in natural uranium is about 0.6 MBq kg^{-1}.

Uranium is present in the earth's crust in the form of different minerals. About a hundred minerals with a uranium abundance of 50% (by mass) or more have been identified. The most uranium-rich mineral is uraninite (UO_2), which contains 88.15% uranium (10.6 MBq kg^{-1}). The abundance of uranium varies with the type of bedrock, for example, granite has an activity concentration of 37–124 Bq kg^{-1} (3–10 ppm) and black shales 37–15 500 Bq kg^{-1} (3–1250 ppm). The average activity concentration of uranium in the earth's crust is 12–37 Bq kg^{-1} (1–3 ppm), and in water 1.2–12 mBq kg^{-1}) (0.0001–0.001 ppm). The median value and population-averaged mean value of the activity concentrations of ^{238}U in soil are 35 Bq kg^{-1} and 33 Bq kg^{-1}, respectively (UNSCEAR 2000).

1.1.3.3 Thorium

Thorium is found in large quantities in the earth's crust, for example, in minerals such as thorite ($ThSiO_4$), and thorianite ($ThO_2 + UO_2$), and in fractions of the mineral monazite, e.g. (Sm, Gd, Ce, Th)PO_4. The internal heat in the earth is to a large extent due to energy released in the decay of the most abundant isotope, ^{232}Th. The median and population-averaged mean values of the activity concentration of ^{232}Th in soil are 30 Bq kg^{-1} and 45 Bq kg^{-1}, respectively (UNSCEAR 2000).

1.1.3.4 Radium

According to UNSCEAR, the median and population-averaged mean values of the activity concentration for ^{226}Ra are 35 Bq kg^{-1} and 32 Bq kg^{-1}, respectively (UNSCEAR 2000). These values are almost equal to the values found for uranium above, since ^{226}Ra belongs to the uranium series. However, the activity concentration of ^{226}Ra can show local deviations

TABLE 1.6 The thorium series.

Radionuclide	Half-life	Decay
^{232}Th	$1.4 \cdot 10^{10}$ y	α
^{228}Ra	5.75 y	β^-
^{228}Ac	6.15 h	β^-
^{228}Th	1.91 y	α
^{224}Ra	3.66 d	α
^{220}Rn	55 s	α
^{216}Po	0.145 s	α
^{212}Pb	10.6 h	β^-
^{212}Bi	60.6 m	α, β^-
^{212}Po	$2.99 \cdot 10^{-7}$ s	α
^{208}Tl	3.1 m	β^-
^{208}Pb	Stable	–

TABLE 1.7 The actinium series.

Radionuclide	Half-life	Decay
^{235}U	$7.0 \cdot 10^8$ y	α
^{231}Th	25.52 h	β^-
^{231}Pa	$3.28 \cdot 10^4$ y	α
^{227}Ac	21.7 y	β^-
^{227}Th	18.7 d	α
^{223}Fr	22.0 m	β^-
^{223}Ra	11.4 d	α
^{219}Rn	3.96 s	α
^{215}Po	$1.78 \cdot 10^{-3}$ s	α
^{211}Pb	36.1 m	β^-
^{211}Bi	2.14 m	α
^{207}Tl	4.77 m	β^-
^{207}Pb	Stable	–

from that of ^{238}U due to weathering and other environmental factors, although the median and mean values are equal to the activity concentration of ^{238}U. The highest activity concentrations of ^{226}Ra, exceeding 2 kBq kg^{-1}, are generally found in shale (IAEA 2014a), but several areas on earth have much higher activity concentrations. Some examples of these are given in Table 1.8.

1.1.3.5 Radon

Although each of the decay chains includes at least one radon isotope, those mainly associated with health effects are ^{220}Rn (also called *thoron*) and ^{222}Rn, which occur in the thorium and uranium decay chains, respectively. The activity concentration in soil gas can reach hundreds of kBq m^{-3}, but the concentration above ground will depend on transport processes and on the half-life of the radon isotope. It has been estimated that the average rate of transport from soil to air is 0.017 Bq m^{-2} s^{-1} for ^{222}Rn and 1.5 Bq m^{-2} s^{-1} for

TABLE 1.8 Areas with high activity concentrations of ^{226}Ra in soil. The enhanced activity concentration in Kerala is due to the high abundance of thorium, while thermal waters are the source in Ramsar. Central Bohemia is located in an area rich in both uranium and thorium. The world median activity concentration is given for comparison.

Country	Site	Act. conc. (Bq kg^{-1})
India	Kerala	7.8–2500
Islamic Republic of Iran	Ramsar	80–50 000
Czech Republic	Central Bohemia	76–275
World median		17–60

Source: Data from IAEA (2014a) and UNSCEAR (2000).

^{220}Rn (NCRP 1988). This leads to outdoor activity levels ranging from 1–>100 Bq m^{-3}. Assuming a typical activity concentration of 10 Bq m^{-3}, UNSCEAR has estimated the annual effective dose from radon to be 0.095 mSv y^{-1} (UNSCEAR 2000).

Since radon is a rare gas, it will interact poorly with its surroundings and will not be incorporated into chemical compounds. ^{222}Rn has a half-life of 3.8 days, and is readily exhaled from the ground, so it may enter the living areas of dwellings (as will be discussed in more detail below). Measurements of radon exhalation have indicated that diffusion is not the only mechanism of importance in radon transport in the ground, as radon has been found in air exhaled from such depths that it should have decayed if diffusion were the only mode of transport (Malmqvist et al. 1989). Radon must therefore also be transported by gas or fluid flow through cracks in the ground. Thus, non-local sources of ^{226}Ra can have a considerable effect on the radon concentration at ground level.

The concentration of radon near the ground is affected by meteorological factors such as temperature, pressure and wind speed. For example, the exhalation rate is strongly affected by the moisture content in soil, and is higher in the summer, leading to a higher concentration at ground level. It has also been shown that decreased air pressure leads to an increase in the exhalation rate (Tanner 1980).

Several attempts have been made to predict earthquakes by studying radon exhalation from the ground. For example, Hayashi et al. (2015) found that a sinusoidal model could be fitted to radon concentration data during a normal period with no seismic activity, and that anomalies in the data could be an indication of increased seismic activity. Such anomalies were indeed found when the data prior to the 2011 Tohoku-Oki Earthquake off the east coast of Japan were reanalysed.

1.1.4   Series Decay and Equilibria

The activity of each radionuclide in a decay chain can be determined by solving the decay equations. For the first radionuclide in a series, the activity, A_1, varies with time according to

$$\frac{dA_1}{dt} = -\lambda_1 A_1, \tag{1.6}$$

where λ is the decay constant, which is inversely proportional to the half-life, $T_{1/2}$, and is given by

$$\lambda = \frac{\ln 2}{T_{1/2}}. \tag{1.7}$$

The solution to Eqn. 1.6 is

$$A_1 = A_1(0) \cdot e^{-\lambda_1 t}, \tag{1.8}$$

where $A_1(0)$ is the activity of the first radionuclide at time $t = 0$. For the second and third radionuclides in the series, the activity is given by

$$\frac{dA_2}{dt} = \lambda_1 A_1 - \lambda_2 A_2 \tag{1.9}$$

and

$$\frac{dA_3}{dt} = \lambda_2 A_2 - \lambda_3 A_3. \tag{1.10}$$

Inserting the expression for $A_1(t)$ (Eqn. (1.8)) into Eqn. (1.9) gives

$$\frac{dA_2}{dt} = \lambda_1 A_1(0) \cdot e^{-\lambda_1 t} - \lambda_2 A_2. \tag{1.11}$$

The activity of each radionuclide can thus be calculated if the activity of the previous radionuclide in the series is known. The solution to these differential equations for an arbitrary radionuclide in a decay series is given by the *Bateman equations*

$$N_n(t) = \prod_{j=1}^{n-1} \lambda_j \sum_{i=1}^{n} \sum_{j=1}^{n} \left(\frac{N_i(0) \cdot e^{-\lambda_j t}}{\prod\limits_{p=1, p \neq j}^{n} (\lambda_p - \lambda_j)} \right), \tag{1.12}$$

where $N_n(t)$ is the number of atoms of radionuclide n at time t. For example, if $n = 2$ and assuming $N_2(0) = 0$,

$$N_2(t) = N_1(0) \frac{\lambda_1}{\lambda_2 - \lambda_1} \left(e^{-\lambda_1 t} - e^{-\lambda_2 t} \right) \tag{1.13}$$

where $N_1(0)$ is the number of atoms of the first radionuclide at time $t = 0$. The activity, A, is related to the number of atoms by

$$A(t) = \lambda \cdot N(t). \tag{1.14}$$

The differential equations can also be solved by a recursive formalism using the Laplace transform. A complete treatment of this formalism is, however, beyond the scope of this book, and the reader is referred to, e.g., Hamawi (1971), Miles (1981), and Samuelsson (1987).

If the half-life of the mother nuclide in a series is much longer (by roughly a factor of 1000) than the half-life of the daughter, secular equilibrium is established after a time corresponding to approximately three half-lives of the daughter. If the half-life of the mother is about ten times longer than the half-life of the daughter, *transient equilibrium* is established, also after a time corresponding to approximately three half-lives of the daughter. In this type of equilibrium, the activities are not equal, and the daughter will decay with the half-life of the mother, as the continuous feeding of daughter nuclei from the decay of the mother will make the daughter appear to decay with the same half-life as the mother nucleus. These two kinds of equilibrium are illustrated in Figure 1.6. If, on the other hand, the half-life of the daughter is longer than the half-life of the mother, no equilibrium will be established.

If it can be assumed that all the radionuclides in a series remain in the bedrock, secular equilibrium will exist at the site of their creation, and *almost* all the radionuclides in the series will have the same activity. The exception is when radionuclides are created through one of the branches in a branched decay, for example, ^{208}Tl in the thorium series (Fig. 1.6). If secular equilibrium can be assumed, this facilitates the determination of the activity of the alpha-emitting radionuclides in a series, which normally have no associated gamma decay, since gamma-emitting radionuclides in the lower part of the chain can be used to determine activity. However, each of the three series includes at least one isotope of the rare gas radon, which can disturb the equilibrium by leaving the site of creation.

Chemical or mechanical processes such as weathering can also affect the equilibrium. This is the case in areas without previous periods of glaciation, where the upper crust has been exposed to weathering, and disequilibrium therefore prevails. In areas subjected to glaciation, the time since the recession of the ice has been too short to permit substantial weathering, which affects the equilibrium.

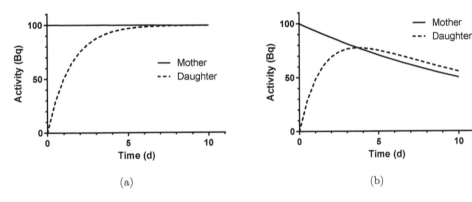

FIGURE 1.6   (a) The half-lives of the mother and daughter nuclides are 1000 days and 1 day, respectively. It can be seen that approximate secular equilibrium is reached after about three half-lives of the daughter (3 d). (b) When the half-lives of the mother and daughter are 10 d and 1 d, respectively, transient equilibrium is reached, also after about three half-lives of the daughter (3 d). Note that equilibrium applies only to the activities; the number of nuclei can differ by several orders of magnitude.

Although equilibrium cannot be achieved for the whole series, parts of the series can fulfil the requirements for equilibrium. The uranium series can thus be divided into five sub-series: ^{238}U – ^{234}U, ^{230}Th, ^{226}Ra, ^{222}Rn – ^{214}Po, and ^{210}Pb – ^{206}Pb. The equilibrium in the first sub-series is established by the short half-lives of the radionuclides between ^{238}U and ^{234}U, although some fractionation and separation can take place. This fractionation, and the fact that thorium may not migrate with the uranium, means that the second sub-series contains only ^{230}Th. The long half-life of ^{230}Th will give it sufficient time to migrate before it decays, and no equilibrium will be established at the site where ^{230}Th was formed. Radium also behaves differently from uranium, and forms its own sub-series for the same reasons as ^{230}Th. Based on arguments similar to those above, the thorium series can be divided into three sub-series: ^{232}Th, ^{228}Ra – ^{224}Ra, and ^{220}Rn – ^{208}Pb (Samuelsson 1994).

1.2 TECHNOLOGICALLY ENHANCED NATURALLY OCCURRING RADIOAC-TIVE MATERIAL (NORM AND TENORM)

The acronyms NORM and TENORM denote *Naturally Occurring Radioactive Material* and *Technologically Enhanced Naturally Occurring Radioactive Material*, respectively. Although radioactive material is not produced in the processes associated with TENORM, the abundance or concentration of naturally occurring radionuclides is increased. This may be the result of concentration or redistribution of the radioactive material, for example, as a result of combustion, uranium ore mining, or the construction of buildings. Below, we briefly describe some processes leading to technological enhancement of radioactive materials, starting with radon in indoor air.

1.2.1 Radon and Radon Exposure Enhanced by Man

The decay products of ^{222}Rn, called *radon daughters*[2], can be divided into two groups. The first group contains short-lived daughters with half-lives of less than 30 minutes: ^{218}Po, ^{214}Pb, ^{214}Bi and ^{214}Po. For historical reasons, these are sometimes denoted RaA, RaB, RaC and RaC′. The second group is composed of the more long-lived daughters and includes ^{210}Pb, ^{210}Bi and ^{210}Po; also denoted RaD, RaE and RaF. In indoor environments, radon and its daughters can be in various degrees of disequilibrium, depending on the balance between the inflow of radon and its removal by ventilation. However, there are usually three fractions of radon and radon daughters in indoor air: radon gas, radon daughters attached to aerosols, and free radon daughters in the air. Since the radon daughters are often formed as positively charged ions, they become attached to surfaces in a process called *plate-out*. This can be used to measure the degree of equilibrium by collecting radon daughters on a wire or plate at a negative electric potential of a few kV. Analysis of the gamma radiation from the attached radon daughters provides an estimate of the equilibrium factor, F, defined as the activity of the short-lived radon daughters at the location under investigation, divided by the corresponding activity that would result if the daughters were in equilibrium with radon.

Due to its migration properties and its sufficiently long half-life, ^{222}Rn can be transported from the ground into indoor environments. The indoor radon concentration will depend not only on the activity concentration of radon in each source, but also to a large extent on the characteristics of the building and the ventilation rate.

The main health risk associated with inhaling radon is an increase in the risk of developing lung cancer. Based on a meta-study including thirteen European studies, Darby et al. (2006) concluded that the cumulative risk of death from lung cancer depends on the radon exposure, and is considerably higher for smokers than for non-smokers; see Figure 1.7. Although the relative risks are approximately equal for smokers and non-smokers, the higher risk for smokers in the absence of radon will cause the cumulative risk to differ greatly between smokers and non-smokers.

1.2.1.1 Potential Alpha Energy

The health effects resulting from the inhalation of radon depend on the amount of energy transferred to the tissues by the emitted α-particles. The radiation dose to the lungs is therefore largely dependent on the concentration of radon daughters in the inhaled air. The quantity *potential alpha energy concentration* is used to quantify this concentration, taking

---

[2]Radon daughters are also called radon progeny.

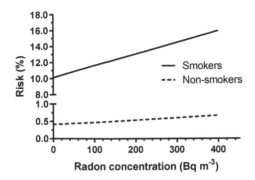

FIGURE 1.7 Cumulative risk of death from lung cancer by the age of 75 years as a function of radon concentration in indoor air. Non-smokers refers to lifelong non-smokers and Smokers to continuing smokers of 15–24 cigarettes per day. Note the gap in the vertical axis. Data from Darby et al. (2006).

into account the energy transferred by α-particles. The *potential alpha energy*, E_{pj}, of one atom of a radon daughter, j, is defined as the sum of the energies of all the α-particles emitted when that daughter and all its progenies decay to ^{210}Pb. Note that only the short-lived radon daughters are included in the potential alpha energy since the mean retention time in the lungs is sufficiently long to permit these to decay before they are absorbed into the blood. The health effects of the much longer-lived radon daughters are minor in comparison to the effects of the short-lived daughters.

The potential alpha energy resulting from all atoms of radon daughter j in a volume of air is then

$$N_j E_{pj} = \frac{A_j}{\lambda_j} E_{pj}, \tag{1.15}$$

where A_j is the activity and λ_j is the decay constant of radon daughter j. The potential alpha energy for ^{222}Rn is 19.2 MeV since this is the sum of the α-particle energies of ^{222}Rn (5.5 MeV) and the α-emitting progenies ^{218}Po (6.0 MeV) and ^{214}Po (7.7 MeV). Note that the β-emitting daughters ^{214}Pb and ^{214}Bi can also be assigned a potential alpha energy since they eventually give rise to the alpha emitter ^{214}Po. The potential alpha energy of each of these β emitters is thus 7.7 MeV.

However, instead of referring to the number atoms in an air volume, it is more practical to determine the concentration of radon daughters. We can then define the potential alpha energy concentration, C_p, as the potential alpha energy per unit volume of air. If the activity concentration of radon daughter j is C_j, this will give rise to a potential alpha energy concentration C_{pj}. For a given mixture of radon daughters, the total potential alpha energy concentration can be obtained by summing over all radon daughters j:

$$C_p = \sum_j C_{pj} = \sum_j \frac{C_j}{\lambda_j} E_{pj}. \tag{1.16}$$

The SI units of potential alpha energy concentration are J m^{-3}, and have replaced the older unit *working level* (WL, 1 WL = $1.3 \cdot 10^5$ MeV dm^{-3} of air, or $2.08 \cdot 10^{-5}$ Jm^{-3}). The working level was originally defined as the potential alpha energy concentration resulting from an activity concentration of 100 pCi ^{222}Rn per litre of air, in secular equilibrium with its short-lived daughters. If we assume that the short-lived daughters of ^{222}Rn are in secular equilibrium with their mother, the potential alpha energy concentration will equal 1 WL if the activity concentration of ^{222}Rn is 3.7 Bq per litre of air.

TABLE 1.9   Potential alpha energy for ^{222}Rn in secular equilibrium with its short-lived daughters. The total potential alpha energy from 3.7 Bq of ^{222}Rn in equilibrium with its short-lived daughters is defined as 1 WL, given by adding the contributions from the number of atoms, N, that will result in 3.7 Bq of each radionuclide. The decay constant, λ, and α-particle energy, E_α, for radionuclide j are given together with the potential alpha energy, E_{pj}, per atom and per 3.7 Bq, $E_{pj,3.7Bq}$. The half-lives are given in Table 1.5.

	$\lambda_j (s^{-1})$	E_α (MeV)	E_{pj} (MeV)	N	$E_{pj,3.7Bq}$ (MeV)
^{222}Rn	$2.10 \cdot 10^{-6}$	5.5	–	$1.76 \cdot 10^6$	–
^{218}Po	$3.78 \cdot 10^{-3}$	6.0	6.0+7.7=13.7	979	$1.34 \cdot 10^4$
^{214}Pb	$4.31 \cdot 10^{-4}$	–	7.7	8585	$6.61 \cdot 10^4$
^{214}Bi	$5.86 \cdot 10^{-4}$	–	7.7	6314	$4.86 \cdot 10^4$
^{214}Po	$4.23 \cdot 10^3$	7.7	7.7	$8.75 \cdot 10^{-4}$	0.007
Sum					$1.28 \cdot 10^5$

The relation between the air concentration, expressed in WL, and the equilibrium concentration, C_{eq}, normally expressed in the units Bq m^{-3}, is thus:

$$C_{p,Rn}[\text{WL}] = \frac{C_{eq,Rn}[\text{Bq m}^{-3}]}{3\ 700} \tag{1.17}$$

for ^{222}Rn (Rn), since 3.7 Bq per litre equals 3700 Bq m^{-3}. For ^{220}Rn (thoron, Tn), the corresponding relation is

$$C_{p,Tn}[\text{WL}] = \frac{C_{eq,Tn}[\text{Bq m}^{-3}]}{275} \tag{1.18}$$

since the different alpha energies in this decay chain yield 1 WL = 275 Bq m^{-3} (remembering that 1 WL was defined based on 100 pCi ^{222}Rn per litre of air).

The quantity *equilibrium-equivalent concentration*, C_{eq}, is used to describe the potential alpha energy concentration of an arbitrary mixture of radon daughters (not necessarily in equilibrium). C_{eq} is defined as the activity concentration of ^{222}Rn (or ^{220}Rn) in equilibrium with its short-lived daughters that has the same potential alpha energy concentration as the mixture (Porstendörfer 1994). This quantity can be calculated from

$$C_{eq} = \sum_j k_{pj} C_j \tag{1.19}$$

where

$$k_{pj} = \frac{E_{pj}/\lambda_j}{\sum_j E_{pj}/\lambda_j}. \tag{1.20}$$

The potential alpha energy of 1 Bq of radon daughter j can be determined by dividing the potential alpha energy per atom, E_{pj}, by the decay constant, λ_j. The coefficient k_{pj} thus represents the fraction of the total potential alpha energy that can be attributed to this daughter nuclide.

To verify Eqn. (1.19) we first note that at equilibrium, the activity concentrations of the short-lived radon daughters are equal, and thus $C_j = C_{eq}$ for all daughter nuclides j. Eqn. (1.16) can then be written

$$C_p = \sum_j \frac{C_j}{\lambda_j} E_{pj} = C_{eq} \sum_j \frac{E_{pj}}{\lambda_j}, \tag{1.21}$$

since the definition of C_{eq} implies that C_p is the same regardless of whether equilibrium prevails or not, i.e.,

$$\sum_j \frac{C_j}{\lambda_j} E_{pj} = C_{eq} \sum_j \frac{E_{pj}}{\lambda_j}. \tag{1.22}$$

Solving for C_{eq} and recalling that the sum of all E_{pj}/λ_j is constant for a given mixture, we find that

$$C_{eq} = \frac{\sum_j \frac{C_j}{\lambda_j} E_{pj}}{\sum_j \frac{E_{pj}}{\lambda_j}} = \sum_j \frac{E_{pj}/\lambda_j}{\sum_j E_{pj}/\lambda_j} C_j = \sum_j k_{pj} C_j. \tag{1.23}$$

Using data from Table 1.9, we can derive an expression for C_{eq} [Bq m^{-3}] for ^{222}Rn with the short-lived daughters $j = 1$ to 4.

$$C_{eq,Rn} = 0.105 C_1 + 0.516 C_2 + 0.379 C_3 + 6 \cdot 10^{-8} C_4 \tag{1.24}$$

Finally, the *equilibrium factor*, F, is used to describe the degree of disequilibrium between the mixture of radon daughters and their mother nuclide in air:

$$F = \frac{C_{eq}}{C_0} \tag{1.25}$$

where C_0 is the actual concentration of the mother nuclide. The difficulties involved in determining the equilibrium factor have contributed to a modification of the method used to measure radon concentration in indoor air. The equipment is usually designed so that only radon gas (and no radon daughters) can enter the detector. This will be discussed later in this section.

The potential alpha energy concentration, described above, is a measure of the energy of the α-particles, but in order to estimate the health effects of radon and radon daughters, the exposure time must also be taken into account. The *exposure* can be expressed as either the time-integrated activity concentration [Bq h m^{-3}], or as the time-integrated potential alpha energy concentration [J h m^{-3}]. The latter may also be expressed in terms of the *working level month*, WLM, which corresponds to the exposure to 1 WL during a working period of one month. Assuming that a working month is equivalent to 170 h and using the relation that 1 WL equals an equilibrium-equivalent concentration of 3700 Bq m^{-3}, we find that 1 WLM corresponds to an *equilibrium-equivalent activity exposure*, $E_{eq,Rn}$ of $6.3 \cdot 10^5$ Bq h m^{-3}.

1.2.1.2 Radon in the Indoor Environment

Elevated radon concentrations in dwellings are often caused by inadequate ventilation. In response to the energy crisis in the 1970s, several countries started to promote the construction of more "energy-efficient" buildings to reduce the need for heating. However, improved insulation may also lead to the retention of radon and moisture in buildings. Five main sources of radon in dwellings can be identified: soil or bedrock, building materials, water, outdoor air, and natural gas.

The soil or bedrock is generally the largest contributor of radon in indoor air due to the large concentrations of radon in soil gases. Activity concentrations in excess of tens of kBq m^{-3} are not uncommon. The most important radon isotope in this respect is ^{222}Rn, because of its relatively long half-life of 3.8 days, although ^{220}Rn can also be important. Radon in soil gases is transported into the building by the pressure gradient that normally exists between outdoor and indoor air. This pressure gradient is caused by ventilation systems and temperature differences, together with wind. Heating of indoor air causes convective flow towards the ceiling inside the building, which will create an underpressure at the floor. Thus, a large temperature gradient between indoor and outdoor air may enhance the inflow of radon. When the indoor radon concentration is monitored at close intervals, a diurnal cycle is seen, with a maximum during night-time when the temperature gradient is greatest. The radon concentration decreases as the outdoor air warms up and the building is ventilated during the day. This diurnal variation means that the measurement strategy for indoor radon is very important. Reliable measurements must be carried out over a time span of two to three months, preferably during the coldest season.

Radon in natural gas is seldom a problem in indoor environments due to the long delay between the production of the gas and its delivery to households. The same is true of the water from municipal water systems, where the transport time allows the radon to decay below significant concentrations. Outdoor air also contains a certain amount of radon, but this is insignificant due to dilution. This leaves two sources which may be significant contributors to radon exposure: water from drilled wells and building materials.

1.2.2 Sources Generated by Industrial and Technological Processes

Many human activities involve processes that redistribute and/or enhance naturally occurring radioactive elements in nature. This will inevitably also affect the radiation environment of humans compared to the undisturbed human habitats in which man lived thousands of years ago. In Section 1.1 we learned that man has always been subject to radiation from natural sources. However, in the past 10 000 years, during which fixed domiciles and agricultural practices became established, new kinds of radiation exposure have arisen, increasing the global human exposure to radiation. Early techniques using the mineral content of various ores also led to new pathways for external and internal exposure of individuals working in mines. The advent of industrialization in the latter half of the 19th century led to a substantial increase in industrially enhanced radioactivity, and increased the global radiation dose burden to humans even further. One example is the use of oil and coal. These fossil fuels originate from organic matter formed more than 100 million years ago, and contain primordial radionuclides emanating from the two principle decay chains (the ^{238}U and ^{232}Th series), which are omnipresent in the earth's crust.

1.2.2.1 Radioactivity Associated with Fossil Fuels

Typical ranges for the activity concentrations of uranium and thorium in refined coal are 4–370 Bq kg^{-1} (0.3–30 ppm) and 12–410 Bq kg^{-1} (3–100 ppm), respectively. At a given site, the concentration of thorium is often about three times higher than the concentration of uranium. Table 1.10 presents examples of NORM concentrations in coal from some countries. However, the concentrations can be much higher in some areas.

Upon combustion, these elements may be released into the atmosphere through gaseous and volatile effluents, leading to exposure of workers at the facility and other people in the

TABLE 1.10 Activity concentration of some naturally occurring radionuclides in coal. Single values represent averages.

Country	Activity concentration (Bq kg^{-1})					
	^{238}U	^{226}Ra	^{210}Pb	^{210}Po	^{232}Th	^{40}K
Australia	8.5–47	19–24	20–33	16–28	11–69	23–140
Brazil	72	72	72		62	
Greece	117–390	44–206	59–205			
Romania	80	126	210	262	62	
UK	7–19	7.8–21.8			7–19	55–314
USA	6.3–73	8.9–59.2	12.2–77.7	3.3–51.8	3.7–21.1	

Source: Data from IAEA (2003a).

surrounding area. After combustion, the radionuclides in the uranium and thorium series are concentrated in the ash, which has a concentration about ten times that of the raw material. The *bottom ash* will remain in the plant but the *fly ash* may follow the hot effluents through the stack, although this can be mitigated to some extent. Table 1.11 gives some examples of typical activity concentrations in fly ash in some countries. However, the concentrations may vary depending on the mineral composition of the coal and, hence, the location of the mine. In recent years, attempts have been made to extract uranium from waste coal ash. This is considered commercially profitable at levels above 200 ppm U per kg of ash.

A well-known feature of fly ash is that the radionuclides are enriched in the smaller particles; the enrichment factor varies between the radionuclides. Sahu et al. (2014) found that ^{210}Po in fly ash from Indian coal-fired power stations had the highest enrichment factor, followed by ^{238}U. The activity concentration of ^{210}Po in 2-μm particles was found to be about three times that in 18-μm particles. A good correlation was also found between the sulphur content of the coal and the activity concentration of ^{210}Po, as well as the enrichment factor for this radionuclide.

TABLE 1.11 Activity concentration of some naturally occurring radionuclides in coal ash.

Country	Activity concentration (Bq kg^{-1})		
	^{238}U series	^{232}Th series	^{40}K
USA	100–600	30–300	100–1200
Egypt	16–41	9–11	
Germany	6–166	3–120	125–742

Source: Data from IAEA (2003a).

The combustion of coal and the mining process give rise to radiation exposure and the accumulation of radioactive waste. The increased radon concentration in underground mines has a substantial impact on the risk of developing lung cancer in underground workers. It is therefore important that appropriate mitigation measures, such as ventilation, are employed. Waste rock and soil often have elevated concentrations of radium isotopes and ^{40}K, and activity concentrations of ^{226}Ra up to 55 000 Bq kg^{-1} have been measured in sediments in waste water discharged to the environment (IAEA 2003a).

Many estimates have been made of the activity released from coal-fired power plants over the years, indicating that the activity released to the atmosphere exceeds the releases from nuclear power plants. However, modern emission reduction techniques, such as scrubbers, filters and desulphurization, have decreased the emission of radionuclides from coal-fired power plants, and these claims are generally no longer correct.

Example 1.1 *A source term for a coal-fired plant*

A source term for a 1000-MWe coal-fired power plant was published in 1978 by McBride et al. (McBride et al. 1978), and it is of interest to study the details of the calculations.

The power of the plant is given in MWe, which means that the electrical power produced by the plant is 1000 MW. The thermal power output is usually considerably higher and the efficiency depends on the construction of the plant. The calculations were performed using data from a plant that had a peak power of 290 MWe when 96.1 tonnes of coal were consumed per hour.

During one full year of operation (365×24 hours) the unit should thus have consumed $8.42 \cdot 10^5$ tonnes of coal, but if the capacity factor of the plant is assumed to be 80%, then $6.74 \cdot 10^5$ tonnes of coal will be consumed per year. Thus $6.74 \cdot 10^{11}$ g per year is consumed, which equals $2.32 \cdot 10^9$ g MWe^{-1} y^{-1} ($6.74 \cdot 10^{11}$ g y^{-1}/290 MWe). If the concentrations of uranium and thorium in the coal are assumed to be 1 ppm and 2 ppm, respectively, this means that the amounts of uranium and thorium in the consumed coal are $2.32 \cdot 10^3$ g MWe^{-1} y^{-1} and $4.64 \cdot 10^3$ g MWe^{-1} y^{-1}, respectively.

The release to the atmosphere can be estimated if we assume that all the uranium and thorium are contained in the ash, and that 1% of the ash is released to the atmosphere. Thus, the amounts of uranium and thorium released to the air per MWe per year will be 23.2 g and 46.4 g, respectively. For a 1000-MWe plant, these releases will be $2.32 \cdot 10^4$ g uranium and $4.64 \cdot 10^4$ g thorium, and these will also be accompanied by their nonvolatile daughters. Assuming that the radioactive daughters of ^{238}U, ^{235}U and ^{232}Th are in secular equilibrium with their parents, and are released in the same proportions, a source term can be calculated.

McBride et al. calculated an annual release of 296 MBq per radionuclide in the ^{238}U series, 12.95 MBq per radionuclide in the ^{235}U series, and 185 MBq per radionuclide in the ^{232}Th series. However, radon was not included in these calculations since it is assumed that all the radon will be released.

Although ^{238}U, ^{235}U and ^{232}Th are not mobilized in rock formations, their daughters belong to Group II of the periodic table (which also includes radium), and these may dissolve in the groundwater, especially if they form chlorides. If this groundwater comes into contact with oil deposits in an oil field, significant amounts of ^{226}RaCl$_2$ and ^{228}RaCl$_2$, and the chlorides of their daughters, may be dissolved in the oil. This can lead to large deposits in drilling pipes and other oil extraction equipment which are rich in radium and its daughters. These solid deposits, often in the form of insoluble barium, calcium and strontium compounds, referred to as *scale*, exhibit concentrations of ^{226}Ra ranging from 100 Bq kg^{-1} to 15 MBq kg^{-1} (IAEA 2003b).

When mixed with water, these deposits form sludge. Sludge can exhibit ^{226}Ra activity concentrations ranging from 50 Bq kg^{-1} to 800 kBq kg^{-1}. *Produced water* is a term used in the oil industry to describe water that is produced as a by-product together with oil and gas, and contains activity concentrations of ^{226}Ra ranging from 0.001–1200 Bq L^{-1} (IAEA 2003b). Scale may also be deposited from produced water that has been in contact with uranium- or thorium-rich formations in the bedrock at high pressure and temperature. When the pressure and temperature decrease, the solubility decreases and radionuclides can precipitate inside the equipment.

Both scale and produced water constitute a significant radiological waste problem for the oil and gas industry. Radiological concerns are mainly associated with internal contamination of workers by inhalation of dust from scale and sludge containing the ^{226}Ra daughters ^{210}Pb and ^{210}Po, rather than external exposure to these deposits. The United States Environmental Protection Agency (2014) has estimated that each oil well in the United States generates, on average, 100 tonnes of scale per year. This must be disposed of as radiological waste, and is often injected into deposition wells in the bedrock. However, the formation of scale is not only a problem from a radiation protection point of view, since scale formation also reduces the efficiency of the equipment and may disrupt oil production.

1.2.2.2 Radioactivity Associated with the Production and Use of Minerals

The metallic and chemical properties of thorium, being stronger but lighter than many other metals, have made it useful in airborne engines (e.g. aircraft, rockets and missiles). Also, glass containing thorium has advantageous optical properties, such as a high refractive index and low dispersion, making it useful in the design of lenses with minimum chromatic aberration. However, the use of thorium in refined minerals has decreased significantly since the 1970s due to concerns over radioactivity.

The nuclear industry primarily uses uranium with enhanced ^{235}U content; from 1% for heavy-water reactors to 3–5% for light-water reactor fuel. The uranium remaining after enrichment is referred to as *depleted uranium*, which typically has a ^{238}U content of 99.6–99.8% by weight. Depleted uranium can also be recovered from spent nuclear fuel since fission in a nuclear reactor will decrease the abundance of ^{235}U. However, this type of depleted uranium can be identified by the presence of ^{236}U, which is formed during the fission process.

Due to its very high density (19 100 kg m^{-3}) depleted uranium has also found use in the weapons industry, for example, in tank armour and warheads. However, concern has been raised regarding the health effects on personnel resulting from the detonation of weapons containing depleted uranium. Plausible exposure pathways include surface contamination of skin, inhalation, and indirect ingestion of small fragments of depleted uranium. It is also used in radiation shields for sealed, gamma-emitting radiation sources, again due to its high density. When intact, freshly made and free from radium, depleted uranium poses little health hazard, as its gamma dose rate is less than that from natural uranium. Depleted uranium has also been used previously as counterweights (trim weights) in aircraft since its high density provides a large mass in a limited volume. A large aeroplane can contain about 1 500 kg of depleted uranium.

However, the toxic chemical properties of depleted uranium are just as important as its radiological properties. Uptake in the gastrointestinal tract is low, but inhalation leads to accumulation in the lungs. It may also be accumulated in the liver and kidneys, where it may cause severe injury, including kidney failure. Generally, the toxicity of a metal depends on its solubility; the more soluble it is, the more harm it can cause.

In order for mining of uranium to be economically feasible, mass contents in the minerals exceeding 10 ppm are often required, depending on the price of uranium. Although uranium and radium may be present in other types of mined minerals, the tailings of the mineral ores resulting from uranium mines naturally have higher concentrations of ^{238}U, ^{234}U and ^{226}Ra. If not properly contained, uranium daughters will leak from these tailings, or slag heaps, into the groundwater, which may lead to the exposure of local residents in the vicinity of the mining site. Mining may also lead to increased radioactivity in slag heaps and groundwater close to the mine.

High mortality rates and increased occurrence of respiratory diseases among underground workers were first reported among the miners working in the silver mines in the

Czech/German region Erzgebirge (Erz Mountains) by Georgius Agricola in the 16th century (Samet 1994). The disease, then called *bergkrankheit* (mountain disease), probably referred to what we now know as lung cancer, silicosis and tuberculosis.

Many deposits of rare earth metals also contain thorium (^{232}Th). The removal of scale and other deposits from drainpipes and other equipment may also lead to exposure to significant amounts of ^{232}Th and its daughters. As in the coal and oil industry, slag from mineral mining has also been used for landfill, constituting another pathway for the exposure of humans to radiation. Groundwater in disused mines also becomes enriched in ^{238}U, ^{232}Th and ^{226}Ra, and the potential radiological harm must be considered when the mines are drained.

Zircon ($ZrSiO_4$), which is used, for example, in ceramics (glazes) and in glass making, exhibits high concentrations of naturally occurring radionuclides, up to 10 kBq kg^{-1} of ^{238}U or ^{232}Th (IAEA 2003a). The dispersion of dust during processing must be controlled to avoid internal contamination of the workers. Dust filters in zirconium processing facilities have been found to have ^{210}Pb and ^{210}Po activity concentrations as high as 200 kBq kg^{-1} and 600 kBq kg^{-1}, respectively (World Nuclear Association 2015). The removal of radionuclides from zircon is not considered economically feasible.

Monazite is a phosphate mineral that contains rare earth metals. It also contains high activities of U (25–75 kBq kg^{-1}) and Th (41–575 kBq kg^{-1}) (IAEA 2003a).

1.2.2.3 Phosphate Ore and Phosphate Fertilizers

Modern agriculture relies to a large extent on fertilizers to enhance the production of crops. Adding phosphates (mainly P_2O_5) provides growing crops with phosphorus, and the worldwide consumption of P_2O_5 has been estimated to be in excess of $40 \cdot 10^6$ tonnes per year. However, the phosphates used for fertilizers are extracted from minerals that may also contain ^{238}U, ^{232}Th and ^{226}Ra (and their associated daughter products). Activity concentrations of these radionuclides in phosphate fertilizers used in Finland have been reported to be 100–600 Bq kg^{-1} of ^{238}U, 10–30 Bq kg^{-1} of ^{228}Th and 30–110 Bq kg^{-1} of ^{226}Ra (Mustonen 1985). The IAEA (2003a) has reported activity concentrations in phosphoric acid of 1200–1500 Bq kg^{-1} for ^{238}U and 300 Bq kg^{-1} for ^{226}Ra, while the corresponding concentrations for NPK (nitrogen, phosphorus, potassium) fertilizers were 440–470 Bq kg^{-1} and 210–270 Bq kg^{-1}.

Adding phosphates as fertilizers can double the load of these naturally occurring radionuclides, and may thus increase human exposure to them. The use of phosphates containing ^{226}Ra will lead to an uptake of the nuclide in, for example, tobacco plants, leading to the accumulation of the daughter nuclide ^{210}Po. This is an α emitter that will be inhaled when the tobacco is smoked, significantly increasing the risk of lung cancer.

Phosphogypsum (calcium sulphate, $CaSO_4 \cdot xH_2O$) is a waste product of the processing of phosphate to produce various types of phosphate-based fertilizers, and the combined activity of the uranium and thorium series radionuclides may be a few kBq kg^{-1}. The disposal of gypsum together with leaching of the naturally occurring radionuclides in the fertilizers into the groundwater will eventually lead to the release of radioactive material to the marine environment. However, in the North Sea and the North Atlantic, discharges are dominated by ^{226}Ra, ^{228}Ra and ^{210}Pb from oil and gas production, which account for 90% of the total discharges of α-emitting radionuclides (World Nuclear Association 2015).

1.2.2.4 Manufacturing of Equipment and Household Goods

Naturally occurring radioactive materials are used in the manufacture of many common products. Some of these products make use of the physical properties of radioactive materials, and users may be unaware of their radioactive properties. Some radioactive materials provide colour or are luminous. For example, when radium is mixed with metal oxides to form a salt, the emission of the highly ionizing α-particles from ^{226}Ra and β-particles in the subsequent decay chain causes excitation of electrons in the atomic lattice of the salt. When these excited electrons decay back to their ground state, energy is released in the form of light in the visible wavelength range. This process is called *fluorescence*, which is a form of *luminescence*[3].

Radium was first studied in the form of a salt, and the continuous radioluminescence seen in the salt crystals led to radium salts being described as "self-luminous". This property rapidly found applications in conditions of poor light, e.g. for instruments in aircraft and other vehicles. However, the harmful effects of the radioactive properties of radium were revealed in the late 1920s, when young women working with paints containing radium salts developed skeletal conditions such as bone carcinoma, which often proved fatal. These women were painting watch dials and licked their paintbrushes to obtain a fine point. In spite of this, radium continued to be used in radioluminescent paint in watches and other equipment on a broad scale until the 1960s. Radioluminescence can also be achieved by less harmful radionuclides than radium. A common light source in exit signs and watch dials is a glass tube coated with a fluorescent layer and filled with tritium gas. The low energy of the β-particles from tritium (18.6 keV) ensures that no ionizing radiation is emitted from these sources.

In contrast to radium, where the direct radioluminescent properties of the salt are exploited, uranium-rich minerals have long been used as dyes or dye enhancers in the glass and china industry. When uranium in oxide form is added to a glass mix before melting, the glass takes on a distinctive yellow-green colour. The production of glass and tableware using uranium reached its peak in the 1930s, when so-called "depression ware" was produced. In addition to alpha and beta emission, the daughters in the ^{238}U series also emit gamma radiation (e.g. from ^{212}Bi, which emits gamma radiation at 1.76 MeV). Dose rates exceeding 300 times the natural background gamma radiation have been measured on the surface of this kind of glassware using radiation dosimeters. This is equivalent to having a 10- to 50-kBq ^{137}Cs source in the kitchen cupboard (Buckley et al. 1980).

Uranium salts have also been widely used in glazes for ceramic tableware and tiles, for example, sodium uranate $Na_2O(UO_3)_2 \cdot 6H_2O$, which gives products a distinctive orange colour. When used for glazes, the radium component was often removed, because chemically purified uranium was required, and this type of glaze was thus not a source of ^{222}Rn. The demand for uranium minerals increased sharply in the 1940s and 1950s due to the newly discovered fission properties of uranium, making uranium too expensive for use in making glass and ceramics, and its use in this industry ceased almost completely after the Second World War.

Arc welding with tungsten electrodes containing thorium oxide is another source of radiation exposure to humans. A small amount of thorium oxide (also called thoria), 1–2%, is mixed with the tungsten in the manufacturing of electrodes to enhance the stability of the arc. Although the physical strength of the electrode is increased by the addition of thoria, small particles (1–10 μm in diameter) are released from the electrode, which can be inhaled

---

[3]Another form of luminescence is *phosphorescence*, which is the delayed emission of light following excitation. This type of luminescence can be found, for example, in emergency exit signs. These signs continue to glow long after the light that excites the electrons has been turned off.

by the welder. Air activity concentrations of ^{232}Th of about 50 μBq L^{-1} have been found in the breathing zone of welders (Jankovic et al. 1999).

Many types of building materials contain substantial amounts of ^{226}Ra, ^{232}Th and ^{40}K, as can be seen from Table 1.12. Especially aerated concrete based on alum shale exhibits high ^{226}Ra levels, with activity concentrations exceeding 10 000 Bq kg^{-1}. Porous aerated concrete also promotes the escape of the rare gas ^{222}Rn from the decay of its mother ^{226}Ra, which increases the indoor air concentration of the ^{222}Rn daughters in buildings. Granite is a popular material used in the construction industry for exterior cladding of buildings, and in kitchens and bathrooms. It can contain up to 10 ppm of both uranium and thorium, and a tenfold increase in gamma radiation levels, compared with the natural background radiation, has been observed when measured close to the surface of granite fittings.

TABLE 1.12   Activity concentrations of ^{40}K, ^{226}Ra and ^{228}Ra in building materials. The data should be regarded as selective examples, as no systematic survey has been carried out on the radioactivity in building materials.

Material	Activity concentration (Bq kg^{-1})		
	^{226}Ra	^{232}Th	^{40}K
Concrete	1–250	1–190	5–1570
Aerated and lightweight concrete	9–2200	<1–220	180–1600
Clay bricks (red)	1–200	1–200	60–2000
Sand-lime bricks and limestone	6–50	1–30	5–700
Phosphogypsum	4–700	1–53	25–120
Cement	7–180	7–240	24–850
Ceramic tiles	30–200	20–200	160–1410

Source: Data from IAEA (2003a).

Electrostatic discharge brushes, containing the naturally occurring α emitter ^{210}Po, have been widely used in industrial applications where surfaces must be protected from electrostatic discharge. Polonium-based antistatic devices, mainly used in car painting, can contain up to 18 MBq ^{210}Po. Although this is a weak γ emitter (one 0.80-MeV photon per 100 000 decays), it can be extremely hazardous if accidentally ingested or inhaled. After a former USSR agent was assassinated using polonium in 2006, it has become widely recognized that access to polonium should be restricted.

1.2.2.5   Water Treatment and the Provision of Running Water

Drinking water from small private wells may contain high concentrations of uranium and radium. It has been found that the concentration of ^{226}Ra in drinking water is correlated to the uranium content in the bedrock from which the water is taken. However, the ^{226}Ra concentration in water is less strongly correlated with the concentrations of its daughter products, e.g. ^{222}Rn, due to disequilibrium. The ^{226}Ra concentration in water drawn from two adjacent wells can also vary considerably. The WHO (2011) has recommended a provisional guideline value of 30 μBq L^{-1} for the total activity concentration of uranium in drinking water. However, this value is based on the chemical toxicity of uranium.

A market has developed for resins and other water filtration devices to remove radionuclides from water. However, long-lived uranium and radium isotopes can accumulate in filters reaching activity levels at which they are considered radioactive, thus requiring appropriate disposal. Deposits on the inside of water pipes, especially water traps, may contain high levels of radium and uranium. Gamma radiation monitors at scrap recycling sites often detect radioactivity in water pipes and water filters.

In 2001, the WHO presented a summary of the efficiency of some methods of reducing the content of radionuclides in water, which can be found in Table 1.13. The methods included in the study are filtration (sand, activated carbon, ion-exchange resins), semipermeable membranes with an applied pressure, as in reverse osmosis, chemical treatment, such as lime-soda ash for softening and subsequent precipitation. When water is treated, the chemicals react with the minerals in the hard water to form insoluble compounds, which precipitate and can be removed by filtration. The use of ion-exchangers is a common means of removing elements from a sample. To remove radionuclides forming positive ions (*cations*), the resin is prepared with a surplus of hydrogen ions (H^+). The hydrogen ions in the resin will then be replaced by the radionuclide. For the removal of negative ions (*anions*), the resin may have a surplus of hydroxide ions (OH^-). The concentration of the radionuclide in the water is then determined by measuring the activity in the resin and the amount of water passing through it. However, it is important to check the efficiency of the ion-exchange resin by making measurements on the eluent (the water emerging from the filter).

TABLE 1.13 The efficiency of six methods of reducing the content of radionuclides in water: coagulation, sand filtration, filtration through activated carbon, precipitation softening, ion-exchange, and reverse osmosis.

| | Treatment method, % removal | | | | | |
	Coag.	Sand filtr.	Activ. carbon	Precip. softening	Ion exchange	Reverse osmosis
Sr	10–40	10–40	0–10	>70	40–70	>70
I	10–40	10–40	40–70	0–10	40–70	>70
Cs	10–40	10–40	0–10	10–40	40–70	>70
Ra	10–40	40–70	10–40	>70	>70	>70
U	>70	0–10	10–40	>70	>70	>70
Pu	>70	10–40	40–70	0–10	>70	>70
Am	>70	10–40	40–70	0–10	>70	>70

Source: Data from WHO (2011).

1.2.2.6 Exemption and Clearance

All the industrial and technical procedures discussed above create waste, the handling and disposal of which must be regulated by the appropriate authorities. International recommendations by the IAEA suggest exemption levels for waste with activity concentrations ranging from 100 to 4000 Bq kg^{-1}, depending on the radionuclide(s), the origin of the waste, and the probability and risk of public exposure to these radioactive sources.

Exposure of humans to radiation from sources of natural origin is, for regulatory purposes, generally considered to be a so-called *existing* exposure situation (see Section 6.2.1). However, practices where the activity concentration exceeds some specified level should be regulated as *planned* exposure situations. This includes, for example, exposure to material with an activity concentration of any radionuclide in the uranium or thorium decay chains that exceeds 1 kBq kg^{-1} or with an activity concentration of ^{40}K exceeding 10 kBq kg^{-1}; public exposure to discharges resulting from mining and processing involving radioactive material, and exposure to radon and radon progenies in workplaces. Exposure due to naturally occurring radionuclides in, for example, food, water, construction materials and agricultural fertilizers, is considered to be an existing exposure situation, regardless of the activity concentrations of the radionuclides. However, the government should ensure that radiation protection and reference levels are established in such situations (IAEA 2013).

Radionuclides of artificial origin (see Section 1.3) are subject to criteria for exemption and clearance, but radionuclides of natural origin are exempted on a case-by-case basis. The dose criterion is on the order of 1 mSv y^{-1}, which is comparable to the background levels of radiation from natural sources. As discussed above, the exemption criteria do not apply to materials containing less than 1 kBq kg^{-1} of any radionuclide in the uranium or thorium decay series or less than 10 kBq kg^{-1} of ^{40}K.

1.3 ANTHROPOGENIC RADIATION

Radiation that originates from man-made processes is referred to as anthropogenic radiation. The era of anthropogenic radiation started with the advent of the x-ray tube in the late 1890s. This "novel" type of radiation, consisting mainly of photons in the energy range 10–100 keV, was used extensively, resulting in considerable human exposure. In the 1930s, researchers (including Hahn and Fermi, among others) discovered how atoms could be made to disintegrate, thus creating new radioactive substances to which humans had not previously been exposed. A wide range of applications has developed for these artificially produced radioactive substances, which are either produced in accelerators, by protons hitting a target material, or in reactors, where neutrons impinge on atoms that may undergo either activation or fission. Below, we give a brief overview of the sources of radiation that have become important for medical and industrial purposes, but which also constitute health hazards and carry the risk of considerable exposure.

1.3.1 The Nuclear Industry

The discovery of fission in 1938 by Meitner, Hahn and Strassman was the start of a new era in terms of energy production. Although military use initially dominated research efforts to exploit nuclear energy, the civil use of nuclear energy began on a large scale in the 1950s.

1.3.1.1 The Nuclear Fission Process

Fission refers to either an induced nuclear reaction, or a spontaneous nuclear process, in which a heavy nuclide ($Z > 92$) divides into two *fission fragments*, often accompanied by the emission of neutrons. The most commonly known and exploited nuclear fission process is that involving naturally occurring ^{235}U:

$$n +^{235} U \rightarrow^{236} U \rightarrow 2 \cdot f_{Fission} + 2.5n + T_{Kinetic} + \gamma_{Prompt}. \tag{1.26}$$

This reaction is induced by a fluence of neutrons (n) impinging on a lattice of ^{235}U atoms, resulting in the formation of a *compound nucleus* (^{236}U), which splits to produce two *fission fragments* ($f_{Fission}$), and secondary neutrons. The reaction also releases energy, which is distributed as kinetic energy ($T_{Kinetic}$) of the fission fragments and the neutrons, and the emission of prompt gamma radiation (γ_{Prompt}). This process is schematically depicted in Figure 1.8.

The distribution of mass number between the fission fragments is given by the *fission yield*. This is the fractional amount of a given fragment formed in each fission reaction, usually expressed as % per fission. The thermal fission yields for ^{235}U and ^{239}Pu are shown in Figure 1.9. The two peaks show that the fission process usually results in the formation of a lighter and a heavier fragment. The yield can be expressed either as the *independent fission yield*, i.e., the number of atoms of a given nuclide that are produced directly by fission; or as the *cumulative fission yield*, where the production of the nuclide by decay of its precursors is also included. The yields shown in Figure 1.9 are the *chain fission yields*, which is the sum of the cumulative yields of the last nuclide in the decay chain with mass number A, which can be stable or long-lived (see Section 1.3.1.4).

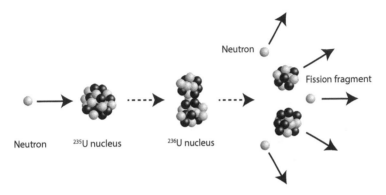

FIGURE 1.8 Fission of ^{235}U by thermal neutrons. The neutron is captured by the ^{235}U nucleus, forming a compound nucleus of ^{236}U, which rapidly develops a "waist". The compound nucleus then splits into two fission fragments, and secondary neutrons are emitted.

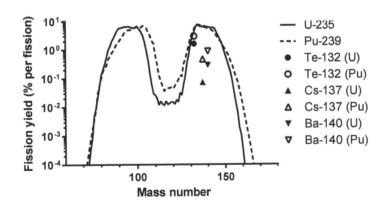

FIGURE 1.9 Chain fission yield for ^{235}U and ^{239}Pu given in % per fission for each mass number. The independent fission yields for some specific radionuclides are also shown. The chain fission yield for the light mass fragments, $A = 1 - 4$, where the highest yield is found for $A = 4$ (about 0.2 % per fission) is not shown in the figure. Data from IAEA (2008).

The cross section, which is the probability of a nuclear reaction taking place per unit flux of impinging particles, depends on the kinetic energy of the neutrons. In a fission reactor, the neutrons must be *moderated*, meaning that their kinetic energy must be reduced, to obtain the optimum cross section for fission of the target ^{235}U atom. The cross section for fission of ^{235}U by neutrons with kinetic energies between 10^3 and 10^6 eV is typically 1–10 barn (1 barn = 10^{-28} m^2), but at thermal energies (0.025 eV) it exceeds 500 barn.

What is remarkable about fission is not only that an atom can be split into two (unequally large) fragments, but that the sum of the rest mass of the two fragments and the neutrons (on average 2.5 neutrons per fission) is substantially less than the rest mass of the original nucleus. Part of the energy in the "missing" rest mass is transformed into kinetic energy of the fragments and neutrons, and γ-radiation. Energy is also released as kinetic energy of β-particles and anti-neutrinos by the decay of the fission fragments. The energy of the fission fragments, γ-rays and β-particles, can in turn be transformed into heat, since the surrounding atoms in the lattice or surrounding medium will slow down and stop them by various electromagnetic and nuclear interactions. The anti-neutrinos, however, will escape without interacting and their energy is thus lost. In addition to the kinetic energies discussed above, nuclear binding energy is also released when neutrons are captured. These reactions are collectively termed *heat dissipation*, and Table 1.14 gives a summary of the "energy budget" of fission in ^{233}U, ^{235}U and ^{239}Pu.

TABLE 1.14 Energy (MeV) released in fission of ^{233}U, ^{235}U and ^{239}Pu. The energy converted into heat in a thermal nuclear reactor is given by subtracting the energy of the anti-neutrinos from the total energy released and adding the binding energy released in the capture of the prompt neutrons that do not take part in fission.

	^{233}U	^{235}U	^{239}Pu
Promptly released energy			
Fission fragments	168.2	169.1	175.8
Prompt neutrons	4.9	4.8	5.9
γ-rays	7.7	7.0	7.8
Protracted energy release			
β-particles	5.2	6.5	5.3
Anti-neutrinos	6.9	8.8	7.1
γ-rays	5.0	6.3	5.2
Total	197.9	202.5	207.1
Total, except anti-neutrinos	191.0	193.7	200.0
Binding energy released	9.1	8.8	11.5
Converted into heat	200.1	202.5	211.5

Source: Data from NPL (2015).

The large amounts of energy released in fission can be explained by the variation in binding energy per nucleon with mass number, A, as shown in Figure 1.10. The binding energy, E_B, of a nucleus is proportional to the difference in the total mass of the nucleons forming the nucleus and the mass of the nucleus, Δm. The proportionality constant is the square of the speed of light, c^2, and according to Einstein's relation, the binding energy is given by $E_B = \Delta m \cdot c^2$. Since the binding energy per nucleon is lower for high A, the splitting of a heavy nucleus results in an increase in binding energy. The total mass of the fission fragments is thus lower than the mass of the fissioning nucleus, and the mass difference is converted into energy that is distributed according to Table 1.14.

In nuclei with fewer nucleons than ^{56}Fe (at the maximum in Fig. 1.10), nuclear reactions can take place in which the nuclides fuse, resulting in a net release of energy. These nuclear reactions are referred to as nuclear *fusion*, and stand in contrast to fission, where the nuclear reactions lead to the fragmentation of the original nuclide. The most common fusion reaction is the fusion of two hydrogen atoms into He, which takes place in the sun and in other

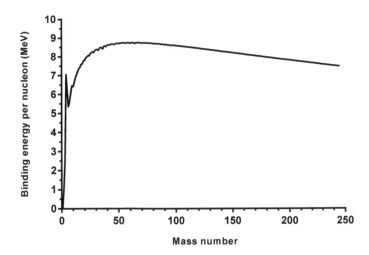

FIGURE 1.10 The binding energy per nucleon is rather constant, around 8 MeV, for most mass numbers. However, the curve decreases towards higher mass numbers and energy can thus be released when a heavy nucleus undergoes fission and is split into two lighter fragments.

stars. However, to achieve fusion, the positively charged nuclei must overcome the Coulomb repulsion between them, which means that they must collide at a considerable speed and hence kinetic energy. These high energies can only be achieved at temperatures of about 10^8 K, which makes it technically complicated to construct a fusion reactor. However, fusion can be achieved in thermonuclear weapons, as will be discussed later in this chapter.

Nuclei such as ^{235}U that can undergo fission are called *fissionable nuclei*. Some fissionable nuclei only undergo fission when bombarded with fast (1–20 MeV) neutrons (e.g. ^{238}U), whereas others undergo fission with neutrons of all energies. The latter are called *fissile nuclei* and among these are ^{235}U and ^{239}Pu. Figure 1.11 shows the cross section for fission of ^{235}U. Note that ^{238}U is thus not fissile, since it will only undergo fission when bombarded with fast neutrons. A third group, *fertile nuclei*, consists of nuclei that do not undergo fission when bombarded with thermal neutrons, but they are transformed into nuclei that are fissile by capturing the neutron. Examples are ^{234}U, ^{238}U and ^{232}Th, which are transformed into ^{235}U, ^{239}Pu and ^{233}U, respectively, by successive beta decay.

Other fission reactions can also be useful for energy production. Vast geological formations of thorium-rich minerals in nature contain the radionuclide ^{232}Th. This radionuclide, which exists in nature in recoverable amounts, can be transmuted into the fissile radionuclide ^{233}U by neutron activation with thermal neutrons in a reactor. ^{233}U has a high fission efficiency, i.e. the ratio of the cross section for fission to that for neutron capture is high, but even if capture should occur, the ^{234}U thus formed can be transformed into fissile ^{235}U by a second neutron capture. The man-made radionuclide ^{239}Pu is produced by neutron capture of ^{238}U followed by two successive beta decays, and also has a substantial cross section for fission by thermal neutrons (peaking at more than 1000 barn at 0.3 eV).

Radionuclides that decay spontaneously into fragments (fissionates) can be produced artificially. The man-made radionuclide ^{252}Cf ($T_{1/2} = 2.64$ y), which is produced by bombarding a curium target with α-particles, decays by spontaneous fission to fission fragments and emits neutrons with a mean energy of about 2 MeV. The distribution of neutron en-

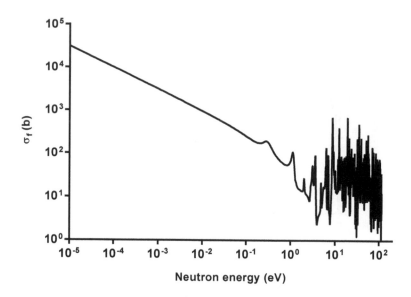

FIGURE 1.11   The cross section for fission of ^{235}U increases rapidly with decreasing neutron energy. Data from Soppera et al. (2012).

ergies from spontaneous decay of ^{252}Cf is shown in Figure 1.12, assuming a Maxwellian distribution described by (Meadows 1967):

$$N(E) \propto E^{1/2} \cdot e^{-E/1.565}. \tag{1.27}$$

^{252}Cf can thus be used as a neutron source, for example, when initiating a nuclear reaction requiring neutrons in a certain energy range, but it cannot be used in a sustainable nuclear chain reaction.

1.3.1.2 Controlled Nuclear Fission

If all the neutrons released during fission are allowed to interact with uranium nuclei, each "primary" fission event will cause, on average, 2.5 "secondary" fission events. Each of these will then cause 2.5 new fission events, and so on. One single fissioning nucleus can thus result in a *chain reaction* of fissions, rapidly releasing large amounts of energy. This may be advantageous in a nuclear weapon, but for energy production in a nuclear power plant, the chain reaction must be controlled. The number of neutrons released per fission is given by the *neutron yield per fission (fission factor)*, ν, which depends on the neutron energy and the fissioning nucleus. Table 1.15 gives the neutron yield for some fertile and fissile nuclides.

A reactor is *critical* if each thermal neutron produces, on average, exactly one new thermal neutron, as depicted in Figure 1.13. The net change in the number of thermal neutrons from one generation to the next is thus equal to 1.0, and under these conditions the chain reaction is self-sustaining. The net change in the number of thermal neutrons is given by the *multiplication factor*, k. If $k < 1$, the reactor is *subcritical* and if $k > 1$ it is *supercritical*.

If the chain reaction is to proceed, it is necessary that $k \geq 1$ and thus at least one neutron must induce a new fission reaction. This requires that at least one of the 2.5 *fast*

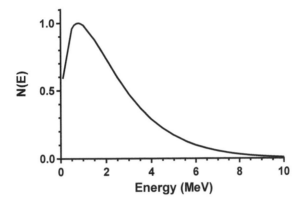

FIGURE 1.12   Normalized number of neutrons at energy E from spontaneous fission of ^{252}Cf.

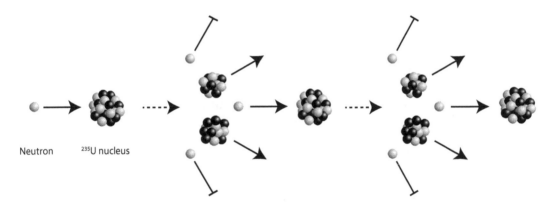

FIGURE 1.13   Neutrons released by fission may in turn cause fission in other fissionable nuclei, which leads to a chain reaction of fission events.

neutrons emitted can be thermalized—i.e., moderated to thermal velocities—in the reactor. Neutrons lose energy in elastic collisions with nuclei, i.e., collisions where the total kinetic energy of the neutron and the nuclei is preserved. It can be shown that the most efficient collisions, where the neutron transfers maximum energy, occurs with light nuclei. Suitable moderator materials are therefore water or carbon.

The multiplication factor depends on several processes and can be calculated by the *four-factor formula*

$$k_\infty = \eta \varepsilon p f, \qquad (1.28)$$

assuming no leakage of neutrons, e.g. for an infinitely large reactor. The four factors will be discussed in more detail below.

Assume that one thermal neutron causes fission in which secondary neutrons are emitted. The mean number of these secondary neutrons, per primary thermal neutron, is determined by the *thermal fission factor*, η, given by

$$\eta = \nu \frac{\sigma_f}{\sigma_f + \sigma_a}. \qquad (1.29)$$

TABLE 1.15 Fission cross section, neutron yield, and fission factor for some neutron energies and nuclides.

	^{233}U	^{235}U	^{238}U	Nat. U	^{239}Pu	^{240}Pu	^{241}Pu
Cross section (b)							
Thermal (0.025 eV)	529.1	582.6		4.19	748.1		1011.1
Fast (0.25 MeV)	2.20	1.28		0.01	1.53	0.10	1.78
Fast (1.0 MeV)	1.90	1.19	0.018	0.03	1.73	1.06	1.56
Neutron yield (ν)							
Thermal (0.025 eV)	2.497	2.436		2.418	2.884		2.948
Fast (0.25 MeV)	2.503	2.426	3.00*	2.426	2.886	2.80	3.01
Fast (1.0 MeV)	2.595	2.522	2.50	2.50	3.001	2.95	3.12
Fission factor (η)							
Thermal (0.025 eV)	2.287	2.068		1.34	2.108		2.145
Fast (0.25 MeV)	2.28	1.99		0.15	2.60	1.13	1.4
Fast (1.0 MeV)	2.52	2.31	0.28	0.44	2.93	2.52	1.4

* For a spectrum of fast neutrons (1–20 MeV).
Source: Data from Choppin et al. (2002) and IAEA (2008).

Note that $\eta < \nu$, the number of neutrons released per fission, since absorption of neutrons will occur in the fuel. The cross section for these absorption processes is given by σ_a, and the cross section for fission is given by σ_f. For thermal neutrons in ^{235}U these cross sections are 99 barn and 583 barn, respectively, giving $\eta = 2.08$ (the number of fast neutrons produced per thermal neutron). For natural uranium, the value of η equals 1.34 since ^{238}U is not fissionable with thermal neutrons ($\sigma_f = 0$), and the cross section for absorption (3.39 b) also differs from that for ^{235}U. Enriched uranium containing 3% ^{235}U will give $\eta = 1.84$. Values of η are given in Table 1.15.

The *fast fission factor*, ε, accounts for the fact that ^{238}U has a small cross section for fission by fast neutrons (≈ 1 b). Some of the fast neutrons may therefore contribute to the overall number of neutrons released in fission. For natural uranium in graphite, $\varepsilon \approx 1.03$. During moderation, the neutrons will lose energy by collisions and gradually be retarded to thermal velocities. However, during retardation they may have energies at which the cross section for absorption by ^{238}U is very large (the *resonance region*), on the order of 1000 b, as shown in Figure 1.14. To avoid excess absorption, the moderator material and the fuel in the reactor pile can be arranged so that the neutrons can be retarded to below the resonance region before leaving the moderator. Some of the neutrons will nevertheless be absorbed, and this is accounted for by the *resonance escape probability*, p, which is typically about 0.9. The last factor in the four-factor formula is the *thermal utilization factor*, f, accounting for processes such as neutron capture in the moderator and in construction materials. The value of f is typically about 0.9. Using the values given above, results in a value of $k_\infty = 1.1$ for natural uranium.

Figure 1.15 shows an example of the neutron cycle in a graphite-moderated, air-cooled reactor, with no losses of neutrons from the reactor. The fuel is assumed to consist of natural uranium. Initially, there are 1000 *thermal* neutrons capable of causing fission, and so releasing *fast* neutrons. The fast neutrons are then brought to thermal energies in the moderator. In this example, the number of thermal neutrons in the second generation is 1118, i.e. the multiplication factor, k_∞, equals 1.12.

Leakage of neutrons from the reactor has so far not been considered, but must be taken

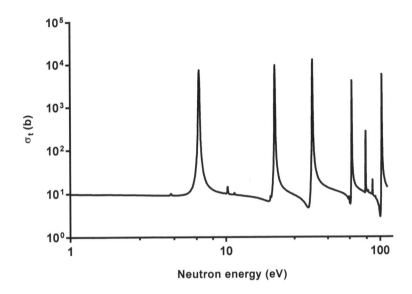

FIGURE 1.14 The region 1–100 eV includes a number of resonances where the cross section for neutron capture is large. During moderation, the neutrons have to pass through this energy region and it is therefore desirable that the neutrons encounter as few ^{238}U nuclei as possible during moderation. Data from Soppera et al. (2012).

into account in a real reactor pile. Both fast and thermal neutrons contribute to the leakage, Λ, which is approximately described by

$$\Lambda = \Lambda_f \Lambda_{th}, \tag{1.30}$$

where Λ_f and Λ_{th} are the fast and thermal *leakage factors*, respectively. Multiplying k_∞ by the leakage factor yields the effective multiplication factor, k_{eff}. If we revert to the data in Figure 1.15, assuming a leakage factor of 0.95, we find that the number of neutrons in the second generation will be 1062, i.e. $k_{eff} = 1.06$.

Leakage of neutrons has a considerable influence on the *critical mass*—the smallest mass of a fissile element that can sustain a chain reaction. The smallest critical mass is obtained for a sphere of the material since the production of neutrons, given by k_∞, is proportional to the volume, and the loss, Λ, is proportional to the surface area. The critical mass of a bare sphere of ^{235}U is 52 kg, and for ^{239}Pu the critical mass is 10 kg. The leakage can be reduced by coating the sphere with a neutron-reflecting material, and if spheres of ^{235}U and ^{239}Pu are coated with 20 cm of natural uranium, the critical masses are reduced to 22.8 kg and 5.6 kg, respectively (Choppin 2002). Although criticality results in large numbers of neutrons being emitted (and hence high radiation doses to nearby people), the chain reaction may proceed at such a slow rate that an explosion is unlikely.

The temperature dependence of k, described by the *temperature coefficient*, has important implications for reactor safety. If the temperature coefficient is positive, k increases with temperature, which leads to an increase in fission rate and hence higher temperatures. The increased temperature will then lead to a further increase in k, leading to further increases in the fission rate and temperature. Under these conditions, the reactor power will increase continuously eventually causing damage to the reactor. For this reason, it is desirable to

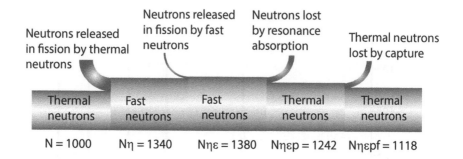

Neutrons released in fission by thermal neutrons

Neutrons released in fission by fast neutrons

Neutrons lost by resonance absorption

Thermal neutrons lost by capture

| Thermal neutrons | Fast neutrons | Fast neutrons | Thermal neutrons | Thermal neutrons |

$N = 1000$ $N\eta = 1340$ $N\eta\varepsilon = 1380$ $N\eta\varepsilon p = 1242$ $N\eta\varepsilon pf = 1118$

FIGURE 1.15 Example of the neutron cycle. For any single generation of neutrons, several processes determine the number of neutrons in the succeeding generation, by either increasing or decreasing the number of neutrons.

operate a reactor with a negative temperature coefficient. An increase in temperature will then lead to a decrease in k, counteracting the power increase.

It is beyond the scope of this book to discuss in detail the processes determining the value of the temperature coefficient. However, the factors in the four-factor formula that are most affected by an increase in temperature are f and p. The fast fission factor usually increases with increasing temperature because more neutrons are captured in ^{239}Pu, as a consequence of the increased average neutron energy. The decrease in the moderator density at higher temperatures also affects f. The increase in f is, however, balanced by a decrease in resonance escape probability, p, due to thermal vibrations which broaden the width of the resonances. This effect increases the probability of a neutron being captured by a ^{238}U nucleus, and increases the number of neutrons captured in the control rods (used to control the reactor, see Section 1.3.1.3). Normally, the increase in p is greater than that in f, and so the temperature coefficient becomes negative.

1.3.1.3 Nuclear Reactors

At the end of 2013, a total of 434 nuclear reactors were in operation around the world, yielding an electrical power of 371 733 MW (MW_e). A total of 72 reactors were under construction, and 2 were undergoing long-term shutdown (IAEA 2014b). A summary of these is given in Table 1.16.

The release of radioactive elements to the environment from nuclear reactors may occur during accidents, but also during normal operation. UNSCEAR has concluded that the maximum annual effective dose per capita to the world's population from the normal operation of nuclear power plants would be below 0.2 μSv, assuming that the present capacity is maintained for the next 100 years (UNSCEAR 2008). However, this dose not only results from releases from nuclear power plants, but also from all other stages in the nuclear fuel cycle, from mining to final disposal. The contributions made by these other stages will be discussed in more detail later.

As discussed above, operating a nuclear reactor produces radioactive fission fragments, which in turn produce other radioactive elements by radioactive decay. In addition, the neutrons released in fission have the capacity to produce radioactive elements by neutron activation of stable elements in the reactor construction and in the moderator material. Some of these processes are depicted in Figure 1.16 and will be further discussed later in this chapter. An understanding of which radioactive elements may be released into the

TABLE 1.16 The status of the world's nuclear reactors as of 31 December 2013. The different types of reactors are abbreviated as follows: PWR: Pressurized light Water moderated and cooled Reactor, BWR: Boiling light Water moderated and cooled Reactor, PHWR: Pressurized Heavy Water moderated and cooled Reactor, LWGR: Light Water cooled, Graphite moderated Reactor, GCR: Gas Cooled, graphite moderated Reactor, FBR: Fast Breeder Reactor, and HTGR: High Temperature Gas cooled Reactor.

	Operational	Under construction	Long-term shutdown	Planned
PWR	273	60	0	78
BWR	81	4	1	9
PHWR	48	5	0	0
LWGR	15	0	0	0
GCR	15	0	0	0
FBR	2	2	1	5
HTGR	0	1	0	0
Total	434	72	2	92

Source: Data from IAEA (2014b).

environment during operation or when an accident occurs therefore requires a basic understanding of the construction of a nuclear power plant.

Referring to Table 1.16, the most common reactor types are those using light water (H_2O) as both moderator and coolant, and among these, the PWR is the most common. We will here give a brief overview of some of these reactor types, focusing mainly on the PWR and BWR.

A PWR, depicted in Figure 1.17, consists mainly of two circuits known as the primary and the secondary circuits. The reactor vessel and the *steam generators* are located in the primary circuit, while the turbine and condenser are located in the secondary circuit. The primary circuit is contained in the *reactor enclosure* (not shown in Fig. 1.17), consisting of about 1-m-thick concrete with an embedded gas-tight steel plate of about 5 mm thickness. Surrounding the reactor vessel is a radiation shield consisting of about 3-m-thick concrete.

The reactor vessel, made of 15- to 20-cm-thick steel, contains water under sufficiently high pressure that it remains in the liquid phase even at high temperatures. As an example, we will consider a PWR producing thermal power of 3160 MW (1100 MWe), operating at 15.5 MPa, with a water temperature at the exit of the reactor vessel of 325 °C. The flow of water through the core of this reactor is about 14 000 kg s^{-1}.

The water in the primary circuit is fed through 2–4 steam generators, where the water is converted into saturated steam. The heat exchange in the steam generators takes place in U-shaped heating tubes, where the water enters the so-called *hot leg*, exchanges heat with water circulating in the secondary circuit, and leaves the heat exchanger tube through the *cold leg*. Only one U-shaped heating tube is shown in Figure 1.17 for clarity, but in reality there may be about 4000 such tubes in each steam generator to maximize the area available for heat exchange. In the example above, the water leaving the steam generator has a temperature of 286 °C, and the steam has a pressure of 6 MPa and a temperature of 276 °C.

A pressurizer is attached to the primary circuit to maintain a constant pressure. This large tank contains both liquid water and steam, and the amount of steam may be adjusted

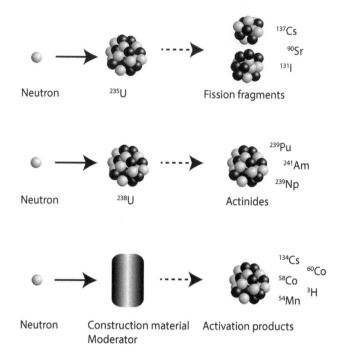

FIGURE 1.16 Production of radionuclides in a reactor. As well as fission products, neutron activation of construction materials and neutron capture also contribute to the production of radionuclides in a reactor system.

by electrically heating the liquid water. Thus, if the pressure in the primary circuit decreases, more steam can be generated in the pressurizer to compensate.

In the secondary circuit, the upper part of the steam generator contains a mixture of steam and water, and the steam generator is equipped with a moisture separator to ensure that only dry steam is fed to the turbine, thus minimizing corrosion. This steam passes through the turbine, and the rotational motion of the turbine is transferred to an electric generator. For practical reasons the turbine system consists of a high-pressure turbine and a low-pressure turbine. As it passes through the turbine, the steam expands and its pressure decreases until it reaches 4 kPa at the entrance of the condenser where the steam is condensed to liquid water.

To achieve efficient condensation, low pressure is desirable in the condenser. However, there is also a lower limit on the pressure, since if it is too low, air will leak into the compressor. The normal ambient air pressure is around 100 kPa, which is considerably higher than the pressure in the condenser. Another reason is that as the pressure of the steam decreases, its temperature will also decrease. If condensation is to take place, there must be a minimum difference between the steam temperature and the temperature of the cooling water, which is typically drawn from the sea, a river, or from a cooling tower. This limits the degree of expansion. Note that the turbine and condenser are totally separated from the water in the reactor, which contains radioactive elements (see below). The primary and secondary circuits are fully isolated from each other.

For a reactor in operation, it is important to know the neutron reproduction factor, k_{eff}, or, rather, the deviation of k_{eff} from 1.0. This is given by the *reactivity*, ρ, defined as

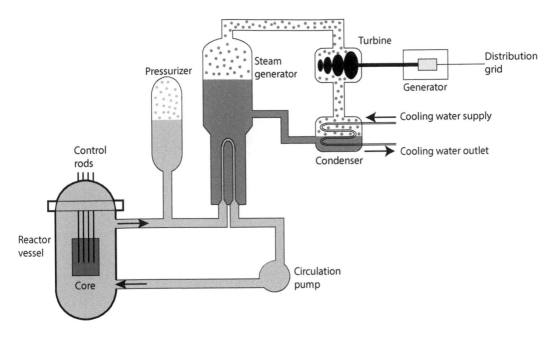

FIGURE 1.17   The basic design of a PWR.

$$\rho = \frac{k_{eff} - 1}{k_{eff}}, \qquad (1.31)$$

and a reactivity of zero corresponds to a reproduction factor of 1. There are several ways to control the reactivity during operation of the reactor. *Control rods* of neutron-absorbing material, such as silver and cadmium, are used to control short-term changes in reactivity. By varying the number of rods inserted into the core, as well as the length of each rod that is inserted, it is possible to regulate the number of neutrons available to cause fission. The reactivity is also controlled by adding boric acid to the water. When fresh fuel is used, the concentration of boric acid can be rather high, but is gradually decreased as the number of fissile nuclei in the fuel decreases due to fission (*burnout*). Varying the concentration of boric acid is, however, a slower process, and it is important for control on a long-term basis. The pH of the reactor water is adjusted by adding LiOH, where the Li is depleted of ^{6}Li to avoid the production of tritium by the reaction

$$^{6}Li + n \rightarrow\, ^{3} H +\, ^{4} He. \qquad (1.32)$$

Similarly, gadolinium (as Gd_2O_3) is added to the fuel to compensate for the higher reactivity of fresh fuel. The isotopes ^{155}Gd and ^{157}Gd have high absorption cross sections ($6.1 \cdot 10^{12}$ barn and $2.5 \cdot 10^{13}$ barn, respectively) and are therefore consumed almost completely. Gadolinium then acts as a burnable absorber or a so-called *burnable poison*.

The general structure of a BWR is depicted in Figure 1.18. The main difference compared with a PWR is that the coolant (water) is allowed to boil in the reactor vessel. The steam produced is then fed through the turbine before condensing in the condenser. A BWR has only a single circuit, and the turbine is therefore not isolated from the radioactive water in this type of reactor.

As an example, let us consider a BWR producing a thermal power of 2540 MW (845

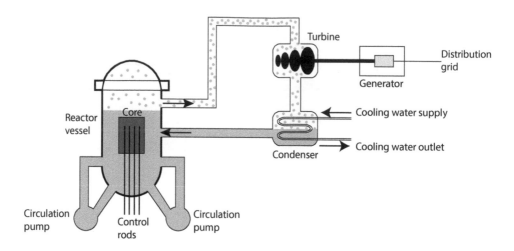

FIGURE 1.18 The basic design of a BWR.

MWe), which operates at 7 MPa and a steam temperature at the exit of the reactor vessel of 286 °C. (The thermal power of BWR's in the world vary between about 500 and 4400 MW.) The flow of water through the core is about 10 000 kg s^{-1}.

To maintain a low pressure in the condenser, residual gases have to be ejected. These gases consist of air that has leaked into the condenser and gases in the steam produced by radiolysis (dissociation of molecules induced by radiation). If fuel damage has occurred, gaseous fission products (such as Xe isotopes) may also be released into the environment through the ejector. Since the steam entering the condenser contains radioactive elements, the condenser is a source of airborne radioactive effluents from the plant.

The reactivity in a BWR may be controlled using control rods made of boron carbide, or by varying the number of steam bubbles, or voids, in the coolant. Both the coolant's ability to absorb neutrons and its moderating capacity are affected by density variations, and thus by the amount of steam it contains, and this will have an effect on the reactivity. The change in reactivity is proportional to the *void coefficient*, which can be either positive or negative, and applies to reactors with liquid coolant or a moderator. If the void coefficient is positive, the reactivity increases with increasing void content, i.e. when less liquid coolant is available for cooling the core. A large positive void coefficient can thus lead to a total loss of coolant. If the void coefficient is negative, as is the case for most BWRs, the reactivity decreases with increasing void content. Conversely, a decrease in void content will then lead to increased reactivity.

The fuel in both PWRs and BWRs consist of cylindrical pellets of uranium dioxide; see Figure 1.19. Each pellet has a diameter of about 8–9 mm and a height of 10–11 mm. The pellets are packed into cladding tubes made of Zircaloy-4, a zirconium alloy consisting of Zr, Sn, Fe, Cr and O. These *fuel rods* are about 4 m long and are bundled into *fuel elements* before being inserted into the core. The arrangement of fuel rods in an element can be, for example, 10 × 10 for BWRs and 17 × 17 for PWRs. The number of fuel elements in the core can vary, depending on the design of the reactor. For example, a BWR may contain 648 elements and a PWR 157 elements. The mass of uranium will then be 125 tonnes and 80 tonnes, respectively. BWR fuel elements are further bundled into supercells, consisting of 4 fuel elements, which allows for the insertion of a cross-shaped control rod between the elements.

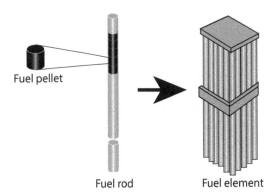

Fuel pellet

Fuel rod   Fuel element

FIGURE 1.19 Fuel pellets of uranium dioxide are packed into fuel rods, which are then bundled into fuel elements.

The use of natural uranium as nuclear fuel is not very efficient due to the low abundance of fissile ^{235}U and the strong absorption of neutrons in the moderator. To increase the amount of fissile material in the fuel, the uranium is therefore enriched to 3–5% ^{235}U by weight. However, if the number of neutrons available for fission could be increased, there would be no need for enrichment of the fuel. This can be achieved by using heavy water (deuterium oxide, D_2O) as moderator and coolant.

The total cross section for both ^{1}H and ^{2}H is dominated by the contribution from elastic scattering, since both nuclei are comparable in mass to the neutron. However, due to the difference in mass between ^{1}H and ^{2}H, the total cross section for thermal neutrons in ^{1}H is 30 barn, while in ^{2}H it is 4 barn (Soppera et al. 2012). In spite of this difference, both can be considered good moderators. However, considering the cross sections for absorption (again, for thermal neutrons), we find that this cross section is far larger for ^{1}H than for ^{2}H: 0.3 barn and $5 \cdot 10^{-4}$ barn, respectively. Thus, heavy water will absorb fewer thermal neutrons, and the amount of ^{235}U in natural uranium will be sufficient to support criticality.

The capture of neutrons in ^{2}H will, however, lead to the production of ^{3}H by the reaction

$$n +^2 H \to^3 H + \gamma, \tag{1.33}$$

and enhanced production of tritium is therefore characteristic of heavy-water reactors.

A reactor that uses fast neutrons to produce fissile material is called a *fast breeder reactor* (FBR). Fissile ^{239}Pu is produced via the absorption of fast, unmoderated neutrons by ^{238}U. Fissile material is thus simultaneously produced (^{239}Pu) and consumed (^{235}U) by the reactor. If the number of neutrons absorbed is sufficiently large, it is possible to produce more fissile material than is consumed. These reactors use liquid sodium as coolant to avoid moderation of the fast neutrons.

Breeding may also be possible using thermal neutrons in a *thermal breeder reactor*. Here, fertile material can be transformed into fissile material by absorption of thermal neutrons, e.g.

$$^{232}Th + n \to^{233} Th \to^{233} Pa + \beta^- + \bar{\nu} \to^{233} U + \beta^- + \bar{\nu}, \tag{1.34}$$

where ^{233}U is fissile, and the half-lives of ^{233}Th and ^{233}Pa are 22.3 m and 27 d, respectively. The high abundance of thorium in the world has led to extensive research, and commercial thermal breeders using thorium as fuel have been proposed.

1.3.1.4 Production of Radionuclides in a Reactor

A self-sustaining neutron flux in a nuclear fuel leads to the production of fission products with long half-lives and considerable amounts of these are thus accumulated in the fuel rods. However, in the absence of fuel damage, these radionuclides remain fixed in the fuel matrix or within the fuel rods. Examples of such fission products are ^{134}Cs , ^{137}Cs, ^{90}Sr, ^{131}I and ^{132}Te. A summary of the fission products produced in a BWR is schematically depicted in Figure 1.20.

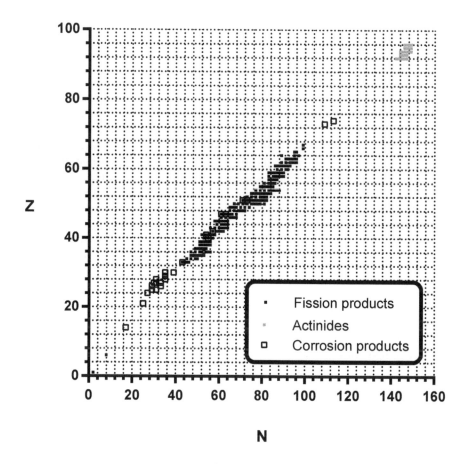

FIGURE 1.20   Fission products, actinides and neutron activated corrosion products found in a BWR.

Tritium (^{3}H) is produced in fuel mainly by *ternary fission,* in which three fission fragments are produced instead of the usual two. Tritium in the coolant is produced by the activation of naturally occurring deuterium in the water, activation of lithium in the water or at the surface of the fuel cladding, as a fission product released during fuel damage, and by activation of boron in the control rods from where it may be released through cracks in the rods.

The production of ^{14}C in fuel is the result of two reactions: ^{17}O (n,α) ^{14}C and ^{14}N (n,p)^{14}C. The ^{14}C present in the coolant has its origin in the activation of naturally occurring

^{17}O and ^{14}N in the water, releases from the fuel during fuel damage, and by activation of construction materials containing nitrogen.

The radionuclide ^{134}Cs is an important example as it is also created by neutron capture in the stable isotope ^{133}Cs, at the end of the mass chain with $A = 133$. Figure 1.21 shows the atomic mass of the nuclides with $A = 133$, and data on the nuclides are given in Table 1.17. The nuclides on the left side of the curve decrease their mass by negative β-decay, while those on the right decay mostly by electron capture. The production of any of the fission products on the left side of the curve will thus eventually lead to the production of ^{133}Cs. The fission products on the right side have low fission yields, and the long half-life of ^{133}Ba prevents the production of ^{133}Cs via this route. As shown in Figure 1.9, the fission yield for mass number 133 is high and these nuclides are thus abundant in the fuel.

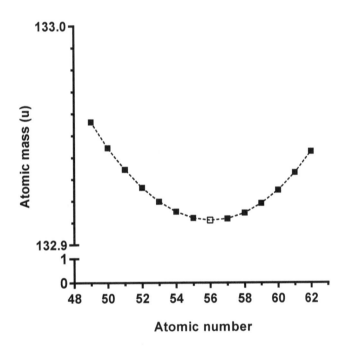

FIGURE 1.21 Atomic masses of nuclides with $A = 133$. The nuclide with the lowest mass, ^{133}Cs, is eventually reached by β-decay or electron capture by nuclei with higher mass.

In addition to fission products and radionuclides produced by their decay, radioactive elements heavier than uranium are produced by neutron capture in ^{238}U and successive β-decays, as shown in Figure 1.22. These elements are referred to as *transuranic elements* or *actinides*, e.g. ^{239}Pu, ^{240}Pu, ^{242}Am and ^{242}Cm. The actinides found in a BWR are shown in Figure 1.20.

The relative abundance of actinides in reactor fuel depends on the age of the fuel, i.e. how long the fuel rod has been in the reactor. For example, the half-lives of ^{239}U and ^{239}Np are 23 m and 2.36 d, respectively, which leads to the rather rapid production of ^{239}Pu. This nuclide is fissile and will therefore also produce fission products. The production of both ^{241}Am and ^{242}Cm will, however, occur at a slower rate due to the rather long half-life of ^{241}Pu (14.3 y). The abundance of any given nuclide will thus depend upon a combination of the capture cross sections and half-lives of the β-emitting nuclides.

TABLE 1.17 Nuclides with mass number 133, their decay mode, and half-lives. β-n stands for delayed neutron emission after β-decay. Nuclear excited states that are populated after β-decay often emit γ-rays, but some are unstable against emission of nucleons, in this case a neutron.

Nuclide	Decay mode	Half-life
^{133}In	β^- (β-n)	165 ms
^{133}Sn	β^-	1.46 s
^{133}Sb	β^-	2.34 m
^{133}Te	β^-	12.5 m
^{133}I	β^-	20.83 h
^{133}Xe	β^-	5.2475 d
^{133}Cs	–	stable
^{133}Ba	ε	10.551
^{133}La	ε	3.912 h
^{133}Ce	ε	97 m
^{133}Pr	ε	6.5 m
^{133}Nd	ε	70 s
^{133}Pm	ε	13.5 s
^{133}Sm	ε	2.89 s

Source: Data from Tuli (2011).

Knowledge of the total inventory (i.e. the total activity) of radionuclides in the core is important in order to make correct decisions regarding, for example, potential accidental releases and radiation shielding during normal operation. Due to their physical and chemical behaviour, the radionuclides can be divided into six groups: ^{3}H & ^{14}C, rare gases, halogens, other soluble fission products, other insoluble fission products, and actinides. The relative contribution to the total inventory for each of these groups in a BWR is shown in Figure 1.23.

From a radiation protection perspective, the rare gases and halogens have the greatest impact, both under normal operation and in the case of accidents involving radioactive releases. However, some of the radionuclides produced are of lesser importance due to low production rate or short half-lives. The most important radioisotopes among the rare gases and halogens are given in Table 1.18 together with their activities in a BWR.

Other fission products, characterized as soluble or insoluble, are summarized in Table 1.19. These may be released into the environment following a serious accident in which the fuel cladding is damaged. Only radionuclides with an activity above 10^{16} Bq in a BWR and half-lives longer than 6 h are given in the table.

However, under normal operation the most important radionuclides in terms of human exposure are the neutron activation products created by neutron capture in metallic corrosion products and in the reactor water. The corrosion products consist of small amounts of nickel, iron or cobalt, which are activated by neutrons to form radionuclides such as ^{58}Co, ^{60}Co, ^{54}Mn, ^{65}Zn and ^{110m}Ag. During transport through the coolant system, these neutron activated metallic impurities can be deposited on the inner walls of the reactor vessel or on other system surfaces. During the replacement of fuel rods, the otherwise hermetically sealed systems are opened, and a substantial amount of activation products may be dispersed to other parts of the reactor building, as well as in spillage or waste water from the washing and handling of contaminated objects. The neutron activated corrosion products typically present in a BWR are shown in Figure 1.20.

TABLE 1.18   Example of an inventory of important rare gas and halogen radionuclides in a BWR.

Rare gases	Inventory (Bq)	Halogens	Inventory (Bq)
^{83m}Kr	$3.15 \cdot 10^{17}$	^{128}I	$2.92 \cdot 10^{16}$
^{85}Kr	$4.50 \cdot 10^{16}$	^{129}I	$1.35 \cdot 10^{11}$
^{85m}Kr	$6.72 \cdot 10^{17}$	^{130}I	$5.21 \cdot 10^{16}$
^{87}Kr	$1.33 \cdot 10^{18}$	^{131}I	$2.58 \cdot 10^{18}$
^{88}Kr	$1.78 \cdot 10^{18}$	^{132}I	$3.77 \cdot 10^{18}$
^{89}Kr	$2.21 \cdot 10^{18}$	^{133}I	$5.33 \cdot 10^{18}$
^{131m}Xe	$3.40 \cdot 10^{16}$	^{134}I	$5.99 \cdot 10^{18}$
^{133}Xe	$5.15 \cdot 10^{18}$	^{135}I	$5.08 \cdot 10^{18}$
^{133m}Xe	$1.65 \cdot 10^{17}$	^{136}I	$2.04 \cdot 10^{18}$
^{135}Xe	$1.73 \cdot 10^{18}$	^{82}Br	$8.04 \cdot 10^{15}$
^{135m}Xe	$1.11 \cdot 10^{18}$	^{83}Br	$3.13 \cdot 10^{17}$
^{137}Xe	$4.80 \cdot 10^{18}$	^{84}Br	$5.70 \cdot 10^{17}$
^{138}Xe	$4.55 \cdot 10^{18}$	^{84m}Br	$2.18 \cdot 10^{16}$
^{139}Xe	$3.36 \cdot 10^{18}$	^{85}Br	$6.69 \cdot 10^{17}$
^{140}Xe	$2.29 \cdot 10^{18}$		

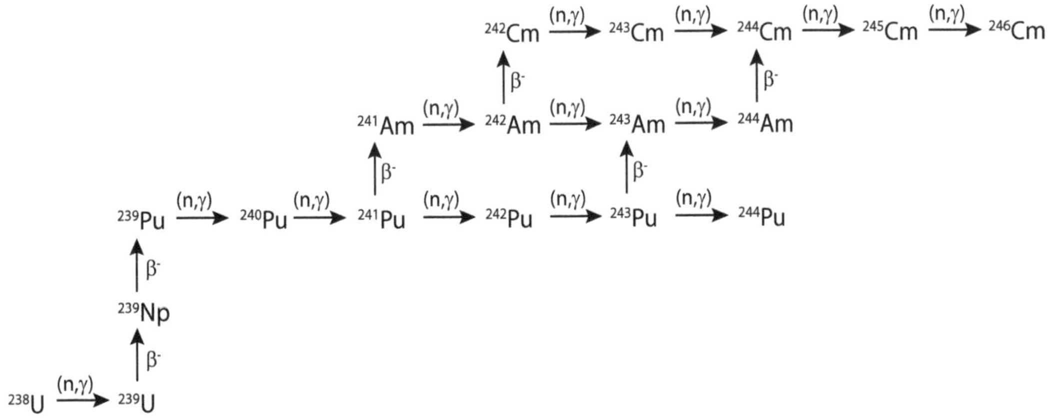

FIGURE 1.22   Production of actinides by neutron capture and β-decay. Several of the actinides (^{239}Pu, ^{240}Pu, ^{241}Pu, ^{243}Pu, ^{242}Am, ^{243}Cm, ^{244}Cm and ^{245}Cm) are fissionable by reactor neutrons.

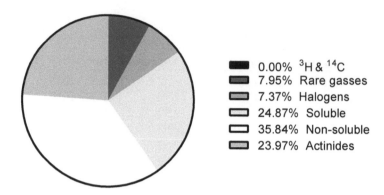

FIGURE 1.23 Relative contribution to the total inventory from radionuclides in the reactor core according to their physical and chemical characteristics.

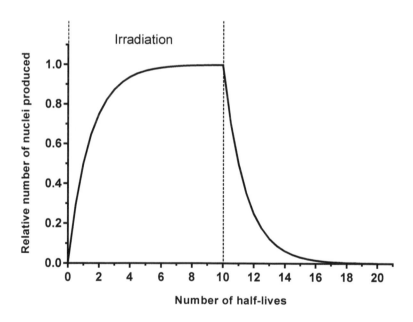

FIGURE 1.24 During activation, the number of nuclei produced increases as a result of simultaneous activation and decay. At the end of activation, production stops and the nuclide decays according to its half-life. Note that the time is given as the number of half-lives for the radionuclide produced.

TABLE 1.19 Soluble and insoluble fission products present in the reactor core of a BWR with an activity above 10^{16} Bq and half-lives longer than 6 h.

Soluble	Inventory (Bq)	Insoluble	Inventory (Bq)
^{99}Mo	$4.84 \cdot 10^{18}$	^{89}Sr	$2.45 \cdot 10^{18}$
^{99m}Tc	$4.29 \cdot 10^{18}$	^{90}Sr	$3.68 \cdot 10^{17}$
^{103}Ru	$4.05 \cdot 10^{18}$	^{91}Sr	$3.13 \cdot 10^{18}$
^{106}Ru	$1.63 \cdot 10^{18}$	^{90}Y	$3.78 \cdot 10^{17}$
^{109}Pd	$8.48 \cdot 10^{17}$	^{91}Y	$3.19 \cdot 10^{18}$
^{112}Pd	$6.63 \cdot 10^{16}$	^{93}Y	$3.83 \cdot 10^{18}$
^{110m}Ag	$1.09 \cdot 10^{16}$	^{95}Zr	$4.32 \cdot 10^{18}$
^{111}Ag	$1.52 \cdot 10^{17}$	^{97}Zr	$4.42 \cdot 10^{18}$
^{115}Cd	$2.30 \cdot 10^{16}$	^{95}Nb	$4.29 \cdot 10^{18}$
^{121}Sn	$1.90 \cdot 10^{16}$	^{95m}Nb	$4.80 \cdot 10^{16}$
^{125}Sn	$2.21 \cdot 10^{16}$	^{105}Rh	$2.63 \cdot 10^{18}$
^{125}Sb	$3.09 \cdot 10^{16}$	^{140}Ba	$4.57 \cdot 10^{18}$
^{127}Sb	$2.33 \cdot 10^{17}$	^{140}La	$4.83 \cdot 10^{18}$
^{128}Sb	$3.62 \cdot 10^{16}$	^{141}Ce	$4.31 \cdot 10^{18}$
^{127}Te	$2.29 \cdot 10^{17}$	^{143}Ce	$4.01 \cdot 10^{18}$
^{127m}Te	$3.80 \cdot 10^{16}$	^{144}Ce	$3.33 \cdot 10^{18}$
^{129m}Te	$1.30 \cdot 10^{17}$	^{142}Pr	$1.71 \cdot 10^{17}$
^{131m}Te	$4.96 \cdot 10^{17}$	^{143}Pr	$3.89 \cdot 10^{18}$
^{132}Te	$3.69 \cdot 10^{18}$	^{145}Pr	$2.73 \cdot 10^{18}$
^{134}Cs	$6.88 \cdot 10^{17}$	^{147}Nd	$1.72 \cdot 10^{18}$
^{136}Cs	$1.52 \cdot 10^{17}$	^{147}Pm	$5.91 \cdot 10^{17}$
^{137}Cs	$5.26 \cdot 10^{17}$	^{148}Pm	$4.26 \cdot 10^{17}$
		^{148m}Pm	$1.16 \cdot 10^{17}$
		^{149}Pm	$1.43 \cdot 10^{18}$
		^{151}Pm	$4.98 \cdot 10^{17}$
		^{153}Sm	$1.18 \cdot 10^{18}$
		^{156}Sm	$6.13 \cdot 10^{16}$

Neutron irradiation of stable or radioactive nuclei produces radioactive nuclei at a rate determined by the number of available nuclei, the neutron fluence rate, and the reaction cross section. The radionuclides present in the reactor core of a BWR decay at a rate determined by their half-life. The simultaneous action of these two processes leads to an increase in the number of nuclei produced that asymptotically approaches an equilibrium value. At the end of neutron irradiation, production stops, and the number of nuclei continues to decrease at a rate determined by their half-life. This process is depicted in Figure 1.24. The time taken to reach equilibrium varies greatly, depending on the half-life of the radionuclide produced. For example, ^{16}N with a half-life of 7.35 s, reaches equilibrium within 1.5 m, while ^{60}Co reaches equilibrium only after 25–30 y because its half-life is much longer (5 y). The number of nuclear reactions per unit time, the production rate, is determined by the number of irradiated nuclei, n, the neutron fluence rate, Φ, and the reaction cross section, σ, according to

$$P = n \cdot \Phi \cdot \sigma. \tag{1.35}$$

If the neutron fluence rate is expressed in s^{-1} cm^{-2} and the reaction cross section in cm^2, the production rate is given in s^{-1}. The net change in the number of radioactive nuclei produced, dN/dt, is determined by the difference between the increase resulting from activation and the decrease resulting from radioactive decay

$$\frac{dN}{dt} = P - \lambda \cdot N,$$
(1.36)

where λ is the decay constant of the radionuclide produced. This equation has the solution

$$N = \frac{P}{\lambda} \left(1 - e^{-\lambda \cdot t}\right)$$
(1.37)

and the activity of the radionuclide produced at a given time, t, is then given by

$$A = \lambda \cdot N.$$
(1.38)

1.3.2 Nuclear Weapons

1.3.2.1 Nuclear Weapons Tests

The first test of a nuclear weapon was performed on 16 July 1945 at the Alamogordo bombing range in New Mexico in the southwestern United States, and was code-named *Trinity*. This bomb was identical to the one called *Fat Man* dropped over Nagasaki in Japan on 9 August 1945, with a yield corresponding to 21 kilotonnes of TNT (trinitrotoluene). This unit is commonly used when rating the output of a nuclear weapon, where 1 tonne of TNT equals 4.184 GJ (originally defined as 1 Gcal). Both Trinity and Fat Man were based on plutonium (see below for a discussion of various designs). However, the first bomb dropped over Japan, on 6 August 1945 was called *Little Boy* and was based on uranium. The yield of this bomb is estimated to have been 15 ktonnes of TNT.

Between 1945 and 1958, when a moratorium, or limited test ban treaty, was established, 272 tests were performed in the atmosphere by the USA, the USSR and the UK. In addition, 22 underground explosions were performed by the USA, together with 6 underwater tests by the USA and the USSR. Thus, during this first test period, 300 tests were conducted, including the first test of a 10-Mt thermonuclear weapon, the *Ivy Mike* test, on 1 November 1952. This hydrogen bomb was tested by the USA at the Enewetak Atoll in the Pacific Ocean. (See below for a discussion of the various types of nuclear weapons.)

The second testing period started in 1960 when France performed three atmospheric tests, followed by one atmospheric and one underground test in 1961. In 1961, the USA and the USSR also started a new series of tests, ending in 1963 with the *Partial Nuclear Test-Ban Treaty*, PTBT, signed by the USA, the USSR and the UK. Due to growing concern about tests of high-yield thermonuclear weapons in the atmosphere, the three countries then agreed to ban nuclear weapons testing in outer space, in the atmosphere, and under water. However, during the period 1963–1980, France and China performed 69 atmospheric tests.

The third testing period, after the PTBT in 1963, involved a large number of underground tests. The number of tests after the test ban is, in fact, three times higher than before the partial ban. However, the total yield of the atmospheric tests is estimated to have been 440 Mt, compared to the total yield of the underground tests, estimated at 80 Mt. The total number of tests performed from 1945 onwards, in the atmosphere and underground, is given in Figures 1.25 and 1.26. Note that, due to the confidential nature of this subject and the difficulties in defining a *test*, the figures are approximate, being based on official government sources and information from research institutes.

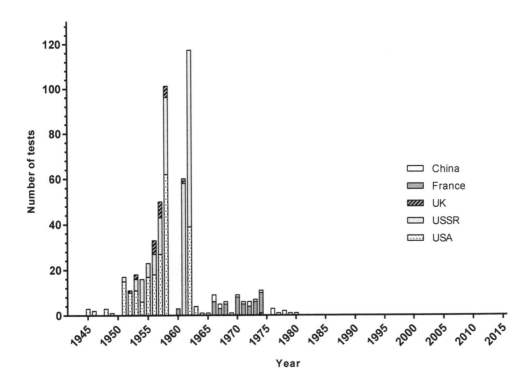

FIGURE 1.25   Atmospheric nuclear weapons tests. Data from CTBTO.

In 1968, the *Nuclear Non-Proliferation Treaty* was signed by the states officially possessing nuclear weapons at that time. These five states (USA, USSR, UK, France and China) declared that they would not transfer nuclear weapons or technology to other states, and that they would decrease the number of their own nuclear weapons. Since then, most of the states in the world have joined or ratified the treaty.

The *Comprehensive Nuclear-Test-Ban Treaty*, CTBT, was opened for signature in 1996. According to this treaty, nuclear explosions are completely prohibited, no matter where or by whom they are conducted. Most of the states in the world have signed and ratified the treaty (the current status can be found at http://www.ctbto.org/map/#status). To establish a verification system to detect nuclear explosions, the CTBTO (*Comprehensive Nuclear-Test-Ban Treaty Organization*) was created by the UN in 1996. The verification system maintained by the CTBTO now includes various kinds of instruments distributed throughout the world. Physical effects such as seismic waves, infrasound, and the detection of radioactive xenon by air filter stations are used to detect the characteristic signs of nuclear explosions.

Figure 1.27 shows the air activity concentration of ^{137}Cs in μBq m^{-3} measured in Stockholm, Sweden, during the period 1957–2013. Also shown is the total number of yearly atmospheric tests. It can be seen that the measured air concentration is correlated with the atmospheric tests, and that even the tests performed in China, on the other side of the world, have a measurable air concentration in Sweden. The two large peaks are due to the releases from the nuclear power plant accidents at Chernobyl in 1986 and at Fukushima in 2011.

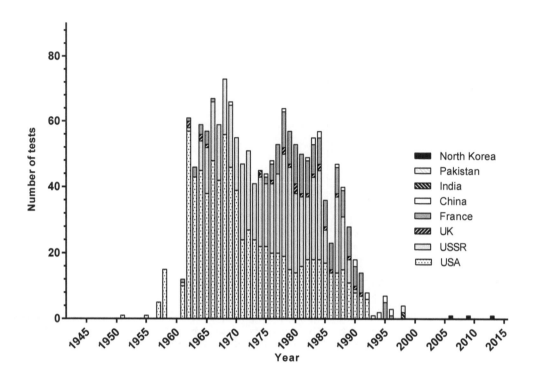

FIGURE 1.26   Underground nuclear weapons tests. Data from CTBTO.

According to the CTBT, detonations of nuclear weapons and all other nuclear explosions are banned, but there is no general agreement on what defines a nuclear explosion. The aim of the treaty is not to ban basic research or peaceful applications and too narrow a definition would allow applications that ought to be banned. In addition to a "scientific" definition, there is also a political debate on the range of such definitions, since a certain degree of testing is deemed necessary to ensure that existing weapons remain safe to handle.

One reason for testing a nuclear weapon is to study how well it functions, depending on its design and construction. The purpose could also be to determine how detonation of the weapon will influence military equipment (*weapons effects tests*). Reactors and accelerators could be used for these tests. *Safety tests*, some of which are included in Figures 1.25 and 1.26, are performed to ensure that a nuclear device cannot be detonated by events such as fire or unauthorized use. A nuclear device should therefore only be detonated intentionally. Some tests have been performed to study the characteristic signs of a nuclear explosion, and so enhance the capability of identifying tests not in compliance with the test ban treaty.

There are, however, several ways to test the construction of a nuclear weapon, without the need to perform a full-scale nuclear explosion. Among these are *hydrodynamic tests*, where the dynamics during detonation can be studied using metals in liquid form. The fissile fuel can then be replaced by a metal such as depleted uranium. The behaviour of fissile material when rapidly compressed is studied in *hydronuclear tests*. These can be performed with varying yield, from zero to a full-scale explosion. Even if the yield is less than the yield of a full-scale explosion, these tests are sometimes called *zero-yield*. These kinds of tests were performed during the moratorium of 1958–61. *Subcritical tests* were

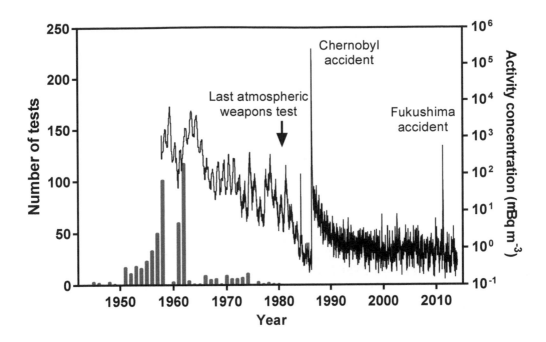

FIGURE 1.27 Activity concentration of ^{137}Cs in the air in Stockholm, Sweden, 1957–2013, and the total number of yearly atmospheric tests. Data from the Swedish Defence Research Agency.

performed by the USA and Russia at the end of the 1990s and beginning of the 2000s. In these tests, the amount of fissile material is intentionally too low to permit a critical mass to be formed. These tests are similar to hydrodynamic tests, but in this case the properties of real plutonium are studied. Material properties may also be studied using powerful lasers.

Modern-day computer capacity enables complicated processes to be simulated within a reasonable calculation time. In 2002, scientists at Los Alamos and Lawrence Livermore National Laboratories in the USA made three-dimensional simulations of a nuclear weapons explosion (NNSA 2002). The simulation time was, however, about 4 months, which can be compared to a real explosion which lasts a few microseconds. With further developments in processor speed, it could be argued that these simulations should be regarded as nuclear weapons tests. However, access to advanced Monte Carlo codes for the calculation of the factors required, such as neutron transport, is highly restricted.

The total yield from all known atmospheric nuclear tests is given in Table 1.20. These tests have contributed to the dispersion of radioactive debris and subsequent fallout that influences the radiation dose to the world's population. UNSCEAR has estimated that the annual per capita effective dose has now declined to <0.005 mSv from its maximum of 0.11 mSv in 1963. At the time of active atmospheric testing, ^{95}Zr and its daughter ^{95}Nb were the main contributors to the external effective dose. However, since 1966, ^{137}Cs has become the main contributor, due to its long half-life, and it is today the only radionuclide from fallout that makes a significant contribution to the external effective dose.

The radionuclides released in fallout and their contributions to the annual per capita effective dose, received before 2000, are given in Table 1.21. Between 2000 and 2100 it is estimated that only ^{3}H (0.1 μSv), ^{14}C (120 μSv), ^{90}Sr (8.6 μSv), ^{125}Sb (0.003 μSv) and

TABLE 1.20 Yield in Mt of atmospheric nuclear tests by country. The total yield is divided into the yields from fission tests and fusion tests.

Country	Yield (Mt)		
	Fission	Fusion	Total
USSR	85.3	162	247.2
USA	81.5	72.2	153.8
China	12.2	8.5	20.72
France	6.17	4.02	10.19
UK	4.22	3.83	8.05

Source: Data from UNSCEAR (2008).

^{137}Cs (124 μSv) will contribute to the total per capita effective dose of 253 μSv. Of these, only ^{137}Cs contributes to the external exposure (114 μSv external and 10 μSv internal) and after 2100 the effective dose (2243 μSv) will be received from ^{14}C (2230 μSv), ^{90}Sr (0.02 μSv) and ^{137}Cs (13 μSv) (UNSCEAR 2008).

However, some tests resulted in severe contamination of the local environment and significant doses to neighbouring populations. The most significant was the *Castle Bravo* test of a thermonuclear weapon at Bikini Atoll on 1 March 1954. A sudden change in wind direction and a miscalculation of the yield led to heavy fallout over inhabited areas and the forced evacuation of nearby islands, including the Rongelap and Rongerik Atolls. The effective doses from external exposure caused by the extended area of fallout and the higher yield was 1.9 Sv on Rongelap, while US service personnel at Rongerik received 0.8 Sv. Equivalent doses to the thyroid reached 200 Sv on Rongelap. In addition to the inhabitants of the islands, a Japanese fishing boat, *Lucky Dragon*, was in the area and 23 fishermen were exposed to the fallout. The external effective doses to the fishermen ranged between 1.7 and 6 Sv (UNSCEAR 2008). The fishermen began to recover during May, but some of the men were infected at the hospital by hepatitis b, which eventually killed one of the men (Lindell 2003).

The hydrogen bomb, which was assumed to have a yield of 5 Mt, contained lithium in the form of LiD, which is an essential compound in the construction of thermonuclear devices since the fuel, ^{3}H, is produced by the reaction ^{6}Li(n,α) ^{3}H. The lithium used, however, consisted of a large amount of ^{7}Li, which produces ^{6}Li by the reaction ^{7}Li(n,2n)^{6}Li. The amount of ^{6}Li, and hence the amount of ^{3}H produced was thus higher than expected, which increased the yield by about a factor of three, to more than 15 Mt.

1.3.2.2 Effects of Nuclear Weapons

The destructive action of a nuclear weapon is mainly due to *shock waves*, and in this respect they are similar to conventional explosives. The physics behind an explosion is that large amounts of energy are released within a limited volume. The resulting sudden increase in pressure and temperature converts the material in the device into gases in a highly compressed state. Since the shell is also converted to gas, there is nothing to prevent rapid expansion, and a shock wave is produced. In air, the term *blast wave* is generally used for this pressure wave.

There are, however, a number of features that distinguish nuclear explosions from conventional explosions:

TABLE 1.21   Radionuclides released in fallout from atmospheric nuclear tests and their contributions to per capita annual effective dose (μSv).

Radionuclide	External	Inhalation	Ingestion	All
^{3}H			24	24
^{14}C			144	144
^{54}Mn	19	0.1		19
^{55}Fe		0.01	6.6	6.6
^{89}Sr		2.6	1.9	4.5
^{90}Sr		9.2	97	106
^{91}Y		4.1		4.1
^{95}Zr	81	2.9		84
^{103}Ru	12	0.9		13
^{106}Ru	25	35		60
^{125}Sb	12	0.1		12
^{131}I	1.6	2.6	64	68
^{140}Ba	27	0.4	0.5	28
^{141}Ce	1.1	0.8		1.9
^{144}Ce	7.9	52		60
^{137}Cs	166	0.3	154	320
^{239}Pu		20		20
^{240}Pu		13		13
^{241}Pu		5		5
Total	353	149	492	994

Source: Data from UNSCEAR (2008).

- The energy released in a nuclear explosion can be orders of magnitude higher than in conventional explosions.

- The initiation of the pressure wave is different because in a nuclear weapon a smaller mass is required to release a given amount of energy.

- The higher temperatures reached in a nuclear explosion (about 10^7 °C) will cause a large part of its energy to be emitted as light and heat, e.g. *thermal radiation*.

- The radioactive fission products produced in the chain reaction will give rise to *initial ionizing radiation*. This radiation may cause skin burns and fires at large distances from the detonation site.

- Several radionuclides with relatively long half-lives are produced in the explosion and in subsequent decay of the fission products. These may cause radioactive fallout over large areas, depending on the height of the detonation.

The fraction of the energy emitted as thermal radiation depends on the environment in which the weapon is detonated. For example, an atmospheric detonation below an altitude of about 30 km will release 35–45% of its energy as thermal radiation in the visible and infrared wavelengths. This is due to degradation of a large part of the initial thermal radiation, for example, through the absorption of soft x-rays and heating of the air close to the burst. Below about 12 km, about 50% of the energy is released as a blast wave. At this altitude the shock energy thus corresponds to the air shock produced by a conventional explosive with half the explosive yield (a 20-kt nuclear weapon corresponds to a conventional weapon containing 10 kt of TNT). The portion released in the blast wave decreases with altitude because of the decreasing density of the air, and the proportion of energy released as thermal radiation increases correspondingly (Glasstone and Dolan 1977).

The fraction of the energy released as shock and thermal radiation from a nuclear fission weapon is about 85%, regardless of the height of detonation. The rest of the energy, 15%, is released as nuclear radiation, i.e. γ-radiation, particles and x-rays. One third of this energy (5% of the total energy released) is the *initial radiation*, mainly γ-radiation, which is emitted within some minutes after the explosion. The rest is *delayed radiation* resulting from the decay of various radionuclides produced in the fission process. This radiation is emitted over a considerably longer period of time.

A thermonuclear weapon (see below) also contains a fission device, but this makes a small contribution to the energy released. The smaller fissile mass and the fact that the main thermonuclear reaction produces no fission products will lead to less production of radionuclides. Thus, only about 5% of the energy released in the explosion is carried by delayed nuclear radiation. The fraction of the energy delivered by delayed radiation is often not included when stating the yield of a weapon.

The effects of a nuclear weapon are largely determined by the location of the explosion in relation to the surface of the earth, and five main types of explosions can be identified (Glasstone & Dolan 1977): *air bursts, high-altitude bursts, underwater bursts, underground bursts* and *surface bursts*. These can be regarded as typical cases, but real explosions may be combinations of the above.

The definition of an air burst is an explosion which is below about 30 km in altitude, but which is sufficiently high that the *fireball* created by the explosion will not reach the ground. This roughly spherical fireball is formed by the incorporation of surrounding material in the high-temperature region of the exploding core. The surface temperature of the fireball (and hence the luminance) does not vary significantly with the yield, and thus the observed brightness of a fireball will be approximately the same regardless of the yield. Since the fireball formed by an explosion of a 1 Mt weapon may be more than 1500 m in diameter, the minimum height of an air burst in practice is about 750 m. The qualitative aspects of an air burst are essentially the same regardless of the yield, in that the air blast contains almost all of the shock energy. Part of the energy is, however, also transferred to the ground, to an extent determined by the height of the explosion.

The fallout from an air burst is of minor concern, unless the burst occurs near the earth's surface. The fission products may then adhere to soil particles, which will become contaminated and return to the ground close to the site of the explosion. If the weapon is detonated at higher altitudes, the fission products will be dispersed in the lower atmosphere and follow air movements prior to deposition. The subjects of atmospheric dispersion and deposition will be discussed in Section 3.2.

An explosion that occurs in the atmosphere above about 30 km altitude is defined as a high-altitude burst. The fraction of the energy contributing to the shock wave decreases with increasing altitude and is less than the shock wave portion of an air burst. The thermal energy radiated is also affected by this decrease in shock wave energy, since more energy will be converted to thermal radiation. In addition, the thermal energy is able to travel a longer distance due to the lower density of the surrounding air, since there are fewer air molecules that can interact with the radiation.

The amount of nuclear radiation emitted from an explosion of a nuclear weapon depends on the yield, but the fraction of the energy released as ionizing radiation is independent of the altitude of the explosion. However, the relative fluence of γ-radiation and neutrons changes with altitude because of different interaction mechanisms in the atmosphere (e.g. γ-radiation is produced when neutrons interact with nitrogen nuclei in the lower atmosphere). Since the fraction of the released energy converted to nuclear radiation in a high-altitude burst is higher than in an air burst, and the fact that much of the γ-radiation is produced at rather low altitudes, the fluence of initial nuclear radiation will be higher. The delayed nuclear radiation (the fallout) is, however, of little concern.

Another important effect of a high-altitude burst is the *electromagnetic pulse*, EMP. This is an effect of the free electrons released when prompt γ-radiation interacts with the surrounding air. Although this effect is also produced in surface bursts and bursts in the lower atmosphere, the affected area is considerably larger for high-altitude bursts. An EMP is similar to radio waves, although with much higher strengths and a wide range of frequencies and amplitudes. As they resemble radio waves, they can be picked up by antennas and generate high voltages and strong currents in electrical equipment. An EMP thus has the capacity to disrupt power generation systems, communications and computers at large distances from the site of the explosion.

The prompt γ-radiation will interact with the surroundings of the burst, mainly by Compton scattering, producing electrons and positively charged ions. Due to the mass difference, the electrons will move away from the burst faster than the heavier ions, and a region of charge separation is created. This results in an electrical field between the positive region nearer to the burst and the negative region further away, which can reach a maximum in 10^{-8} s. In perfectly homogeneous surroundings, this field would be static and no energy would be radiated, but since the conditions are not perfect, an electron current will be generated. This current is generally in the upward direction since most of the disturbances in homogeneity occur in the vertical direction (because of density differences and proximity to the ground). The EMP is emitted perpendicular to the current.

For surface bursts (see below), the conductivity of the ground can enhance the EMP by forming a return path for the electrons. Thus, current loops are formed, which generate large magnetic fields running clockwise around the burst point (seen from above). This process is equivalent to a current flowing in a solenoid, producing a magnetic field at its centre. Both the electric and magnetic fields developed in the ground contribute to the EMP.

An underground burst or an underwater burst is defined as an explosion that occurs such that its centre is below the ground or under the water surface, respectively. If the explosion occurs at a rather shallow depth, some of the energy is released as an air blast, but most of the energy appears as shock. The thermal radiation is generally absorbed very locally and in some cases may melt the surrounding ground minerals. However, some thermal radiation may escape into the atmosphere. Contamination of the explosion site will be considerable, since almost all of the radionuclides produced will be retained in the ground or dispersed in the water.

An explosion that occurs at, or close to, the surface is referred to as a surface burst. The heat generated in the explosion will vaporize some of the ground material, which will then be mixed with the fireball. Depending on the structure of the ground below the explosion and the explosion height, the updraft of air into the cloud (see below) may also cause debris from the ground to be transported up into the cloud. A surface blast will result in a large amount of debris being drawn into the cloud, while the fission products are still vaporized. The fission products will then condense on dirt particles, producing a large amount of fallout. An air burst producing a small updraft of debris will produce less contaminated dirt particles since the particles reach the cloud after the fission products have condensed. The fallout from a surface burst is therefore considerable, as is the EMP. In addition, a crater will be formed by vaporization and removal of ground material.

For example, a 1-kt surface burst will create a crater 9 m deep, with a radius of 20 m. Most people within a radius of about 700 m will die, if not shielded, due to various effects, depending on the distance from the explosion site. Within a radius of 450 m, thermal radiation causes third-degree burns (extending through the entire skin, or dermis) and within a radius of 700 m, the burns will be first-degree (extending through the outer layer of skin, the epidermis). At this distance, the initial ionizing radiation will cause lethal damage that results in death within a few weeks. Further away, at 960 m, the radiation dose from the initial ionizing radiation is at the limit where tissue effects will develop. The shock wave will account for half of the deaths within a radius of 200 m and cause lung damage almost 300 m from ground zero. The delayed radiation from the radioactive fallout can be expected to cause an effective dose rate of 3 Sv during the first hour after the explosion within an approximately elliptically shaped area with the dimensions 200 m by 7000 m.

The fireball, which consists of hot air and gaseous weapon materials, is developed within 10^{-6} s. After its formation, it immediately starts to grow in size and its temperature decreases, partly because it incorporates a larger mass of air and partly because energy escapes as emitted radiation. The fireball from a 1-Mt weapon attains a diameter of about 130 m within 0.7 ms, and after 10 s will have grown to its maximum size, about 1700 m. The heat of the fireball also causes it to rise at a rate of about 75–105 m s^{-1} to its maximum height of about 7 km from the explosion point. It has then cooled to a temperature where visible radiation is no longer emitted. During this inflation, the vapours in the interior of the fireball start to condense, forming solid particles and water droplets (Glasstone and Dolan 1977).

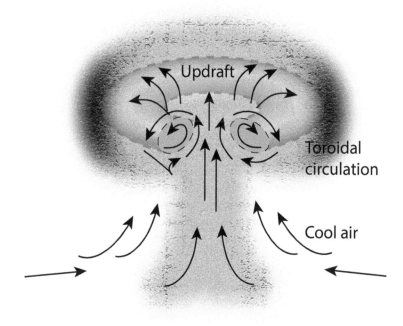

FIGURE 1.28 Schematic view of the toroidal circulation within the radioactive cloud, and the air currents, resulting from a nuclear explosion.

As the rising fireball is cooled and rises through the surrounding air, its shape is gradually changed from the initial spherical form to a toroidal shape. This is depicted in Figure 1.28, which also indicates that the toroidal circulating matter is more or less hidden by the cloud of water droplets and debris. The fireball is further cooled by an updraft of colder air through the centre of the toroid, and this cooling will eventually stop the toroidal motion. The cloud turns white due to scattering of light by water droplets (as in cumulus clouds). The reddish colour seen at the beginning of the formation of the cloud is due to chemical interactions taking place at higher temperatures.

The maximum height of the cloud is reached after about 10 minutes, after which it continues to grow laterally. This lateral growth gives the cloud its characteristic mushroom shape. The radius of the head of the cloud is about twice the radius of the stem for a burst with a yield less than 20 kt, and at higher yields, the cloud may be much wider than the stem. For example, for a burst of a few Mt, the cloud may be five times as wide as the stem.

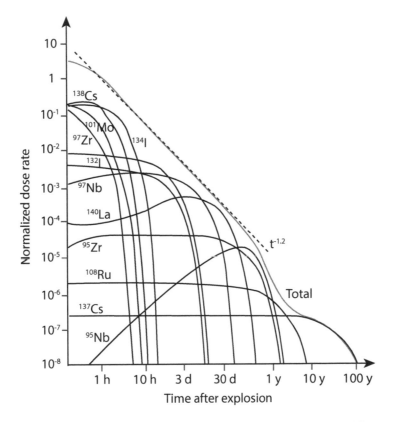

FIGURE 1.29 The total dose rate from fallout decreases approximately as $t^{-1.2}$. The curves have been normalized to unit dose rate at 1 h after the explosion. This is also discussed in Section 7.1.

The vast amounts of fission products produced in the explosion will result in a time-dependent total activity that depends on the decay of the fission products and simultaneous production of decay products, as depicted in Figure 1.29. Although the activity of the individual radionuclides shows a complicated pattern of growth and decay, the decrease in the total activity of the fallout is approximately proportional to $t^{-1.2}$, where t is the time after the explosion. This approximate proportionality holds for times between about 0.5 and 5000 h after the explosion. Note that the expression is only valid if the fallout has ceased and no removal, for example by weather, occurs.

The curve shown in Figure 1.29 is normalized to unit dose rate at unit time, i.e. 1 h after the explosion. Thus, by measuring the dose rate at any time, the dose rate at any other time can be found by applying the $t^{-1.2}$ dependency.

Example 1.2 *Estimating the dose rate from fallout*

Assume that a measurement performed 10 h after a nuclear explosion yields the dose rate 31.5 mSv h^{-1}. Using data from Table 1.22 we find that the fraction of the reference dose rate at 10 h is 0.063 and the reference dose rate at 1 h is then 31.5/0.063 = 500 mSv/h.

This value of the dose rate at 1 h after the explosion may be used to estimate doses to the population, but may also be used to calculate the dose rate at any other time. Assume that we want to estimate the dose rate at the site 36 h after the explosion. According to Table 1.22 the fraction of the reference dose rate is now 0.015 and the dose rate 36 h after the explosion is then 500 × 70.015 = 7.5 mSv/h.

TABLE 1.22   Dose rate relative to a reference dose rate of 1000 mSv/h 1 h after a nuclear explosion.

Time (h)	Relative dose rate	Time (h)	Relative dose rate
1	1000	36	15
1.5	610	48	10
2	400	72	6.2
3	230	100	4.0
5	130	200	1.7
6	100	400	0.69
10	63	600	0.40
15	40	800	0.31
24	23	1000	0.24

Source: Data from Glasstone and Dolan (1977).

By expressing the $t^{-1.2}$ dependency as

$$R_t = R_1 \cdot t^{-1.2}, \tag{1.39}$$

where R_1 and R_t are the instrument readings at unit time and at time t, respectively. If $R_t = 0.1 \cdot R_1$, we find that $t = 0.1^{-1/1.2} = 6.8$, i.e. the dose rate has decreased by approximately a factor of 10 after 7 h. If we assume that $R_t = 0.01 \cdot R_1$, we find that $t = 0.01^{-1/1.2} = 46$ and we can conclude that, as a good approximation, the dose rate decreases by a factor of 10 when the time increases by a factor of 7. This can also be seen in Table 1.22 by comparing, for example, the dose rates given at 1, 6 and 48 h after the explosion, or the dose rates at 2, 15 and 100 h after the explosion.

The dose received during a given time interval, t_a to t_b, can be found by integrating Eqn. 1.39

$$D \approx R_1 \int_{t_a}^{t_b} t^{-1.2} dt = 5R_1 \left(t_a^{-0.2} - t_b^{-0.2} \right), \tag{1.40}$$

which is valid for times between 0.5 h and 5000 h.

1.3.2.3 Nuclear Fission Bombs

An important difference between the fission process in a nuclear reactor and in a fission bomb is the speed of the chain reaction. The lifetime of a neutron generation is about 0.01 μs and most of the energy is released between the 51st and the 58th generation, i.e. during 0.07 μs (for a 100-kt weapon), starting 0.5 μs after the first initiating neutron. The power output of the nuclear weapon is thus $1.8 \cdot 10^{12}$ times that of a nuclear power plant.

The operation of a reactor relies on well-controlled criticality in which delayed neutrons contribute to controlling the chain reaction. Variations in power output thus occur on a rather slow timescale. However, if the criticality is allowed to increase, the prompt neutrons can maintain the chain reaction, resulting in rapid multiplication of the number of nuclei undergoing fission. The vast amounts of energy released will cause a rapid increase in temperature, and thus expansion of the core. This expansion causes an outward acceleration of the core material, resulting in a loss of material from the core surface and a pressure decrease in its interior. The loss of material from the surface is due to the high pressure gradient at the surface, which is analogous to the string breaking when a weight tied to its end is rotated at a high speed releasing the weight. Eventually, the density in the core will be too low to prevent neutrons from leaking and the prompt chain reaction will terminate. At this point, the core is no longer supercritical but passes rapidly through a state of criticality before becoming sub-critical. However, even when the core is sub-critical, the chain reaction continues to some extent, as does heat generation (albeit decreasing rapidly).

To slow down the acceleration, most bombs are equipped with a *tamper* of a high-Z material. The tamper, which surrounds the core, is heated by the thermal radiation and may reduce the temperature decrease in the core by re-emitting part of this heat back to the core. The combined effect of this re-heating of the core and the pressure built up in the tamper due to its increase in temperature will reduce the outward acceleration of the core contents and prolong the time available for the chain reaction. Thus, the fissile material in the core will be used more effectively.

Generally, there are two methods of achieving a critical mass of fissile material in a nuclear weapon. In a *gun-type* weapon, two sub-critical masses of the fissile material are brought together by a conventional explosive, as shown in Figure 1.30. In an *implosion-type* weapon, the fissile material is in the form of a hollow sphere that is compressed by an explosive. For good efficiency, compression must be uniform over the whole surface of the sphere, and the explosive is therefore placed around the sphere, as shown in Figure 1.31. The bomb used in the Trinity test and the bomb dropped over Nagasaki (Fat Man) were of the implosion type with plutonium as the fissile material, while the bomb dropped over Hiroshima (Little Boy) was based on uranium and had a gun-type construction. Critical masses for a bare sphere of ^{235}U and ^{239}Pu are 46 kg and 16 kg, respectively. If the sphere is surrounded by a neutron-reflecting material, these critical masses decrease to 15 kg and 6 kg.

It is very important that the chain reaction starts just before the material becomes overcritical and this is achieved by neutrons emitted from a neutron source placed inside the weapon. If neutron-emitting radionuclides are present in the core, the chain reaction may start too early, giving rise to a lower yield. One such radionuclide is ^{240}Pu, which always accompanies the fissile ^{239}Pu. Weapons-grade plutonium (enriched to more than 90% ^{239}Pu) is therefore produced in such a manner as to keep the amount of ^{240}Pu as low as possible, for example, by reactors with a short fuel cycle that limits the production of ^{240}Pu. Because ^{240}Pu is present, it is of vital importance that the fissile material is brought together as

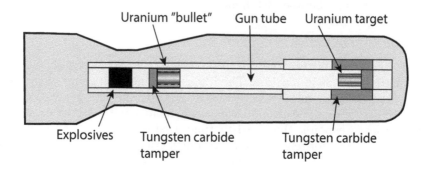

FIGURE 1.30 Schematic view of "Little Boy", the 15-kt uranium bomb used at Nagasaki on 6 August 1945. The length of the bomb was 3 m and its diameter was 0.7 m. This weapon was of the gun-type construction, where the hollow uranium "bullet" was shot towards the uranium target using a conventional explosive (cordite). In total, the weapon contained 64 kg of highly enriched uranium. Fission was initiated by four Po-Be sources (initiators), placed around the base of the uranium target. Most of the tamper, made of tungsten carbide, was placed around the target and the tamper "lid" was carried with the "bullet".

rapidly as possible, and since the implosion method is faster than the gun method, this type is often used for plutonium bombs. Both methods may, however, be used for uranium (enriched to more than 90% ^{235}U).

1.3.2.4 Thermonuclear Weapons

The first test of a fusion bomb was conducted by the USA in 1952 at the Enewetak Atoll in the Pacific Ocean. The yield was 10.4 Mt, but since the bomb contained liquid deuterium requiring cooling, it was not suitable as a practical weapon due to its size. Later developments have led to a significant decrease in size, and thermonuclear weapons now fit into missile warheads. Since thermonuclear weapons make use of the fusion of hydrogen, they are often referred to as *hydrogen bombs.*

The hydrogen is often in the form of deuterium, D, and tritium, T, which may undergo fusion by the reaction

$$^{2}H + ^{3}H \rightarrow ^{4}He + n, \tag{1.41}$$

in which the energy released (Q) is 17.6 MeV. This reaction requires a kinetic energy of the reactants corresponding to a temperature of about $4 \cdot 10^{7}$ K to overcome the Coulomb barrier and achieve the high temperatures necessary. This may be achieved using a fission bomb.

Details of the construction of the internal parts of thermonuclear weapons are classified, but most weapons are constructed according to the *Teller–Ulam design*[4]. This kind of bomb consists of two stages: a primary stage (fission) and a secondary stage (fusion), as depicted in Figure 1.32. The primary stage consists of an implosion-type fission bomb, similar to the one depicted in Figure 1.31, where explosives, arranged in a soccer ball pattern, detonate, and compress a hollow core of plutonium. However, the centre of the core is filled with D

---

[4]Edward Teller (1908–2003), Hungarian-born American theoretical physicist; Stanisław Ulam (1909–1984), Polish-born American mathematician.

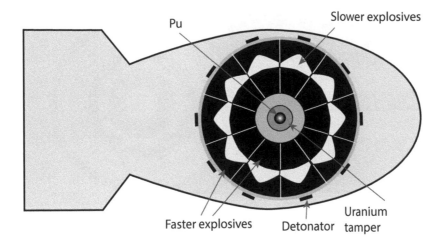

FIGURE 1.31 Overview of the construction of "Fat Man", used at Hiroshima on 9 August 1945. The bomb, 3.3 m long and 1.5 m in diameter, was of the implosion type, and contained 6.2 kg Pu in two hemispheres, surrounded by a tamper of uranium. The tamper also serves as a neutron reflector and increases the yield since fission of the ^{238}U is stimulated by the fast neutrons produced in the chain reaction. Criticality was achieved by compressing the two hemispheres together with explosives mounted so as to give a spherically symmetric detonation front. The explosion was initiated by detonators, triggered with a precision of μs. Initiation of the chain reaction was achieved with a Po-Be source, centrally placed between the plutonium hemispheres. The Po and Be, initially separated, were mixed by the shock wave.

and T gas, which fuse at the high temperatures reached during fission, and the neutrons released during D–T fusion gives rise to additional fission, known as *boosting*.

The secondary stage contains the fusion fuel, usually ^{6}LiD (lithium deuteride), which will produce tritium when bombarded by neutrons from the fission reaction in the primary stage

$$^6Li + n \rightarrow^3 H +^4 He \tag{1.42}$$

with a value of Q of 4.78 MeV. Fusion can then be induced in the mixture of D and T in the secondary stage provided the temperature is sufficiently high. However, the temperature required for fusion must be reached before the shock wave from the fission reaction destroys the secondary stage, and therefore a much faster process must be used. It is generally believed that the photons released in the primary explosion deliver energy to the secondary stage by ablation of the tamper surrounding the D-T fuel. At the temperatures prevailing in the fission core, almost 95% of the energy is released as thermal energy, with a mean photon energy of about 10 keV.

The photon flux is directed through a porous medium and will heat the interior of the casing surrounding the secondary stage. As the tamper is heated, it will be removed by vaporization (ablated) at such a rate that the recoil of the rest of the tamper will exert pressure on the fusion fuel and the fissile spark plug. This pressure is sufficient to compress the fissile material so that it becomes critical. The energy released by fission contributes to the heating, and the temperature increases sufficiently to enable fusion. The neutrons released in fission of the spark plug will also produce additional T from the ^{6}LiD in the fuel. A large part of the yield can also be attributed to fission of ^{238}U in the tamper. Although

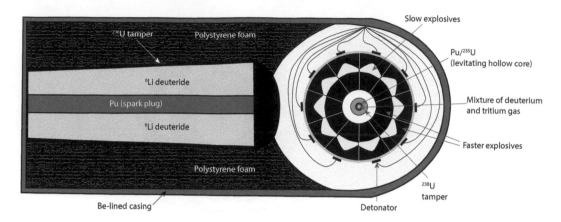

FIGURE 1.32   Illustration of the construction of a thermonuclear weapon of Teller–Ulam design.

^{238}U is not fissile (i.e. cannot undergo fission with thermal neutrons, and therefore cannot sustain a chain reaction), it can undergo fission due to the fast neutrons released by fusion.

A *neutron bomb* can be regarded as a thermonuclear weapon with a smaller amount of fissile material. In addition, the parts otherwise designed to keep the neutrons inside the weapon are made transparent to neutron radiation, and the tamper is replaced by a material that does not undergo fission. It is a common misconception that a neutron bomb kills people and leaves buildings intact. Some structures may collapse at blast pressures that are survivable for humans. The original intention was to use neutron bombs to destroy tank formations, since although tanks offer good protection against the blast, they are penetrated by fast neutrons.

1.3.3   Radioisotopes Used in Medicine

The medical application of radiation started in 1896, only months after Conrad Röntgen's first publication in 1895 of his observation of the radiation emitted when electrons hit a positively charged electrode in vacuum. Although the benefits of x-rays for medical purposes soon became evident, it took longer to realise that there were risks associated with exposure to ionizing radiation, not only in terms of overexposure of the patient, but also in terms of unnecessary exposure of the staff involved. However, electron accelerator tubes do not cause irradiation of people outside the laboratory, or the environment, and hence the environmental aspects of exposure by medical devices are not dealt with in detail here.

Radionuclides have been used for non-invasive visualization of anatomic structures and physiological processes in the body since the late 1940s, and have found a wide range of diagnostic applications. Tens of millions of nuclear medical diagnostic examinations are performed globally each year. Initially, iodine, especially ^{131}I, was widely used for radiotherapy of thyroid diseases, due to its high uptake in the thyroid. Today, the most commonly used isotope in medical diagnostics is ^{99m}Tc, and most procedures for testing physiological function use ^{99m}Tc. ^{99m}Tc is obtained at the clinic from a so-called generator containing the mother nuclide ^{99}Mo. The soluble ^{99m}Tc produced in the generator is then eluted by saline solution before use.

Effects on the environment were seen when radionuclides came into medical use. It is interesting to note that the amount of ^{131}I released to the environment as the result of treatment of thyroid disease exceeds the amount released to the Irish Sea by the Sellafield nuclear fuel reprocessing plant. Obsolete medical radiation sources, used in the 1920s and onwards for intracavitary treatment of, for example, cervical carcinoma, also constitute a liability in terms of radiological safety and radiation protection in many countries.

Radionuclides used in medicine are either produced in reactors (not only as fission products, but also neutron activation products of various target materials), or in accelerators (mostly cyclotrons), where charged particles collide and interact with target materials to produce well-defined amounts of artificial radionuclides with high chemical purity. Figure 1.33 shows the principle of a cyclotron.

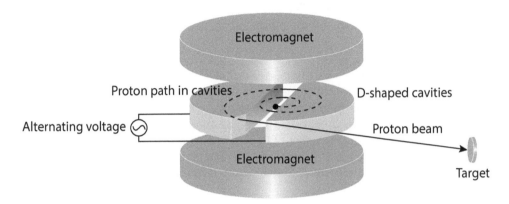

FIGURE 1.33 When protons are emitted from the source in the air gap between the two D-shaped cavities, they are forced to move in a circular path by the magnetic field. The electric potential of each cavity alternates, as they are connected to an alternating voltage. When one cavity is at a positive potential, the other is at a negative potential. When the protons enter the air gap, they are thus accelerated by the potential difference between the cavities. Since the time taken for a particle to travel one semicircle is independent of the radius, the protons will be accelerated every time they pass the air gap if the half-period of the alternating voltage is equal to the time taken to travel one semicircle. After gaining energy, the protons can be guided to a target.

A considerable commercial market has developed for radionuclides intended for various applications, including not only medical use, but also for research and industrial applications. The world's supply of ^{99}Mo is produced in nine research reactors, most of which use highly enriched uranium (45–90% ^{235}U), in Belgium (BR-2), the Netherlands (HFR), Canada (NRU), South Africa (Safari-1), the Czech Republic (LVR-15), Poland (MARIA), France (OSIRIS), Australia (OPAL) and Argentina (RA-3) (OECD/NEA 2012IAEA 2010b). Neutron-depleted radionuclides are generally produced in cyclotrons, while fission products and neutron-rich nuclei are generally produced in reactors. Tables 1.23 and 1.24 list some of the radionuclides used in medicine. Due to the rather rapid development of new substances for diagnostics and treatment, these should be regarded as examples of common medical radionuclides.

Cyclotrons are used to produce short-lived positron emitters, such as ^{16}O, ^{18}F and ^{13}C, which are commonly used in cancer diagnostics (especially in tomography with positron emitters in combination with conventional computed tomography, PET-CT). In 2006, over 260 cyclotron facilities producing medical radionuclides were in operation throughout the world, and the number is growing rapidly due to the increased use of PET-CT examinations (IAEA 2006). The number of cyclotrons used for the production of PET tracers in 2010 was 671, distributed around the world as follows (IAEA 2010a):

- Africa: 6

- Asia: 211

- Australia: 11

- Europe: 183

- Middle East: 18

- North America and Mexico: 220

- South America: 22

In some systemic radionuclide therapies, the organ-specific properties of the radionuclide itself are used: for example, the administration of ^{131}I to patients with goitre or thyroid carcinoma, the use of bone-seeking radionuclides such as ^{89}Sr, ^{153}Sm and ^{189}Rh for pain relief in patients with bone metastases, and the use of ^{32}P for the treatment of the blood disease polycythaemia vera. In so-called targeted therapies, the radionuclide is guided in body fluids by a pharmaceutical agent. The aim of this kind of treatment is to damage the cancer cells while sparing healthy cells, and radionuclides emitting particles with a short range in tissue are therefore preferred, since short-range particles will deliver most of their energy in the target cells.

Suitable radionuclides for therapy are α emitters, pure β-emitting radionuclides or radionuclides emitting Auger electrons. However, since imaging is often needed, it is desirable that the radionuclides also emit γ-radiation. Imaging with, for example, gamma cameras can help to estimate organ-specific uptake for use in calculations of the activity that should be administered to obtain a certain therapeutic effect, as well as for calculations of the dose to healthy tissue. Examples of these are the α emitter ^{213}Bi and the low-energy β emitter ^{177}Lu, which are being used increasingly in therapy. Another example is ^{90}Y, where bremsstrahlung can be used for gamma camera imaging of the uptake in the body.

Imaging is also widely used in the diagnostics of physiological processes in the body, for example, locating metastases in the skeleton, studies of the filtration rate in the kidneys, or lung perfusion studies. The activity administered in diagnostics is, however, much lower than in therapies: an examination of thyroid uptake requires a few hundred kBq, while several GBq may be administered in the treatment of thyroid cancer.

The radionuclides used in nuclear medicine generally have the following features.

- They are relatively short-lived, since delayed and long-term uptake of the radionuclide in the body would lead to high accumulated absorbed doses.

- They emit gamma radiation, either directly by nuclear disintegration, or secondarily from positron emission leading to annihilation radiation with gamma energies of 0.511 MeV. This enables the emitted γ-rays to be detected using external detectors, often NaI(Tl) crystals.

- They should preferably have no radioactive daughters to limit the radiation dose. ^{99m}Tc has a purely β-emitting daughter, ^{99}Tc, but because of its long half-life ($T_{1/2} = 2.1 \cdot 10^5$ y) compared to its mother, ($T_{1/2} = 6.0$ h), there is negligible activity, and hence radiation dose contribution, from ^{99}Tc per unit administered activity of ^{99m}Tc. Nevertheless, when ^{99}Tc is released from reprocessing plants, it has an environmental impact because of its high accumulation in certain marine organisms.

TABLE 1.23 Some cyclotron-produced radionuclides used in medical applications. The main decay mode is given in cases where the probability of the alternative mode is less than 10%. The applications given are only examples, and some radionuclides are used in a number of different applications.

Nuclide	Half-life	Decay	Applications
^{11}C	20 m	ε	PET, myocardial metabolism
^{13}N	10 m	ε	PET, brain physiology and pathology
^{15}O	122 s	ε	PET, regional cerebral blood flow
^{18}F	110 m	ε	PET, glucose metabolism
^{57}Co	272 d	ε	Organ size markers; Calibration source
^{64}Cu	13 h	ε, β^-	Copper metabolism
^{67}Cu	62 h	β^-	Therapy
^{67}Ga	3.3 d	ε	Tumour imaging
^{68}Ga	68 m	ε	PET, tumour localization
^{68}Ge	271 d	ε	Production of ^{68}Ga
^{81m}Kr	13 s	IT	Imaging, pulmonary function
^{82}Rb	1.3 m	ε	PET, myocardial perfusion
^{82}Sr	25 d	ε	Production of ^{82}Rb
^{111}In	2.8 d	ε	Tumour localization
^{123}I	13 h	ε	Imaging, receptors in the brain
^{201}Tl	3.0 d	ε	Heart diagnostics
^{211}At	7.2 h	ε, α	Targeted therapy, cancer

- The gamma energies should preferably be around 100–150 keV for planar imaging. This is the gamma energy at which most gamma camera imaging systems exhibit the highest detection efficiency and, when equipped with low-energy collimators, the maximum image quality. The exception to this is the positron emitters, which produce two annihilation photons emitted in coincidence. This coincidence is used in dual-head gamma camera systems to achieve high spatial image resolution.

- They should have versatile chemical properties, making it possible to combine them with biomarkers and molecules so that they can be transported to specific organs, or bound to receptors on cancer cells. Tc and I are examples of such elements, and isotopes of these are commonly used in nuclear medicine.

In addition to diagnostics and therapy, the low-energy beta emitter ^{14}C has been used within medical research to study the metabolism of various pharmaceuticals. This radionuclide used to be a biomarker for the diagnosis of the commonly occurring *Helicobacter pylori* (now known to cause stomach ulcers), but is now used less often to avoid long-term radiation doses. ^{14}C has too long a physical half-life to be practical for common physiological studies in nuclear medicine.

It has become routine in cancer diagnostics to use PET-CT for patients who have been given tumour-seeking agents labelled with a positron emitter, and due to the short half-life of these radionuclides, they must be produced just before they are used. Since many clinics do not have cyclotrons, and cannot produce their own radionuclides, there is a constant need for transport. It is therefore important to bear in mind the radiological safety aspects, including possible accidents. The safety of transportation is also important for ^{99m}Tc generators that are distributed by air and road from a few suppliers to clinics all over the world.

Releases of medical radionuclides into the environment are mainly the result of patients excreting radionuclides during and after medical procedures. These radionuclides will pass through the sewage system and reach water recipients, such as coastal waters or rivers. UNSCEAR (2008) estimated that the global annual usage of ^{131}I was 600 TBq, which would result in an annual collective dose of 0.009 manSv from releases to the environment. Attempts have been made to reduce environmental contamination by medical radionuclides by using closed sewage systems in hospitals. However, this can actually result in higher exposures to humans since the sewage tanks must be maintained and emptied, requiring human operators who will be occupationally exposed.

It is important to recognize that medical radionuclides contribute to the radionuclides found in environmental samples, e.g. ^{131}I in air filters and in seaweed in coastal waters. This is especially important when interpreting the radionuclide content in bioindicators from sites that may be affected by both water treatment plants and nuclear power plants. The radionuclide content in sewage sludge is usually the best indicator of the release rate of relatively long-lived medical radionuclides from urban settlements.

In external radiation therapy, the irradiation source is placed outside the body, in a traditional so-called *teletherapy* geometry, using the beam from an accelerator or from a γ-emitting radioactive source that is shielded and collimated so that a beam of radiation is directed towards the patient. A sealed source can also be placed in close contact with the body surface, e.g. skin for the treatment of skin carcinoma or other skin diseases known to respond to irradiation.

Teletherapy sources were introduced in the 1950s and used to consist of ^{137}Cs or ^{60}Co, the latter nuclide being beneficial for target volumes of cancerous tissue located deep inside the patient, due to its higher average gamma energy (1.25 MeV). However, ^{137}Cs was also used, despite its lower gamma-ray energy (0.662 MeV), because it has a longer physical half-life (30.2 y compared with 5.2 y for ^{60}Co). The strength of these sources was often in the range 10–500 TBq, i.e., sufficient to provide an average absorbed dose rate of 0.1 to 1 Gy per minute for a typical patient treatment geometry, often with a source–target distance of 0.5 to 1 m.

In *brachytherapy*, the source is inserted inside the body cavities close to the tumor site (e.g. for the treatment of cervical or rectal cancer), or invasively placed in the tumour-bearing organ (e.g. for prostate cancer). This kind of therapy should not be confused with systemic therapy, where the radionuclide is taken up by the blood. Traditional radiation sources used for brachytherapy consisted of rods of ^{226}Ra inserted into the body. This type of treatment began in the USA and Europe as early as the 1920s. Later, ^{137}Cs sources were used. Today, the most commonly used intracavitary irradiation source is ^{192}Ir which, like ^{226}Ra, has a very complex gamma spectrum, but is less radiotoxic[5].

A typical sealed source for teletherapy consists of a ceramic disc in which radioactive ions are implanted, or a canister (about 20 cm^3) filled with a salt of the radioactive substance, e.g. ^{137}CsCl. If the canister or source containment is broken, the radioactive substance can leak out and be ingested or inhaled. Sealed radioactive sources generally have little value on the scrap metal market. However, sealed sources are often stored together with their original collimators, and the scrap value of the lead can be high enough to be attractive for burglars. Precautions must therefore be taken to ensure secure storage of these sources. There are now safety regulations governing the documentation and physical storage of these types of sources, which are dealt with in more detail later. There is growing concern among national authorities that hospitals have accumulated a significant number of old teletherapy

---

[5] ^{226}Ra is an α emitter with high specific activity. Following intake, radium is to a large extent retained in the skeleton from which it is very slowly excreted.

sources. These could constitute a radiation hazard if the use of the source is not adequately documented, or if appropriate measures are not taken for their secure long-term storage. Twice in recent decades, radiotherapy sources for clinical use have been disposed of inappropriately , causing irradiation of several people. In 1987, in Goiânia in Brazil, an old ^{137}Cs source was stolen by burglars who were unaware of its radiation hazard. This resulted in substantial external and internal exposure of several individuals resulting in at least four fatalities. A similar event took place in Thailand in 2000, involving an abandoned ^{60}Co source. However, this incident led to fewer exposed people and no fatalities.

TABLE 1.24 Some reactor-produced radionuclides used in medical applications. The main decay mode is given in cases where the probability of the alternative mode is less than 10%. The applications given are only examples and some radionuclides are used in a number of various applications.

Nuclide	Half-life	Decay	Applications
^{24}Na	15 h	β^-	Studies of electrolytes in the body
^{32}P	14 d	β^-	Treatment of polycythaemia vera
^{42}K	12 h	β^-	Studies of exchangeable potassium
^{51}Cr	28 d	ε	Studies of kidney clearance
^{59}Fe	44 d	β^-	Metabolic studies
^{60}Co	5.2 y	β^-	Sterilization; calibration source
^{75}Se	120 d	ε	Studies of digestive enzymes
^{89}Sr	51 d	β^-	Pain relief, cancer
^{90}Y	64 h	β^-	Therapy, liver cancer; pain relief, e.g. arthritis
^{99}Mo	66 h	β^-	Production of ^{99m}Tc
^{99m}Tc	6.0 h	IT	Renal studies; Imaging of skeleton metastases
^{103}Pd	17 d	ε	Brachytherapy source
^{125}I	59 d	ε	Localization of thrombosis; renal studies
^{131}I	8.0 d	β^-	Therapy, thyroid diseases; thyroid imaging
^{133}Xe	5.2 d	β^-	Studies of lung ventilation
^{137}Cs	30 y	β^-	Sterilization of blood; calibration source
^{153}Sm	120 d	ε	Pain relief, cancer
^{165}Dy	2.3 h	β^-	Therapy, arthritis
^{166}Ho	27 h	β^-	Diagnosis and therapy, liver cancer
^{169}Er	9.4 d	β^-	Pain relief in arthritis
^{169}Yb	32 d	ε	Studies of cerebrospinal fluid in the brain
^{177}Yb	1.9 h	β^-	Production of ^{177}Lu
^{177}Lu	6.6 d	β^-	Pain relief, cancer; therapy, endocrine tumours
^{186}Re	3.7 d	β^-	Pain relief, cancer
^{188}Re	17 h	β^-	Coronary artery irradiation (restenosis prevention)
^{192}Ir	74 d	β^-	Brachytherapy source
^{212}Pb	11 h	β^-	Targeted therapy, cancer
^{213}Bi	46 m	β^-	Targeted therapy, cancer

1.3.4 Radiation Sources in Industry and Research

Radiation sources have found various applications in industry and research. Table 1.25 lists some radionuclides used in industrial and research applications, some of which will be discussed below. Further information on nucleonic gauges can be found in IAEA-TECDOC-1459, see Section 1.6.

TABLE 1.25 Radionuclides used in industrial applications. The main decay mode is given in cases where the probability of the alternative mode is less than 10%. Examples of common uses of the radionuclides are given.

Nuclide	Half-life	Decay	Applications
^{3}H	12 y	ε	EC detectors; Tracer studies; Lights
^{14}C	5700 y	β^-	Radiocarbon dating
^{36}Cl	0.3 My	β^-	Dating of water
^{51}Cr	28 d	ε	Tracer studies
^{54}Mn	312 d	ε	Tracer studies
^{55}Fe	2.7 y	ε	X-ray fluorescence analysis
^{57}Co	272 d	ε	X-ray fluorescence analysis
^{60}Co	5.2 y	β^-	Sterilization; Radiography; Calibration; Thickness gauges
^{63}Ni	101 y	β^-	EC detectors
^{65}Zn	244 d	ε	Tracer studies
^{75}Se	120 d	ε	Radiography
^{85}Kr	11 y	β^-	Thickness gauges
^{90}Sr	29 y	β^-	RTGs; Thickness gauges; Calibration
^{99m}Tc	6.0 h	IT	Tracer studies
^{109}Cd	461 d	ε	X-ray fluorescence analysis
^{137}Cs	30 y	β^-	Sterilization; Calibration; Level gauges; Tracer studies
^{147}Pm	2.6 y	β^-	Thickness gauges
^{170}Tm	129 d	β^-	Radiography
^{169}Yb	32 d	ε	Radiography
^{192}Ir	74 d	β^-	Radiography
^{198}Au	2.7 d	β^-	Tracer studies
^{204}Tl	3.8 y	β^-	Thickness gauges
^{210}Pb	22 y	β^-	Dating of soil and sediments
^{210}Po	138 d	α	Static eliminators
^{226}Ra	1600 y	α	Level gauges
^{232}Th	14 Gy	α	Lights
^{238}Pu	88 y	α	Radioisotope thermoelectric generators
^{241}Am	433 y	α	Calibration; Well logging; Smoke detectors
^{244}Cm	18 y	α	Thickness gauges
^{252}Cf	2.6 y	α	Soil moisture gauges; Well logging

Neutron activation analysis is widely used to investigate the elemental composition of a sample by the activation of stable nuclei, which may then be analysed by gamma spectroscopy. These applications make use of both prompt γ-radiation, and γ-radiation from decaying radionuclides produced by activation. Activation by neutron capture has been discussed in Section 1.3.1.4, and the activity of the sample was given by Eqns. 1.37 and 1.38. Inelastic scattering of neutrons can also be used as an analytical tool, especially for elements with a low cross section for capture of thermal neutrons, such as C and Si. For example, ground moisture can be determined by inserting a probe containing a neutron source and a neutron-sensitive detector into the ground. Since neutrons are effectively scattered by hydrogen, the detector signal will be a function of the ground's hydrogen content, and thus its water density. The neutron source may be a reactor, a neutron generator, or a radionuclide that decays by spontaneous fission, such as ^{252}Cf. Neutron sources consisting of an α emitter and a nuclide that emits neutrons when bombarded with α-particles, such as ^{241}Am-Be (the neutrons interact with the ^{9}Be), are also used, for example, during the start-up of research reactors. A similar neutron source is the combination ^{239}Pu-Be, which is used for calibration. Tritium is used as a target for neutron production in neutron generators. These generators produce nearly monoenergetic neutrons by the reaction T(d,n)^{4}He, in which a tritium target is bombarded by deuterons.

The transmission of γ-rays is used in various applications to determine the thickness of materials such as sheet metals. Since attenuation of the γ-radiation depends on the thickness of the material, the detector signal varies accordingly. Feedback of the detector signal to the press on the production line thus enables automatic adjustment of the thickness. Radioisotope gauging with β-emitting sources is often used in the production of paper and plastic films. Coating thickness can be determined by studying the backscattered radiation. These thickness gauges often contain a radioactive source of ^{85}Kr, ^{90}Sr, ^{241}Am, ^{147} or ^{244}Cm.

A very common application of radiation sources in industrial activities is the level gauge (^{60}Co or ^{137}Cs) for determining the level of a substance in a container. This device is used similarly to the thickness gauge. When the filling height is below the line of sight between the source and the detector, the detector signal is high, but as the container is filled, the signal decreases. These gauges are also used to monitor the flow of material in pipes, depicted in Figure 1.34. The activity of the radioactive source is typically a few hundred GBq.

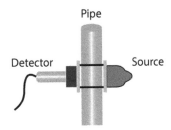

FIGURE 1.34 Changes in the flow (or level) inside the pipe can be monitored by measuring the radiation from a radioactive source, placed on the pipe opposite the detector.

X-ray fluorescence (XRF) analysis is a widely used technique to determine the composition of samples (and of various body organs *in vivo*). The material is irradiated by a radioactive source, ionizing some of the atoms. Vacancies created in the electron shells are then filled by electrons from outer shells with the emission of characteristic x-rays. The energies of these x-rays depend on the element emitting the radiation, since the energies of the electron shells are Z-dependent. By analysing the energies and the fluence of the emitted radiation, the amount of each element can be determined. It is preferable that the energy of the radiation emitted from the radioactive source is higher than the K shell electron binding energy of the elements to be analysed, to obtain a high yield of K x-rays from the sample. XRF can also be performed using an x-ray tube, but equipment with a radioactive source is more suitable for field use.

Gamma radiography is very similar to x-ray imaging, but a radioactive source is used instead of an x-ray unit. This enables the equipment to be transported, for example, to construction sites where it can be used to detect faulty welded joints. Sources used for radiography have activities on the order of 1 TBq, and are enclosed in a shield of tungsten or depleted uranium when not in use. Before radiography, a collimator connected to a hose is positioned close to the weld and the source is then remotely conveyed to the correct position by a cable in the hose. This procedure ensures that the radiographer does not come into close contact with the source.

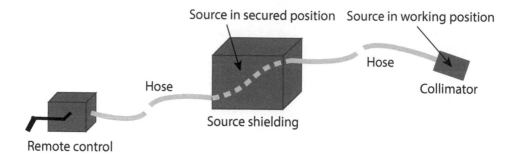

FIGURE 1.35 The radiography source is moved within the hose to the collimator by means of a remotely controlled wire. After exposure, the source is withdrawn and secured in the shielded container.

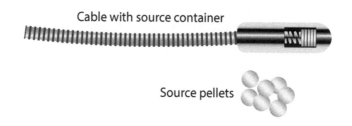

FIGURE 1.36 The source container (a few cm long) contains the radioactive source in the form of small pellets, which are held in place by a spring. The wire at the end of the source is connected to the remote control shown in Figure 1.35.

Irradiation with γ-rays is a common method of sterilizing medical equipment (see also Section 1.6). The high absorbed dose required to kill bacteria (several kGy) is achieved mainly by using ^{60}Co sources with a high activity, on the order of 150 000 TBq. These sources are thus the strongest sealed sources used in society today. Foods may also be sterilized to extend their shelf life, but the use of radiation for sterilization of food is restricted in many counties. Sources of ^{137}Cs (a few hundred TBq) are often used to irradiate blood before transfusions to prevent graft-versus-host disease or to inactivate microbes. Blood irradiation units can be located at the clinic as they do not require large facilities.

Small amounts of radionuclides are often used in so-called *tracer studies*. Very low levels are detectable due to the high sensitivity of radiation detectors, and thus even elements that are otherwise toxic can be used. The activity per unit mass of a substance is inversely proportional to the product of the atomic weight and the half-life. For example, 1 pg (10^{-12} g) of ^{14}C will have an activity of 0.16 kBq, while 1 pg of ^{137}Cs will have an activity of 3.2 kBq. The corresponding activities for ^{65}Zn and ^{60}Co are 300 kBq and 530 kBq, respectively. Tracer studies are employed in many environmental, industrial and medical applications, for example, tracing liquid effluents using tritium; studying erosion processes with ^{137}Cs, ^{51}Cr or ^{198}Au; tracing heavy-metal effluents using ^{54}Mn or ^{65}Zn; and tracing industrial waste by ^{198}Au or ^{99m}Tc.

Radioactive sources of high activity have been used for electricity production, for example, to power spacecraft and lighthouses at remote locations. These *radioisotope thermo-electric generators* (RTGs), may contain about 700 TBq ^{90}Sr or 10 TBq ^{238}Pu. The energy transferred by the particles emitted to the surrounding material during decay is enough to raise the temperature to about 1000 °C. This heat is converted into an electrical voltage using an array of thermocouples, and an RTG therefore has no moving parts. A suitable radionuclide for an RTG should have a relatively long half-life for sustainability, and a high power density. The energy density of ^{238}Pu is 0.54 W g^{-1}, which is higher than for ^{90}Sr (0.46 W g^{-1}).

Energy density can be illustrated by a simple example: consider ^{90}Sr, which decays 100% by β^--decay, emitting 546 keV electrons, with a half-life of 29 y. The activity of 1 g ^{90}Sr is 5 TBq and thus $5 \cdot 10^{12} \cdot 546 \cdot 10^3 \cdot 1.6 \cdot 10^{-19} = 0.4$ J s^{-1} is released as kinetic energy. The α emitter ^{238}Pu has a lower specific energy, about 0.6 TBq g^{-1}, but the energy of the emitted α-particles is about 5500 keV. The energy emitted as kinetic energy is thus 0.5 J s^{-1}. These values agree with the energy densities given above, although in this example, several processes that influence the power density have been omitted. For example, the exact form of the β^--spectrum from ^{90}Sr has not been considered, nor has the fact that some α-particles with an energy lower than 5500 keV are emitted by ^{238}Pu.

Other miscellaneous applications of radionuclides are tritium used in lightning preventers, ^{14}C for dating of organic matter, ^{36}Cl for groundwater dating, and ^{210}Pb for soil and sediment dating. Weak β-sources are used, for example, in chromatography as sources for electron capture detectors. The electrons emitted by β-decay are picked up by atoms in the sample, which can then be collected by an electrical potential difference. The combined use of ^{46}Sc, ^{60}Co, ^{110m}Ag, ^{140}La and ^{198}Au to monitor the performance of blast furnaces has resulted in small amounts of ^{60}Co in almost all steel products, and therefore steel manufactured before World War II is preferred for shielding in measurements where a very low radiation background is needed, such as whole-body counting.

The use of spallation reactions at research facilities constitutes a potential source of radionuclides in the environment. When a target is bombarded by neutrons, a large number of radionuclides is produced in the target, as well as in the moderator and neutron reflector, some of which are not found in nature.

1.4   REFERENCES

- Atwood, D. A. 2010. *Radionuclides in the Environment*. Singapore: John Wiley & Sons Ltd.

- Bé, M-M., Dulieu, C., Mougeot, X. and Kellett, M. 2014. *HALF-LIVES Table of Recommended Values*. Note Technique LIST/LNHB/2014/18, Laboratoire National Henri Becquerel. Paris.

- Beeman, J. W., Biassoni, M., Brofferio, C. et al. 2012. First measurement of the partial widths of ^{209}Bi decay to the ground and to the first excited states. *Phys Rev Lett.*. 108: 062501.

- Bouville, A. and Lowder, W. M. 1988. Human population exposure to cosmic radiation. *Radiation Protection Dosimetry*. 24: 293–299.

- Buchholz, B. A. 2007. Carbon-14 bomb-pulse dating. *Wiley Encyclopedia of Forensic Science*. Lawrence Livermore National Laboratory, UCRL-BOOK-237334.

- Buckley, D. W., Belanger, R., Martin, P. E. et al. 1980. *Environmental Assessment of Consumer Products Containing Radioactive Material.* U.S. Nuclear Regulatory Commission. NUREG/CR-1775.

- Choppin, G., Liljenzin, J-O. and Rydberg, J. 2002. *Radiochemistry and Nuclear Chemistry.* Woburn: Butterworth-Heinemann.

- Darby, S., Hill, D., Deo, H. et al. 2006. Residential radon and lung cancer: Detailed results of a collaborative analysis of individual data on 7148 persons with lung cancer and 14 208 persons without lung cancer from 13 epidemiologic studies in Europe. *Scandinavian Journal of Work, Environment and Health.* Vol. 32, Suppl. 1:1–83.

- Doering, C. and Saey, P. 2014. Hadley cell influence on ^{7}Be activity concentrations at Australian mainland IMS radionuclide particulate stations. *Journal of Environmental Radioactivity.* Vol. 127:88–94.

- Glasstone, S. and Dolan, P. J. 1977. *The Effects of Nuclear Weapons.* United States Department of Defence and United States Department of Energy. USA.

- Hamawi, J.N. 1971. A useful recurrence formula for the equations of radioactive decay. *Nucl. Techn.* 11:84–88.

- Hayashi, K., Yasuoka, Y., Nagahama, H. et al. 2015. Normal seasonal variations for atmospheric radon concentration: A sinusoidal model. *Journal of Environmental Radioactivity.* Vol. 139:149–153.

- IAEA, International Atomic Energy Agency. 2003a. *Extent of Environmental Contamination by Naturally Occurring Radioactive Material (NORM) and Technological Options for Mitigation.* Technical Reports Series No. 419. International Atomic Energy Agency. Vienna.

- IAEA, International Atomic Energy Agency. 2003b. *Radiation Protection and the Management of Radioactive Waste in the Oil and Gas Industry.* Safety Reports Series No. 34. International Atomic Energy Agency. Vienna.

- IAEA, International Atomic Energy Agency. 2006. *Directory of Cyclotrons Used for Radionuclide Production in Member States 2006 Update.* IAEA-DCRP/2006. International Atomic Energy Agency. Vienna.

- IAEA, International Atomic Energy Agency. 2008. *Handbook of Nuclear Data for Safeguards. Database Extensions, August 2008.* INDC(NDS)-0534. International Atomic Energy Agency. Vienna.

- IAEA, International Atomic Energy Agency. 2010a. *Nuclear Technology Review 2010.* GC(54)/INF/3. International Atomic Energy Agency. Vienna.

- IAEA, International Atomic Energy Agency. 2010b. *Nuclear Technology Review 2010.* GC(54)/INF/3. Production and Supply of Molybdenum-99, NTR2010 Supplement. International Atomic Energy Agency. Vienna.

- IAEA, International Atomic Energy Agency. 2014a. *The Environmental Behaviour of Radium: Revised Edition.* Technical Reports Series No. 476. International Atomic Energy Agency. Vienna.

- IAEA, International Atomic Energy Agency. 2014b. *Nuclear Power Reactors in the World*. Reference Data Series No. 2. International Atomic Energy Agency. Vienna.

- Jankovic, J. T., Underwood, W. S. and Goodwin, G. M. 1999. Exposures from thorium contained in thoriated tungsten welding electrodes. *Am Ind Hyg Assoc J.*. 60(3):384–389.

- Levin, I. and Hesshaimer, V. 2000. Radiocarbon: A unique tracer of global carbon cycle dynamics. *Radiocarbon*, 42:69–80.

- Lindell, B. 2003. *Herkules Storverk*. Atlantis, Stockholm. [In Swedish]

- Malmqvist, L., Isaksson, M. and Kristiansson, K. 1989. Radon migration through soil and bedrock. *Geoexploration* 26:135–144.

- McBride, J. P., Moore, R. E., Witherspoon, J. P. and Blanco, R. E. 1978. Radiological impact of airborne effluents of coal and nuclear plants. *Science*. Vol. 202, No. 4372: 1045–1050.

- Meadows, J. W. 1967. ^{252} Fission neutron spectrum from 0.003 to 15.0 MeV. *Phys. Rev.*. 157: 1076–1082.

- Miles, R.E. 1981. An improved method for treating problems involving simultaneous radioactive decay, buildup, and mass transfer. *Nucl. Sci. Eng.*. 79:239–245.

- Mustonen, R. 1985. Radioactivity of fertilizers in Finland. *The Science of the Total Environment*. 45: 127–134.

- NCRP, National Council on Radiation Protection and Measurements. 1988. *Measurements of Radon and Radon Daughters in Air*. NCRP Report No. 97. Bethesda, USA.

- NNSA. 2002. NNSA News March 7, 2002. National Nuclear Security Administration, U.S. Department of Energy.

- NPL (National Physical Laboratory). 2015. Section 4.7.1: Nuclear fission. In. *Kaye & Laby Table of Physical and Chemical Constants*. National Physical Laboratory, UK.

- O'Brien, K., Friedberg, W., Sauer, H. H. and Smart, D. F. 1996. Atmospheric cosmic rays and solar energetic particles at aircraft altitudes. *Environment International*. 22: S9–S44.

- OECD/NEA 2012. *A Supply and Demand Update of the Molybdenum-99 Market*. http://www.oecd-nea.org/med-radio/docs/2012-supply-demand.pdf (accessed 2015-05-05)

- Portstendörfer, J. 1994. Tutorial/Review. Properties and behaviour of radon and thoron and their decay products in the air. *J. Aerosol Sci.*. 2: 219–263.

- Sahu, S. K., Tiwari, M., Bhangare, R. C., and Pandit, G. G. 2014. Enrichment and particle size dependence of polonium and other naturally occurring radionuclides in coal ash. *Journal of Environmental Radioactivity*. Vol. 138:421–426.

- Samet, J. M. 1994. Health Effects of Radon. In *Radon: Prevalence, Measurements, Health Risks and Control*, ed. N. L. Nagda, 33–48. Philadelphia, USA: American Society for Testing and Materials.

- Samuelsson, C. 1987. Application of a recursion formula to air sampling of radon daughters. *Nuclear Instruments and Methods in Physics Research.* A262:457–462.

- Samuelsson, C. 1994. Natural radioactivity. In *Radioecology. Lectures in Environmental Radioactivity*, ed. E. Holm, 3–20. Singapore: World Scientific.

- Soppera, N., Dupont, E., and Bossant, M. 2012. *JANIS Book of Neutron-Induced Cross-Sections.* OECD NEA Data Bank.

- Tanner, A. B. 1980. Radon migration in the ground: A supplementary review. In *Natural Radiation Environment III*, ed. T. F. Gesell and W. M. Lowder, USDOE Rept. CONF-780422, NTIS, USA.

- Tuli, J. K. 2011. *Nuclear Wallet Cards.* Brookhaven National Laboratory. Upton, New York.

- UNSCEAR, United Nations Scientific Committee on the Effects of Atomic Radiation. 1982. *Ionizing Radiation: Sources and Biological Effects, Annex B.* New York: United Nations.

- UNSCEAR, United Nations Scientific Committee on the Effects of Atomic Radiation. 1993. *Sources and Effects of Ionizing Radiation, Annex A.* New York: United Nations.

- UNSCEAR, United Nations Scientific Committee on the Effects of Atomic Radiation. 2000. *Sources and Effects of Ionizing Radiation, Vol.I: Sources, Annex A.* New York: United Nations.

- UNSCEAR, United Nations Scientific Committee on the Effects of Atomic Radiation. 2008. *Sources and Effects of Ionizing Radiation, Vol.I: Sources of Ionizing Radiation, Annex B.* New York: United Nations.

- United States Environmental Protection Agency. 2014. http://www.epa.gov/radiation/tenorm/oilandgas.html (accessed 2016-06-28).

- WHO, World Health Organization. 2011. *Guidelines for Drinking-Water Quality*, 4th Ed. World Health Organization, Geneva.

- World Nuclear Association. 2015. http://www.world-nuclear.org/info/Safety-and-Security/Radiation-and-Health/Naturally-Occurring-Radioactive-Materials-NORM/ (accessed 2016-06-28).

1.5 EXERCISES

1.1 Knowing that the neptunium series starts with ^{237}Np and ends with ^{205}Tl, determine the other elements in this decay chain, from the relation $4n+1$, described in Section 1.2.

1.2 The activity of ^{14}C in living organisms is assumed to be 13.2 disintegrations per minute per gram carbon. When the organism dies, no more ^{14}C is assimilated and the amount of ^{14}C in the organism decreases due to radioactive decay. An activity measurement on 200 g carbon from an old piece of bone gave the value 1780 disintegrations per minute per gram carbon. How old was the piece of bone?

1.3 The specific activity of ^{14}C in carbon from an old camp fire was 182 Bq/kg. The specific activity in living trees is 220 Bq/kg and the half-life of ^{14}C is 5730 years. How many years have passed since the camp fire was in use?

1.4 One of the decay series starts with a uranium isotope and ends with a lead isotope. The half-life of the uranium isotope is $4.5 \cdot 10^9$ years and the intermediate radionuclides have comparatively short half-lives. Calculate the time needed for a mineral to contain atoms of lead and uranium at the ratio 1:2 if the mineral originally contained only uranium.

1.5 The present-day abundances of the two main uranium isotopes ^{238}U and ^{235}U are 99.27% and 0.72%, respectively, and their half-lives are $4.5 \cdot 10^9$ years and $7.13 \cdot 10^8$ years. If it can be assumed that the abundance ratio could never have been greater than unity, what is the maximum possible age of the earth's crust? (Rutherford derived a rather good estimate of the age of the earth, in about 1904, using this method).

1.6 Either transient or secular equilibrium will exist depending on the half-lives of the mother and daughter radionuclides in a radioactive decay. Use Eqn. (3.1) to find the relation between the half-lives of mother and daughter for

a. transient equilibrium, where both radionuclides decay at a rate given by the half-life of the daughter, and

b. secular equilibrium, where both radionuclides have the same activity.

1.7 Calculate the specific activities in Bq kg^{-1} for the radionuclides ^{238}U, ^{235}U and ^{234}U.

1.8 Knowing that the relative abundance of ^{40}K is 0.012%, show that the specific activity of naturally occurring K is 30.6 kBq kg^{-1}.

1.9 Show that 1 WL = 275 Bq m^{-3} of ^{220}Rn when in secular equilibrium with its daughters.

1.10 Use the data from Table 1.9 to show that the equilibrium-equivalent concentration, C_{eq}, for ^{222}Rn with the short-lived daughters ($j = 1$ to 4) is given by the expression $C_{eq,Rn}$ [Bq m^{-3}] = $0.105C_1 + 0.516C_2 + 0.379C_3 + 6 \cdot 10^{-8}C_4$.

1.11 Assume that the activity of ^{137}Cs ($T_{1/2} = 30$ y) per unit area is 3 kBq m^{-2}. Calculate the mass of ^{137}Ba per unit area, produced by the decay of ^{137}Cs, after 10 y.

1.12 Show that the energy released in fission of ^{235}U is about 200 MeV.

1.13 A thin foil of ^{115}In is irradiated by neutrons for 3.0 h. The activity of the ^{116}In produced ($T_{1/2} = 54.2$ minutes) was then measured 8.0 minutes after the end of irradiation and found to be 100 kBq. Calculate the number of ^{116}In nuclei produced per unit time during irradiation, and the neutron fluence rate. The mass of the foil was 2.5 g and it can be assumed that the foil consisted of 100% ^{115}In. The cross section for the reaction ^{115}In(n,γ)^{116}In is 202 barn.

1.14 ^{211}At ($T_{1/2} = 7.2$ h) is produced by the reaction ^{209}Bi(α,2n)^{211}At. Calculate the activity of ^{211}At after 15 h irradiation if the production rate is 10^6 s^{-1}. By how much will the activity increase if the irradiation time is doubled?

1.15 Estimate the dose received by a person who enters a fallout area 36 h after an explosion and leaves the area after 12 h based on the information that the dose rate measured at the site 40 h after the explosion was 25 mSv h^{-1}.

1.6 FURTHER READING

IAEA. 2005. *Technical Data on Nucleonic Gauges.* IAEA-TECDOC-1459. International Atomic Energy Agency. Vienna.

IAEA. 2008. *Trends in Radiation Sterilization of Health Care Products.* STI/PUB/1313. International Atomic Energy Agency. Vienna.

Neeb, K. H. 1997. *The Radiochemistry of Nuclear Power Plants with Light Water Reactors.* Berlin: Walter de Gruyter & Co.

Malley, M. C. 2011. *Radioactivity: A History of a Mysterous Science.* New York, USA: Oxford University Press

Tuniz, C. 2012. *Radioactivity: A Very Short Introduction.* Gosport, Hampshire: Oxford University Press.

CHAPTER 2

Radiation Biology and Radiation Dosimetry

CONTENTS

T HIS chapter starts with a brief review of the mechanisms governing the interaction of charged and uncharged radiation with matter, with the aim of presenting some of the key concepts used in dosimetry and the operation of radiation detectors. Simple relations are then presented between the absorbed dose and kerma in terms of interaction coefficients. Cavity theories are also introduced to prepare the reader for the more comprehensive discussion of radiation detectors later in this book. The section on radiation biology gives an overview of the biological effects of radiation, and includes the basic knowledge required to follow subsequent discussions on radiation protection and intervention levels in emergency preparedness. Quantities related to the harmful effects of exposure to ionization radiation, such as the effective dose, are then discussed, and calculations of the absorbed dose and kerma from external sources with various geometries are presented. The chapter ends with a description of the models developed by the International Commission on Radiological Protection for calculating the absorbed dose from radioactive sources inside the human body.

2.1 INTERACTION OF RADIATION WITH MATTER

2.1.1 The Interaction of Charged Particles with Matter

Charged particles interacting with matter may lose energy either by *collisions* with (mainly) atomic electrons, or by *radiative interactions*, in which electromagnetic radiation is emitted by the impinging charged particle. Collisions result from the action of the Coulomb force, and a fraction of the energy of the charged particle may then be transferred to an atomic electron. This may lead to the ejection of the atomic electron from the atom, i.e. to *ionization* of the atom, as shown on the left in Figure 2.1. However, if the energy transferred to the atomic electron is not sufficient to cause ionization, the atomic electron may instead be transferred to a higher energy state, i.e. the interaction causes *excitation* of the atom. The excited atom will then emit radiation when it returns to its original state.

In a radiative interaction, the charged particle is deflected (*scattered*) by the Coulomb force due to the charge on the atomic nucleus. Charged particles that are forced to change their velocity or direction will emit electromagnetic radiation, an example of which is radio antennas. This is also the cause of emission of radiation from x-ray tubes, where the rapid slowing down of electrons as they hit the anode causes the kinetic energy of the electrons to be dissipated as heat (by collisions with electrons in the anode material) or emitted as electromagnetic radiation. This type of electromagnetic radiation is called *bremsstrahlung* and is depicted on the right in Figure 2.1.

Both these processes lead to a loss of kinetic energy of the charged particle, which can be described as the energy loss per unit length travelled in the material, dT/dx. This energy loss depends on the mass and the charge of the impinging particle, as well as on its energy and on the properties of the material (e.g., the atomic number, Z) through which it passes. This is called the *stopping power*, S, defined as

$$S = -\frac{dT}{dx},$$
(2.1)

where the minus sign indicates that the energy of the charged particle decreases as it passes through the material. For heavy charged particles, such as protons and α-particles, the *collisional stopping power* of a particle with charge ze (where e is the elementary charge) is given by (Anderson 1984)

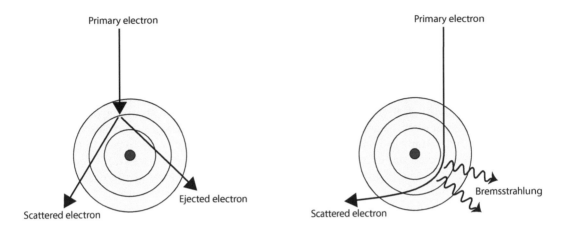

FIGURE 2.1   Charged particles, in this case electrons, can interact with matter by collisions (left) or by radiative interactions (right).

$$S_c = 2z^2 \pi r_0^2 \frac{m_e c^2}{\beta^2} n_v Z \ln \left(\frac{2m_e v^2}{I} \right)^2,$$ (2.2)

where r_0 is the classical electron radius ($\approx 2.8 \cdot 10^{-15}$ m), m_e is the electron rest mass ($\approx 9.1 \cdot 10^{-31}$ kg), and c is the speed of light in vacuum ($\approx 3 \cdot 10^8$ m s^{-1}). The parameter β is given by the ratio of the speed of the charged particle, v, and the speed of light in vacuum, i.e. v/c. The number of atoms per m^3 in the medium with atomic number Z is denoted n_v, and depends on Avogadro's number, N_A, the density, ρ, and the molar mass, M, of the medium as $\rho N_A / M$.

The parameter I is called the *average excitation energy*, and is a measure of the energy transferred from the charged particle to an atomic electron in a collision. In deriving Eqn. 2.2 it is necessary to consider the minimum and maximum energy transferable to an atomic electron in a collision. The maximum energy can be obtained by applying classical laws of energy and momentum conservation, since it can be assumed that the charged particle is nonrelativistic, i.e. moving with a speed considerably less than the speed of light in vacuum. However, the minimum energy is not easily determined due to the discrete energy levels of the bound electron states in an atom. The binding energy can vary significantly depending on the atomic electron participating in the collision, since it differs between the shells and subshells within the atom. Thus, the minimum amount of energy transferred varies with the target material. The average excitation energy is then an average value weighted by the probabilities of the different energy transitions between the electronic states.

Electrons can also lose energy by the emission of *Cherenkov* (or *Čerenkov*) radiation. An electron moving with higher velocity than the speed of light in the particular medium will emit energy in the form of a bluish light. The speed of light in a medium, c_m, depends on the index of refraction, n, as $c_m = c/n$, where c is the speed of light in vacuum. For water, the index of refraction is 1.33 and an electron with kinetic energy >0.6 MeV will then move faster than light in water and thus emit Cherenkov radiation. This radiation is emitted in a cone around the moving electron, where the half-angle of the cone, θ, is given by $\cos \theta = c/nv$, where v is the electron velocity.

Heavy charged particles may also lose energy by nuclear collisions, but this makes only a very slight contribution to the stopping power. The greatest contribution thus comes from collisions with electrons, especially in the outer shells, and the stopping power thus depends mainly on the electron density of the medium, given by the product of the number of atoms per unit volume and the number of electrons per atom, i.e. $n_v Z$ in Eqn. 2.2.

The stopping power for protons in water is shown in Figure 2.2, where it can be seen that there is a peak at a proton energy of about 0.2 MeV. This is called the *Bragg peak*, and is the energy at which the charged particle most effectively transfers its kinetic energy to the material. At particle energies above the Bragg peak, the stopping power decreases smoothly with increasing energy, while at energies below the Bragg peak the stopping power decreases rapidly, and the charged particle will then essentially come to rest in the material. If we consider a material externally irradiated with charged particles, the Bragg peak will appear at a certain depth in the medium, since the kinetic energy of the charged particles decreases gradually in the medium. This means that the range of heavy charged particles in a medium is rather well defined. For example, the range of a 1-MeV proton in water is about 2.5 mm.

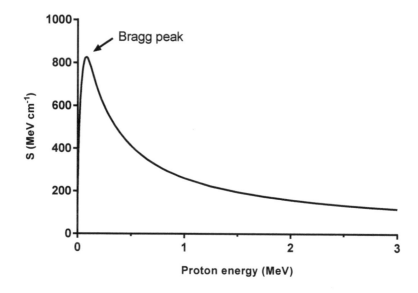

FIGURE 2.2   Stopping power, S, for protons in water. Data from http://physics.nist.gov.

The maximum energy that can be transferred in a collision between a heavy charged particle and an electron can be approximated by the expression (Anderson 1984)

$$E_{max} = 2m_e v^2. \tag{2.3}$$

Assuming a proton with a kinetic energy of 1 MeV and using the classical relation between kinetic energy, T, and velocity, $T = mv^2/2$, where m is the proton rest mass, then the speed of the proton will be $1.4 \cdot 10^7$ m s^{-1}. E_{max} will then be about $2 \cdot 10^{-3}$ MeV, i.e. the proton loses only a small proportion of its kinetic energy in each collision. As can be seen in Figure 2.2, the stopping power of a proton with a kinetic energy of 1 MeV is about 250 MeV cm^{-1}, or 25 MeV mm^{-1}. Although the stopping power is not constant due to the loss of energy of the proton, we can make an approximate estimate of the number of collisions along a path of 1 mm. The proton must make $25/2 \cdot 10^{-3} = 12\ 500$ collisions in order to lose 25 MeV, if we assume that the maximum possible energy is transferred in each collision. Thus, heavy charged particles are characterized as densely ionizing particles. The tracks of heavy charged particles in a medium are thus almost equal in length for particles of the same energy, and are straight due to the large difference in mass between the heavy charged particle and the much lighter electron with which it collides.

The collisional energy losses for light charged particles, such as electrons and positrons, can be described in a similar way as for heavy charged particles. However, some of the assumptions made when deriving Eqn. 2.2 are not valid for light charged particles. For example, an electron may lose all its kinetic energy in a single collision with an atomic electron, and events in which the incident electron loses a large fraction of its kinetic energy in a single collision are quite frequent. Also, in contrast to the straight tracks of heavy charged particles, a collision between an incident electron and an atomic electron may result in a substantial change in direction of the incident electron. The range of light charged particles in a medium is therefore less well defined.

The derivation of an expression for the collisional stopping power, S_c, as given in Eqn. 2.2, will require both quantum mechanical and relativistic treatment, and is outside the scope of this book. However, the radiative stopping power, S_r, is given by

$$S_r = Z^2 r_0^2 \alpha n_v R(T, Z) T, \tag{2.4}$$

where α is given by

$$\alpha = \frac{e^2}{4\pi\epsilon_0 \hbar c} \tag{2.5}$$

and includes the fundamental constants vacuum permittivity, ϵ_0, electron charge, e, and the reduced Planck constant, $\hbar$. The function $R(Z, T)$ is the spectrum average of the intensity function, and accounts for the distribution of photon energies in the bremsstrahlung. For high-energy electrons, $R(Z, T)$ can be approximated by $4\ln(183 \cdot Z^{-1/3})$ (Anderson 1984). It follows from the Z^2 dependence in Eqn. 2.4 that bremsstrahlung is more significant when electrons impinge on materials with high atomic numbers.

Figure 2.3 shows the contribution from collisional and radiative stopping power to the total stopping power for electrons in water. The collisional stopping power for protons is shown for comparison, demonstrating the difference in collisional stopping power for light and heavy charged particles. Since water has a low effective atomic number, the radiative part of the stopping power is less important than the collisional part below about 100 MeV, i.e. in all practical situations concerning electrons emitted by β-decay, which typically have energies of about 1 MeV. In elements with high atomic numbers, exemplified by lead in Figure 2.4, the radiative losses become important at significantly lower energies.

The stopping power is proportional to the density of the medium. Although not explicitly stated in Eqn. 2.2, the density is used in the determination of the number of atoms per unit volume, n_v. The stopping power can therefore differ considerably between different phases (solid, liquid, gas) of the same element. To overcome these difficulties when tabulating values of stopping power, the quantity *mass stopping power*, S/ρ, is used. It is also possible to calculate the mass stopping power for media consisting of more than one element using Bragg's additivity rule. According to this rule, the mass stopping power is calculated as a weighted sum of the mass stopping powers for each element in the medium as

$$\frac{S}{\rho} = \sum_i \epsilon_i \left(\frac{S}{\rho}\right)_i, \tag{2.6}$$

where ϵ_i is the fraction of element i, by weight, in the medium.

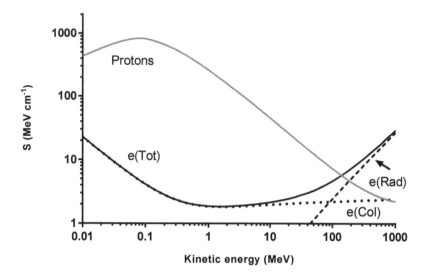

FIGURE 2.3 Stopping power for electrons (e) in water, divided into collisional (col), radiative (rad) and total (tot) stopping power. The stopping power for protons is shown for comparison. Data from http://physics.nist.gov.

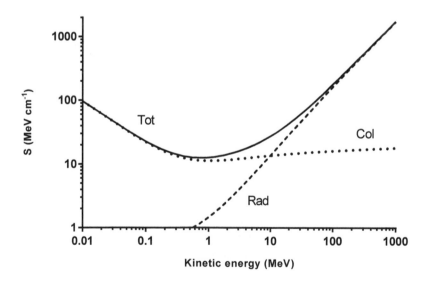

FIGURE 2.4 Stopping power for electrons in lead, divided into collisional (col), radiative (rad) and total (tot) stopping power. Data from http://physics.nist.gov.

Collisional stopping power is closely related to an important quantity in applied radiation protection, the *linear energy transfer* (LET), L_Δ, defined in ICRU Publication 60 (ICRU 1998) as

$$L_\Delta = \frac{dE_\Delta}{dl}. \tag{2.7}$$

For charged particles, L_Δ is defined as the energy transferred by electronic collisions, dE_Δ, per unit length, dl, within a medium, with the restriction that the energy transferred by each interaction is less than a certain amount, denoted Δ. The LET can thus be described as a restricted collisional stopping power, where the collisions generate secondary charged particles with such low energies that it can be assumed that their energy will be locally deposited due to the short range of these particles. The stopping power is sometimes referred to as an unrestricted linear energy transfer, where the value of Δ is chosen so that no interactions are excluded (theoretically, $\Delta \to \infty$).

Collisions resulting in secondary electrons with energies greater than Δ are thus not included in L_Δ, but it is assumed that these secondary electrons will cause further ionization along their paths. These particles are denoted *δ-particles*. The energy limit Δ cannot be assigned a specific numerical value, but must be determined in each case. For example, the energy transferred in a collision depends on the electronic binding energies in the irradiated medium. Therefore, the energy of an impinging electron required to ionize the atoms in the medium will depend on the elemental composition of the medium.

It has been found that electrons emitted by β-decay can be treated similarly to photons (see Eqn. 2.14) when traversing a medium, i.e. the continuous energy distribution results in an almost exponential decrease in fluence with increasing thickness of the absorbing material (Nilsson 2015). The relation between electron fluence and the thickness, x, of an absorber, assuming an incident fluence, ϕ_0, is then

$$\phi(x) = \phi_0 e^{-(\beta/\rho) \cdot \rho \cdot x}, \tag{2.8}$$

where β/ρ is the mass attenuation coefficient of the β-particles. This coefficient varies with the energy, T, of the β-particles and the atomic number of the absorber, and values for β/ρ in the units m^2 kg^{-1} can be approximated by

$$\beta/\rho = \frac{3.5 \cdot Z}{A \cdot T^{1.14}} \tag{2.9}$$

for absorbers with low atomic numbers, where (A) is the mass number of the absorber. For absorbers with high atomic numbers, the relation is

$$\beta/\rho = \frac{0.77 \cdot Z^{0.31}}{T^{1.14}}. \tag{2.10}$$

Note that these expressions are only approximate, although they are sufficiently accurate for radiation protection purposes.

2.1.2   The Interaction of Uncharged Radiation with Matter

Uncharged radiation consists mainly of photons and neutrons. Although other elementary particles, such as neutrinos and neutral pions, belong to this class of radiation, we will restrict the discussion here to photons (x-rays and γ-rays) due to their importance in radiation protection. The basic interaction mechanisms of neutrons have been discussed previously in Section 1.3.1. Uncharged radiation may also be called indirectly ionizing radiation since the interactions do not occur as a result of the Coulomb force. However, in most cases, the interactions will give rise to the liberation of charged particles, i.e. directly ionizing radiation, as discussed above.

Four kinds of interactions between photons and matter are relevant for understanding the characteristics of indirectly ionizing radiation in the field of radiation protection. Two of them involve the scattering of photons (*coherent* scattering and *incoherent* scattering, where the latter is denoted *Compton scattering* if scattering is assumed to take place from a free electron), and the third involves the absorption of the photons (*photoelectric absorption*). The fourth involves *pair production*, which leads to the creation of two new photons. These four processes will be discussed below.

Let us first assume that a beam containing a flux of N_0 photons is impinging on a medium. For each length interval, dx, into the medium, a proportion of the particles, dN, is absorbed or scattered out of the beam by a process called *attenuation*. The probability of these interactions is characterized by the *reaction cross section*, σ, which can be defined as the probability per atom per unit area, i.e.

$$\sigma = \frac{-d\phi}{\phi} \frac{1}{n_v dx}, \tag{2.11}$$

where $-d\phi$ is the fluence (the number of particles per unit area impinging on the medium) removed from the incident fluence in a slab of thickness dx. The number of atoms per unit area is given by $n_v dx$ since n_v is the number of atoms per unit volume. The unit of σ is the barn (b), where 1 b equals 10^{-28} m^2, and although expressed as an area, the cross section may be significantly different from the actual cross-sectional area of, for example, an atom. Rearranging this expression gives

$$\frac{-d\phi}{\phi} = n_v \sigma dx. \tag{2.12}$$

The product $n_v \sigma$ is called the *linear attenuation coefficient*, μ, which can also be written $\mu = n_m \sigma \rho$, where n_m is the number of atoms per unit mass. The linear attenuation coefficient thus depends on the energy of the incident radiation (through σ), as well as on the irradiated material (through ρ and n_m). It follows that the dimension of μ is inverse length, often given in the unit cm^{-1}.

Returning to the flux of N_0 photons impinging on a medium, we find that

$$\frac{dN}{N} = -\mu dx, \tag{2.13}$$

since dN/N is equivalent to $-d\phi/\phi$ due to cancellation of the area. Solving for N and explicitly stating that N is a function of distance, x, gives

$$N(x) = N_0 e^{-\mu x}. \tag{2.14}$$

$N(x)$ is thus the number of photons *remaining* at depth x.

The attenuation coefficient is dependent on the cross sections of the four interaction processes mentioned above, all of which have a significant probability of gamma photons in the energy range 30–3000 keV. Since the linear attenuation coefficient also depends on the density of the medium, it is often desirable to use the so-called *mass attenuation coefficient*, μ/ρ (for the same reasons as discussed above for the mass stopping power). This quantity is related to the probability of an interaction per unit mass depth, ρx, and is often given in the units cm^2 g^{-1}:

$$N(x) = N_0 e^{-(\mu/\rho)\cdot\rho\cdot x}. \tag{2.15}$$

When counting the number of photons with a given primary energy, the above relationship, Eqn. 2.15, describes how the flux depends on the thickness, x, of the attenuating material. However, the radiant energy imparted per unit length does not always follow this simple relation, since the interaction processes yield secondary photons that can contribute to the total flux of photons in the beam, creating so-called *build-up* (see Section 2.7).

In elastic scattering of photons, the direction of motion of the photon is changed, without any change in the photon energy. Conversely, when a photon is inelastically scattered, some of the photon energy is transferred to the medium, resulting in the emission of an atomic electron. Elastic scattering can be described by regarding the photon radiation as an electromagnetic wave that causes an electron in the irradiated medium to oscillate. This oscillating electron will then radiate since it is an accelerating electrical charge. This type of scattering was first described by J. J. Thomson and is thus referred to as *Thomson scattering*. For low photon energies, the wavelength of the electromagnetic wave is comparable to the size of an atom, and the whole atom will take part in scattering. For this reason, this type of scattering is called *coherent scattering* (or *Rayleigh scattering*).

Inelastic scattering, or Compton scattering, is conveniently described by regarding the electromagnetic radiation as particles, i.e., photons. The fact that electromagnetic radiation can be described as both waves and particles (wave–particle duality) is reflected in the relation between the photon energy, E_γ, and the frequency, ν, of the electromagnetic wave: $E_\gamma = h\nu$, where h is Planck's constant. We will hereafter use the notation $h\nu$ for the photon energy.

In Compton scattering, a photon with energy $h\nu$ collides inelastically with a loosely bound atomic electron, i.e. $h\nu \gg E_b$, where E_b is the binding energy of the atomic electron, as illustrated in Figure 2.5. The electron is ejected at an angle ϕ to the direction of the incident photon, and a photon is scattered at an angle θ. The energy of the scattered photon, $h\nu'$ is given by

$$h\nu' = \frac{h\nu}{1 + \alpha(1 - cos\theta)}, \tag{2.16}$$

where α is given by

$$\alpha = \frac{h\nu}{m_e c^2}, \tag{2.17}$$

m_e being the rest mass of an electron. Eqn. 2.16 follows from the laws of energy and momentum conservation.

The scattered photon will thus have lower energy than the incident photon, the difference in energy being kinetic energy of the ejected electron. Since it is assumed that the ejected electron is at rest prior to scattering, and that the binding energy can be neglected (loosely bound, or free electron), the electron energy, T_e, is given by $h\nu - h\nu'$. The angle of the ejected electron is related to the angle of the scattered photon by

$$\phi = cot^{-1}\left[(1 + \alpha)tan\frac{\theta}{2}\right]. \tag{2.18}$$

The two extremes for the photon scattering angle, θ, are 0° (no scattering) and 180° (backscattering), and the corresponding angles for the ejected electron are $\phi = 90$° and $\phi = 0$°, respectively. At the limit where there is no scattering, we can also conclude that

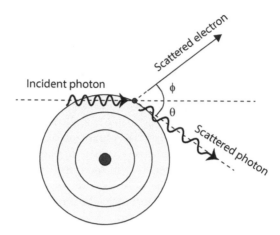

FIGURE 2.5 Compton scattering of an impinging photon with energy $h\nu$ from an atomic electron, leading to a scattered photon with energy $< h\nu$ and a recoil electron. The relation between the energies of the scattered photon and the recoil electron is determined by the scattering angles ϕ and θ.

$h\nu' = h\nu$ and that $T_e = 0$. Correspondingly, for the backscattering situation, we have $h\nu' = h\nu/(1 + 2\alpha)$ and $T_e = 2\alpha h\nu/(1 + 2\alpha)$.

The photon scattering angle depends on the energy of the incoming photon, in a relation mathematically described by Klein and Nishina in 1929. The Klein–Nishina differential cross section per unit solid angle describes the probability of a photon being scattered in a solid angle between θ and $\theta + d\theta$, i.e. within a cone around θ. The results of normalized calculations of this cross section are shown in Figure 2.6. At low photon energies, the angular distribution is almost symmetrical, and forward scattering and backscattering are more probable than lateral scattering. With increasing energy of the incident photons, more photons are scattered in the forward direction.

The Klein–Nishina differential cross section can be used to derive an expression for the total cross section for Compton scattering, σ_c, if the effects of electron binding energies can be neglected. If the parameter α (see Eqn. 2.16) is much greater than unity, σ_c is given by

$$\sigma_c = \pi r_0^2 \left(\frac{1 + 2\ln 2\alpha}{2\alpha} \right). \tag{2.19}$$

The mass attenuation coefficient, defined above as $\mu/\rho = n_m\sigma$, accounts for all types of interactions and can be regarded as being composed of contributions from the different interaction processes. The part attributable to Compton scattering is therefore given by $(\mu/\rho)_c = n_m\sigma_c$. The energy dependence of this coefficient is implicit in the parameter α, which is proportional to $h\nu$ and, due to the natural logarithm in the numerator of Eqn. 2.19, it follows that the mass attenuation coefficient for Compton scattering decreases monotonically with increasing incident photon energy. This dependence cannot be seen in the experimental data shown in Figure 2.7, but the shape can be re-created by taking the effects of electron binding energies into account. It should also be noted that the cross section is independent of the atomic number of the scattering medium, although the mass attenuation coefficient for Compton scattering depends on Z through n_m.

Photoelectric absorption refers to the absorption of a photon with energy $h\nu$ by an atomic electron in the inner shell of an atom, often the K shell. This leads to the emission

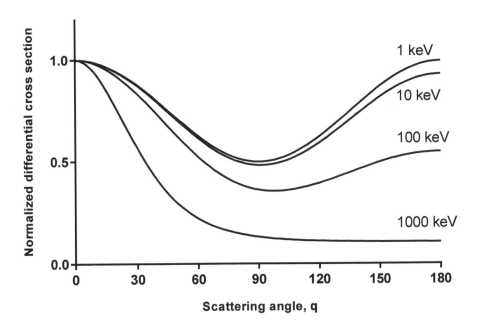

FIGURE 2.6 Normalized plots of the Klein–Nishina differential cross section per unit solid angle for different photon energies (1–1000 keV). The scattering angle θ is defined in Figure 2.5.

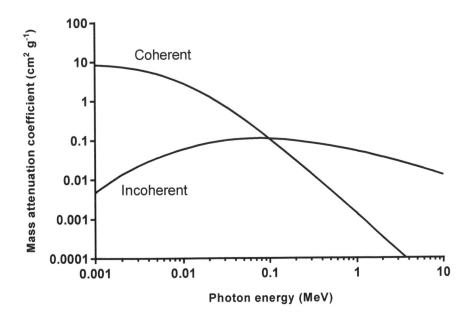

FIGURE 2.7 Mass attenuation coefficients for coherent and incoherent (Compton) scattering in iodine. Data from http://physics.nist.gov.

of an electron with energy $h\nu - E_K$, where E_K is the binding energy of the K electron, followed by a characteristic x-ray photon when the vacancy in the K shell is filled by an electron from an outer shell. This outer electron will also give rise to a vacancy when it moves to the K shell, which will also be occupied by an outer electron, thereby creating a sequence of x-rays of different energies.

The radiation emitted when the vacancies are filled depends on the atomic number of the medium and is thus called characteristic x-radiation, as the energies of the x-rays are characteristic of each element. A vacancy in the K shell can be filled by an electron from any of the outer shells: N, M, etc. If the vacancy is filled by an electron from the L shell, the emitted radiation is denoted K_α, while if the electron is from the M shell, the radiation is denoted K_β. Similarly, vacancies in the L shell give rise to L_α and L_β x-rays.

The vacancies may also be filled without the emission of characteristic x-rays. In this case, the energy lost by the electron when moving from a higher energy level to a lower one is transferred directly to an outer electron, which is ejected from the atom. The prerequisite for the emission of an electron is that the energy difference between the initial and the final state of the electron filling the vacancy is greater than the binding energy of the ejected electron. These so-called *Auger electrons* are monoenergetic, in contrast to the continuous energy spectra observed when electrons are emitted by β-decay. The relative abundance of characteristic x-rays in relation to the number of Auger electrons can be found using the *fluorescence yield*, ω, which is the probability of the emission of a characteristic x-ray photon when a vacancy is filled. The K fluorescence yield, i.e. the probability of emission of a characteristic x-ray photon per vacancy in the K shell, is shown in Figure 2.8 for elements with $Z = 1$ to $Z = 110$. It can be seen that the emission of characteristic x-rays is most important for heavier elements.

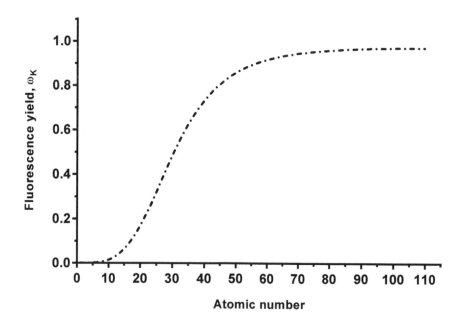

FIGURE 2.8 The K fluorescence yield gives the probability of emission of a characteristic x-ray photon per vacancy in the K shell.

Photoelectric absorption occurs mostly via the electrons in the K shell, and in this case the total cross section for photoelectric absorption is given by

$$\sigma_K = 4\sqrt{2}\frac{8\pi r_0^2}{3}\alpha^4 Z^5 \left(\frac{m_e c^2}{h\nu}\right)^{7/2} \tag{2.20}$$

and the total cross section for the entire atom can be approximated by (Anderson 1984)

$$\sigma = \sigma_K \left(1 + 0.01481(\ln Z)^2 - 0.00079(\ln Z)^3\right). \tag{2.21}$$

Since Eqn. 2.20 does not include relativistic effects, the dependence on $h\nu$ and Z is not exactly as stated in the equation above. The atomic nucleus is screened by the atomic electrons, which introduces an energy dependence into Eqn. 2.20, i.e. the dependence on Z varies with the photon energy. This effect leads to an average value of the Z exponent between 4 and 5. However, the dependence on Z is still very large, and the probability of photoelectric absorption in heavy elements, such as lead ($Z = 82$), will be significantly higher than in tissue-equivalent material ($Z = 7.1$). For a given medium with an atomic number Z, the cross section for photoelectric absorption decreases smoothly with photon energy as $(h\nu)^{-3}$ for low energies, and as about $(h\nu)^{-2}$ for higher energies, as can be seen in Figure 2.9.

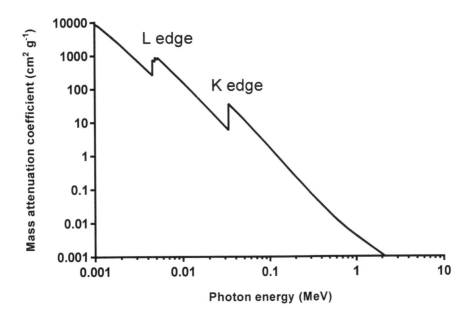

FIGURE 2.9   Mass attenuation coefficients for photoelectric absorption in iodine. The sharp discontinuities in the curve are explained in the text. Data from http://physics.nist.gov.

The sharp discontinuities in the curve occur at the electron binding energies in the atomic shells, and are called the K edge and L edge. They can be explained by following the curve from higher energies to lower. As the photon energy decreases, the mass attenuation coefficient (or the cross section) increases according to Eqn. 2.20. However, when the photon energy is lower than the binding energy of electrons in the K shell, these electrons cannot take part in photoelectric absorption, and the cross section decreases. A further decrease in photon energy leads to an increase in cross section, involving electrons in higher shells, until the photon energy equals the binding energy of the electrons in the L shell. Note that the L edge actually has three edges due to the subshells of the L shell.

The fourth and final interaction mechanism for photons discussed here is pair production, i.e. the creation of an electron–positron pair from the photon's energy. This process can only occur in the vicinity of an atomic nucleus or an atomic electron due to the requirement that an electric field is present, and the fact that energy and momentum must be conserved in the reaction. Since the photon energy must be sufficient for the creation of an electron–positron pair, the lower energy limit for pair production is 1.022 MeV (twice the energy equivalent to the rest mass of an electron or positron, 511 keV). Pair production occurring in the vicinity of an atomic electron is denoted *triplet production*, and the threshold energy for this reaction is 2.044 MeV ($4m_e c^2$) since the momentum is shared between three particles of equal mass (in contrast to pair production where the momentum is shared with the much heavier nucleus).

The cross section for pair production is given by

$$\sigma = Z^2 r_0^2 \alpha h(h\nu, Z), \tag{2.22}$$

where $h(h\nu, Z)$ is a function that increases monotonically with photon energy. Figure 2.10 shows the mass attenuation coefficient for pair production together with mass attenuation coefficients for the other interaction mechanisms discussed above. Photoelectric absorption is the dominating interaction mechanism at low photon energies, while pair production dominates at high energies. At intermediate energies, most of the interactions occur by incoherent scattering. Although Figure 2.10 shows the case for iodine, the main features are similar for other elements. The boundaries between the dominating processes are, however, different for different elements. Generally, the energy interval in which scattering dominates is broader in light elements, while the interval in which photoelectric absorption and pair production dominate is wider in heavy elements.

2.2   RADIATION DOSIMETRY

2.2.1   Absorbed Dose and Kerma

Absorbed dose is defined by the International Commission on Radiation Units and Measurements (ICRU) in Publication 60 (ICRU, 1998), and is a measure of how much of the energy in the radiation field is retained per unit mass in a small volume. The SI unit of absorbed dose is the Gy (gray), but in fundamental SI units 1 Gy equals 1 J kg^{-1}. The absorbed dose rate [Gy s^{-1}] to air in free air (meaning that no scattering material is present) can be defined in terms of the photon fluence rate, $\dot{\Phi}$, the photon energy, E, and the *mass energy absorption coefficient*, μ_{en}/ρ, for air. This relation, given in Eqn. 2.23, is valid for monoenergetic photons of energy E.

$$\dot{D} = \dot{\Phi} \left(\frac{\mu_{en}}{\rho} \right)_E E \tag{2.23}$$

If the radiation field consists of photons of different energies, the contribution from each photon energy must be included in the calculation of the absorbed dose rate by integration over the photon energy, similar to Eqn. 2.25.

Kerma is an acronym for the kinetic energy released per unit mass, denoted K, and is also defined in ICRU Publication 60. It is a measure of how much of the photon energy in the radiation field is transferred, via interactions, to the kinetic energy of charged particles,

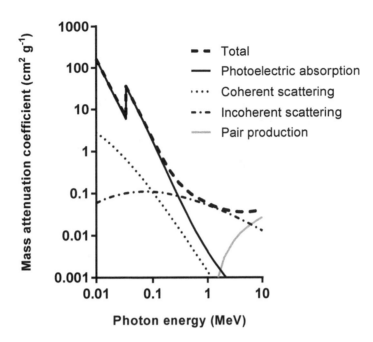

FIGURE 2.10 Mass attenuation coefficients for various interaction mechanisms in iodine. Data from http://physics.nist.gov.

per unit mass in a small volume. The unit of kerma is also Gy. Some of the kinetic energy transferred to charged particles may be dissipated by bremsstrahlung, thus leaving the volume, and will therefore not contribute to the absorbed dose, as shown in Figure 2.11. This energy loss is taken into account when defining the kerma in terms of the *mass energy transfer coefficient*, μ_{tr}/ρ, instead of μ_{en}/ρ. The kerma rate in free air from monoenergetic photons can thus be written

$$\dot{K} = \dot{\Phi}\left(\frac{\mu_{tr}}{\rho}\right)_E E. \qquad (2.24)$$

In the general situation, where the radiation field consists of photons of various energies, the kerma rate is given by the relation

$$\dot{K} = \int \dot{\Phi}_E E\,(\mu_{tr}/\rho)\,dE, \qquad (2.25)$$

where $\dot{\Phi}_E$ is the differential fluence rate, and the subscript for μ_{tr}/ρ has been omitted for increased readability. Since the energy fluence rate, $\dot{\Psi}$, equals $\dot{\Phi} \cdot E$, the kerma rate is also given by

$$\dot{K} = \int \dot{\Psi}_E\,(\mu_{tr}/\rho)\,dE. \qquad (2.26)$$

If radiative losses (bremsstrahlung and annihilation in flight) are neglected, the energy transferred from photons to charged particles is given by the *collision kerma*, K_c:

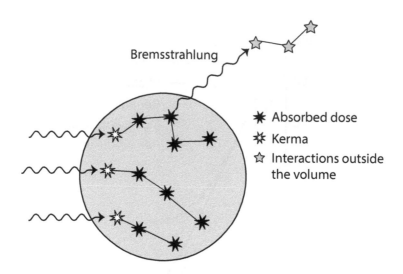

FIGURE 2.11 Some of the energy transferred from photons to charged particles may be dissipated as bremsstrahlung. The dissipated energy will thus not contribute to the absorbed dose.

$$K_c = \int \Psi_E \left(\mu_{en}/\rho \right) dE. \tag{2.27}$$

The relation between (μ_{tr}/ρ) and (μ_{en}/ρ) is given by

$$(\mu_{en}/\rho) = (\mu_{tr}/\rho) \left(1 - g \right). \tag{2.28}$$

The term g depends on the photon energy and on the cross sections for Compton scattering and pair production, i.e. processes that lead to the creation of secondary photons with sufficient energy to escape the volume element under consideration.

2.2.2 Charged Particle Equilibrium and Cavity Theory

The collision kerma is thus a measure of the energy that *may* be locally deposited in a volume. However, a prerequisite is that *charged particle equilibrium (CPE)* prevails in the volume. In this case, the energy from charged particles that enter the volume is equal to the energy carried by charged particles leaving the volume. For example, the total kinetic energy of the electrons liberated by photons per unit mass in a volume is given by the kerma. The part of this energy that is dissipated by collisions is given by the collision kerma. If all the electrons dissipate their energy within the volume, then the absorbed dose will equal the collision kerma. However, some of the electrons may escape the volume and their energy will be dissipated outside the volume, and thus the absorbed dose will be less than the collision kerma. If CPE prevails, it can be assumed that these escaping electrons are replaced by electrons liberated outside the volume (and thus not included in the collision kerma), as shown in Figure 2.12. Therefore, under the condition of CPE, the absorbed dose is equal to the collision kerma. In a medium externally irradiated by photons, CPE can often be assumed at a depth exceeding the range of the liberated electrons (strictly speaking, *transient CPE*, since the photon fluence decreases with depth due to attenuation), as shown in Figure 2.13. These concepts are further discussed in several books on dosimetry (e.g. Attix 1991 and McParland 2010).

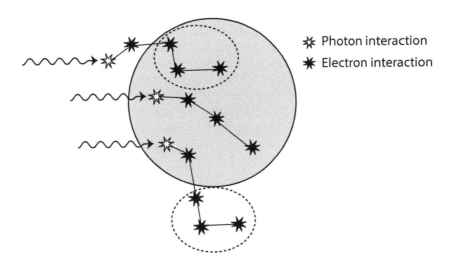

FIGURE 2.12 At charged particle equilibrium, the energy deposited within a volume by electrons liberated by photon interactions outside the volume is balanced by the energy deposited outside the volume by electrons liberated by photon interactions within the volume. These interactions are shown in the dashed ellipses.

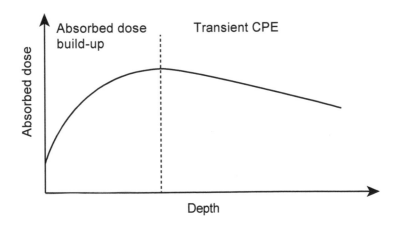

FIGURE 2.13 The absorbed dose in a medium irradiated by photons reaches a maximum at a depth corresponding to about the range of the electrons set in motion by the photon interactions in the medium. The depths below the maximum are characterized by transient charged particle equilibrium. Note that the absorbed dose at the surface will not equal zero due to electrons liberated in the surrounding medium, and electrons backscattered within the medium.

Measurements of absorbed dose require some kind of detector, which often has a composition different from the medium in which the absorbed dose is to be determined. For example, the absorbed dose to water can be measured by placing an air-filled detector at a specified point in the water. The detector will then give a response that is proportional to the absorbed dose to air (the detector cavity), and in order to determine the absorbed dose to water corrections are made using *cavity theories*. These corrections depend on the size of the detector in relation to the range of the charged particles in the detector material, and two cases will be discussed here: a large detector and a small detector.

Let us consider a water-filled tank, irradiated by a photon field. The size of the tank is large enough for CPE to prevail in the water. The dose, D_w, to the water is then given by

$$D_w = K_{c,w} = \int \Psi_{E,w} \left(\mu_{en}/\rho\right)_w dE, \qquad (2.29)$$

where $\Psi_{E,w}$ and $(\mu_{en}/\rho)_w$ are the differential energy fluence in water and the mass energy absorption coefficient for water, respectively. If a large detector is placed at the same point in the water, it can be assumed that CPE also prevails in the detector, and the mean absorbed dose to the detector is then given by

$$\bar{D}_d = \int \bar{\Psi}_{E,d} \left(\mu_{en}/\rho\right)_d dE, \qquad (2.30)$$

where the subscript d denotes the detector material. If it can be assumed that the presence of the detector in the water does not disturb the energy fluence, which will then be equal in the water and in the detector, the absorbed dose in the water is given by

$$D_w = \bar{D}_d \frac{(\mu_{en}/\rho)_w}{(\mu_{en}/\rho)_d}. \qquad (2.31)$$

Thus, for a *large detector*, the relation between the absorbed dose in the two media depends on (μ_{en}/ρ), i.e. the difference in *photon interaction* in the media.

For a *small detector*, the relation will instead depend on how the *electrons* interact in the different media, i.e. it will depend on the mass stopping powers, $(S(E)/\rho)$ of the media. This follows from the Bragg–Gray cavity theory, and small detectors are often referred to as *Bragg–Gray cavities*. The Bragg–Gray cavity theory assumes that the fluence of the charged particles (liberated by the photons) in the medium does not change when the cavity is inserted into the medium. It further assumes that the fluence of photons is uniform throughout the volume of the cavity, and that no photons interact in the cavity. Under these conditions, the absorbed dose to the medium, e.g. water, is given by

$$D_w = \int \Phi^c_{E,w} \left(S(E)/\rho\right)_w dE \qquad (2.32)$$

and the absorbed dose to the detector by

$$D_d = \int \Phi^c_{E,d} \left(S(E)/\rho\right)_d dE, \qquad (2.33)$$

where $\Phi^c_{E,d}$ is the fluence of charged particles (electrons). The absorbed dose to the water is thus given by

$$D_w = D_d \frac{(S(E)/\rho)_w}{(S(E)/\rho)_d}. \qquad (2.34)$$

Example 2.1 *Bragg–Gray cavity*

A small, air-filled, detector (e.g. an ionization chamber) is inserted into a tank of water and irradiated by photons. The reading from the detector shows that the absorbed dose rate to the detector material (i.e. air) is 8.76 mGy s^{-1}. Assuming that the ratio of the mass stopping power of water to the mass stopping power of air is 1.11, what is the absorbed dose rate to the water?
The dose rate to water is given by Eqn. 2.34, giving

$$D_w = 1.11 \cdot D_d = 1.11 \cdot 8.76, \tag{2.35}$$

which equals 9.72 mGy s^{-1}.
A requirement for a Bragg–Gray cavity is that the absorbed dose depends only on the energy deposited by interactions of charged particles that pass through the cavity, i.e. the charged particles are generated outside the cavity. If the detector is replaced by a large detector, e.g. a LiF thermoluminescent detector, the ratio between μ_{en}/ρ for water and the detector material must be used instead of the mass stopping power. Whether a detector is regarded as small or large depends on the range of the secondary electrons and hence on the photon energy.

2.3   BASIC RADIATION BIOLOGY

Radiation biology is a broad field of research concerned with the effects of ionizing radiation on living matter. The size of the objects studied ranges from single molecules to the organs of the body, and many of the methods used are highly specialized. The effects of high radiation doses, such as those delivered in radiation therapy, i.e. several Gy, as well as the effects of low doses, i.e. a few mSv, on the human body are studied. The aim of this section is to present the basic mechanisms behind the effects on cells and tissues of exposure to ionizing radiation, and their medical consequences.

2.3.1   Effects on Cells and Tissues

2.3.1.1   Animal Cells

Animal cells contain a number of internal components, *organelles*, some of which vary depending on the type of cell, and others that are common to all types of cells. The organelles are suspended in a gel-like substance called the *cytoplasm*, and about 80% of the cell mass consists of water. Animal cells are enclosed in a cell membrane, composed mainly of proteins and lipids, which separates the intracellular fluids from the extracellular fluids. The cell membrane also controls the transport of various molecules (e.g. nutrients) into and out of the cell.

The DNA molecule, located in the cell nucleus, consists of two strands that form a double helix; see Figure 2.14. These strands are composed of *nucleotides*, which are smaller units consisting of a *nucleobase*, a sugar molecule and a phosphate group. The nucleobase can be any of the four nucleobases *thymine* (T), *adenine* (A), *cytosine* (C) or *guanine* (G). The strands are formed when the nucleotides are connected by covalent bonds between the sugar and phosphate, thereby creating an alternating sequence of sugar and phosphate. The two strands in the double helix of the DNA molecule are joined by hydrogen bonds between the bases. The joining of bases follows certain pairing rules, i.e. A will only bind to T, and C will only bind to G. This pairing is an important prerequisite for the repair of lesions caused by ionizing radiation, as will be discussed later. The DNA molecule is about 2 m long, and is packed in the cell nucleus in a chemical complex called *chromatin*, which consists of DNA, RNA (a nucleic acid like DNA), and proteins.

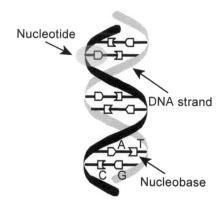

FIGURE 2.14 The double-stranded helix of the DNA molecule is composed of nucleotides consisting of a sugar, a phosphate group and a nucleobase.

A sequence of 10^3 to 10^6 base pairs constitutes a *gene*, which contains information on the production of protein sequences in the cell. The complete set of genetic material in DNA is called the *genome*, and consists of the genes, as well as noncoding DNA (parts that do not encode for any protein). Due to the regular pairing of the bases, the genetic information can be transferred to the daughter cells by *cell division*. Doubling of the DNA is achieved by longitudinal splitting of the double helix by breaking of the hydrogen bonds between the bases. Each strand then acts as a template of the original molecule allowing the accompanying strand to be created, forming a new double helix. This DNA molecule is thus an exact copy of the one in the mother cell.

Proliferation of cells by cell division follows a regular pattern, described by the cell cycle illustrated in Figure 2.15. Cell division takes place during the phases of *mitosis* and *cytokinesis*, resulting in the formation of two new daughter cells. These cells then enter the *interphase*, where they prepare for a new division. This period is further subdivided into three phases: G1, S and G2.

During the G1 (Gap 1) phase, or the growth phase, the cell increases in size and metabolic action is resumed. At the end of this phase, the cell reaches the *G1 checkpoint* where internal, as well as external conditions are checked to ensure that the cell is ready for entry into the S phase. Possible damage to the DNA is also assessed at this point. Depending on the conditions, the cell can enter the S phase, remain in the G1 phase or enter a resting phase, denoted G0, where the cell can remain for some time before re-entering the cell cycle.

The onset of DNA replication marks the start of the S (synthesis) phase. During this phase, the amount of DNA is doubled, and DNA damage is detected and repaired. Two DNA molecules are needed to enable the formation of two identical daughter cells, containing all the genetic information of the mother cell. If no errors are detected, the cell passes the *G2 checkpoint* and enters the G2 phase. Cell growth continues and the cell is prepared for the next division. If the cell fails to pass the G2 checkpoint, it will remain in the G2 phase and attempts will be made to repair the damage.

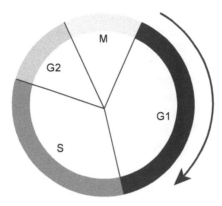

FIGURE 2.15   The cell cycle describes the various phases between cell division, which occurs during mitosis, M. The period between cell divisions is called the interphase and consists of three phases: G1, S and G2. Replication of DNA takes place during the S phase.

During the first part of mitosis, the *prophase*, the chromatin condenses further into *chromosomes*. The chromosomes are now visible in a light microscope. All human cells (apart from egg cells and sperms) contain 46 chromosomes, arranged as 23 pairs, where only the 23rd pair differs between males and females. This pair contains the sex chromosomes, consisting of two X chromosomes in females and one X and one Y chromosome in males. After replication, each chromosome consists of two identical *sister chromatids* that are joined at the *centromere*; see Figure 2.16.

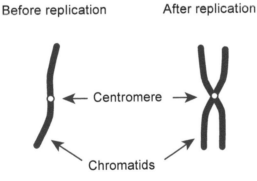

FIGURE 2.16   Before replication, a chromosome consists of a single chromatid, which contains DNA in a coiled structure. After replication, the two sister chromatids are joined at the centromere.

The chromosomes then align at the equator of the cell during the next phase, the *metaphase*, where a structure called the *mitotic spindle* connects to each chromosome centromere by microtubules. The mitotic spindle then contracts and separates the sister chromatids to opposite poles of the cell. This separation takes place during the *anaphase*. During the *telophase*, the nuclear membrane is reformed and the chromosomes unwind into chromatin. Due to the separation of the sister chromatids during the anaphase, each of the two cell nuclei now contains the same genetic material as the mother cell (before replication of the DNA). Finally, the cytoplasm is divided between the two daughter cells.

In *meiosis*, the number of chromosomes in each daughter cell is reduced by half. This occurs in the formation of cells in the reproductive organs, since the offspring would otherwise have twice the number of chromosomes as the parents. In humans, 23 chromosomes are inherited from the mother and 23 from the father, resulting in a total of 46 (23 pairs) chromosomes. During meiosis, the sister chromatids are not separated, but each sister chromatid is transferred to one of the two daughter cells, which then contains half of the DNA (duplicated) from the dividing cell. During the ensuing mitotic cell division, the sister chromatids are separated, resulting in four cells, each containing half of the genetic information.

2.3.1.2 DNA Lesions and Repair Mechanisms

The effects of radiation on living organisms start with the initial physical reactions, excitation and ionization, about 10^{-18} s after irradiation, but may extend over several months, years or decades. After about 10^{-12} s, these initial reactions lead to the formation of radicals, and physical and chemical reactions then take place between the radicals and the DNA molecule in the cell nucleus. Damage to the DNA molecule will then initiate cell repair processes about a microsecond after irradiation. Depending on the outcome of the repair, various biological effects may occur at different times after irradiation. These can be summarized as *early effects*, occurring within approximately 90 d after irradiation, *late effects*, occurring months to years after exposure (by definition, later than 90 d after exposure (ICRP 2012a)), and *carcinogenesis*. It should be emphasized that, provided the radiation dose and the distribution of the dose are similar, the biological effects do not differ between external and internal exposure.

Carcinogenesis is regarded as a *stochastic effect*, i.e. the probability, and not the severity, of cancer development is proportional to the radiation dose. Stochastic effects will be discussed later in this section. Early and late effects were previously called deterministic effects, since it was thought that the effects were proportional to the radiation dose, and that the outcome of a certain exposure caused the same effect in all the individuals exposed. However, these effects can be altered by biological response modifiers and are not necessarily predetermined (ICRP 2012a). They are therefore now referred to as early or late organ or tissue reactions, and will be further discussed later in this chapter.

Exposure of cells to ionizing radiation can result in a direct hit on the DNA molecule, i.e. a molecular bond is affected through one or more of the interaction mechanisms discussed in Section 2.1. This is denoted the *direct effect*. However, exposure of a cell to radiation also leads to the formation of radicals by radiolysis of water molecules in the cell. These radicals may then cause damage to the DNA molecule through chemical reactions, and this is denoted as the *indirect effect*. Direct and indirect effects are depicted schematically in Figure 2.17.

The interaction of ionizing radiation with water molecules will produce short-lived radical cations, free electrons and excited water molecules through ionization and excitation processes. The formation of water radicals and electrons by ionization is described by

$$\text{Ionizing radiation} + H_2O \rightarrow H_2O^+ + e^- \tag{2.36}$$

and excited water molecules are formed by the reaction

$$H_2O \rightarrow H_2O^*. \tag{2.37}$$

Both the radical cations and the excited water molecule are unstable and dissociate to form hydroxyl and hydrogen radicals:

$$H_2O + H_2O^+ \rightarrow H_3O^+ + OH^{\cdot} \tag{2.38}$$

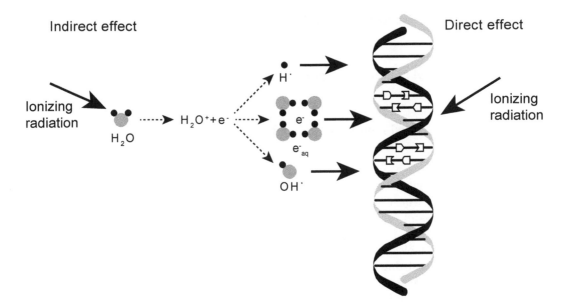

Indirect effect

Direct effect

Ionizing radiation

H_2O

$H_2O^+ + e^-$

$H^{\cdot}$

e^-

e^-_{aq}

$OH^{\cdot}$

Ionizing radiation

FIGURE 2.17 Damage to the DNA molecule in the cell nucleus can be caused by an ionizing particle (direct effect) or by chemical reactions caused by radicals, produced when ionizing radiation interacts with water molecules in the cell (indirect effect).

and

$$H_2O^* \rightarrow OH^{\cdot} + H^{\cdot} \tag{2.39}$$

within a period of about 10^{-13} s. Excited water molecules may also be de-excited by emitting a photon.

Both these reactions lead to the formation of hydroxyl radicals $(OH^{\cdot})$, which are highly reactive, due to the unpaired electron, and oxidize the surrounding medium. The short lifetime of the hydroxyl radical (about 1 μs in pure water) means that it can only diffuse a short distance before it reacts, producing another radical.

The electron ejected when the water molecule is ionized will also contribute to the formation of radicals. It may produce a hydrogen radical (i.e. a hydrogen atom, which has an unpaired electron) by interacting with a water molecule, or it may be thermalized (lose energy) by collisions in the medium. The thermalized electron will then attract water molecules and form e^-_{aq}, a solvated electron that is surrounded by water molecules. The solvated electron reacts by

$$e^-_{aq} + H^+ \rightarrow H^{\cdot} \tag{2.40}$$

and

$$e^-_{aq} + H_2O \rightarrow H_2O^- \rightarrow OH^{\cdot} + H^{\cdot} \tag{2.41}$$

to produce hydrogen and hydroxyl radicals.

In the presence of oxygen, the solvated electron reacts with the oxygen molecule by

$$e^-_{aq} + O_2 \rightarrow O_2^-, \tag{2.42}$$

which can react with hydrogen ions by

$$O_2^- + H^+ \rightarrow HO_2^{\cdot}. \tag{2.43}$$

The hydroperoxyl radical is also produced by

$$H^{\cdot} + O_2 \rightarrow HO_2^{\cdot}, \tag{2.44}$$

and two hydroperoxyl radicals may react to form hydrogen peroxide by

$$2HO_2^{\cdot} \rightarrow H_2O_2 + O_2. \tag{2.45}$$

The radiosensitivity of oxygenated cells is thus higher, which is reflected in the *oxygen enhancement ratio*, which will be discussed below.

The interactions of ionizing radiation and radicals with the DNA molecule are able to cause various types of lesions in different constituents of the molecule, e.g. *crosslinks* between the DNA molecules and proteins, as well as the destruction of hydrogen bonds. The *nucleobases*, consisting of the four compounds *thymine, adenine, cytosine* and *guanine*, may be damaged or removed from the DNA molecule. Damage can also occur in one or both of the two strands of the DNA molecule: *single strand breaks* (ssb) and *double strand breaks* (dsb). An absorbed dose of 1 Gy to the cell will induce damage to about 3000 base pairs, 1000 single strand breaks and 40 double strand breaks, i.e., approximately 5000 lesions per cell per Gy. The various lesions are illustrated in Figure 2.18.

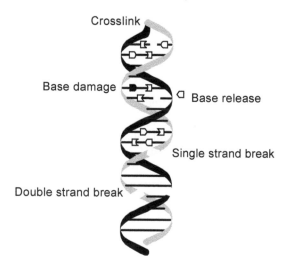

FIGURE 2.18   Various types of DNA lesions.

DNA lesions can be repaired by various mechanisms, the success of which depends on the type of lesion. Lesions affecting one of the DNA strands, e.g. base damage and single strand breaks, are repaired by *excision repair*, in which various enzymes recognize and remove the damaged part, and restore the DNA molecule. If both strands of the DNA molecule are damaged, it can be restored by *homologous recombination* (HR) or *non-homologous end joining* (NHEJ). HR is only possible in the late S phase or G2 phase of the cell cycle since it uses an undamaged copy of the region for sequence homology. If it is not necessary to remove any part of the DNA molecule in the repair process, HR is error free and leads to complete repair of the DNA molecule. NHEJ is most probable in the G1 and S phases, although it can also operate in the other phases. This can give rise to errors such as sequence changes and chromosome aberrations. Sequence changes occur when nucleotide pairs are lost, while chromosome aberrations are induced by the joining of non-homologous ends. Double strand breaks may also be repaired by *single strand annealing*, in which the damaged part is removed and the ends are rejoined. However, this process always results in loss of genetic material.

Unrepaired DNA lesions can be observed during mitosis as *chromosome aberrations*; the type of aberration depending on the stage of the cell during exposure. Exposure during the G1 phase can lead to *dicentric* chromosomes and rings, as well as *acentric* fragments, as illustrated in Figure 2.19. A dicentric chromosome can be formed when a chromatid is broken, resulting in one part containing the centromere and the other only a fragment. If two broken chromatids with centromeres are joined, the resulting chromatid will then contain two centromeres, and after the S phase a dicentric chromosome will be formed. A ring is formed when a chromatid is broken at two sites. The broken ends may join, giving rise to an overlapping ring after the S phase. The fragments formed when a chromatid is broken may rejoin correctly with the original chromosome, and in this case no chromosome aberration will occur. However, the fragments may join another chromosome, leading to exchange of genetic material between the chromosomes, or they may be lost during mitosis due to the lack of a centromere. This results in loss of genetic material from the cell.

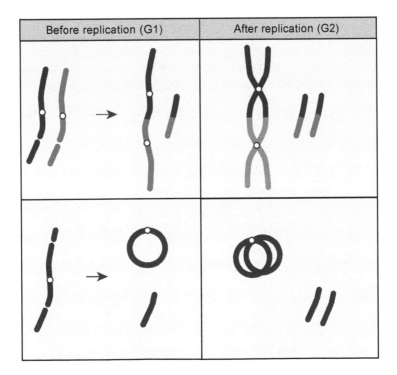

FIGURE 2.19 A dicentric chromosome is formed when two broken chromatids join (top). An overlapping ring is formed from a chromatid broken at two sites (bottom).

The cell is generally regarded as "dead" when it has lost the ability to divide indefinitely (*clonogenic cell death*). The cell may thus still be able to synthesize DNA, produce proteins and even go through a number of cell divisions. Several types of cell death can be identified, e.g. *mitotic catastrophe*, *apoptosis* and *necrosis*, which all occur in response to exposure to ionizing radiation. Mitotic catastrophe, or mitotic death, occurs during mitosis, and a damaged cell may undergo several divisions before it loses its ability to divide. Apoptosis (interphase cell death) is a form of programmed cell death, when the cell is instructed to break down into fragments that can be removed from the body without inducing an inflammatory response. This is in contrast to necrosis, which is characterized by cell oedema (excess fluid), an increase in membrane permeability (leading to the release of the cell contents into the surrounding tissue), and inflammatory response.

2.3.1.3 Radiosensitivity of Cells

The response of cells upon exposure to ionizing radiation can be studied by *colony forming assays*. A specified number of cells are seeded in each of two culture dishes, and the cells in one dish are exposed to a certain absorbed dose (often 1–10 Gy), while the cells in the other dish serve as a control. The cells are then allowed to divide and form colonies for about two weeks. The effect on the cells is quantified by the *surviving fraction*, defined as the ratio of the number of colonies formed in the exposed sample to the number of colonies formed in the unexposed (control) dish. It is important to consider the fact that the number of colonies formed is normally less than the number of seeded cells. This is described by the *plating efficiency*, defined as the ratio of the final number of colonies to the number of seeded cells (in the unirradiated sample). The plating efficiency is typically between 50 and 100%.

A plot of the surviving fraction at different levels of exposure (mean absorbed dose to the cells) results in a *cell survival curve*, such as that shown in Figure 2.20. The shape of the cell survival curve is commonly modelled by the *linear–quadratic model*, describing the decrease in surviving fraction, SF, with increasing dose, D, and is given by

$$SF = e^{-\left(\alpha D + \beta D^2\right)}. \tag{2.46}$$

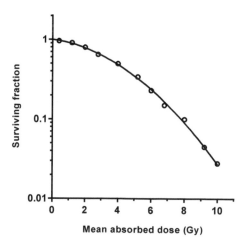

FIGURE 2.20 Cell survival curve showing the fraction of surviving cell colonies (log scale) after exposure to various mean absorbed doses. Experimental data can be modelled by a linear-quadratic model, given by Eqn. 2.46 (solid line).

Cell survival after exposure is affected by several parameters, such as:

- intrinsic radiosensitivity,

- cell cycle phase,

- radiation quality,

- oxygen level and

- exposure parameters, as discussed below.

The *intrinsic radiosensitivity* reflects the difference in cell survival of different cell types, as shown in Figure 2.21. This effect may be due to differences in apoptotic signalling between different cell types. Cells with decreased apoptotic signalling will be more resistant to DNA lesions, and hence less sensitive to exposure to ionizing radiation. The cell survival curve for these cells has a broad shoulder (i.e. a moderate decrease in surviving fraction over a rather large dose interval), in contrast to more sensitive cells, where the shoulder may be hardly visible. In these cells, apoptotic signalling is very active, and apoptosis occurs even at low absorbed doses.

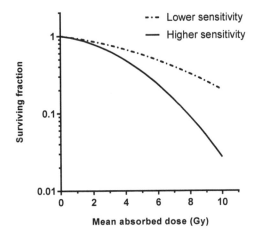

FIGURE 2.21 The cell survival curve for cell types that are less sensitive to irradiation has a broad shoulder, while cell survival curves for highly sensitive cells exhibit a narrow, or barely visible shoulder.

The cell is most sensitive to radiation during the M and G2 phases, and somewhat less sensitive during the late G1 and early S phases. Cells in the early G1 phase are less sensitive than cells in the late S phase, and cells are least sensitive in the G0 phase. The high radiosensitivity during mitosis implies that cells that are highly proliferating, i.e. that divide often, are more sensitive to radiation than cells that divide more seldom. This is called the *law of Bergonie and Tribondeau*, who found that the radiosensitivity is proportional to the rate of proliferation, and inversely proportional to the degree of differentiation (i.e. the degree of specialization regarding the functioning of the cell). However, it should be noted that cell cycle effects are not relevant for the response of tissue to radiation since the cells in a particular organ or tissue are in different phases at any given moment in time.

The effect of radiation is highly related to the LET of the radiation. Low-LET radiation, e.g. electrons, is sparsely ionizing, i.e. the probability of several interactions in the cell nucleus is rather low. On the other hand, when exposed to high-LET radiation, e.g. α-particles or neutrons, the cell DNA may suffer several interactions, and double-strand breaks will be highly probable. The difference between low-LET and high-LET radiation is reflected in the cell survival curves shown in Figure 2.22.

The difference in radiation sensitivity for different radiation qualities is quantified by the *relative biological effectiveness*, RBE. The RBE is defined as the ratio between the absorbed dose resulting from γ-radiation, D_γ, and the absorbed dose resulting from the radiation in question, D_r, that produces the same biological effect. Thus $RBE = D_\gamma/D_r$. Examples of reference radiation are 250 kV x-rays or photons from the decay of ^{60}Co.

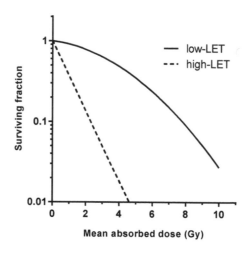

FIGURE 2.22 Exposure to high-LET radiation will lead to more serious damage to the DNA. Since this is more difficult to repair, cell survival decreases rapidly with increasing absorbed dose. The cell survival curve for low-LET radiation exhibits a shoulder and a less steep decrease.

The RBE varies with the chosen biological endpoint, e.g. the cell survival fraction. It also varies with the absorbed dose, as shown in Figure 2.22, where the low-LET curve can be assumed to represent the reference radiation. The ratio between the absorbed doses producing the same effect will vary depending on the survival fraction chosen as the biological endpoint. Due to the shoulder in the low-LET curve, reflecting the lower efficacy at low absorbed doses, the RBE will be higher at low doses.

The variation in RBE with LET is schematically depicted in Figure 2.23. RBE increases slowly with increasing LET to about 10 keV μm^{-1}, followed by a a more rapid increase as the LET increases above 10 keV μm^{-1}. For many cell types, the RBE reaches a maximum at about 100 keV μm^{-1}, and then starts to decline. Low-LET radiation causes fewer lesions in the DNA, and the RBE is therefore low. When LET equals 100 keV μm^{-1}, the distance between interactions in DNA is approximately 2 nm, which coincides with the "width" of the DNA molecule, i.e. the distance between the two strands. Thus, the probability of double strand breaks is high. At higher values of LET, the distance between interactions decreases and more lesions can occur in the DNA. However, since one double strand break can be sufficient to kill the cell, further ionization does not cause more lethal damage.

Direct effects, i.e. ionization of the DNA molecule, often lead to the loss of a hydrogen atom from the DNA strand, thus producing a free radical site. In the presence of oxygen, a single strand break can be created by the interaction between an oxygen molecule and the

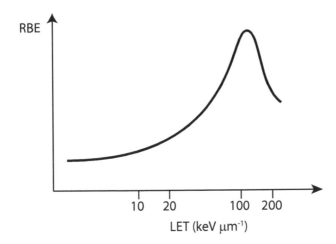

FIGURE 2.23 RBE vs. LET shown in a semilogarithmic diagram. The RBE reaches a maximum at a LET of about 100 keV μm^{-1}.

free radical site. Oxygen can also take part in the formation of free radicals in the cytoplasm, as described in Eqn. 2.42. The effect of oxygen is quantified by the oxygen enhancement ratio, OER, defined similarly to the RBE, i.e. the ratio of the absorbed doses that produce a given biological effect, without and with oxygen. The OER varies rapidly with oxygen partial pressure between 0 and about 20 mm Hg (IAEA 2010), where it reaches a plateau at a value just below 3.0. Oxygenated cells can thus be almost three times more radiosensitive than anoxic cells. The effect of oxygen can be substantial in tumour cells due to poor blood supply, resulting in a low oxygen partial pressure of 5 mm Hg in some regions. In normal tissue, the oxygen partial pressure is often between 10 and 80 mm Hg and most of these cells are well within the plateau. Variations in oxygen pressure in normal tissue within this range will therefore not cause any observable variations in radiosensitivity.

Exposure parameters that affect the sensitivity of cells to radiation include the duration of the exposure and *dose fractionation*. Dose fractionation means that the total absorbed dose is delivered in fractions. The effect of fractionation on cell survival is shown in Figure 2.24. Consider, for example, an absorbed dose of 8 Gy delivered as a single dose, resulting in a survival fraction of about 0.09 (solid line in Fig. 2.24). If the same absorbed dose is delivered in two fractions of 4 Gy each, the survival fraction increases to about 0.25 due to the shoulder of the cell survival curve (dashed line). Thus, the same absorbed dose, if given in several, smaller fractions, will result in higher cell survival. This example also shows that cells can tolerate a higher absorbed dose if it is given in smaller fractions.

Extending the exposure by increasing the time between fractions, or exposing the cells at a low dose rate, will further increase cell survival due to the proliferation of cells between the exposures. The next fraction will then expose a greater number of cells, which will respond according to the cell survival curve for that particular cell type. This effect is depicted by the dashed curve in Figure 2.25.

2.3.1.4 Effects on Tissue

The effects on cells, discussed above, can lead to loss of function in organs or tissues. Some tissues, e.g. bone marrow, are continuously renewed from stem cells, and radiation

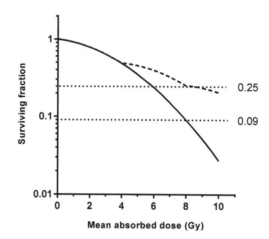

FIGURE 2.24 The survival fraction of the cells increases if the total absorbed dose is given in smaller fractions. This effect is more pronounced if the cell survival curve has a broader shoulder, and is thus dependent on the intrinsic radiosensitivity of the particular cell type. The dotted lines illustrate that an absorbed dose of 8 Gy, delivered as a single dose, results in a survival fraction of about 0.09, while the survival fraction will increase to 0.25 if the same absorbed dose is given in two equal fractions.

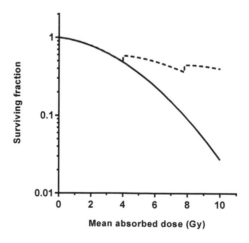

FIGURE 2.25 The survival fraction of the cells increases if they are allowed to proliferate between exposures, or when exposed at a low dose rate (dashed line).

may decrease the proliferation of these cells. Other tissues may be affected by damage to supporting structures, blood vessels or to the cells in the tissue itself (different kinds of damage can also occur in combination). As mentioned in Section 2.3.1.2, the response of tissue to radiation can be divided into early (acute) tissue effects and late tissue effects, depending on the time between exposure and the manifestation of damage. The biological endpoints considered are mainly changes that can be seen when studying tissue samples in a microscope (i.e. by histopathology), or changes in the function of the tissue. Note that late effects should not be confused with stochastic effects, although these terms are sometimes used synonymously.

Tissues with rapid cell turnover, e.g. bone marrow, oral and intestinal mucosa, and the epidermis (the outermost layer of the skin), are likely to show an early response to radiation. Due to the continuous renewal of these cells, a large fraction will be in some stage of mitosis, and are therefore more sensitive to radiation. Loss of these cells, which are vital for the renewal of the tissue, will result in early clinical symptoms, i.e. within 3 months of exposure. The length of the latent period, between exposure and clinically relevant effects, is independent of the absorbed dose and varies with the type of tissue exposed. However, both the severity of the damage and the time required for complete restoration are proportional to the absorbed dose.

Late effects occur in tissues whose cells divide less frequently, e.g. the lungs, the kidneys, the central nervous system and the dermis (the layer of skin below the epidermis). The expression of radiation damage will then be delayed until the cells enter mitosis. These tissues can also suffer from damage to connective tissue and/or blood vessels. A loss of capillaries will affect perfusion of the tissue or organ, and hence its oxygen supply. Late effects can also be affected by an early effect in the same tissue. The latent time for late effects depends on the absorbed dose, in contrast to early effects. A lower absorbed dose results in a longer latent time, and a long follow-up period (months to years) is therefore important for those exposed to absorbed doses where tissue effects can be expected.

A certain number of cells must be lost before the function of a tissue or organ is affected, and there may thus be no tissue effect if only a few cells are lost, i.e. the absorbed dose is below a certain *threshold dose*. Threshold doses for some organs and effects are given in Table 2.1. As the absorbed dose increases, more and more cells will be lost and the severity will increase until maximum severity is reached. This is described by the *dose–effect* relationship, shown in Figure 2.26.

The threshold dose for the formation of cataracts has been substantially reduced in recent years. In the ICRP 1990 and 2007 recommendations, a threshold dose of 5 Gy was given for acute exposure. However, this value was based on data from studies with rather short follow-up periods. Based on the statistical variation in the data from studies on cataract formation, it is now believed that the threshold dose may be even lower than 0.5 Gy (ICRP 2012a).

The corresponding effects on skin, including threshold doses and latent times, are given in Table 2.2. Exposure to radiation causes three kinds of erythema (a rash, or redness of the skin). Early transient erythema may occur within a few hours of exposure to doses exceeding 2 Gy, and is caused by changes in the permeability of the blood vessels. Approximately 10 days after exposure, the main erythematous reaction begins, as a result of inflammation following damage to the epithelium, the thin tissue lining both the outside and inside cavities of the body, for example, the alimentary canal, organs (including skin) and blood vessels. Later, damage to the blood vessels in the dermis, restricting the blood supply, may lead to ischaemia. This results in oxygen and glucose deficiency which affects the cell's metabolism.

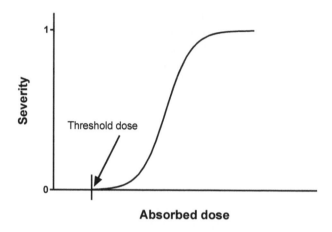

FIGURE 2.26 The severity of the effect on a tissue or an organ increases with increasing absorbed dose, provided the absorbed dose exceeds the threshold dose for that particular type of tissue.

TABLE 2.1 Threshold organ/tissue doses resulting in ≈1% incidence of morbidity, assuming acute exposure. The doses given are for adults, unless otherwise stated. The effects on skin are given in Table 2.2.

Organ/tissue	Effect	Latent time	Absorbed dose (Gy)
Testes	Temporary sterility	3–9 weeks	≈ 0.1
Testes	Permanent sterility	3 weeks	≈ 6
Ovaries	Permanent sterility	<1 weeks	≈ 3
Bone marrow	Depression of haematopoiesis[1]	3–7 days	≈ 0.5
Eye	Cataract	>20 years	≈ 0.5
Lung	Acute pneumonitis	1–3 months	6–7
Kidneys	Renal failure	>1 years	7–8
Bladder	Fibrosis[2]/necrosis	>6 months	15
Brain (adult)	Cognitive defects	Several years	1–2
Brain (infants <18 months)	Cognitive defects	Several years	0.1–0.2

[1] Formation and development of blood cells.
[2] Formation of excess connective tissue (scarring).
Source: Data from ICRP (2012a).

TABLE 2.2   Threshold doses to skin resulting in ≈1% incidence of morbidity, assuming acute exposure. The doses given are for adults, unless otherwise stated.

Effect	Latent time	Absorbed dose (Gy)
Early transient erythema	2–24 hours	2
Main erythema reaction	≈ 1.5 weeks	6
Temporary hair loss	≈ 3 weeks	3
Permanent hair loss	≈ 3 weeks	7
Dry desquamation	≈ 4 − −6 weeks	14
Moist desquamation	≈ 4 weeks	18
Secondary ulceration[1]	>6 weeks	24
Late erythema	8–10 weeks	15
Dermal necrosis	>10 weeks	18
Dermal atrophy[2]	>1 years	10
Telangiectasia[3]	>1 years	10

[1] Progressive loss of dermal tissue.
[2] Breakdown of tissue.
[3] Dilated blood vessels.
Source: Data from ICRP (2012a).

It is important to consider the rather long latent time for shedding of the skin (desquamation), ulceration and necrosis after an accidental exposure, e.g. improper handling of a highly active radioactive source. The absorbed dose may be unknown, and the initial signs of skin erythema may seem rather harmless. However, the absorbed dose may have been sufficiently high to cause irreparable damage.

Exposure of the growing embryo or foetus to ionizing radiation can result in similar effects to those discussed above, i.e. damage to a large number of cells, affecting the integrity or the function of an organ or tissue. The damage depends on the stage of development of the embryo/foetus when exposed. The main effects are malformations, mental retardation or death of the embryo/foetus. There is strong evidence for a threshold dose of 0.1 Gy for intrauterine damage to the embryo or foetus (IAEA 2010).

The embryo is very sensitive to radiation during the pre-implantation phase (8–10 d after conception), since it consists of very few cells. A lesion in some of these cells may thus lead to death of the embryo. However, if the embryo survives the exposure, it will develop normally. Malformations, characteristic of the stage of pregnancy, may occur during the development of limbs and organs during the formation of the organs (organogenesis), which takes place up to about 8.5 weeks after conception. These effects have been demonstrated in animal studies, but no dose-dependent malformations were found after the atomic bombings of Hiroshima and Nagasaki. However, mental retardation found among children exposed *in utero* in weeks 8–15, close to the hypocentre, could be correlated to the dose. This period, 8–15 weeks after conception, is the most sensitive period for mental retardation, followed by weeks 16–25.

2.3.1.5   *Acute Radiation Syndrome, ARS*

The effects on tissue described above are mainly related to the local exposure of tissues and organs. However, in the case of a radioactive accident it is likely that the whole body will be exposed, and damage to several organs will contribute to the overall medical outcome. The response to whole-body irradiation can be divided into the *haematological, gastrointestinal, mucocutaneous* and *neurovascular* syndromes, depending on the organ groups affected.

The onset of ARS is called the prodromal phase, and is characterized by nausea, headache, fatigue and vomiting. The time before symptoms develop depends on the absorbed dose. The prodromal phase is followed by a latent period, the duration of which is approximately inversely proportional to the absorbed dose. The time to onset of prodromal symptoms and the length of the latent period can be used as the basis for a prognosis of the outcome for the exposed subject. The latent period is followed by one or several of the four syndromes listed above, depending on the radiation dose, possibly in combination with infections. The exposed subject may then recover, provided the absorbed dose did not exceed the lethal level and that appropriate medical care has been administered.

The haematological syndrome occurs in the dose range 1–10 Gy, and has a latency period of about three weeks. Due to the killing of precursor cells in the bone marrow, the blood is depleted of leukocytes and thrombocytes, leading to bleeding and infections. With the appropriate treatment, i.e. bone marrow transplantation, the prognosis is good. Without medical treatment, the survival time is less than two months.

The gastrointestinal syndrome, which can be expected between 5 and 20 Gy, manifests as damage to the intestinal mucosal lining. Damage to this barrier results in the loss of electrolytes and fluid, as well as infections that will be severe due to the haematological syndrome. The symptoms that arise, after a latency period of 3–5 days, are diarrhoea, fever, cramps and a decrease in blood plasma volume. The survival time is about two weeks after exposure, and although medical care may prolong survival, the prognosis is poor due to damage to other organs, e.g. the kidneys and lungs.

The mucocutaneous syndrome is seen as ulceration of the skin and mucosa and occurs in the dose range 5–20 Gy. After a latency period of 5–12 days, symptoms such as pain, skin loss, fever, hair loss and infections become apparent. The prognosis for exposed subjects is poor, and the survival time is expected to be 2–3 weeks, depending on the area exposed; a larger exposed area makes the symptoms more severe and reduces the survival time.

The neurovascular (or cerebrovascular) syndrome involves effects on the brain and central nervous system at doses above 20 Gy. The exposed subject will die within a few days due to cardiovascular and neurological damage, e.g. cerebral inflammation and oedema. The latent time depends on the dose, but is generally between 0.5 and 3 hours.

The rate of mortality after exposure to ionizing radiation increases with increasing absorbed dose, above the threshold dose, as shown in Figure 2.26. The dose resulting in the death of 50% of the exposed individuals is denoted as the mean (or median) lethal dose, LD50. In order to distinguish death due to exposure to radiation from other causes of death, a time period after exposure is specified, usually 30 or 60 days. The mean lethal dose is then denoted LD50/30 or LD50/60.

The mean lethal dose varies considerably between species, and is between about 2 and 9 Gy for mammals. A value of 3–4.5 Gy is often used for humans, but since those exposed to such high doses are almost always given medical care, this does not represent a true mean lethal dose. The mortality rate after whole-body exposure therefore depends more on medical care and other complications, than on the absorbed dose. An alternative to LD50 is to define dose ranges with different prognoses, such as: ≤ 2 Gy survival very likely, 2–8 Gy survival possible with medical care, ≥ 8 Gy survival unlikely despite medical care (IAEA 2010).

The prognosis for subjects exposed to radiation can be assessed by using the criteria for triage given in Table 2.3. These criteria were used after the Chernobyl accident, and it was found that the onset of vomiting, the lymphocyte count and hair loss were the most important indicators for medical decision making. The cytogenetic radiation dose is assessed from biological dosimetry, mostly by studying the frequency of chromosome aberrations in lymphocytes.

TABLE 2.3 Triage criteria used to assess the prognosis for subjects exposed to radiation after the Chernobyl accident. Lymphocyte counts were obtained on day 3 after exposure, and hair loss refers to the first 2 weeks after exposure. The dose given refers to the cytogenetic dose, assessed by biological dosimetry. The last column gives the number of actual deaths, including those due to skin burns, in relation to the number of subjects exposed in the particular dose range. See also Section 7.4.

Severity	Vomiting	Lymphocyte count $/mul^{-1})$	Hair loss	Dose (Gy)	Mortality
Mild	No	> 600	No	< 2	0/105
Intermediate	After 1–2 h	300–600	No	2–4	0/53
Severe	After 30–60 m	100–300	Yes	4–6	6/23
Very severe	Immediate	< 100	Yes	6–16	19/22

Source: Data from IAEA (2010).

Other scoring systems have been proposed, for example, by the European Society for Blood and Marrow Transplantation (EBMT). This system is discussed in the *TMT* (Triage, Monitoring and Treatment) *Handbook* (Rojas-Palma et al. 2009); see also Section 7.4.3. The severity and time of onset of some symptoms, as well as lymphocyte counts, can be used in scoring for the first 48 hours after exposure to determine the severity of ARS. Some of the symptoms listed are erythema, nausea, vomiting, diarrhoea, elevated body temperature and low blood pressure. The combined assessment of the symptoms then results in a score of I, II or III, indicating the need for outpatient monitoring, hospitalization for curative treatment or hospitalization with treatment for *multiple organ failure* (MOF), respectively. Note that the reliability of the diagnosis requires that a large part of the body is exposed and that the duration of the exposure is less than a few hours.

Another diagnostic tool used to assess the appropriate medical treatment of an exposed subject is the *METREPOL System*. The damage to critical organ systems is quantified and the grading for one or more of the organ systems are combined to give a so-called response category (RC), which determines the kind of medical care required. The organ systems included are the neurovascular (N), haematopoietic (H), cutaneous (C) and gastrointestinal (G). In addition to damage to these organ systems, the risks of *multiple organ dysfunction* (MOD) and multiple organ failure are also considered.

Although medical examination and the determination of a response category are the tasks of medical professionals, it is instructive to briefly describe the steps involved in determining the RC.

- The signs and symptoms of the four organ systems are first assessed by rating them between 1 and 4, in increasing order of severity. The ratings can be found in tables, listing several symptoms for each organ system. As an example, let us consider the symptom *ulcer/necrosis* for organ system C. The degrees of severity are: *epidermal only* (1), *dermal* (2), *subcutaneous* (3) and *muscle/bone involvement* (4). If a symptom is absent, it is assigned the value zero.

- The maximum degree of severity found for an organ system is then assigned as an index to that organ system, e.g. C_3 if subcutaneous damage was the most severe of the symptoms found for the cutaneous system.

- The above steps are repeated for the other organ systems, resulting in a grading code, e.g. C_3 H_2 N_1 G_1.

- The response category, at a certain time after exposure, is determined by the highest severity index, i.e. $RC3_{4d}$ in this example, assuming that the assessment was made 4 days after exposure.

- The above steps are repeated periodically, at intervals determined by the RC and the presence of clinical complications (e.g. bleeding, infections).

The RC provides a measure of the severity and the probability of autologous recovery, and will trigger a certain therapeutic strategy. For RC1 and RC2, autologous recovery is certain and likely, respectively. Autologous recovery is possible for RC3, while it is most unlikely for RC4. The METREPOL system is further discussed in the *TMT Handbook* (Rojas-Palma et al. 2009) and at http://www.remm.nlm.gov/.

2.3.2 Stochastic Effects

Stochastic effects differ from tissue effects in several important respects. Stochastic effects arise from lesions in a few cells (or a single cell) and may result in cancer or hereditary effects, depending on whether the damaged cell is a somatic cell (tissue cell) or a germ cell (develops into egg cells or sperms). In contrast to tissue effects, the *severity* of the stochastic effect in an individual is independent of the radiation dose; either a stochastic effect will appear, or the exposure will cause no harm to the exposed individual. However, the *frequency* of the effect in an exposed *population* increases with radiation dose. Moreover, there is probably no threshold dose for stochastic effects (the subject of a threshold dose and the shape of the dose–response curve for stochastic effects will be further discussed below).

2.3.2.1 Carcinogenesis

Carcinogenesis and cancer growth are very complex mechanisms, involving changes in DNA and chromosomes, as well as the formation of blood vessels in a tumour. The aim of this section is to discuss only the basics of this broad subject.

The human genome is composed of a large number of genes, some of which are activated and some of which are inactivated. The cell's ability to activate and inactivate genes is one way of increasing its flexibility in response to different kinds of stimuli. This regulation of genes can be caused by ionizing radiation. Cancer is the result of the activation of *oncogenes* and inactivation of *tumour suppressor genes*, and various combinations of oncogenes and tumour suppressor genes have been assigned to different types of cancer in humans. Oncogenes may be activated by chromosome changes, such as deletions, translocations and gene amplification (a disproportionate increase in the number of copies of a gene).

The time at which cancer develops after exposure to ionizing radiation has been studied extensively among the survivors of the atomic bombings of Hiroshima and Nagasaki in the Second World War. It was found that the incidence of solid cancers (i.e., other than leukaemia) increased with absorbed dose and age at exposure, and that the latent period was about 10–15 years and independent of the radiation dose. For leukaemia, the incidence peaked at 6–7 years after exposure, with a latency of about 2–3 years.

However, an increased incidence of thyroid cancer was found among children exposed after the Chernobyl accident in 1986 (UNSCEAR 2000a). The occurrence of these cases could be related to the consumption of milk, contaminated by ^{131}I, which, like stable iodine, is concentrated in the thyroid. Fortunately, this type of cancer is curable, and most of the children survived.

Mutations and chromosome changes caused by ionizing radiation do not necessarily lead to the development of cancer. Several mechanisms are involved, and the development of cancer is generally described by a multistep theory, characterized by three steps. The first step is *initiation*, where a cell is subject to irreversible changes (mutations). If these changes are to develop into cancer, they must be followed by *promotion*. The latent cancer cell then develops into a proliferating cancer cell through interactions with other mutations and stimuli. During the third step, *progression*, the cancer cells proliferate further and develop invasive and metastatic behaviour. Ionizing radiation can be of importance in all three of these steps, for example, by activating oncogenes (initiation) and inactivating tumour suppressor genes by chromosomal deletions (progression).

When discussing the risk of developing cancer following exposure to ionizing radiation, several other contributing factors must be taken into account. Apart from factors related to exposure, i.e. dose and dose rate, radiation quality and the parts of the body exposed, other factors such as gender, age and genetic disposition play a role, as do lifestyle and habits (e.g. smoking, diet and stress).

The effect of dose rate has been briefly discussed in Section 2.3.1.3. Not only cell death, but also the frequency of mutations and subsequent risk of cancer decrease with decreased dose rate. This is believed to be due, in part, to increased recovery from sub-lethal DNA lesions. The effects of radiation quality on cell survival (also discussed in Section 2.3.1.3) also apply to carcinogenesis, and the variation of RBE with LET (see Fig. 2.23) is the basis for the *radiation weighting factors* defined by the ICRP. The relation between the part of the body exposed and the risk of stochastic effects is given by *tissue weighting factors*, also defined by the ICRP. (The concepts of radiation weighting factors and tissue weighting factors are discussed later, in Section 2.4.)

The incidence of cancer varies considerably among different populations. For example, the incidence of liver cancer in Japanese women (Osaka) is 11.5 cases per 100 000 population, which can be compared with 0.3 in one region of Canada (Prince Edward Island). The corresponding values for men are 46.7 and 0.7, respectively. The incidence of thyroid cancer among females is highest in the Filipino population in Hawaii (25.5); while among males, the highest incidence (6.1) is found in Iceland (UNSCEAR 2000b). These values exemplify variations in cancer incidence due to gender and geographical location. The incidence of leukaemia, however, does not vary to the same extent as solid cancers between different populations.

The effect of radiation on cancer incidence has been estimated from epidemiological studies on several groups of people, or *cohorts*, as they are often called. One large cohort is the Japanese atomic bomb survivors, studied in the Life Span Study (LSS). In addition to the LSS, risk estimates are based on cohort and *case–control* epidemiological studies (see below) on survivors of atomic bombings exposed *in utero*, patients treated with radiation for malignant as well as benign diseases, patients who have undergone diagnostic examinations, and people exposed at work or by natural sources. These studies have been extensively discussed by the UNSCEAR (2000b and 2006).

Cohort studies and case–control studies are two common types of epidemiological studies. In a cohort study, a population is chosen and the occurrence of various endpoints concerning health effects in the cohort is investigated. The number of individuals in the cohort may decrease with time due to death, but no new subjects are added to the cohort. A cohort study can be either prospective or retrospective, i.e. either a cohort is defined and followed over time (prospective) or a cohort of people is selected from a register and followed in time from some moment in the past (retrospective). The LSS is a combined prospective and retrospective cohort study. In a case–control study, a number of subjects with a specific disease (cases) are compared with a group of subjects without the disease (controls) who are as similar as possible to the cases regarding all other factors (e.g. age, gender, etc.).

The excess cancer risk arising from exposure to radiation can be expressed as the *excess absolute risk* (EAR) or the *excess relative risk* (ERR). The ERR is a measure of the relative increase in cancer incidence rate compared with an unexposed group of people. The ERR can also be expressed as the relative risk (RR) minus 1. If, for example, ERR equals 1, the incidence rate in the exposed population is twice that in the unexposed population. EAR is a measure of the absolute increase, for example, the extra number of cancers occurring annually per 100 000 people in the exposed population. (Risk estimates and the relation between exposure to ionizing radiation and risk will be further discussed in Section 6.2.3.)

For solid cancers (i.e. other than leukaemia), the sex-averaged ERR decreases with *age at exposure*. For example, exposure to 1 Gy (photons, whole-body exposure) at age 10 y will give an ERR of about 0.9 at an attained age (age at observation) of 60 y, while the same exposure at age 50 y will result in an ERR of about 0.2 at age 60. The EAR also decreases with age at exposure (UNSCEAR 2006).

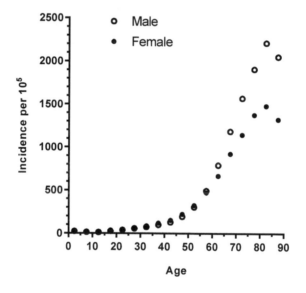

FIGURE 2.27 Cancer incidence for all cancers except skin cancer (melanoma is, however, included) per 100 000 inhabitants in the Nordic countries in 2012. Data from Engholm et al. (2010) and Engholm et al. (2015).

The ERR for solid cancers decreases with attained age after exposure, while the EAR increases with attained age (UNSCEAR 2006). Since the cancer incidence rate increases with attained age in the unexposed population (see Fig. 2.27), the excess increase in cancer incidence rate in the exposed population (given by the EAR) will contribute comparatively less to the total cancer incidence rate at higher attained age. This is reflected by the decrease in ERR.

The risk of leukaemia due to exposure to ionizing radiation has been studied extensively, showing a rather high relative risk compared to other cancer types. There are also differences in the dose–response, in that the risk of leukaemia increases nonlinearly with increasing dose, showing a decreasing slope at low doses. Regarding the influence of time after exposure, it has been found that the EAR decreases with increasing time after exposure if the subjects were exposed during childhood. For those exposed in adulthood, the EAR varies only slightly with time after exposure.

2.3.2.2 Hereditary Effects

Radiation-induced changes in germ cells may be transferred to the next generation and give rise to hereditary effects. Transgenerational mutations, which may be dominant or recessive, have been studied, for example, in fruit flies (*Drosophila*) and mice, and been found to increase with radiation dose. However, no transgenerational mutations have been detected in the large study on the Japanese atomic bomb survivors.

When assessing the risk of hereditary effects, it is important to consider that mutations causing effects in the offspring also occur spontaneously. In order to take this spontaneous mutation rate into account, the *doubling dose* was defined as the absorbed dose to the gonads that will produce a number of mutations equal to those that occur spontaneously in an average generation (30 y). Data from mouse studies have been used to determine induced mutation rates, while human data were used to determine spontaneous mutation rates. The doubling dose, DD, was estimated to be 0.82 Gy (UNSCEAR 2001), although the ICRP retains a value of 1 Gy, based entirely on mouse data (ICRP 2007). The expected increase in the frequencies of genetic diseases, i.e. the risk per unit dose, is calculated using the doubling dose method by

$$P \cdot \frac{1}{DD} \cdot MC \cdot PCRF, \qquad (2.47)$$

where P is the baseline frequency of the class of genetic disease under study and MC is the *mutation component* (see Eqn. 2.48), specific for that disease class. The last factor in Eqn. 2.47, the *potential recoverability correction factor*, $PCRF$, accounts for the fact that spontaneous mutations causing disease differ from radiation-induced mutations, and that recoverability of mutations in live births is possible (ICRP 2007). The numerical value of the $PCRF$ is thus less than 1. The above relation is based on the postulate that the disease frequency in a population is due to a balance between the rate at which spontaneous mutations enter the gene pool, per generation, and the rate at which these mutations are removed by natural selection. Equilibrium is assumed between the introduction and removal of mutations in an unexposed population (ICRP 2007). The mutation rate will increase in an exposed population, and the equilibrium will be disturbed. However, it is predicted that this population will reach a new equilibrium over a number of generations.

The mutation-class-specific component, MC, is defined by

$$MC = \frac{\Delta P/P}{\Delta m/m}, \qquad (2.48)$$

where ΔP is the change in baseline frequency, P, due to a change, Δm, in mutation rate, and m is the spontaneous mutation rate.

Recent estimates of probability coefficients for heritable diseases up to the second generation given by the ICRP (ICRP 2007) are $0.2 \cdot 10^{-2}$ Sv^{-1} for the whole population and $0.1 \cdot 10^{-2}$ Sv^{-1} for adult workers.

2.4 DOSIMETRIC QUANTITIES USED IN RISK ESTIMATION

Broadly speaking, the absorbed dose is a measure of the radiation energy transferred to a small volume, divided by the mass of the volume. The absorbed dose is mathematically defined at a point, but the mean absorbed dose to a volume is usually considered. This information is, however, not sufficient to describe the relation between exposure to radiation and its effects on living tissue. The biological effects also depend on the type of radiation, i.e. α-particles, β-particles, photons (x-rays or γ-rays) or neutrons. This dependence is accounted for by multiplying the mean absorbed dose to a tissue by a *radiation weighting factor*, w_R (ICRP 2007). The weighted quantity is then called the *equivalent dose*, H_T, and is expressed in the unit Sv (sievert). The equivalent dose is thus defined as

$$H_T = \sum_R w_R \cdot D_{T,R}, \tag{2.49}$$

where $D_{T,R}$ is the mean absorbed dose to a tissue, or an organ, T, caused by radiation of type R. The equivalent dose is a *protection quantity* used in regulatory guidelines to avoid tissue effects and to limit stochastic effects (we will later encounter some *operational quantities*). The values of the radiation weighting factor for different kinds of radiation are given in Table 2.4. The radiation weighting factor as a function of neutron energy is also shown in Figure 2.28.

A similar quantity is the *RBE-weighted absorbed dose*, defined as the product of the absorbed dose in an organ or tissue and the RBE of the radiation (IAEA 2005), i.e.,

$$AD_T = \sum_R D_{T,R} \cdot RBE_{T,R}. \tag{2.50}$$

This quantity is intended to account for tissue effects due to differences in biological effectiveness in different organs or tissues. The unit for the RBE-weighted absorbed dose is the *gray-equivalent*, Gy-Eq.

TABLE 2.4 Radiation weighting factors, w_R. Note that the neutron energy, E_n, must be inserted (in MeV) in the expressions for w_R.

Radiation type	w_R
Photons	1
Electrons & muons	1
Protons and charged pions	2
α-particles, fission fragments & heavy ions	20
Neutrons, $E_n < 1$ MeV	$2.5 + 18.2 \cdot e^{-[\ln(E_n)]^2/6}$
Neutrons, 1 MeV $\leq E_n \leq 50$ MeV	$5.0 + 17.0 \cdot e^{-[\ln(2 \cdot E_n)]^2/6}$
Neutrons, $E_n > 50$ MeV	$2.5 + 3.25 \cdot e^{-[\ln(0.04 \cdot E_n)]^2/6}$

Source: Data from ICRP (2007).

It can be seen from the values of the radiation weighting factors, that the absorbed dose from γ-radiation must be twenty times higher than the absorbed dose from α-particles to cause the same degree of biological damage to living tissue. The value of a radiation weighting factor should ideally be derived from stochastic effects investigated *in vivo*, reflecting the RBE. This is not always possible, and the ICRP has therefore also considered the LET when defining these weighting factors. Although the values are given without any uncertainty, there are of course variations between photons of different energies, etc. These

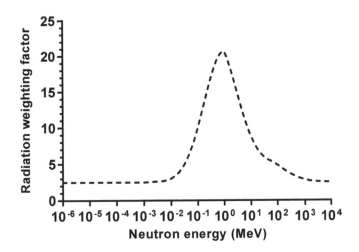

FIGURE 2.28 Radiation weighting factor for neutrons as a function of energy, calculated from the expressions given in Table 2.4.

differences have been acknowledged by the ICRP, but the differences in weighting factors are considered to be small. An extensive discussion of radiation weighting factors can be found in ICRP Publication 103 (ICRP 2007).

The equivalent dose is calculated for a specific organ or tissue, but exposure of a single body organ is rarely encountered in environmental situations (except for the intake of radioactive iodine, which is almost exclusively accumulated in the thyroid). External exposure may also be combined with internal exposure, and a quantity is therefore required that can describe different exposure situations. The quantity recommended by the ICRP for this purpose is the *effective dose*, E, defined as

$$E = \sum_{T} w_T \cdot H_T, \tag{2.51}$$

where the equivalent dose H_T for each exposed tissue T is multiplied by the tissue weighting factor for that tissue, w_T (Jacobi 1975, ICRP 2007). The sum of the tissue weighting factors over all tissues and organs that are believed to be sensitive to stochastic effects is 1, and the contribution of each of these organs or tissues to the total *detriment* is reflected by the value of w_T. The ICRP describes radiation detriment as "...a concept used to quantify the harmful effects of radiation exposure in different parts of the body. It is determined from nominal risk coefficients, taking into account the severity of the disease in terms of mortality and years of life lost. Total detriment is the sum of the detriment for each part of the body (tissues and/or organs)" (ICRP 2007). Table 2.5 presents the values of w_T recommended by the ICRP; the values from both Publications 60 and 103 are given since many published dose coefficients are still based on ICRP Publication 60.

The steps involved in developing the tissue weighting system have also been described by the ICRP (ICRP 2007). For the 14 organs or tissues in Table 2.5, risk estimates were determined for the incidence of cancer, based on cancer incidence data obtained from studies on the survivors of the atomic bombings of Hiroshima and Nagasaki. The reason for using incidence data is that the records of the causes of a disease are often more complete than

TABLE 2.5   Tissue weighting factors recommended by the ICRP in Publication 103 (ICRP 2007) and Publication 60 (ICRP 1991) (in brackets). Values for brain and salivary glands were not included in Publication 60.

Tissue	w_T	Tissue	w_T
Red bone marrow	0.12 (0.12)	Liver	0.04 (0.05)
Colon	0.12 (0.12)	Thyroid	0.04 (0.05)
Lungs	0.12 (0.12)	Bone surface	0.01 (0.01)
Stomach	0.12 (0.12)	Brain	0.01 (–)
Breasts	0.12 (0.05)	Salivary glands	0.01 (–)
Gonads	0.08 (0.20)	Skin	0.01 (0.01)
Bladder	0.04 (0.05)	Remainder tissues	0.12 (0.05)
Oesophagus	0.04 (0.05)	Total	1.00 (1.00)

Source: Data from ICRP (1991) and ICRP (2007).

records of mortality. These risk estimates were then divided by a factor of 2 (the *dose and dose-rate effectiveness factor*, DDREF) to account for the assumed lower risk at low doses and low dose rates. To account for different baseline risks in different populations, weighting was performed between ERR and EAR, resulting in nominal risk coefficients. It should be borne in mind that not all forms of cancer are lethal, but non-lethal cancers may reduce the quality of life. This is accounted for by adjusting for the lethality fraction and for quality of life. In addition, an adjustment was made for years of life lost, as the age distribution differs between different cancer forms. Because of the uncertainties in all these adjustments and in basic cancer incidence data, only four different values (0.01, 0.04, 0.08 and 0.12) were used for the tissue weighting factors.

The remainder tissues in the recommendations in Publication 103 (ICRP 2007) are the adrenal glands, the extrathoracic region of the lungs, the gall bladder, heart, kidneys, lymphatic nodes, muscle, oral mucosa, pancreas, prostate (male), small intestine, spleen, thymus and uterus/cervix (female). The remainder tissues in the previous recommendations in Publication 60 (ICRP 1991) were the adrenal glands, brain, upper large intestine, small intestine, kidney, muscle, pancreas, spleen, thymus and uterus, and were treated according to the *splitting rule*. This rule states that if an organ or tissue in the remainder group receives the highest equivalent dose in the body, the tissue weighting factor for the remainder tissues (0.05) should be divided between this tissue and the rest of the remainder tissues, meaning that the most affected tissue is assigned a w_T value of 0.025. The splitting rule is not used in the present recommendations and the value of w_T for the remainder tissues is applied to the arithmetic mean of the radiation dose to the organs and tissues in the remainder group.

The SI units of effective dose are J kg^{-1} but this has been assigned the name sievert (Sv). Both equivalent dose and effective dose are thus given in sievert, and it is important that the quantity is clearly stated to avoid misunderstanding. The effective dose, E, is also a protection quantity that can be assigned a risk coefficient for different detriments (e.g. cancer incidence), and can be used for regulatory purposes concerning populations. Since the values of w_T are determined based on averages over all ages and both sexes, the effective dose is not applicable to individuals.

The effective dose, E, has replaced the previously used quantity *effective dose equivalent*, H_E (ICRP 1977, ICRP 1991), and although the effective dose has been in use for many years, dose conversion factors published before the replacement, and thus given in terms of effective dose equivalent, are still useful (see, for example, Jacob et al. 1988). Relations between E and H_E have been published by the ICRP (ICRP 1996b) and the difference is less than 12% for photon energies above 100 keV in all irradiation geometries (see Fig. 2.29). For rotational symmetry and for photon energies above 100 keV, E/H_E is between 0.95 and 1.0. In environmental applications the two quantities can therefore often be considered interchangeable.

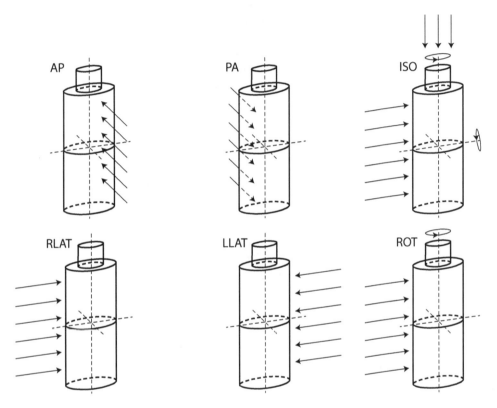

FIGURE 2.29 Six different standardized geometries used when determining the effective dose. The abbreviations for the various geometries are explained in the text.

The effective dose can be related to the physical quantities photon fluence, absorbed dose and kerma by conversion coefficients. The ICRP gives conversion coefficients from both photon fluence and air kerma to effective dose (ICRP 1996b, ICRP 2010, ICRP 2012b). As mentioned above, the IAEA has also published conversion coefficients relating effective dose to source activity. In addition, conversion coefficients have been derived from phantom measurements in contaminated environments (see, for example, Golikov et al. 2007). To determine the effective dose from a known value of the air kerma, detailed information is required on the energy deposition in the human body, which means that the equivalent dose to each organ, associated with the appropriate organ weighting factor, must be known. This can, however, be achieved through Monte Carlo simulations using mathematical phantoms, which are digitalized phantoms based on images from computerized tomography or magnetic resonance imaging.

Due to the distribution of the organs in the body, the effective dose will depend on the irradiation geometry. Hence, the simulations will yield different conversion factors for different irradiation geometries. Since the geometrical variation can be considerable in real exposure situations, conversion coefficients are given for some idealized geometries that are approximations of real situations (Fig. 2.29).

Six different standardized geometries are used when determining the effective dose, taking into account the different dose distributions in organs and tissues in different irradiation situations. The anterior-posterior (AP) geometry is used for radiation incident on the front of the body, and the posterior-anterior (PA) is used when the radiation is incident on the back of the body. When the radiation is incident from the side of the body, the right lateral (RLAT) or left lateral (LLAT) geometry is used, depending on the direction of the radiation. Isotropic (ISO) geometry is used to simulate an isotropically incident radiation field, and rotational (ROT) geometry is applicable when the irradiation is rotationally symmetrical. The most useful geometries for environmental applications are thus the ROT and ISO geometries. The conversion coefficients E/K_a for these two geometries are shown in Figure 2.30 for different photon energies.

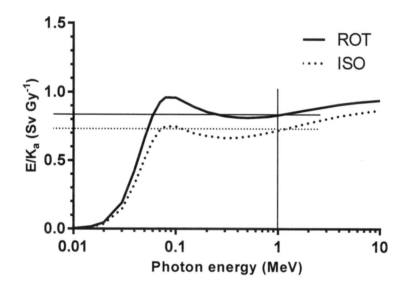

FIGURE 2.30 Effective dose per unit air kerma for rotationally (ROT) and isotropically (ISO) symmetrical irradiation. The conversion coefficients for the mean photon energy of natural background (vertical line at1 MeV) is indicated by the horizontal lines. Data from ICRP (1996b).

According to Figure 2.30, the conversion coefficient giving the relation between the effective dose to the air kerma in rotationally symmetrical irradiation resulting from naturally occurring radionuclides is about 0.8 Sv Gy^{-1}. Golikov et al. (2007) gave values of 0.71 Sv Gy^{-1} for adults and 1.05 Sv Gy^{-1} for a 1-year-old child, based on phantom measurements in an environment contaminated with ^{137}Cs (γ-ray energy 662 keV). This demonstrates that the value will vary with the size of the body. Since the conversion coefficient is based on the absorbed dose to irradiated organs, both the distance from the source and the degree of self-shielding by the body depend on the size of the body.

Equivalent dose and effective dose, discussed above, refer to a reference person, defined by the ICRP. However, in order to optimize radiation protection for groups of people, collective dose quantities have been introduced (ICRP 2007). These quantities are the *collective equivalent dose*, S_T, and the *collective effective dose*, S, both given in the unit *man sievert* (man Sv). The collective equivalent dose, which refers to an organ or a tissue, T, is only used in special situations, and will not be discussed further in this context.

The collective effective dose is defined as the mean effective dose to a group of people, multiplied by the number of people in the group. Note that this is not the same as adding the individual doses, since individual doses cannot be added. This is analogous to the fact that if equal amounts of water, with temperatures of 20 °C and 30 °C, respectively, are mixed, the temperature of the mixture will not be 50 °C. According to the ICRP, to avoid misuse of the collective effective dose, it is important to consider the number of exposed individuals, the age and sex of the exposed individuals, the range of individual doses and their distributions in time, as well as the geographical distribution of the exposed individuals (ICRP 2007).

2.5   OPERATIONAL QUANTITIES

Since the protection quantities equivalent dose and effective dose cannot be measured, some operational (measurable) quantities must be defined. These quantities should provide an accurate estimate of the protection quantities and be defined in such a way as to not underestimate the protection quantity in question. Physical quantities, such as the absorbed dose, are used to calculate protection quantities using weighting factors, as described above, or by using conversion coefficients. The operational quantities are then derived from the physical quantities through definitions given by the ICRU for photon and neutron irradiation, respectively (ICRU 1993, ICRU 2001). These definitions utilize the so-called ICRU sphere: a tissue-equivalent sphere with a 30-cm diameter, and the absorbed dose at different depths in the sphere. The relations between protection and operational quantities can be found in ICRP Publications 74 and 116 (ICRP 1996b, ICRP 2010).

Instruments used for radiation protection purposes are often designed to show an operational quantity, and guidelines for the calibration of these instruments have been issued by the IAEA (IAEA 2000b). Spectroscopic instruments, incorporating, for example, sodium iodide detectors, that are designed to give the dose rate are also available. The transformation of spectroscopic data into dose rates is then performed by spectrum analysis and application of a software algorithm (see Chapter 4).

Several operational quantities have been defined by the ICRU for monitoring both personal exposure and the radiation environment. The primary quantity of interest in environmental monitoring is the *ambient dose equivalent*, $H^*(10)$, which is used to estimate the effective dose to individuals. The ambient dose equivalent at a point in a real radiation field is defined as the *dose equivalent* (see below) at a depth of 10 cm in the ICRU sphere when the sphere is homogeneously irradiated from one direction by a hypothetical radiation field, such that the fluence and energy distribution are the same as in the real field. The point at a depth of 10 cm in the ICRU sphere should then be on a radius in the opposite direction to the hypothetical field.

The dose equivalent, H, is defined as

$$H = Q \cdot D, \tag{2.52}$$

where D is the absorbed dose at a point in tissue, and Q is a quality factor, defined as a function of the unrestricted (i.e. $\Delta \to \infty$) linear energy transfer, L, of charged particles in water (ICRP 1991). Values of Q are given in Table 2.6.

Although the relation between biological effects and L can be rather complex, the correlation is strong, and justifies the use of this simple relationship in radiation protection.

The relation between $H^*(10)$ and E is shown in Figure 2.31, where it can be seen that

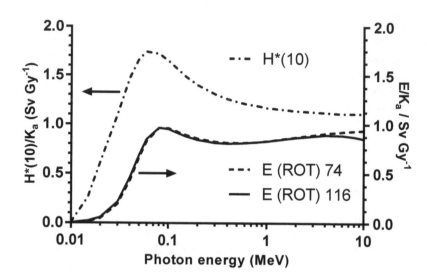

FIGURE 2.31   Ambient dose equivalent, $H^*(10)$, per unit air kerma, and effective dose, E, per unit air kerma, for a rotationally symmetric (ROT) irradiation geometry. (The curves for effective dose differ slightly due to a change in the computational phantom and a more elaborate dose calculation in ICRP Publication 116.) The comparison shows that the definition of ambient dose equivalent overestimates the effective dose at all photon energies shown. The data for $H^*(10)$ and E (ROT) 74 were taken from ICRP Publication 74 (ICRP 1996b), and the data for E (ROT) 116 were taken from ICRP Publication 116 (ICRP 2010).

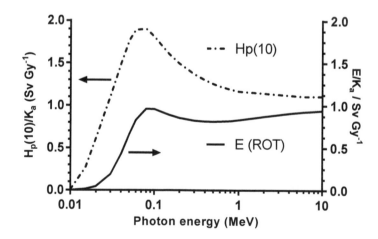

FIGURE 2.32   Personal dose equivalent, $H_p(10)$, per unit air kerma, and effective dose, E, per unit air kerma, for a rotationally symmetric (ROT) irradiation geometry in a perpendicularly incident radiation field. It can be seen that the definition of personal equivalent dose overestimates the effective dose at all photon energies shown. Data from ICRP (1996b).

TABLE 2.6 Values of the quality factor, $Q(L)$, for different intervals of linear energy transfer (LET).

L	$Q(L)$
$L < 10$ keV/μm	1
$10 \leq L \leq 100$ keV/μm	$0.32L$-2.2
$L > 100$ keV/μm	$300/\sqrt{L}$

Source: Data from ICRP (1991).

an instrument designed to measure the ambient dose equivalent overestimates the effective dose for all photon energies in this geometry. This overestimation will cause deviations when the effective dose is estimated from field gamma-ray spectrometry measurements and compared to the reading of a dose rate meter that has been designed to give the ambient dose equivalent. The fact that the dose rate meter is calibrated at a specified photon energy may also cause a deviation, which will be discussed in Example 2.2 below.

To monitor the effective dose to an individual, the operational quantity *personal dose equivalent*, $H_p(10)$, is used. The personal dose equivalent denoted $H_p(0.07)$ is used to monitor doses to the skin, hands, wrists and feet, while $H_p(3)$ is used to monitor the dose to the lens of the eye. (The numbers in brackets denote depths of 10 mm, 0.07 mm and 3 mm.) When the effective dose is estimated using detectors such as *thermoluminescence dosimeters* worn by individuals, special attention must be paid to their calibration. These detectors are calibrated to show the personal dose equivalent, and if the reading is to be used for comparisons with measurements made with other types of instruments, the relation between the value and air kerma must be known. Figure 2.32 shows the personal dose equivalent per unit air kerma, compared to the effective dose per unit air kerma.

Example 2.2 *Comparison of the reading from a dose rate meter with the effective dose estimated from spectroscopic measurements*

Let us assume that we have made field gamma measurements using a calibrated detector allowing the activity of the radionuclides in the ground to be determined. The naturally occurring radionuclides can be assumed to be homogeneously distributed with depth, and the inventory can thus be given in Bq kg^{-1}. However, for ^{137}Cs, a plane source will be assumed with the activity given in Bq m^{-2}.

This is a common assumption when the depth distribution is unknown, although the resulting equivalent surface deposition will underestimate the true inventory due to the absorption of photons in the ground. The photon fluence rate above the ground is, however, reasonably well described by this quantity.

Since we now know the activity of all the significant radionuclides present at the site, we can calculate the effective dose to a human at the site using conversion factors from, for example, IAEA-TECDOC-1162 (IAEA 2000). If we then measure the ambient dose rate with a dose rate meter and subtract the contribution from cosmic radiation, we will find that the dose rate meter gives a higher value than that obtained from the field gamma measurements.

This discrepancy is due to the fact that the dose rate meter is designed to show the ambient dose equivalent, which always overestimates the effective dose for ROT geometry. However, there may be other sources of discrepancy, such as a difference in angular sensitivity of the two measurement systems, and the fact that the dose rate meter is calibrated in a radiation field from ^{137}Cs (662 keV), while the mean photon energy at the site is higher due to the naturally occurring radionuclides.

We can correct for the difference in photon energy used when calibrating the instrument and the photon energy at the site by using the relation between the ambient dose equivalent and air kerma (Fig. 2.31).

Assuming that the mean photon energy at the site is 1.5 MeV, and since the air kerma at the measurement site must be independent of the instrument used, we find that

$$\frac{H^*(10)_{1.5}}{H^*(10)_{0.7}} = \frac{1.15}{1.2} = 0.96, \tag{2.53}$$

where the subscripts indicate the photon energies (0.662 MeV is rounded up to 0.7 MeV). The ambient dose equivalent at the site is thus

$$H^*(10)_{1.5} = 0.96 \cdot H^*(10)_{0.7}. \tag{2.54}$$

In this case, a rather small correction is obtained. However, if we estimate the activity of an ^{241}Am source by measuring the ambient dose equivalent rate with a dose rate meter calibrated for ^{137}Cs, the result would be quite different.

The ambient dose equivalent per unit air kerma for 59.5 keV photons (main photon energy for ^{241}Am) is 1.7 Sv Gy^{-1}, and the correction factor would thus be 1.4. In this case the measured ambient dose equivalent would be underestimated.

2.6 FLUENCE RATE FROM VARIOUS SOURCE GEOMETRIES

The primary photon fluence consists of photons that pass through a medium without interacting with the source or the medium between the source and the detector. However, as discussed above, the total fluence will also consist of photons that were not originally directed towards the detector but whose direction was changed through Compton scattering. This scattered radiation will also contribute to the detector signal. The relation between primary and scattered radiation is given by the build-up factor.

2.6.1 Volume Sources

Figure 2.33 shows the geometry used for the calculation of the primary fluence rate at a point P in air at a distance h above a homogeneous volume source. The number of photons emitted per unit time from a volume element dV at a depth z in the soil is denoted $S_V(z, r, \eta)dV$, where dV is given by $drdRd\rho$ (the azimuthal angle η is not shown in the figure). S_V is thus a measure of the number of photons emitted per unit volume of the source, and is called the source strength. Taking into account self-attenuation in soil (μ_s)[1] and attenuation in air (μ_a), the differential (primary) photon fluence rate at the point P, is given by

$$d\dot{\phi}_p = \frac{S_V dV \cdot e^{-\mu_s(R-h/\cos\theta) - \mu_a h/\cos\theta}}{4\pi R^2}. \tag{2.55}$$

Here we have simply treated the contribution from the volume element dV as the primary fluence rate from a point source. The exponential factor is the attenuation in the media (soil and air) between the source and the point P. The number of photons emitted per unit time from the volume element can be calculated from the activity, A, using the relation $S_V dV = A \cdot f$, where f is the probability of the emission of a photon (γ-ray) with a certain energy in each decay of the radionuclide.

---

[1]The *linear attenuation coefficient* is a measure of the probability of the interaction of a photon with the irradiated medium per unit length. If we ignore the scattered radiation, the decrease in photon fluence, ϕ, after traversing a medium of thickness x is given by $\mu \cdot \phi$.

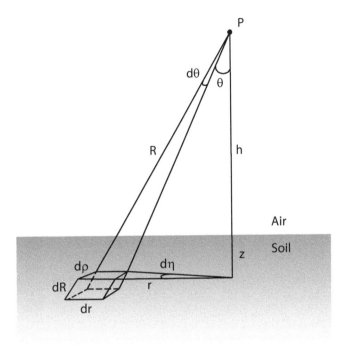

FIGURE 2.33 Geometry used for the calculation of the fluence from a volume source in the soil. It is assumed that the radionuclide is homogeneously distributed in the soil. Modified from Isaksson (2011).

The total primary photon fluence rate can be determined by summing the contributions from each volume element dV, which is equivalent to integrating the differential fluence rate over the whole volume. However, this calculation can be simplified by changing to spherical coordinates, which yields $dV = drdRd\rho = R^2 \sin\theta d\theta d\eta dR$. The primary photon fluence is thus given by

$$\dot{\phi}_p = \int\limits_0^{2\pi} \int\limits_{h/\cos\theta}^{\infty} \int\limits_0^{\pi/2} \frac{S_V \cdot R^2 \cdot \sin\theta \cdot e^{-\mu_s(R-h/\cos\theta)-\mu_a h/\cos\theta}}{4\pi R^2} d\eta dR d\theta. \qquad (2.56)$$

There are unfortunately only a few special cases of distributions for which Eqn. 2.56 can be solved analytically. Two examples will be given here: an infinite homogeneous volume source and an infinite homogeneous slab source of finite thickness. An infinite volume source distribution (horizontally and vertically) can often be assumed for radionuclides occurring naturally in the ground, since the mean free path of photons from naturally occurring radionuclides is about 10 cm in soil (based on 3 MeV photons and a soil density of 1500 kg m^{-3}).

Assuming an *infinite homogeneous volume source* and integrating Eqn. 2.56 gives

$$\int\limits_0^{2\pi} d\eta = [\eta]_0^{2\pi} = 2\pi \qquad (2.57)$$

and

$$\int\limits_{h/\cos\theta}^{\infty} e^{-\mu_s(R-h/\cos\theta)}dR = e^{\mu_s \cdot h/\cos\theta}\left[-\frac{1}{\mu_s}e^{-\mu_s \cdot R}\right]_{h/\cos\theta}^{\infty} = \frac{1}{\mu_s}. \tag{2.58}$$

Eqn. 2.56 then becomes

$$\dot{\phi}_p = \frac{S_V}{2\mu_s}\int\limits_{0}^{\pi/2} \sin\theta \cdot e^{-\mu_a h/\cos\theta}d\theta \tag{2.59}$$

The integral in Eqn. 2.59 can be evaluated by using the properties of the exponential integral, $E_n(x)$, defined by

$$E_n(x) = x^{n-1}\int\limits_{x}^{\infty} \frac{e^{-t}}{t^n}dt. \tag{2.60}$$

and the substitution $t = \mu_a h/\cos\theta$ thus enables Eqn. 2.59 to be rewritten as

$$\dot{\phi}_p = \frac{S_V}{2\mu_s}\int\limits_{0}^{\pi/2} \sin\theta \cdot e^{-\mu_a h/\cos\theta}d\theta = \frac{S_V}{2\mu_s}\int\limits_{\mu_a h/\cos 0}^{\mu_a h/\cos(\pi/2)} \sin\theta \cdot e^{-t}\frac{\cos^2\theta}{\mu_a h\sin\theta}dt =$$

$$\frac{S_V}{2\mu_s}\mu_a h\int\limits_{\mu_a h}^{\infty} \sin\theta \cdot e^{-t}\frac{\cos^2\theta}{(\mu_a h)^2\sin\theta}dt = \frac{S_V}{2\mu_s}\mu_a h\int\limits_{\mu_a h}^{\infty} \frac{e^{-t}}{t^2}dt. \tag{2.61}$$

Inserting the expressions $x = \mu_a h$ and $n = 2$ into Eqn. 2.60 gives the expression for the primary fluence rate.

Primary fluence rate from an infinite homogeneous volume source

$$\dot{\phi}_p = \frac{S_V}{2\mu_s}E_2(\mu_a h) \tag{2.62}$$

The value of the exponential integral $E_2(x)$ can be found in standard mathematical tables and graphs, but a useful feature of $E_2(x)$ is that it equals 1 for $x = 0$ and decreases monotonically to 0 as x approaches infinity. This can be utilized for the special case when the detector is placed on the surface of an infinite volume source, i.e. when $h = 0$. In this case, there is no attenuation in air and $\mu_a h = 0$. Since $E_2(0) = 1$, Eqn. 2.62 simplifies to

$$\dot{\phi}_p = \frac{S_V}{2\mu_s}. \tag{2.63}$$

The next type of source considered is the *infinite homogeneous slab source*. Ploughing of arable land after the deposition of radionuclides may lead to a homogeneous distribution of the radionuclides in the upper layer of the soil, while the deeper soil layers may be depleted of radionuclides resulting from the deposition. The fluence rate from such a slab source (of infinite lateral extension) with thickness z can be found by subtracting one volume source from another: one extending from the surface of the ground to infinite depth and the other extending from depth z to infinite depth. The fluence rate from each source can be found using Eqn. 2.62, and the difference is given by

$$\dot{\phi}_p = \frac{S_V}{2\mu_s} \left[E_2 \left(\mu_a h \right) - E_2 \left(\mu_s z + \mu_a h \right) \right].$$ (2.64)

If we once again consider a point at the surface of the soil, the attenuation in the air vanishes from the calculation and Eqn. 2.64 reduces to

$$\dot{\phi}_p = \frac{S_V}{2\mu_s} \left[1 - E_2 \left(\mu_s z \right) \right]$$ (2.65)

2.6.2 Area Sources

A source commonly encountered in field measurements is a two-dimensional area source, or *plane source*. In this case, the number of photons emitted per unit time from an area element dA, is considered. This is expressed as $S_A(r, \eta)dA$, where dA is $drd\rho = dr \cdot r \cdot d\eta$ (see Fig. 2.34). The source strength, S_A, is thus a measure of the number of photons emitted per unit area of the source.

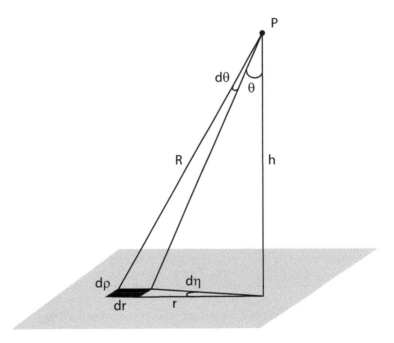

FIGURE 2.34 Geometry used for the calculation of the fluence from an area source in which it is assumed that the radionuclide is distributed homogeneously over the surface. Modified from Isaksson (2011).

Using the geometry shown in Figure 2.34, the differential (primary) photon fluence rate at point P, taking attenuation in air (μ_a) into account, can be written

$$d\dot{\phi}_p = \frac{S_A dA \cdot e^{-\mu_a h / \cos\theta}}{4\pi R^2} = \frac{S_A dr \cdot r \cdot d\eta \cdot e^{-\mu_a h / \cos\theta}}{4\pi R^2},$$ (2.66)

where the differential fluence rate at P is assumed to originate from a point source. Integrating over the azimuthal angle, η, from 0 to 2π yields

$$d\dot{\phi}_p = \frac{S_A dr \cdot r \cdot e^{-\mu_a h / \cos\theta}}{2R^2}. \tag{2.67}$$

To find the total primary photon fluence rate, it is necessary to integrate the contribution from all "point sources" in the plane, dA, which is assumed to be infinite. However, three variables are involved in this calculation (R, r and θ), and in order to integrate the expression, it is necessary to find a relation between at least two of them. e.g. r and R. From simple geometry, $r = R\sin\theta$ and $dr\cos\theta = Rd\theta$. Substituting these relations into Eqn. 2.67 gives an expression that can be integrated over the vertical angle θ. By making the same variable substitution as for the volume source, $t = \mu_a h / \cos\theta$, the primary fluence rate can be expressed as an exponential integral.

$$\dot{\phi}_p = \int_0^{\pi/2} \frac{S_A \sin\theta \cdot e^{-\mu_a h / \cos\theta}}{2\cos\theta} d\theta = \frac{S_A}{2} \int_{\mu_a h}^{\infty} \frac{e^{-t}}{t} dt \tag{2.68}$$

The exponential integral in this case is $E_1(\mu_a h)$, and the fluence rate from a two-dimensional source of infinite extent is thus given by:

Primary fluence rate from area source of infinite extent

$$\dot{\phi}_p = \frac{S_A}{2} E_1(\mu_a h) \tag{2.69}$$

Calculating the fluence rate from an area source of finite size is quite complicated, and is beyond the scope of this book (detailed calculations can be found, for example, in Isaksson 2011 and Kase and Nelson 1972). Therefore, only the resulting expressions are presented below. The primary fluence rate at a point vertically above the centre of a circular source can be calculated from Eqn. 2.70. In this expression, which assumes that there is no attenuation in the air, r_{max} is the radius of the source and h is the height of the detector above the source.

$$\dot{\phi}_p = \frac{S_A}{4} \ln\left(\frac{r_{\max}^2 + h^2}{h^2}\right) \tag{2.70}$$

If attenuation in air is taken into account, the expression becomes slightly more complicated, and will include an exponential integral; see Eqn. 2.71. The angle α is the angle between the detector and the perimeter of the (circular) source. In the limit where r_{max} approaches infinity, α will approach 90° and hence $1/\cos\alpha \longrightarrow \infty$. The exponential integral E_1 will then be zero, and Eqn. 2.71 reduces to the expression for an area source of infinite extent (Eqn. 2.69).

$$\dot{\phi}_p = \frac{S_A}{2} \left[E_1(\mu_a h) - E_1\left(\frac{\mu_a h}{\cos\alpha}\right) \right] \tag{2.71}$$

In practical situations, an area source is often approximated by a point source to simplify the calculations, especially when making the first estimate of the activity of a source. The inverse square law (which says that the fluence rate decreases inversely proportionally to the square of the distance between the source and detector) is not valid for an extended source, but is often a good approximation. However, the accuracy of this approximation depends on the radius of the source in relation to the distance to the detector. An extended area source can be replaced by a point source in the calculations when the distance to the detector exceeds a certain value. It can be shown that a circular source can be approximated by a point source when this distance from the source to the detector is greater than 2.2 times the radius of the source. It is thus acceptable to make this approximation when the distance between the source and the detector is greater than the diameter of the source.

The calculations can also be simplified by treating the source as an infinite source. The dimensions at which the source can be regarded as being infinite can be investigated by comparing Eqn. 2.71 and Eqn. 2.69, and will depend on the distance between the source and the detector, as well as the photon energy. For 662 keV photons from ^{137}Cs the fluence rate from a source with a radius of approximately 500 m is equivalent to that from a source of infinite extent.

2.6.3 Spherical Sources

An example of the calculation of the fluence rate in this geometry is given by calculating the radiation dose to a person surrounded by a radioactive plume. Figure 2.35 shows the geometry used to calculate the fluence at the centre of a sphere.

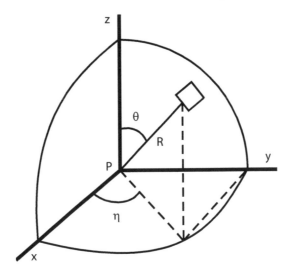

FIGURE 2.35   Geometry for the calculation of the primary fluence rate inside a homogeneous spherical source. The source extends to the radius r_{max} (not shown in the figure). Modified from Isaksson (2011).

The differential fluence rate from volume element dV is given by

$$d\dot{\phi}_p = \frac{S_V dV \cdot e^{-\mu_a R}}{4\pi R^2},$$
(2.72)

where μ_a is the linear attenuation coefficient of air. If the source is composed of a medium other than air, μ_a should be replaced by the attenuation coefficient of the medium in question. In spherical coordinates, the fluence rate at P can be written as

$$\dot{\phi}_p = \int\limits_0^{2\pi} \int\limits_0^{r_{\max}} \int\limits_0^{\pi} \frac{S_V \cdot R^2 \cdot \sin\theta \cdot e^{-\mu_a R}}{4\pi R^2} d\eta \, dR \, d\theta \qquad (2.73)$$

and the integrals can be readily evaluated as

$$\dot{\phi}_p = \frac{S_V}{4\pi} \cdot 2\pi \cdot \left[-\frac{1}{\mu_a} e^{-\mu_a \cdot R} \right]_0^{r_{\max}} \cdot [\cos\theta]_0^\pi = \frac{S_V}{\mu_a} \left(1 - e^{-\mu_a r_{\max}} \right). \qquad (2.74)$$

In environmental applications, a person completely enveloped in a radioactive plume is a relatively rare event. So, we will instead assume that the source is a cloud or plume of air containing radioactive isotopes in the form of gases or particles, and that the point at which we want to calculate the dose is at ground level. This means that the person is surrounded by a half-sphere of air containing the radionuclides. The primary fluence rate will then be half the value calculated from Eqn. 2.74.

Primary fluence rate from a radioactive cloud of infinite extent

$$\dot{\phi}_p = \frac{S_V}{2\mu_a} \left(1 - e^{-\mu_a r_{\max}} \right) \qquad (2.75)$$

If the source can be considered infinite, Eqn. 2.75 will be reduced to Eqn. 2.63.

In relation to infinite sources, it is interesting to see how large a radioactive plume must be before the primary fluence rate reaches equilibrium. Using the linear attenuation coefficient for 662 keV gamma radiation from ^{137}Cs in air, we find that the factor $(1 - e^{-\mu_a r_{\max}})$ equals 0.99 for a plume radius, $r_{\max}$, of 495 m. Thus, for a plume with a radius greater than a few hundred metres, approximate calculations can be performed by assuming the source to be infinite. The radius will be shorter in a denser medium, and in water, for example, it is a few tens of centimetres.

2.6.4 Line Sources

Line sources can be encountered, for example, in the form of pipes containing a radioactive liquid, and the method for calculating the fluence rate is similar to those above. In this case, however, a small length interval is integrated along the source, as depicted in Figure 2.36. The source strength, S_L, gives the number of photons emitted per unit length of the source. The fluence rate from a line element, dy, at a point, P, located at a perpendicular distance h from the source, is calculated by treating the line element as a point source. Integrating along the line source then yields

$$\dot{\phi}_p = \frac{S_L}{4\pi} \int\limits_{-y_1}^{y_2} \frac{dy}{r^2}. \qquad (2.76)$$

Integration is facilitated by replacing the integration limits with the absolute values of the angles θ_1 and θ_2 (in radians), i.e.,

$$\dot{\phi}_p = \frac{S_L}{4\pi} \int\limits_{-|\theta_1|}^{|\theta_2|} \frac{h \sec^2\theta}{h^2 \sec^2\theta} d\theta \qquad (2.77)$$

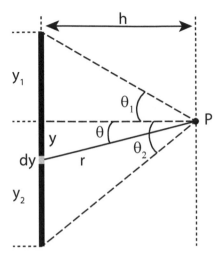

FIGURE 2.36 Geometry for the calculation of the fluence rate from a line source.

yielding:

Primary fluence rate from a line source

$$\dot{\phi}_p = \frac{S_L}{4\pi}\left(|\theta_2| + |\theta_1|\right) \tag{2.78}$$

In deriving Eqn. 2.78, it was assumed that the source was not shielded and that attenuation in the air could be neglected. For a shielded source, as illustrated in Figure 2.37, the attenuation in the shielding material introduces an exponential factor into the integral, i.e.

$$\dot{\phi}_p = \frac{S_L}{4\pi h}\int_{-|\theta_1|}^{|\theta_2|} e^{-\sum_i \mu_i t_i \sec\theta}\,d\theta. \tag{2.79}$$

This integral cannot be solved analytically, but can be expressed as a *Sievert integral*

$$\dot{\phi}_p = \frac{S_L}{4\pi h}\int_{0}^{|\theta_2|} e^{-\sum_i \mu_i t_i \sec\theta}\,d\theta + \int_{0}^{|\theta_1|} e^{-\sum_i \mu_i t_i \sec\theta}\,d\theta, \tag{2.80}$$

which is written

$$\dot{\phi}_p = \frac{S_L}{4\pi h}\left\{ F\left(|\theta_2|, \sum_i \mu_i t_i\right) + F\left(|\theta_1|, \sum_i \mu_i t_i\right)\right\}. \tag{2.81}$$

Numerical values for the Sievert integral are shown in Figure 2.38. Note that two values must be taken from the graph in order to calculate the fluence rate according to Eqn. 2.81.

There are several other source geometries for which the fluence rate must sometimes be calculated, such as cylindrical sources (e.g. a waste tank). These geometries have been extensively discussed by, for example, Kase and Nelson (1972) and Shultis and Faw (2000).

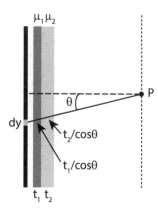

FIGURE 2.37 A line source, shielded by two different materials of thickness t_1 and t_2, and with attenuation coefficients μ_1 and μ_2, respectively.

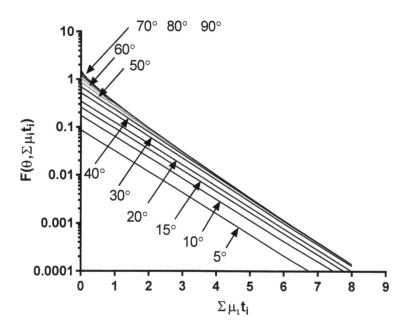

FIGURE 2.38 Values of the Sievert integral for various angles as a function of the thickness of the attenuating material. For angles between 70° and 90°, the curves virtually coincide.

2.7   ABSORBED DOSE AND KERMA FROM EXTERNAL RADIATION SOURCES

2.7.1   Calculations of Absorbed Dose and Kerma

The air kerma rate from a point source can be calculated using the *air kerma rate constant,* Γ, defined in ICRU Publication 60 as

$$\Gamma_\delta = \frac{l^2 \dot{K}_\delta}{A}, \tag{2.82}$$

where the subscript δ specifies that photons with energies higher than δ are included in the calculation of Γ_δ, and A is the activity. Detailed calculations of the air kerma rate constant from basic radiation quantities can be found, for example, in Ninkovic et al. (2005). The kerma rate from a point source is thus given by

$$\dot{K} = \frac{\Gamma A}{l^2}, \tag{2.83}$$

where the subscript δ has been dropped and l is the distance from the point source to the detector.

TABLE 2.7   Values of the conversion factor for some selected radionuclides. The values give the absorbed dose rate at 1 m distance from a 1-kBq source.

Radionuclide	Conversion factor $(mGy h^{-1} kBq^{-1})$
^{40}K	$2.2 \cdot 10^{-8}$
^{60}Co	$3.6 \cdot 10^{-7}$
^{99m}Tc	$2.1 \cdot 10^{-8}$
^{131}I	$6.2 \cdot 10^{-8}$
^{137}Cs[1]	$9.5 \cdot 10^{-8}$
^{133}Ba	$9.3 \cdot 10^{-8}$
^{192}Ir	$1.4 \cdot 10^{-7}$
^{226}Ra	$2.2 \cdot 10^{-9}$
Nat. U[2]	$1.5 \cdot 10^{-8}$
^{241}Am	$3.7 \cdot 10^{-8}$

[1] Including ^{137m}Ba
[2] Also valid for depleted U
Source: Data from IAEA (2000a).

The units of the air kerma rate constant are thus Gy m^2 Bq^{-1} s^{-1}, but other units are also used in published tables. For example, Ninkovic et al. (2005) calculated air kerma rate constants for a large number of radionuclides, which were given in the units μGy m^2 GBq^{-1} h^{-1}. The *Radionuclide and Radiation Protection Data Handbook* (Delacroix et al. 2002) instead presents constants expressed as the radiation dose in mSv h^{-1} from a point source with an activity of 1 MBq at a distance of 30 cm. This kind of conversion factor, giving the absorbed dose (or other quantities describing radiation dose), is often referred to as a *gamma constant*. The reason for not strictly adhering to SI units is that the values can be very small. These constants should therefore be used with great care, and the units checked before using a published constant. Table 2.7 gives the values of the conversion constant for some radionuclides.

Example 2.3 *Estimation of the activity of a point source*

A measurement is made with an accurately calibrated dose rate meter at a distance of 1 m from a ^{137}Cs point source. The instrument shows that the dose rate is 3.5 $\mu Sv\,h^{-1}$. A background measurement is made at a large distance from the Cs source and shows 0.2 $\mu Sv\,h^{-1}$. What is the activity of the source?

After subtracting the background dose rate, the net dose rate from the source at 1 m is 3.3 $\mu Sv\,h^{-1}$. Consulting the Radionuclide and Radiation Protection Data Handbook (Delacroix et al. 2002), we find that a point source of 1 MBq will lead to a dose rate of $1.07 \cdot 10^{-3}$ $mSv\,h^{-1}$ at a distance of 30 cm from the source. It may appear to be simply a question of performing some mental arithmetic to obtain the activity, but the conversion of the distance from 30 cm to 100 cm while applying the inverse square law is likely to cause problems. We will therefore start by calculating Γ in the units $mSv\,m^2\,MBq^{-1}\,h^{-1}$, according to Eqn. 2.82, dropping the subscript, and using the dose rate instead of the kerma rate:

$$\Gamma = \frac{0.3^2 \cdot 1.07 \cdot 10^{-3}}{1 \cdot 10^6} = 9.6 \cdot 10^{-11} \ \mathrm{mSv\,m^2\,Bq^{-1}\,h^{-1}}. \tag{2.84}$$

The activity of the source can now be found by using Eqn. 2.83

$$3.3 \cdot 10^{-3} = \frac{9.6 \cdot 10^{-11} \cdot A}{1^2}, \tag{2.85}$$

which can be rewritten as

$$A = \frac{3.3 \cdot 10^{-3} \cdot 1^2}{9.6 \cdot 10^{-11}}, \tag{2.86}$$

giving a value of 34 MBq.

An alternative to using the air kerma rate constant is presented in IAEA-TECDOC-1162 (IAEA 2000a). In this document, procedures are given for calculating the effective dose (see below) or the absorbed dose rate for a variety of source geometries: point sources, line sources and spill, ground contamination, skin contamination and plume immersion (as well as procedures for calculating internal contamination). To demonstrate the use of the IAEA procedure when estimating the activity of a point source, Example 2.3 is repeated, this time for a shielded source, to demonstrate the use of the half-value layer.

Example 2.4 *Estimation of the activity of a shielded point source using the IAEA procedure*

A measurement is made with an accurately calibrated dose rate meter 1 m from a ^{137}Cs point source that is shielded by 3 mm of iron. The instrument shows that the dose rate is 3.5 $\mu Gy\,h^{-1}$. A background measurement made at a large distance from the Cs source shows 0.2 $\mu Gy\,h^{-1}$. What is the activity of the source?

As before, the net dose rate from the source 1 m away is 3.3 $\mu Gy\,h^{-1}$. According to the IAEA procedure, the absorbed dose rate is given by

$$\dot{D} = \frac{A \cdot CF_7 \cdot (0.5)^{\frac{d}{d_{1/2}}}}{X^2}, \tag{2.87}$$

where $\dot{D}$ is the absorbed dose rate in mGy h^{-1}, A is the activity of the source in kBq, X is the distance from the source in m, and d is the thickness of the shielding [cm]. The quantity $d_{1/2}$ is the half-value layer [cm] of the shielding material, given for some specified radionuclides and tabulated in IAEA-TECDOC-1162 together with the values of the conversion factor CF_7 [mGy h^{-1} kBq^{-1}]. Note that the half-value layer does not take into account scattered radiation (build-up).

Rearranging Eqn. 2.87 gives the activity

$$A = \frac{\dot{D} \cdot X^2}{CF_7 \cdot (0.5)^{\frac{d}{d_{1/2}}}} \tag{2.88}$$

and inserting the tabulated values for ^{137}Cs gives

$$A = \frac{3.3 \cdot 10^{-3} \cdot 1^2}{9.5 \cdot 10^{-8} \cdot (0.5)^{\frac{0.3}{1.19}}} \tag{2.89}$$

giving the value 41.1 MBq.

If we neglect the shielding we arrive at the same result as in Example 2.3, although that result was assumed to give the effective dose. The relation between the effective dose and the absorbed dose will depend on the distance from the body and on the position of the source relative to the body. Therefore, the assumptions made for tabulated values should always be carefully checked.

The effective dose is given by:

$$E_{ext} = \frac{A \cdot CF_6 \cdot T_e \cdot (0.5)^{\frac{d}{d_{1/2}}}}{X^2}, \tag{2.90}$$

where T_e is the duration of the exposure, and the conversion factor CF_6 [mSv h^{-1} kBq^{-1}] is that tabulated in IAEA-TECDOC-1162.

The use of the kerma rate constant (or the corresponding constant for dose) requires knowledge of the air kerma rate. There are, however, many situations in which the air kerma rate is unknown, but the activity per unit area may have been measured instead. *Dose rate conversion coefficients*, relating the activity per unit mass in the ground to the absorbed dose rate in air 1 m above the ground, have been published, for example, by Clouvas et al. (2000) and Quindos et al. (2004). These factors, given in the units nGy h^{-1} per Bq kg^{-1}, can be applied when the activity in the ground has been determined, for example, from an *in situ* (field gamma-ray spectrometry) measurement or gamma spectrometric soil sample analysis. For anthropogenic radionuclides that may be deposited on a surface, it would be convenient to have a coefficient relating the effective dose (from gamma radiation) to the activity per unit area. Such coefficients have been published by the IAEA (IAEA 2000a) and McColl and Prosser (2002). Examples are given in Table 2.8.

A useful rule of thumb concerning ^{137}Cs is that 1 MBq m^{-2} will result in 17 mSv y^{-1} if the Cs is deposited on the ground surface. If it has penetrated into the ground and is distributed in the top 5 cm, the effective dose will be 5.2 mSv y^{-1}.

Example 2.5 *Calculation of the absorbed dose rate from uranium series radionuclides*

As an example of the use of conversion coefficients, let us consider the naturally occurring uranium

TABLE 2.8   Ambient dose rate in mSv h^{-1} from a deposition of 1 kBq m^{-2}.

Radionuclide	Conversion factor (mGyh^{-1}kBq^{-1})
^{60}Co	$8.3 \cdot 10^{-6}$
^{90}Sr	$1.0 \cdot 10^{-9}$
^{103}Ru	$1.6 \cdot 10^{-6}$
^{131}I	$1.3 \cdot 10^{-6}$
^{137}Cs[1]	$2.1 \cdot 10^{-6}$
^{140}Ba	$6.4 \cdot 10^{-7}$
^{140}La	$7.6 \cdot 10^{-6}$
^{239}Pu	$1.3 \cdot 10^{-9}$

[1] Including ^{137m}Ba

Source: Data from IAEA (2000a).

decay series starting with ^{238}U and ending with the stable lead isotope ^{206}Pb. The bismuth isotope ^{214}Bi is often used to determine the activity of the radionuclides in the series due to its easily identified peak at 609.4 keV in the gamma spectrum. From the conversion coefficients given by Clouvas et al. (2000) a coefficient of 0.05348 nGy h^{-1} per Bq kg^{-1} can be used to calculate the absorbed dose rate to air from ^{214}Bi. If secular equilibrium can be assumed, the activities of all the members of the series will be equal, and the dose rate from all the radioactive elements in the series can be calculated using the conversion coefficient for the whole uranium series, which is 0.38092 nGy h^{-1} per Bq kg^{-1} (Clouvas et al. 2000), and the activity of the measured ^{214}Bi.

Calculating the activity of the radionuclides in a decay series utilizing secular equilibrium, as in the previous example, requires that the measured radionuclide does not have a branched decay. One example where this is not the case can be found in the thorium series.

Example 2.6 *Calculation of the absorbed dose rate from thorium series radionuclides*

One member of the thorium series (^{232}Th to ^{208}Pb) is ^{208}Tl, which has a convenient γ-ray decay at 583.1 keV. However, the measured activity of this radionuclide cannot be used in the same manner as previously for ^{214}Bi in the uranium series. The mother radionuclide of ^{208}Tl, which is ^{212}Bi, decays either by α-decay (35.9%) or β-decay (64.1%) to ^{208}Tl and ^{212}Po, respectively. Therefore, the activity of ^{208}Tl is only 35.9 % of the activity at equilibrium.

Both the thorium series and the uranium series contain isotopes of radon, which easily escape from the ground or from a sample. Since the calculations above assume secular equilibrium, the degree of equilibrium must also be checked by determining the activity of a member of each series that is above radon.

2.7.2 Build-Up

So far, only the absorbed dose and kerma from the *primary* radiation field, i.e. the radiation emerging directly from the radiation source, has been considered. However, the radiation field consists of primary, as well as scattered radiation due to coherent or incoherent scattering of photons in the media surrounding the radiation source. The total absorbed dose rate from both primary (p) and scattered (s) photons can then be written

$$\dot{D}_E = \dot{D}_{p,E} + \dot{D}_{s,E} = B_E \cdot \dot{D}_{p,E}, \qquad (2.91)$$

where B_E is the *build-up factor*, defined as

$$B = 1 + \frac{\dot{D}_s}{\dot{D}_p}. \qquad (2.92)$$

The build-up factor is generally calculated using Monte Carlo simulations, but various analytical approximations have also been developed by curve fitting (see, for example, Shultis and Faw 2000). These analytical expressions are convenient when the radiation field consists of photons of several energies, as in the case of environmental measurements. The contribution from each energy can then be integrated to yield the total absorbed dose (or total fluence rate). However, since the build-up factor depends on the quantity being studied, e.g. the fluence or absorbed dose, it is important to choose the correct parameters in these analytical expressions. For example, the build-up factor for fluence will account only for the increase in the number of photons due to scattering, while the build-up factor for absorbed dose is affected by the energy distribution of the scattered photons, since the mass energy absorption coefficient, μ_{en}, is a function of the photon energy. It should also be noted that published build-up factors may overestimate the build-up behind, for example, a radiation shield. Most calculations assume that the scattering medium is of infinite thickness and the build-up factor determined at a point within such a medium will then include the contribution from backscattered photons. The fluence rate behind an absorber will, however, not be affected by backscattered photons in the medium.

An example of the analytical expressions for the build-up factor is the Berger approximation, in which the build-up factor is expressed in the form

$$B(E_0, \mu t) \approx 1 + a\mu t e^{b\mu t}. \qquad (2.93)$$

The parameters a and b both depend on the incident photon energy E_0 and on the attenuating material, e.g. water or lead, and parameter values can be found, for example, in Shultis and Faw (2000). The parameters will also depend on the quantity for which the build-up is calculated. For example, the absorbed dose will depend on the exposed medium, as well as on the energy spectrum of the photons, which will change due to attenuation and build-up in the medium. For the same reasons, the build-up factor for the kerma will be different from that for the absorbed dose. The build-up factor given by Eqn. 2.93 also depends on the (energy-dependent) linear attenuation coefficient, μ, and the thickness, t, of the medium. The product of these is often denoted the *mean free path*, i.e. the mean distance a photon travels in a medium before interaction. Figure 2.39 shows the build-up factor for some media commonly encountered in radiation protection.

Several parameters, e.g. photon energy and the composition of the medium traversed by the photons, will influence the build-up factor. It will generally increase with decreasing photon energy in elements with low Z since more scattered photons will be generated due to the comparatively high probability of Compton scattering in these elements. In elements

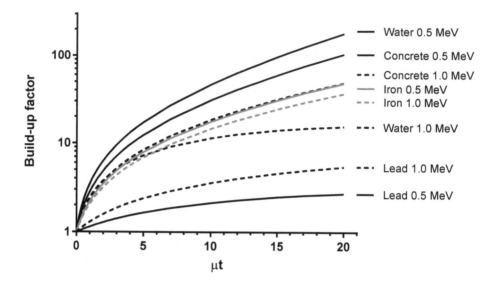

FIGURE 2.39 Build-up factors for air kerma, calculated with the Berger approximation. The thickness of each material is given as the number of mean free paths, μt.

with high Z, the build-up factor will instead decrease with decreasing photon energy since photoelectric absorption dominates. Thus, electrons are generated at the expense of scattering. Increasing the distance traversed by the photons will lead to a continuous decrease in the build-up factor. This energy-dependence of the build-up factor can be seen in Figure 2.39. For elements (or compounds) with low Z, such as water, concrete and iron, the build-up factor increases with decreasing photon energy, while the reverse is seen for lead, which is an element with high Z.

TABLE 2.9 Linear attenuation coefficient, μ, and distance, t, corresponding to 20 mean free paths for photons of energy 0.5 MeV and 1 MeV in various materials.

Material	Energy (MeV)	$\mu(cm^{-1})$	t (cm)
Lead	0.5	1.83	11
Lead	1.0	0.803	25
Iron	0.5	0.662	30
Iron	1.0	0.472	42
Concrete	0.5	0.201	100
Concrete	1.0	0.147	136
Water	0.5	0.0964	207
Water	1.0	0.0705	284

Build-up is also affected by the atomic number, and will be lower in elements with high Z than in elements with low Z. This is an effect of the increasing probability of both photoelectric absorption and pair production with increasing atomic number, neither of which generate secondary photons (apart from annihilation photons and characteristic x-

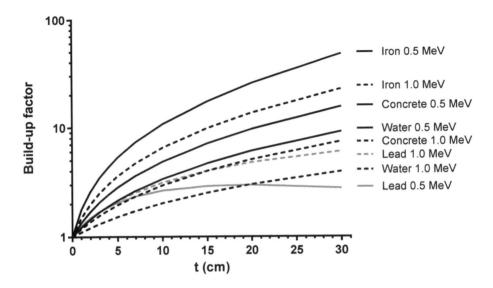

FIGURE 2.40 Build-up factors for air kerma, calculated with the Berger approximation. The thickness of each material, t, is given as the distance penetrated.

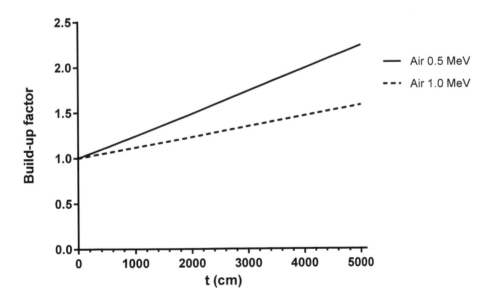

FIGURE 2.41 Build-up factors in air for air kerma, calculated with the Berger approximation. The thickness, t, is given as the distance penetrated.

rays). This effect can also be seen in Figure 2.39 as the build-up factor for lead is lower than for the other materials, for both photon energies shown in the figure.

Note that although all the build-up factors in Figure 2.39 are given for penetration depths up to 20 mean free paths, this corresponds to large variations in the thickness of the materials, as can be seen in Table 2.9. The variation in the build-up factor with material thickness is shown in Figure 2.40, where it can be seen that the ordering of the curves has changed compared to Figure 2.39. This is because the number of mean free paths corresponding to a given distance in cm varies considerably depending on the material. For example, 30 cm in water corresponds to 3 mean free paths for 0.5-MeV photons, while it corresponds to 20 mean free paths in iron.

Since the density of air is significantly lower than that of the solid materials discussed above, the mean free path will be considerably longer. For example, 20 mean free paths in air is 1900 m for 0.5-MeV photons and 2600 m for 1.0-MeV photons. At these distances from the source, the photons will have become attenuated by a factor of $2 \cdot 10^{-9}$. The build-up factor in air is shown in Figure 2.41.

Example 2.7 *Attenuation and build-up*

Assume that a point source emits 1-MeV photons. The photon fluence rate 1.0 m (t_0) from a point source, Φ_0, is $1.0 \cdot 10^9 \ m^{-2} \ s^{-1}$. Calculate the fluence rate at a distance, t, 2000 m from the source.

If we neglect attenuation and build-up in the air, the fluence rate at 2000 m is given by the inverse square law:

$$\dot{\Phi} = \dot{\Phi}_0 \frac{t_0^2}{t^2} = 1.0 \cdot 10^9 \frac{1^2}{2000^2}, \tag{2.94}$$

giving $\dot{\Phi} = 250 \ m^{-2} \ s^{-1}$.
Since the linear attenuation coefficient, μ, equals $7.65 \cdot 10^{-3} \ m^{-1}$, the attenuating factor will be:

$$e^{-7.65 \cdot 10^{-3} \cdot 2000}, \tag{2.95}$$

which equals $2.3 \cdot 10^{-7}$ and the fluence rate will then be $5.8 \cdot 10^{-5} \ m^{-2} \ s^{-1}$.
The build-up factor is 40 and the resulting fluence rate is thus $40 \cdot 5.8 \cdot 10^{-5}$, which equals $2.3 \cdot 10^{-3} \ m^{-2} \ s^{-1}$.

2.8 ABSORBED DOSE FROM INTERNAL RADIATION SOURCES

2.8.1 Modelling the Behaviour of Radionuclides in the Human Body

Estimating the radiation dose from ingested, inhaled, or otherwise internalized radionuclides in the human body is a more complicated task than estimating the radiation dose from an external source. The main difference is that it is impossible to measure the internal dose by physical means, whereas the dose from external radiation therapy, for example, can be determined with an accuracy of about ±5%. However, for the purpose of radiation protection, an accuracy within a factor of two may be sufficient for a decision to be made regarding protective measures. Despite the rather large uncertainties in some cases, the external radiation dose can be estimated by direct measurements. However, to estimate the internal dose, we have to rely on models and calculations, often in combination with some kind of measurement of the activity of a given radionuclide in, for example, blood or urine. Another problem is that internal exposures are protracted in time and the radiation dose must therefore be integrated over time.

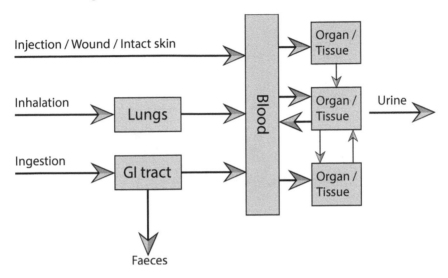

FIGURE 2.42 Pathways for internal contamination by radionuclides. After entering the body through one of the intake routes, the radionuclide will be transported by the blood to other organs or tissues, depending on the physical and chemical characteristics of the element.

The transport of radionuclides in the body is schematically depicted in Figure 2.42, showing the main intake routes: transfer through wounds or intact skin, inhalation and ingestion. Inhaled radionuclides will be deposited in the lungs and their distribution will be governed by the size distribution of the inhaled particles. Once deposited in the body, they will be transferred to the blood at a rate depending on their solubility, which is independent of the isotope, and is determined solely by the chemical characteristics of the element in question[2]. Intake by ingestion will also eventually lead to uptake in the blood when the radionuclide reaches the small intestine. Inhaled radionuclides may also find their way from the oesophagus or the lungs into the gastrointestinal (GI) tract. Entry into the body through wounds will lead to faster transfer to the blood, since neither the lungs nor the GI tract are involved. Once in the blood, radionuclides find their way to different organs in the body. In medical terms, this is referred to as *systemic uptake*. For example, strontium will be incorporated into the skeleton due to its chemical similarity to calcium. Radionuclides are eventually excreted in the urine or faeces. Some excretion may also take place through sweat, saliva or tears.

Calculations of internal dose require substantial knowledge on the behaviour of the radionuclide in the body: 1) the organ or tissue in which the radionuclide is retained and the change in retention with time, 2) the number of radioactive decays during the time the radionuclide is retained by that organ or tissue, 3) the amount of energy released in each decay, 4) which organs or tissues can be affected by the radiation, and the amount of energy

---

[2]There are a few exceptions to this rule. Especially for small nuclei, such as tritium, the difference in atomic mass between the isotopes influences the probability of reaction with other biomolecules; hence the metabolic behaviour (and thus the biokinetics) may also differ slightly between the isotopes.

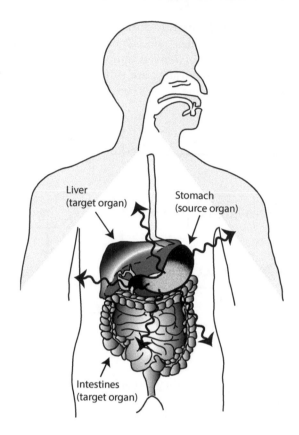

FIGURE 2.43 Following the ingestion of a radioactive substance, the stomach will be the source organ. The radiation dose to other (target) organs depends on the distance between the source and target organs, the attenuating characteristics of the organs, and the type of radiation.

that is deposited in each organ or tissue, and 5) the mass of each of these organs or tissues. Some of these are purely physical in nature, such as the energy released in each decay of a radionuclide, while others will require some kind of approximation, for example, the organ or tissue mass. The organ containing the radionuclide is described as the *source organ* and the organ affected by the radiation emitted by the radionuclides in the source organ is called the *target organ*, as illustrated in Figure 2.43. Different radionuclides will be accumulated in different source organs.

The transport of radionuclides in the body is described by mathematical models that take the time dependence of the activity in each source organ into account. The target organs, on the other hand, and the amount of energy deposited in each target organ will depend on the location of the organs within the body, and on the type of radiation released from the radionuclides in the source organ. Furthermore, since the radionuclides may be retained in more than one source organ, all combinations of source organs and target organs must be considered in order to obtain an accurate estimate of the internal dose. Thus, in order to be able to calculate the internal radiation dose, information must be available on the radioactive element that has entered the body, the route of intake and the transfer rates to and from the different organs or tissues within the body.

Extensive research carried out over the years has resulted in two main paradigms for estimating the internal radiation dose in man: the MIRD formalism[3] and the ICRP concept. Despite the use of different symbols and notations, these two paradigms are based on the same physical and physiological processes. The MIRD formalism is mostly used in nuclear medicine, where radioactive substances, i.e. radiopharmaceuticals, are administered to patients for diagnosis or treatment. The main task is then to estimate the absorbed dose to healthy tissue, for the purpose of protection in diagnostic nuclear medicine, or to estimate the absorbed dose to a tumour with the aim of destroying the cancer cells while limiting the effect on normal tissue during therapy with radiopharmaceuticals. In the ICRP concept, the intention is to estimate the equivalent dose to target tissues and to estimate the effective dose. However, since the incorporation of radioactive elements leads to protracted exposure, it is more correct in this context to use the quantities *committed equivalent dose* and *committed effective dose*. We will leave the MIRD formalism to the nuclear medicine community and concentrate on the methods used for the calculation of internal dose developed by the ICRP.

2.8.1.1 The Human Respiratory Tract

Since inhalation involves all the main organs of the body, i.e. the lungs, GI tract and blood, it is appropriate to start with this intake route. The ICRP human respiratory tract model, HRTM (ICRP 1994a), describes the fate of inhaled radionuclides by specifying compartment models and transfer parameters for deposition in, and clearance from, the lungs through absorption into the blood, transport to the GI tract and to regional lymph nodes. (Compartment models will be discussed in more detail in Section 3.1.)

Good knowledge of the kinetics of clearance is important when estimating doses to the lung tissues and to other body organs due to systemic uptake. In the ICRP model, clearance from the respiratory tract is described by particle transport processes and absorption. These particle transport processes are responsible for transferring the substance to the GI tract and to the lymph nodes, as well as between different regions of the respiratory tract. Absorption includes all types of transfer from the respiratory tract to the blood.

It must first be determined where the inhaled radionuclides are deposited in the lungs, both for dosimetric purposes and for the calculation of clearance and absorption by the blood. Since the radionuclides in the inhaled air may be in vapour form as well as attached to aerosols, the model must be able to deal with a wide range of particle sizes, from the size of an atom, with a diameter of about 0.5 nm, to large aerosols 0.1 mm in size. Aerosol particles are usually described by their *activity median aerodynamic diameter* (AMAD), which is the diameter of a sphere of unit density that has the same settling velocity in air as the particle in question.

The human respiratory tract (Fig. 2.44) can be divided into two regions: the *extrathoracic airways* and the *thoracic airways*. The extrathoracic airways are further divided into the anterior nasal passage (ET_1) and the rest of the extrathoracic airways (ET_2) due to their different clearance behaviour, which will be described below. The thoracic airways are also subdivided into several regions, the trachea and bronchi (BB), the bronchioles (bb) and the alveoli (AI), for the same reason. The fraction of the inhaled aerosols deposited in each of these regions will depend on the particle size.

The transport of particles in the respiratory tract is depicted schematically in Figure 2.45, where the shaded circles indicate regions where particles may be deposited initially. Once deposited, the particles may then be relocated or retained in the walls of the airways.

---

[3]Medical Internal Dose Committee, USA.

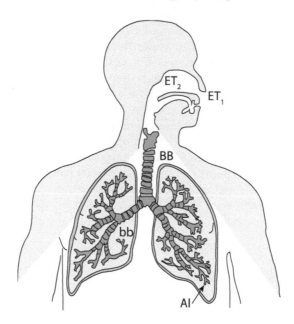

FIGURE 2.44 The human respiratory tract and the regions identified in the ICRP HRTM.

The (rather small) retained fractions of the deposited substance are shown as the regions denoted ET_{seq}, BB_{seq} and bb_{seq}, where the subscript seq denotes sequestration.

It is assumed that the removal of particles from the anterior nasal passage (ET_1) takes place only by wiping or blowing the nose, while the ET_2 region, which is covered with a mucosal layer, can only be cleared by rinsing with a fluid. The thoracic regions BB and bb are cleared by the action of the cilia covering the surfaces of these airways. However, the difference in the rate at which particles are cleared motivates further division of the BB and bb regions into two sub-regions (Fig. 2.45), where it is assumed that clearance is faster in regions BB_1 and bb_1 than in regions BB_2 and bb_2. Experimental data suggest that the AI region should also be divided into sub-regions; in this case, three different clearance rates are assumed.

Gases and vapours are subject to deposition mechanisms other than particles, and their uptake is determined by both the solubility and the reactivity of the gas. In the HRTM, this is described by three classes of uptake: SR-0, SR-1 and SR-2. SR-0 includes (almost) insoluble and nonreactive gases, such as H_2, He and N_2. The gases and vapours included in class SR-1 are denoted soluble or reactive, but since both the solubility and the reactivity can range from moderate to high, many different combinations are possible. The uptake may then be described by an uptake fraction, f_r, which is the fraction of the inhaled gas or vapour that is retained during normal breathing. Class SR-1 gases and vapours are assumed to be taken up in the BB, bb and AI regions. An example of an SR-1 class gas is oxygen, which is relatively insoluble and has a moderate reactivity in the conducting airways, but is highly reactive in the alveoli. The absorption of oxygen in the alveoli is almost complete, where it reacts with haemoglobin and is transported to the rest of the body via the blood. Class SR-2 contains highly soluble and reactive gases, such as SO_2, and these are only deposited in the extrathoracic airways.

Inhaled substances that are deposited in the respiratory tract are removed by three main routes: absorption into the blood, particle transport to the GI tract and particle transport

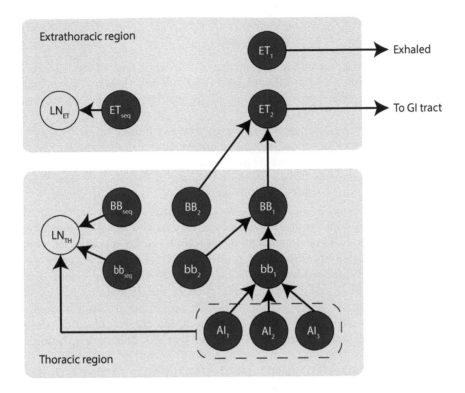

FIGURE 2.45 Inhaled radionuclides, initially deposited in the regions depicted by the shaded circles, are further transported in the respiratory tract at a rate depending on the characteristics of each region. The regions shown in the figure (including the lymph nodes, denoted LN) also constitute the compartments in the ICRP particle transport compartment model, and reference values for the parameters specifying the transfer rates can be found in ICRP Publication 66 (ICRP 1994a).

to regional lymph nodes, see Figure 2.45. Generally, the clearance rates depend on where the particles are located in the respiratory tract, the physical and chemical form of the substance, and the time since deposition. The clearance kinetics is very important since this determines both the amount of substance retained in the respiratory tract (and hence the dose to the tissues therein), and the amount transported to other organs via absorption into the blood, which will determine the doses to these organs. In addition, clearance rates will also influence the rates of faecal and urinary excretion.

Clearance from the respiratory tract is modelled by first-order kinetics, and the amount of the substance in region i at time t is thus given by

$$\frac{dR_i(t)}{dt} = -\lambda_i(t) R_i(t), \tag{2.96}$$

where the clearance rate, $\lambda_i(t)$, is the sum of the clearance rates due to particle transport, $m_i(t)$, and absorption, $s_i(t)$. The overall clearance rate can thus be written

$$\lambda_i(t) = m_i(t) + s_i(t). \tag{2.97}$$

This relation may be rewritten to explicitly include the particle transport rate to the GI tract, $g_i(t)$, and to the regional lymph nodes, $l_i(t)$

$$\lambda_i(t) = g_i(t) + l_i(t) + s_i(t). \tag{2.98}$$

The particle clearance rates are assumed to be independent of the element, while the absorption rates are different for different kinds of elements. However, the absorption rates are assumed to be the same for all regions in the respiratory tract, apart from the ET_1 region.

As the clearance rates generally change with time, as indicated in the equations above, this may cause difficulties in modelling the clearance since time-dependent functions must be found for the rates. To overcome this problem, the time dependence is instead modelled by combining compartments within a region, as shown in Figure 2.45. The time-dependent clearance of a region can then be modelled using constant rate parameters.

Absorption by the blood is regarded as a two-stage process. In the first stage, the particles are dissociated so they can be absorbed more easily. In the second stage, the element is absorbed into the blood. Time-dependent *dissociation* can be modelled by assuming that a fraction of the particles are dissociated and absorbed at a specified rate, while the rest of the particles are dissociated at another (slower) rate. Time-dependent *uptake*, on the other hand, is modelled by assuming that a fraction of the dissociated (dissolved) particles are retained and absorbed at a specified rate. The remaining fraction is then assumed to be transferred instantaneously to the blood. To simplify the calculations, the amounts of deposited radionuclides reaching the blood are given by three default absorption parameters: F, M and S.

Elements that are absorbed quickly into the blood belong to type F (fast absorption), although some of the radioactive substance in regions ET_2 and BB will be swallowed instead of absorbed. Type M (moderate) includes radionuclides that are absorbed at intermediate rates, and type S (slow) includes radionuclides with poor solubility.

2.8.1.2 The Human Alimentary Tract

As discussed above, some of the inhaled substances will eventually reach the GI tract (Fig. 2.46), as do ingested radionuclides. Therefore, the transport of radionuclides through the GI tract must be modelled. A GI tract model was published by the ICRP in 1979 as part of ICRP Publication 30 (ICRP 1979). This model, shown schematically in Figure 2.47, was developed for dosimetric purposes (for the calculation of radiation doses to workers). Although an extended model, the human alimentary tract model (HATM), was later developed and presented in ICRP Publication 100 (ICRP 2006), the ICRP 30 model still forms the basis for the dose coefficients for intakes of radionuclides, used to calculate radiation doses to workers and members of the public, published in the ICRP Publications 68 (ICRP 1994b) and 72 (ICRP 1996a), respectively. Dose coefficients for the intake of radionuclides and conversion coefficients for external radiation have also been compiled in ICRP Publication 119 (ICRP 2012b). Tables of dose coefficients can also be found in the International Basic Safety Standards, published by the IAEA (IAEA 2011).

As shown in Figure 2.47, the model consists of four regions: the stomach (ST), the small intestine (SI), the upper large intestine (ULI) and lower large intestine (LLI). The transport of radionuclides through these four regions is described by first-order kinetics in a chain-like compartment model. Radionuclides from the substance that is transported through the GI tract will be absorbed into the blood, and it is assumed that absorption takes place only from the small intestine. The fractional uptake is specified by the rate parameter f_1, which depends on the chemical characteristics of the radionuclide. The rate parameter for the transfer from the small intestine to the blood is given by

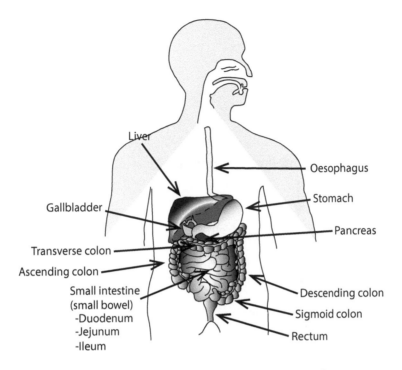

FIGURE 2.46 A schematic view of the organs constituting the human GI tract.

$$f_1 = \frac{\lambda_B}{\lambda_{SI} + \lambda_B}. \tag{2.99}$$

Re-arranging the above equation gives

$$\lambda_B = \frac{f_1 \lambda_{SI}}{1 - f_1}. \tag{2.100}$$

As can be seen from Figure 2.48, the choice of the value of f_1 determines not only the amount of the radioactive substance entering the blood, but also the fraction of the ingested activity that is transported further along the GI tract. Thus, the radiation dose to the intestines, as well as the activity excreted via the faeces, depends on f_1.

The more elaborate model of the GI tract in ICRP Publication 100 was developed because new data had become available on the transit of radionuclides through the gut, and on the location of sensitive cells. The new model can be used for children and adults, and takes into account absorption and retention, as well as excretion. The new model also includes the mouth and oesophagus. In the former ICRP 30 model, the dose coefficients for children were calculated simply by using a smaller mass of the GI tract. Another problem associated with the ICRP 30 model is that neither the colon nor the oesophagus was explicitly included in the model since these organs were not assigned tissue weighting factors (for the calculation of effective dose) before the 1990 recommendations from the ICRP (ICRP 1991). Some approximations therefore had to be made when calculating the dose coefficients given in ICRP 68; for example, the dose to the colon was calculated as the mean of the doses to the upper and the lower large intestine, weighted by the masses of the two organs.

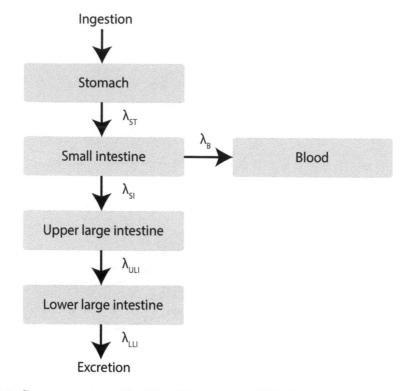

FIGURE 2.47 Compartment model of the GI tract from ICRP Publication 30. The values of the rate parameters in the model are $\lambda_{ST} = 24\ d^{-1}$, $\lambda_{SI} = 6\ d^{-1}$, $\lambda_{ULI} = 1.8\ d^{-1}$ and $\lambda_{LLI} = 1\ d^{-1}$. The rate parameter for absorption to the blood, λ_B, is given by Eqn. 2.100.

2.8.1.3 Biokinetic and Metabolic Models

Biokinetic models for calculations of the amount of radioactive substance retained in body organs have been published in ICRP Publication 30 (ICRP 1982). However, some of these models have since been replaced, and the newer models have been used when calculating dose conversion factors. The elements for which new models have been developed include tritium, iodine, polonium, radium, uranium and americium. For the actinides and the alkaline earth metals, the new models are based on physiological features. The new models have been published in ICRP Publications 56, 67 and 69 (ICRP 1990, ICRP 1993 & ICRP 1995a).

As an example of these biokinetic models, we can study the model for caesium. The new model for caesium, enabling the calculation of age-dependent dose coefficients, was published in ICRP Publication 67, and is schematically illustrated in Figure 2.49[4]. This compartment model is an example of the use of different compartments in the same volume or organ. The two compartments Total body A and B both refer to the whole body, but describe different rates of transport of the radioactive substance.

Using these kinds of models it is possible to estimate the retention of a radionuclide in different body organs, and they therefore form the basis for estimating the equivalent doses to the organs, and the effective dose. It is also possible to estimate the rate of excretion, which is important when performing *bioassays*. The term bioassay refers to some kind of

---

[4]The figure shows an adaptation of the generalized compartment model for hydrogen, cobalt, ruthenium, caesium and californium, presented in ICRP Publication 78 (ICRP 1997).

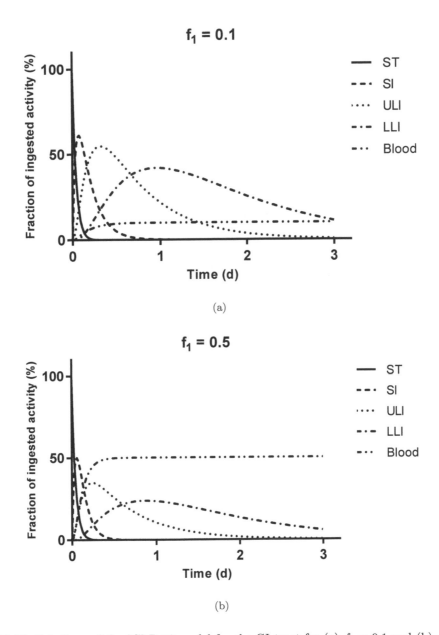

FIGURE 2.48 Solutions of the ICRP 30 model for the GI tract for (a) $f_1 = 0.1$ and (b) $f_1 = 0.5$. The fractional amount of activity absorbed by the blood, determined by the value of f_1, affects the amount of the radioactive element available for systemic uptake in the body organs, as well as the amount transported through the colon.

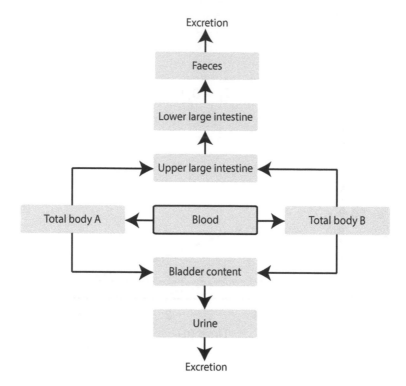

FIGURE 2.49 Compartment model describing the biokinetics of Cs. Radioactive material that has entered the blood is distributed throughout the whole body. However, the biokinetics of Cs is characterized by initial rapid removal followed by slow removal. This is modelled by dividing the whole body into two compartments called Total body A and Total body B. The biological half-times in Total body A and Total body B for an adult are defined as 2 and 110 days, respectively. The fractional uptake by the two compartments from blood is 10% in Total body A and 90% in Total body B. It is further estimated that 80% of the radioactive material leaves the body via the urine and 20% via faeces.

measurement intended to provide an estimate of the amount of a substance that has entered the body. Examples of bioassays are measurements of the activity concentration in urine or faeces, and whole-body measurements. In the case of Cs, the retention function, $R(t)$, where t is given in days, is described by the fairly simple expression

$$R(t) = 0.1e^{-\frac{ln2}{2}} + 0.9e^{-\frac{ln2}{110}}. \qquad (2.101)$$

2.8.2 Dose Calculations for Internal Exposure

In order to calculate the retention of a radionuclide in a body organ, or determine the excretion rate at a certain time after intake, all the relevant models discussed above must be combined. Figure 2.50 shows the total body retention for the intake of 1 Bq ^{137}Cs by inhalation, ingestion and injection. The retention curves for ingestion and injection coincide due to complete absorption in the small intestine ($f_1 = 1$). In the case of inhalation, the HRTM, the GI tract model and the systemic model are utilized, whereas in the case of ingestion there is no need to include the HRTM in the calculations.

Figure 2.51 shows the excretion rate for an intake of 1 Bq ^{137}Cs. These kinds of curves (or a corresponding table) can be used to estimate the inhaled or ingested activity *if the time of intake is known*. Similarly, the retention curve shown in Figure 2.50 could be used to estimate the intake based on measurements of the whole-body activity.

The activities given in Table 2.10 can be used as a useful rule of thumb to estimate the committed effective dose from the ingestion of Sr, I or Cs. In emergency situations, such approximate rules may be quite useful.

TABLE 2.10 Approximate intake (ingestion) of radionuclides giving a committed effective dose of 1 mSv.

Radionuclide	Activity (kBq)
^{89}Sr	400
^{90}Sr	40
^{131}I (adult)	70
^{131}I (1 y)	7
^{134}Cs	50
^{137}Cs	75

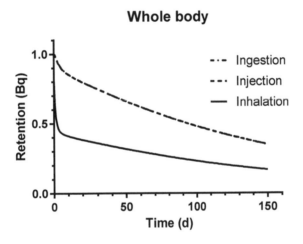

FIGURE 2.50 Retention of an intake of 1 Bq ^{137}Cs in the total body after inhalation, ingestion and injection. Only two curves are visible since the retention curves for ingestion and injection are almost identical. The retention of inhaled Cs was calculated assuming fast absorption (type F) and $f_1 = 1$. Note that the initial rapid removal is followed by slower removal characterized by a biological half-time of about 100 d.

Example 2.8 *Inhalation of ^{131}I*

During thyroid monitoring at a research facility where ^{131}I is used, an activity of 2 MBq was measured in the thyroid of one of the researchers. By consulting the laboratory journal, it was concluded that *contamination had probably occurred three days earlier. The route of intake was most certainly inhalation of iodine vapour. Estimate the committed effective dose and the committed equivalent dose to the thyroid.*

First, we must determine the inhaled activity. The activity measured in the thyroid three days after intake will have been subject to both physical decay and removal by physiological processes. In addition, only a fraction of the inhaled activity reaches the thyroid. Simply correcting the measured activity for physical decay and applying a dose coefficient will not give the correct answer. ICRP Publication 78 (ICRP 1997) presents bioassay calculations at different times after intake and, referring to Table A.6.17, we find that 20% of the inhaled activity is assumed to be found in the thyroid three days after intake. Thus, the inhaled activity is estimated to be 10 MBq. The dose coefficient, $e(50)$, can also be found in ICRP Publication 78, and is $2.0{\cdot}10^{-8}$ Sv Bq^{-1}. The committed effective dose is then estimated to be $2.0 \cdot 10^{-8} \cdot 1 \cdot 10^7 = 0.2$ Sv.

Dose coefficients for the calculation of the committed equivalent dose can be found in ICRP Publication 71 (ICRP 1995b), and for elemental iodine (Table 5.19.4(d)) the dose coefficient is $3.9 \cdot 10^{-7}$ Sv Bq^{-1}, which yields a committed equivalent dose of $3.9 \cdot 10^{-7} \cdot 1 \cdot 10^7 = 3.9$ Sv.

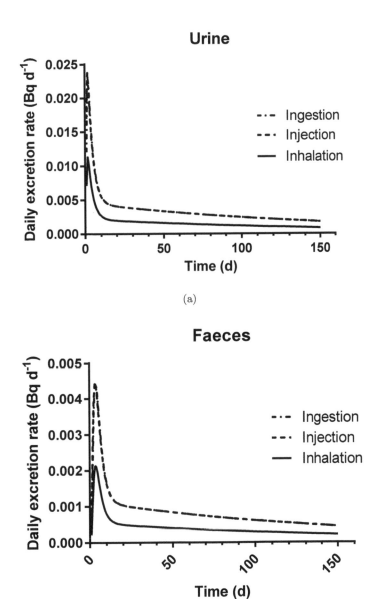

FIGURE 2.51 Excretion rates in Bq d^{-1} in urine (a) and faeces (b) following an intake of 1 Bq ^{137}Cs. The excretion rate from inhaled Cs was calculated using absorption type F and $f_1 = 1$.

2.9   REFERENCES

- Anderson, D. W. 1984. *Absorption of Ionizing Radiation.* University Park Press, Baltimore, USA.

- Attix, F. H. 1991. *Introduction to Radiological Physics and Radiation Dosimetry.* Wiley-VCH Verlag GmbH, Germany.

- Clouvas, A., Xanthos, S., Antonopoulos-Domis, M. and Silva, J. 2000. Monte Carlo calculation of dose rate conversion factors for external exposure to photon emitters in soil. *Health Physics*, Vol. 78, No. 3, 295–302.

- Delacroix, D., Guerre, J. P., Leblanc, P. and Hickman, C. 2002. *Radionuclide and Radiation Protection Data Handbook 2002.* Radiation Protection Dosimetry, Vol. 98, No. 1.

- Engholm, G., Ferlay, J., Christensen, N., Bray, F., Gjerstorff, M. L., Klint, A., Køtlum, J. E., Ólafsdóttir, E., Pukkala, E. and Storm, H. H. 2010. NORDCAN: A Nordic tool for cancer information, planning, quality control and research. *Acta Oncol.* Vol. 49, No. 5, 725–736.

- Engholm, G., Ferlay, J., Christensen, N., Kejs, A. M. T., Johannesen, T. B., Khan, S., Leinonen, M., Milter, M. C., Ólafsdóttir, E., Petersen, T., Stenz, F. and Storm, H. H. 2015. *NORDCAN: Cancer Incidence, Mortality, Prevalence and Survival in the Nordic Countries, Version 7.1 (09.07.2015).* Association of the Nordic Cancer Registries. Danish Cancer Society. Available from http://www.ancr.nu, accessed 27 July 2015.

- Golikov, V., Wallström, E., Wöhni, T., Tanaka, K., Endo, S. and Hoshi, M. 2007. Evaluation of conversion coefficients from measurable to risk quantities for external exposure over contaminated soil by use of physical human phantoms. *Radiation and Environmental Biophysics*, Vol. 46, 375–382.

- IAEA 2000a. *Generic Procedures for Assessment and Response during a Radiological Emergency.* IAEA-TECDOC-1162. International Atomic Energy Agency. Vienna.

- IAEA 2000b. *Calibration of Radiation Protection Monitoring Instruments.* IAEA Safety Reports Series No. 16. International Atomic Energy Agency. Vienna.

- IAEA 2005. *Generic procedures for medical response during a nuclear or radiological emergency.* EPR-MEDICAL (2005). International Atomic Energy Agency. Vienna.

- IAEA 2010. *Radiation Biology: A Handbook for Teachers and Students.* IAEA Training Course Series 42. International Atomic Energy Agency. Vienna.

- IAEA 2011. *Radiation Protection and Safety of Radiation Sources: International Basic Safety Standards.* IAEA Safety Standards Series No. GSR Part 3 (Interim). International Atomic Energy Agency. Vienna.

- ICRP 1977. *Recommendations of the ICRP.* ICRP Publication 26. Ann. ICRP 1 (3).

- ICRP 1979. *Limits for Intakes of Radionuclides by Workers.* ICRP Publication 30 (Part 1). Ann. ICRP 2 (3–4).

- ICRP 1982. *Limits for Intakes of Radionuclides by Workers.* ICRP Publication 30 (Index). Ann. ICRP 8 (4).

- ICRP 1990. *Age-Dependent Doses to Members of the Public from Intake of Radionuclides—Part 1.* ICRP Publication 56. Ann. ICRP 20 (2).

- ICRP 1991. *1990 Recommendations of the International Commission on Radiological Protection.* ICRP Publication 60. Ann. ICRP 21 (1–3).

- ICRP 1993. *Age-Dependent Doses to Members of the Public from Intake of Radionuclides—Part 2 Ingestion Dose Coefficients.* ICRP Publication 67. Ann. ICRP 23 (3–4)

- ICRP 1994a. *Human Respiratory Tract Model for Radiological Protection.* ICRP Publication 66. Ann. ICRP 24 (1–3).

- ICRP 1994b. *Dose Coefficients for Intakes of Radionuclides by Workers.* ICRP Publication 68. Ann. ICRP 24 (4).

- ICRP 1995a. *Age-Dependent Doses to Members of the Public from Intake of Radionuclides—Part 3 Ingestion Dose Coefficients.* ICRP Publication 69. Ann. ICRP 25 (1).

- ICRP 1995b. *Age-Dependent Doses to Members of the Public from Intake of Radionuclides—Part 4 Inhalation Dose Coefficients.* ICRP Publication 71. Ann. ICRP 25 (3–4).

- ICRP 1996a. *Age-Dependent Doses to the Members of the Public from Intake of Radionuclides—Part 5 Compilation of Ingestion and Inhalation Coefficients.* ICRP Publication 72. Ann. ICRP 26 (1).

- ICRP 1996b. *Conversion Coefficients for Use in Radiological Protection against External Radiation.* ICRP Publication 74. Ann. ICRP 26 (3–4).

- ICRP 1997. *Individual Monitoring for Internal Exposure of Workers.* ICRP Publication 78. Ann. ICRP 27 (3–4).

- ICRP 2006. *Human Alimentary Tract Model for Radiological Protection.* ICRP Publication 100. Ann. ICRP 36 (1–2).

- ICRP 2007. *The 2007 Recommendations of the International Commission on Radiological Protection.* ICRP Publication 103. Ann. ICRP 37 (2–4).

- ICRP, 2010. *Conversion Coefficients for Radiological Protection Quantities for External Radiation Exposures.* ICRP Publication 116, Ann. ICRP 40(2–5).

- ICRP, 2012a. *ICRP Statement on Tissue Reactions / Early and Late Effects of Radiation in Normal Tissues and Organs: Threshold Doses for Tissue Reactions in a Radiation Protection Context.* ICRP Publication 118. Ann. ICRP 41(1/2).

- ICRP 2012b. *Compendium of Dose Coefficients Based on ICRP Publication 60.* ICRP Publication 119. Ann. ICRP 41 (Suppl.).

- ICRU 1993. *Quantities and Units in Radiation Protection Dosimetry.* ICRU Publication 51. ICRU Publications, Bethesda, USA.

- ICRU 1998. *Fundamental Quantities and Units for Ionizing Radiation.* ICRU Publication 60. ICRU Publications, Bethesda, USA.

- ICRU 2001. *Determination of Operational Dose Equivalent Quantities for Neutrons.* ICRU Publication 66. ICRU Publications, Bethesda, USA.

- Isaksson, M. 2011. Environmental dosimetry: Measurements and calculations. In *Radioisotopes–Applications in Physical Sciences*, Ed. N. Singh. Intech. DOI: 10.5772/22731. (http://www.intechopen.com/books/radioisotopes-applications-in-physical-sciences/environmental-dosimetry-measurements-and-calculations)

- Jacob, P., Paretzke, H. G., Rosenbaum, H. and Zankl, M. 1988. Organ doses from radionuclides on the ground. Part I. Simple time dependencies. *Health Physics.* Vol. 54, No. 6, 617–633.

- Jacobi, W. 1975. The concept of the effective dose: A proposal for the combination of organ doses. *Rad. and Environm. Biophys.* Vol. 12, 101–109.

- Kase, K. R. and Nelson, W. R. 1972. *Concepts of Radiation Dosimetry.* Stanford Linear Accelerator Center. SLAC-153. http://www.slac.stanford.edu/cgi-wrap/getdoc/slac-r-153.pdf (accessed 13 June 2014).

- McColl, N. P. and Prosser, S. L. 2002. *Emergency Data Handbook.* NRPB-W19. National Radiological Protection Board, UK.

- McParland, B. J. 2010. *Nuclear Medicine Radiation Dosimetry.* Springer-Verlag, London, UK.

- Nilsson, B. 2015. *Exercises with Solutions in Radiation Physics.* De Gruyter Open Ltd. Warsaw/Berlin.

- Ninkovic, M. M., Raicevic, J. J. and Adrovic, F. 2005. Air kerma rate constants for gamma emitters used most often in practice. *Radiation Protection Dosimetry*, Vol. 115, No. 1–4, 247–250.

- Quindos, L. S., Fernández, P. L., Ródenas, C., Gómez-Arozamena, J. and Arteche, J. 2004. Conversion factors for external gamma dose derived from natural radionuclides in soils. *Journal of Environmental Radioactivity.* Vol. 71, 139–145.

- Rojas-Palma, C., Liland, A., Naess Jerstad, A., Etherington, G., del Rosario Pérez, M., Rahola, T. and Smith, K. (Eds.). 2009. *TMT Handbook, Triage, Monitoring and Treatment of People Exposed to Ionising Radiation Following a Malevolent Act.* SCK-CEN, NRPA, HPA, STUK, WHO.

- Shultis, J. K. and Faw, R. E. 2000. *Radiation Shielding.* American Nuclear Society, La Grange Park, IL, USA.

- UNSCEAR, United Nations Scientific Committee on the Effects of Atomic Radiation. 2000a. *Sources and Effects of Ionizing Radiation, Vol. II: Effects, Annex J.* New York: United Nations.

- UNSCEAR, United Nations Scientific Committee on the Effects of Atomic Radiation. 2000b. *Sources and Effects of Ionizing Radiation, Vol. II: Effects, Annex I.* New York: United Nations.

- UNSCEAR, United Nations Scientific Committee on the Effects of Atomic Radiation. 2001. *Hereditary Effects of Radiation, Scientific Annex.* New York: United Nations.

- UNSCEAR, United Nations Scientific Committee on the Effects of Atomic Radiation. 2006. *Effects of Ionizing Radiation, Vol. I: Report to the General Assembly, Scientific Annexes A and B, Annex A.* New York: United Nations.

2.10 EERCISES

2.1 In an incident at a facility for nuclear fuel fabrication, a female worker was exposed to airborne uranium, which was probably inhaled. In an attempt to estimate the inhaled activity and the radiation dose to the worker, a urine sample and a faeces sample were collected. The samples represent 24 hours' excretion, and the incident probably occurred 2–3 days before sampling. It is also known that the inhaled substance was mainly UO_2, which means that slow absorption (type S) can be assumed.

Measurements at the workplace showed that the estimated particle size was 5 μm (AMAD). The analysis of the urine sample showed the following activities: 0.0147 Bq ^{238}U; 0.0035 Bq ^{235}U, and 0.0818 Bq ^{234}U. The analysis of the faeces sample yielded 51.8 Bq ^{238}U; 12.0 ^{235}U, and 286.2 ^{234}U. The specific activities are therefore $1.24 \cdot 10^4$ Bq g^{-1}, $8.00 \cdot 10^4$ Bq g^{-1} and $2.30 \cdot 10^8$ Bq g^{-1} for ^{238}U, ^{235}U and ^{234}U, respectively.

a. Based on the above measurements, can we assume that an acute intake of uranium has occurred?

b. Is the uranium enriched? If it is, to what extent?

c. Is the estimated time of intake reasonable?

d. Calculate the inhaled activity and the committed effective dose.

e. Estimate the total uncertainty in the calculations.

Dose coefficients and excretion rates calculated from the ICRP models are given in Table 2.11 and Figure 2.52.

TABLE 2.11 Dose coefficients, $e(50)$, for the inhalation of uranium (absorption type S) of 5 μm AMAD.

Radionuclide	f_1	$e(50)$ (Sv Bq^{-1})
^{234}U	0.002	$6.8 \cdot 10^{-6}$
^{235}U	0.002	$6.1 \cdot 10^{-6}$
^{238}U	0.002	$5.7 \cdot 10^{-6}$

Source: Data from ICRP (1994b).

2.2 After the Chernobyl accident, many countries placed restrictions on the intake of activity in various foodstuffs. A common limit for ^{137}Cs in meat was 300 Bq kg^{-1}. Calculate the maximum amount of this meat that can be eaten while ensuring the committed effective dose from meat is below 1 mSv. The dose coefficient for ingestion of ^{137}Cs is $1.3 \cdot 10^{-8}$ Sv Bq^{-1}.

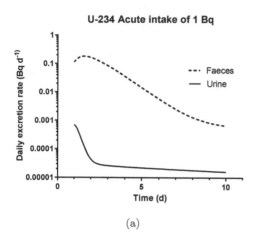

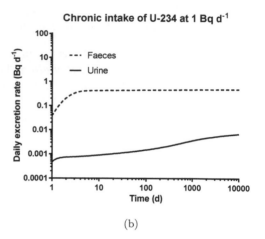

(a)   (b)

FIGURE 2.52 Excretion rates in urine and faeces, expressed in Bq d^{-1}, following an acute and chronic intake of 1 Bq ^{234}U. The excretion rate from inhaled U was calculated assuming absorption type S and $f_1 = 0.002$.

2.3 Show that an area source may be approximated by a point source if the distance from the source is greater than 2.2 times the radius of the source, if attenuation in air is neglected.

2.4 Estimate the minimum distance at which a line source can be approximated by a point source, within an error of 10%.

2.5 A line source consists of a 10.0-cm-long pipe with inner diameter of 5.0 mm. The pipe is enclosed in 1.0 mm Pb and 2.0 mm Cu. The pipe is assumed to carry water containing ^{137}Cs and, in order to estimate the activity, a measurement of the fluence rate is made at the midpoint of the pipe, at a perpendicular distance of 0.70 m from the pipe. The measurement shows that the fluence rate is $2.4 \cdot 10^8$ cm^{-2} s^{-1}. What is the activity of ^{137}Cs in the pipe? Self-absorption in the water may be neglected, but not the attenuation in air.

2.11   FURTHER READING

Hall, E. J. and Giaccia, A. J. 2006. *Radiobiology for the Radiologist*. Lippincott, Williams, & Wilkins.

Joiner, M. and van der Kogel, A. 2009. *Basic Clinical Radiobiology*. CRC Press.

CHAPTER 3

Environmental Exposure Pathways and Models

CONTENTS

T HE exposures of people and other organisms to radiation from many of the sources discussed in Chapter 1 are dependent on how the radioactive contaminants are transported from the source to the organism. Because most radionuclides are released into the atmosphere, the transfer chain usually starts there. However, large amounts of radionuclides may also be released into the environment via inland waterways or the sea. Radionuclides released into the atmosphere follow the movement of the air before being deposited on the ground by precipitation or various dry deposition mechanisms. Some of these radionuclides will find their way into rivers, lakes or the sea through run-off from the ground or by direct deposition onto the water surface. Radioactive material deposited on the surface environment may subsequently be transported to other parts of the ecosystem. In order to quantify the exposure resulting from a particular release of radioactive elements it is therefore important to understand the physical and chemical processes that govern how these elements are transported in the atmosphere, in water and in ecosystems. This is the subject of this chapter.

The models developed for studies of the transport of radionuclides in the environment, and for estimation of radiation doses to the human body, are often of a type called compartment models. We therefore begin with a brief description of compartment modelling. A more elaborate description can be found in, for example, *Compartmental Analysis in Biology and Medicine* by J. A. Jacquez (see Further Reading at the end of this chapter).

3.1 COMPARTMENT MODELS FOR ENVIRONMENTAL MODELLING

Compartment models can be used in a very wide range of contexts: for example, to model fuel transport in engines, chemical reactions, physiological processes, ecosystems and mechanical systems. The *system* to be modelled is divided into discrete regions called *compartments*, each containing a homogeneous distribution of the substance in question. This substance could, for example, be a drug introduced into the blood, which is then transported to different organs in the body. The transport of a substance between compartments is described by *flows*, and models also include *inflows* and *outflows* to and from the system as a whole. These flows are influenced by the concentrations in the compartments, and the compartment model is described mathematically by ordinary differential equations.

3.1.1 The Basics of Compartment Models

An example of a compartment model is the transfer of ^{137}Cs in the food chain from grass to cow's milk, as illustrated in Figure 3.1. Fallout of ^{137}Cs will cause contamination of the grass eaten by grazing cows. Some of the ^{137}Cs in the grass will be ingested and find its way into the milk through physiological processes in the cow's body. Humans who consume this milk will then be exposed to ^{137}Cs. With this kind of model it is possible to estimate the radiation dose to humans if the amount of activity in the grass is known. No attempt will be made here to model the details of the transfers between compartments, such as the physiological processes that transfer ^{137}Cs from a cow's stomach to its milk, but it is possible to model such details.

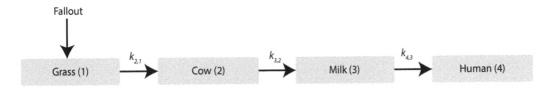

FIGURE 3.1 Simplified compartment model describing the transport of ^{137}Cs from grass to cow's milk and the exposure of humans through the consumption of milk and other dairy products. The transfer coefficients between the compartments are denoted $k_{i,j}$.

The system to be modelled is chosen based on previous knowledge of the radionuclide and how it is transported, for example, in an ecosystem or between body organs. A *closed system* has neither inflows nor outflows, while an *open system* is connected to the environment outside the system being modelled. It is assumed that the content of each compartment in the system is homogeneously mixed within that compartment, and that any material flowing into a compartment is immediately mixed with the material already there. Every molecule in a compartment thus has the same probability of being transported to another compartment in the system.

The flow between two compartments is determined by the concentration or content in the first compartment at a given instant, and a *transfer coefficient*, which has the dimension [time^{-1}]. Transfer coefficients are usually approximated as being time-independent (first-order kinetics), but they may also depend on the concentration or content in a compartment. They may also be time-dependent. If all the transfer coefficients in the model are constant, the system is called a linear compartmental system with constant coefficients. If some of the transfer coefficients are functions of time only, and the remaining coefficients are constants, the system is described as linear with time-varying coefficients. Finally, if at least one of the transfer coefficients is a function of the content in a compartment, the system is called a nonlinear compartmental system. These compartment models are considered to be deterministic, which means that at a given time, t, each transfer coefficient contributing to the model has a specific value. Successive computational runs of the model will then yield the same result.

Due to uncertainties in the transfer parameters, the modelling of some systems may benefit from the use of *stochastic models* in which each transfer coefficient is assigned a distribution of values with a mean value and a variance obtained from an assumed probability density distribution. In each run of the model, the parameter value is chosen at random with a probability given by this distribution. Modern computational techniques have facilitated these kinds stochastic models. They allow greater flexibility in studying how the model output depends on the transfer coefficients.

The mathematics for a simple compartment model can be understood by considering the model for the transport of ^{137}Cs shown in Figure 3.1. Assume that the transfer coefficients, $k_{i,j}$ are known. The flow between the first two compartments is determined by the content in the first compartment (the grass), and the flow between the second and the third compartments is determined by the content in the second compartment (the cow). We can then derive a differential equation for the change in content per unit time, dN/dt, in the second compartment

$$\frac{dN_2}{dt} = k_{2,1} \cdot N_1 - k_{3,2} \cdot N_2, \tag{3.1}$$

where $N = N(t)$ is the (time-dependent) activity.

The change in content per unit time in the cow compartment is thus governed by inflow from the grass compartment (an increase) and outflow to the milk compartment (a decrease). The rate of change in the grass compartment is given by $-k_{2,1} \cdot N_1$, which yields

$$N_1(t) = N_1(0) \cdot e^{-k_{2,1}t}, \tag{3.2}$$

where $N_1(0)$ is the amount of ^{137}Cs in the grass at time $t = 0$. If it can be assumed that the fallout is deposited over a period of time that is short compared to the transfer times, it can be treated as an instantaneous input and defined as a constant $(N_1(0))$ in the model. Inserting the expression for N_1 in Eqn. 3.2 into the right-hand side of Eqn. 3.1 gives a differential equation that can be easily solved, yielding

$$N_2(t) = \frac{k_{2,1}}{k_{3,2} - k_{2,1}} \cdot N_1(0) \cdot (e^{-k_{2,1}t} - e^{-k_{3,2}t}). \tag{3.3}$$

In the above case, we assumed that the input occurred instantaneously, but in some cases it may be necessary to model a continuous input. In such a case, the differential equation describing the activity in the grass compartment would be

$$\frac{dN_1}{dt} = R - k_{2,1} \cdot N_1, \tag{3.4}$$

where R is the rate of input to compartment 1 (grass). Assuming that the grass compartment initially contains no ^{137}Cs, this equation has the solution

$$N_1(t) = \frac{R}{\lambda} \cdot (1 - e^{-k_{2,1}t}). \tag{3.5}$$

Since we are modelling the flows of radioactive substances, we must also take the reduction due to radioactive decay into account. This can simply be done by adding another loss term, $\lambda = ln2/T_{1/2}$, representing the rate of loss resulting from radioactive decay. Furthermore, we can also model the presence of a mother nuclide in a compartment by treating it as an inflow from an external compartment at the rate λ_{mother}. It is very important that the transfer coefficients are expressed in the same units as the decay constant. Since the physical half-life, $T_{1/2}$, can vary by orders of magnitude it is often tabulated in different units: seconds, minutes or years. In calculations where the unit of activity is the becquerel, decay constants and half-lives must be expressed in terms of the SI unit of time, the second.

Example 3.1 *Sedimentation and resuspension in a lake*

The model depicted in Figure 3.2, describing the transfer of a radionuclide within a (simplified) freshwater system, can be used as an example of a compartment model with bidirectional flow. In the model, transfer from the water to the bottom sediment, or sedimentation, occurs at a rate proportional to the concentration of the radionuclide in the water. However, there may be processes driving a flow in the opposite direction, transporting radionuclides from the sediment to the water by resuspension.

The resuspension rate is proportional to the concentration of the radionuclide in the bottom sediment.

Before determining the change in concentrations in the water and sediment, we determine the contents of the radionuclide in the various compartments. The number of radioactive atoms in the water, N_{Water}, will change with time according to

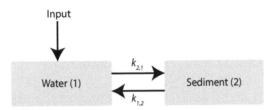

FIGURE 3.2 The transfer of a radionuclide between water and the bottom sediment. The sedimentation rate is determined by the transfer coefficient $k_{2,1}$ and the amount of the radionuclide in the water, while transfer from the bottom sediment back to the water through resuspension is determined by $k_{1,2}$ and the amount of the radionuclide in the sediment.

$$\frac{dN_{Water}}{dt} = -F_{Sedimentation} + F_{Resuspension} = -k_{2,1} \cdot N_{Water} + k_{1,2} \cdot N_{Sediment}, \qquad (3.6)$$

where $F_{Sedimentation}$ and $F_{Resuspension}$ are the flows (number of atoms per unit time) from water to sediment and vice versa. The fraction of atoms transported per unit time is given by the transfer coefficients $k_{2,1}$ and $k_{1,2}$.

If the physical half-life of the radionuclide is of the same order of magnitude as the time intervals studied, radioactive decay of the nuclide in each compartment must be considered when expressing the content of the radionuclide in either of the compartments. The mathematical expression for the time dependence of the content can then be written

$$\frac{dN_{Water}}{dt} = -F_{Sedimentation} + F_{Resuspension} - F_{Decay} =$$
$$= -k_{2,1} \cdot N_{Water} + k_{1,2} \cdot N_{Sediment} - \lambda \cdot N_{Water}. \qquad (3.7)$$

Since we are concerned with a radioactive substance, it is convenient to express Eqn. 3.6 in terms of the activity, A, using the relation $A = \lambda \cdot N$:

$$\frac{dA_{Water}}{dt} = -k_{2,1} \cdot \lambda \cdot N_{Water} + k_{1,2} \cdot \lambda \cdot N_{Sediment}. \qquad (3.8)$$

If radioactive decay is also taken into account, Eqn. 3.8 becomes

$$\frac{dA_{Water}}{dt} = -k_{2,1} \cdot \lambda \cdot N_{Water} + k_{1,2} \cdot \lambda \cdot N_{Sediment} - \lambda^2 \cdot N_{Water}, \qquad (3.9)$$

indicating that there are two processes leading to the loss of the radionuclide from the water compartment: transport through resuspension and radioactive decay. The loss of the radionuclide from the water compartment can be quantified from the values of $(k_{2,1} + \lambda)$. However, it should be noted that the amount transported to the sediment compartment is still governed by $k_{2,1}$ alone.

In practice, it is often easier to determine the concentration of a radioactive nuclide than to measure the amount present, and it is then more convenient to express the compartment model in terms of the activity concentration. If the amount of the radionuclide is expressed as the activity concentration, c [Bq m^{-3}], then, ignoring decay (i.e. assuming a long-lived radionuclide), we have:

$$\frac{dc_{Water}}{dt} = \frac{\lambda}{V_{Water}} \frac{dN_{Water}}{dt} =$$

$$= \frac{\lambda}{V_{Water}}(-k_{2,1} \cdot N_{Water} + k_{1,2} \cdot N_{Sediment}) =$$

$$= \frac{\lambda}{V_{Water}}(-k_{2,1} \cdot \frac{V_{Water}}{\lambda} \cdot c_{Water} + k_{1,2} \cdot \frac{V_{Sediment}}{\lambda} \cdot c_{Sediment}) = \tag{3.10}$$

$$= -k_{2,1} \cdot c_{Water} + k_{1,2} \cdot \frac{V_{Sediment}}{V_{Water}} \cdot c_{Sediment},$$

where V_{Water} and $V_{Sediment}$ are the volumes of the water and the sediment, respectively.

Figure 3.3 shows a solution to the model illustrated in Figure 3.2 with values of $k_{2,1} = 0.04\ \mathrm{d}^{-1}$ and $k_{1,2} = 0.02\ \mathrm{d}^{-1}$. The initial activity in the water is assumed to be 100 Bq. Since the model is bidirectional, the activity in the water and the sediment will reach equilibrium. In this case, the equilibrium activities in the water and sediment are about 33 Bq and 67 Bq, respectively. The ratio between them is explained by the relationship between the transfer coefficients: 0.02 is one-third of the total transfer, while 0.04 is two-thirds. For each period of time, two-thirds of the activity leaves the water while only one-third leaves the sediment, leading to a higher activity in the sediment.

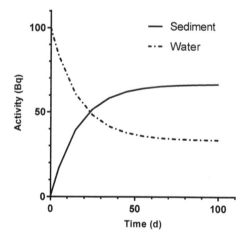

FIGURE 3.3   Since the model is bidirectional, the activity in both the water and the sediment will approach a constant value, determined by the transfer coefficients.

This relation between the activities at equilibrium is the basis for the concept of the *concentration ratio*, C_r (ICRU 2001), which can be used to calculate the activity concentration[1] in, for example, a fish species such as cod, when the activity concentration in the water is known. If Figure 3.2 had shown the activity concentration in water and in cod, instead of the activity in water and sediment, the concentration ratio would have been equal

---

[1] According to ICRU (2001), the recommended name for this quantity is the *volumetric activity density*, A_v. However, since the term *activity concentration* is widely used, it will be used throughout this book.

to 2. This means that if the activity concentration in the water had been 200 Bq kg^{-1}, the cod would have contained 400 Bq kg^{-1} at equilibrium[2].

Although the amount of a radionuclide may be reduced by radioactive decay or transport within the system, the presence of a mother nuclide in either or both of the compartments will increase the amount of the radionuclide present. Instead of using an arrow to denote flow in the compartment model, the presence of a mother nuclide may be treated as a *source term*, calculated as the physical decay constant of the mother nuclide, multiplied by the amount of the mother nuclide present. If we introduce a mother nuclide into the water compartment in the previous example, Eqn. 3.7 becomes

$$\frac{dN_{Water}}{dt} = k_2 \cdot N_{Sediment} - k_1 \cdot N_{Water} - \lambda \cdot N_{Water} + \lambda_M \cdot N_{M,Water}, \qquad (3.11)$$

where λ_M is the decay constant of the mother nuclide, and $N_{M,Water}$ denotes the number of atoms of the mother nuclide present in the water. The product $\lambda_M \cdot N_{M,Water}$ is then equal to the activity of the mother nuclide in the water. If the mother nuclide is also present in the sediment, this will be included in $N_{Sediment}$ according to

$$\frac{dN_{Sediment}}{dt} = k_1 \cdot N_{Water} - k_2 \cdot N_{Sediment} - \lambda \cdot N_{Sediment} + \lambda_M \cdot N_{M,Sediment}. \qquad (3.12)$$

Eqns. 3.11 and 3.12 must be solved simultaneously, since $N_{Sediment}(t)$ must be known in order to solve Eqn. 3.11. However, this can only be done if we know the expression for $N_{Water}(t)$. To solve the system of differential equations, which can be quite complicated for large models, Laplace transform techniques may be used, or numerical techniques such as Euler's method or Runge–Kutta methods. Numerical methods are suitable for computer-aided solutions, and many commercial programs are available for defining and solving compartment models. These programs often allow the user to choose between different numerical solvers, depending on the structure of the model.

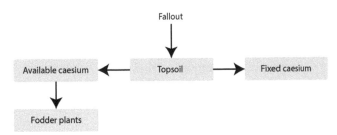

FIGURE 3.4   Simplified compartment model that includes fixation of caesium in the soil after fallout.

Although it is common for a compartment to represent a physical body in the system, such as a lake, a fish or an organ, a compartment may also have a more abstract meaning. Consider, for example, ^{137}Cs fallout over a pasture, where we want to develop a simplified compartment model to predict the activity in the fodder growing in the pasture at a given time after the fallout was deposited, assuming that only the soil becomes contaminated, and that the future activity of the fodder is due to uptake from the soil. We can then identify

---

[2]The degree of equilibrium may be difficult to determine in the field, and the use of the concentration ratio is therefore associated with rather large uncertainties.

two compartments: fodder plants and topsoil. A fraction of the caesium that enters the soil may be fixed in clay minerals, and will therefore not be available for uptake by the fodder plants. This can be taken into account by defining two soil compartments: one containing available caesium, and the other, fixed caesium. Note that the physical counterpart to both these compartments is the soil. The resulting compartment model is illustrated in Figure 3.4.

Example 3.2 *A model for the transfer of iodine to milk*

A number of models have been developed that attempt to describe the transfer of radionuclides from grass to milk (see e.g. Assimakopoulos et al. 1989, Crout and Voigt 1996). In this example, we will discuss a rather simple model for the transfer of iodine which takes some aspects of cow physiology into account (Nilsson 1981). The structure of this model is illustrated in Figure 3.5.

Intake of ^{131}I is assumed to take place by ingestion of contaminated grass, which then enters the cow's stomach. The rate of intake can be modelled for realistic conditions, and depends on the release rate and atmospheric conditions. However, in this example, we will assume that 37 GBq is ingested on the first day. The blood is divided into two compartments, in order to describe the behaviour of iodine in two different chemical forms. The compartment Blood 1 represents inorganic iodine and Blood 2 represents organic iodine. This is an example of the use of compartments to represent not only specific organs, but also different forms of an element in the same organ.

The transfer coefficients in this model are given in Table 3.1, in the unit d^{-1}, together with their abbreviations in the model. Radioactive decay is taken into account by adding the decay constant, λ, 0.086 d^{-1}, to each transfer coefficient. However, for the end compartments, urine and faeces, the decay is explicitly modelled by setting the transfer coefficients equal to the decay constant.

TABLE 3.1 Transfer coefficients used to predict the ^{131}I activity in milk from cows grazing in a contaminated area. The abbreviations of the transfer coefficients are also given.

From	To	Abbr.	Value (d^{-1})
Stomach	Blood 1	S-B1	0.950
Stomach	GI tract	S-GI	0.432
Blood 1	Thyroid	B1-T	0.280
Blood 1	Udder	B1-U	0.200
Blood 1	GI tract	B1-GI	0.518
Blood 1	Kidneys	B1-K	0.371
Thyroid	Blood 1	T-B1	0.066
Thyroid	Blood 2	T-B2	0.043
Udder	Milk	U-M	2.00
GI tract	Faeces	GI-F	1.00
Blood 2	GI tract	B2-GI	1.00
Blood 2	Kidneys	B2-K	0.033
Kidneys	Urine	K-U	3.00

This model results in a number of coupled differential equations, in which the activity of ^{131}I in the thyroid, for example, is given by

$$\frac{dA_{Thyroid}}{dt} = B1\text{-}T \cdot A_{Blood1} - T\text{-}B1 \cdot A_{Thyroid} - T\text{-}B2 \cdot A_{Thyroid}. \qquad (3.13)$$

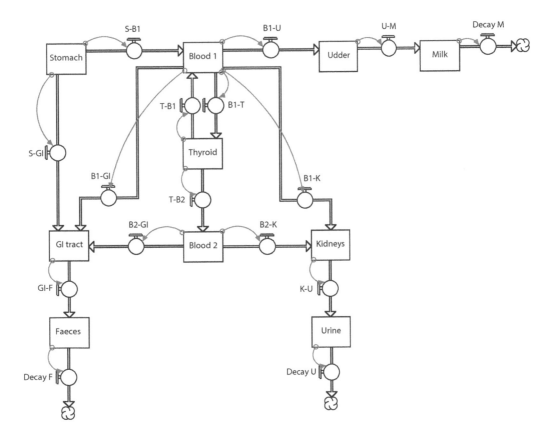

FIGURE 3.5 Structure of a model used to predict the activity in milk at different times after the intake of radioactive iodine. The abbreviations of the transfer coefficients are explained in Table 3.1.

This equation can only be solved if A_{Blood1} is known, and hence all the differential equations have to be solved simultaneously. The details of the numerical procedures available for this are, however, beyond the scope of this text, but can be found in most textbooks on numerical methods. Solving the equations to find the activity in milk results in the data shown in Figure 3.6. From this figure it is evident that the maximum activity in the milk is found about three days after intake. The activity in milk then amounts to about 10% of the total ingested activity.

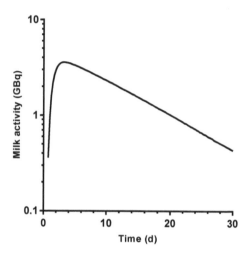

FIGURE 3.6 The time dependence of ^{131}I activity in milk, calculated using the model shown in Figure 3.5.

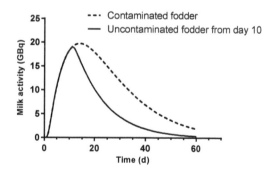

FIGURE 3.7 The time dependence of ^{131}I activity in milk with a continuous input, when the grazing cattle were instead fed uncontaminated fodder from day 10 after the release.

In the case of an accidental release of iodine into the environment, it is possible to decrease the activity in milk by feeding the cattle uncontaminated fodder. For this example, we will assume that the input to the model in Figure 3.5 is a function of time, where the initially deposited ^{131}I decreases only by radioactive decay. The input (GBq) is then given by the expression

$$Input = 37 \cdot e^{ln2 \cdot t/8.02} \tag{3.14}$$

with the time, t, given in days. The resulting activity in milk is shown in Figure 3.7. This figure also shows the activity in milk when uncontaminated fodder is given to the cows from day 10 after the release.

These model calculations are very valuable in determining when the level of contamination of the milk can be considered to be below the action level for consumption.

In a more comprehensive model describing the transfer of a radionuclide through a particular type of ecosystem, the expression for the time dependence of the content in each of the compartments can be calculated using the general equation for an n-compartment model:

$$\frac{dq_i}{dt} = -q_i\left(t\right)\sum_{\substack{j=1\\j\neq i}}^{n} k_{j,i} + \sum_{\substack{j=1\\j\neq i}}^{n} k_{i,j}\cdot q_j\left(t\right) + ex_i, \tag{3.15}$$

where q_i is the concentration or quantity of the substance in compartment i at time t, $k_{j,i}$ is the transfer coefficient for transfer to compartment j from compartment i, $k_{i,j}$ is the transfer coefficient for transfer to compartment i from compartment j, and ex_i is the inflow to compartment i from outside the modelled system. This equation simply expresses a mass balance relation.

We now consider all the flows contributing to inflow of the radionuclide into a given compartment, i, as illustrated in Figure 3.8. There will be at most $i-1$ inflows from the other compartments. If we consider all the individual flows in this system, which can at most be $n\cdot(n-1)$, a matrix can be formed in which each element represents the value of the flow from each compartment to every other compartment in the model, as shown in Figure 3.9. When all possible compartments have been identified, the user must decide which are the most important flows.

Many flows will be physically impossible, or insignificant, and need not be considered in the model. Although computer capacity no longer places severe restrictions on calculations, computing time can be reduced and the robustness of the model improved if insignificant flows and compartments are identified and omitted as early as possible in the modelling process. This is referred to as model reduction. Some simplifications of reality must also be made in order to obtain a model that is implementable, especially when modelling biospheric ecosystems which, like many biological systems, are highly complex. However, if one of the significant transfer coefficients is inaccurately determined, the model may not represent the real situation. For this reason, it may be necessary to compare the model output with data measured over a period of time so as to ensure that the transfer coefficients have been reasonably well determined.

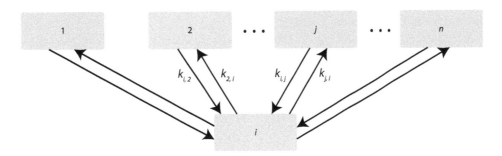

FIGURE 3.8 Flows to and from a compartment in a multi-compartment model.

j \ i	1	2	$\cdots$	n
1	–	$k_{1,2}$	$k_{1,i}$	$k_{1,n}$
2	$k_{2,i}$	–	$k_{2,i}$	$k_{2,n}$
$\cdots$	$k_{j,1}$	$k_{j,2}$	$\cdots$	$k_{j,n}$
n	$k_{n,1}$	$k_{n,2}$	$k_{n,i}$	–

FIGURE 3.9 Matrix showing the transfer coefficients describing the flows between compartments in a multi-compartment model. The transfer coefficients $k_{j,i}$ describe the transport *to* compartment *j from* compartment *i*.

The purpose of a radioecological model is often to predict the radionuclide content in a particular species or material. This can be expressed as the content of the end compartment, N_{end}, which will be related to the content of all the other compartments in the model, and will thus be a function of the model's transfer coefficients $k_{j,i}$. Some examples of the tools and strategies used to optimize radioecological models, in terms of how well the model predicts observed data for a particular end compartment, are presented below.

3.1.2 Variance and Sensitivity of a Compartment Model

When solving a model using transfer coefficients randomly sampled from their individually assigned distributions, a single simulation run will yield single end values in each of the compartments. This end value may, for example, be the activity concentration in freshwater fish from a contaminated lake, which could be used to calculate the absorbed dose to humans when the fish is consumed. However, this does not provide much information unless the simulation is repeated with different parameter values, as many as 1000–1 000 000 times. When the values for compartment i are plotted versus the corresponding values for compartment j, as shown in Figure 3.10, the distribution of the results can reveal a great deal of useful information. The *covariance*, c_{ij}, can be identified by determining the (Pearson) correlation coefficient, r, in the scatter plot of these data, since the correlation coefficient and the covariance are related by

$$r = \frac{c_{ij}}{s_i \cdot s_j}, \tag{3.16}$$

where s_i and s_j are the standard deviations of the distributions of the activity concentration values in compartment i and j, respectively.

A correlation coefficient close to zero indicates that there is no linear relation between the two compartments, which also suggests that the flows connecting them act independently and cannot be simplified. On the other hand, if the correlation coefficient is close to unity, this is a strong indication that there is a covariance or co-dependence between the compartments. A correlation close to unity also means that one of the two compartments is probably redundant in determining the end output, since the values in one compartment simply reflect the values in the other.

Using this procedure it is possible to identify compartments that are not important in determining the radionuclide content, although they may have physiological, biological or ecological relevance.

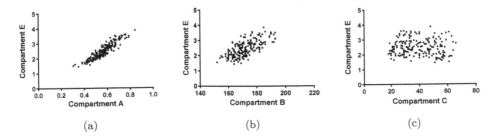

(a)   (b)   (c)

FIGURE 3.10 In (a) the values in compartment E are well correlated with the values in compartment A, indicating that there is a strong linear relation between the activity concentrations in the two compartments. In (b), the correlation between the results obtained in compartments E and B is weaker than in (a), but the activity concentration in the compartments may still be regarded as being linearly related. The scatter plot in (c) indicates that there is no linear relation between the values in compartment E and compartment C.

Example 3.3 *Sensitivity analysis of the model for calculating the transfer of iodine to milk*

In this example, each of the transfer coefficients in the model for the transfer of iodine to milk (Fig. 3.5) was assigned a probability density function (PDF) instead of a single value. The PDF used for each transfer coefficient was a normal distribution with a mean equal to the value given in Table 3.1, and a standard deviation of approximately 20% of the mean. The differential equations are then solved by randomly choosing a value for each transfer coefficient using their respective PDFs. These calculations are repeated 1000 times and the correlation between the variables in the model is calculated.

One way of visualizing the correlations is a tornado chart, which has been so named because of its resemblance to a tornado. Correlations between the activity in milk and the transfer coefficients in this example are shown in Figure 3.11.

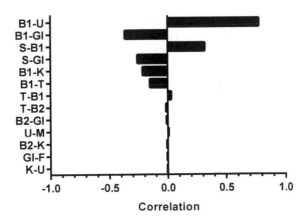

FIGURE 3.11 The dependence of the activity in milk on the transfer coefficients, shown in a tornado chart.

It is clear from Figure 3.11 that the transfer coefficient B1-U is strongly correlated to the activity in milk.

The correlation is positive, which means that an increase in the transfer coefficient will result in an increase in the activity in the milk. The transfer coefficient B1-GI also has a considerable impact on the activity in the milk, although in this case the correlation is negative. The activity in the milk will thus increase as B1-GI decreases.

A *sensitivity analysis* can be also performed by keeping all, or most of, the transfer coefficients constant, apart from the coefficient being studied, $k_{x,y}$. By varying the value of $k_{x,y}$ it is possible to visualize how sensitive the output of the model is to the radionuclide content in the end compartment. It is also important to determine whether there are *interactions* between any of the transfer coefficients.

To describe interactions we first consider a model consisting of a number of variables, X_1, X_2, ...X_n, predicting an outcome variable Y (in our case the content in the end compartment). If all variables except X_i are kept fixed, and X_i is changed by ΔX_i, the outcome will change by an amount ΔY_i. Similarly, if X_j is changed by ΔX_j, the outcome will change by ΔY_j. If there is no interaction between the two variables, a simultaneous change in both X_i and X_j, will give a change in the output variable of $\Delta Y_i + \Delta Y_j$. If there is an interaction between the variables, the output will change by a different amount, which can be written

$$Y(X_1, X_2, \ldots, X_i + \Delta X_i, X_j + \Delta X_j, \ldots, X_n) =$$
$$= Y(X_1, X_2, \ldots, X_i, X_j, \ldots, X_n) + \Delta Y_i + \Delta Y_j + c \cdot \Delta X_i \cdot \Delta X_j. \tag{3.17}$$

If the constant c is a positive scalar, the interaction is said to be positive, and a simultaneous change in the variables may have a multiplicative effect on the change in Y. If $c < 0$, the interaction between the two variables is negative, reducing the change in Y, and if $c = 0$ there is no interaction between the two variables. Identifying interactions between transfer coefficients is essential when constructing a realistic and robust radioecological model.

The important topic of (a priori) *identifiability* of the parameters in a compartment model should also be mentioned (Jacquez 1999). In general, a necessary condition for an unambiguous solution is that the number of independent observations is equal to, or greater than, the number of parameters in the system of differential equations. However, this is not a sufficient condition, since the possibility of achieving an unambiguous solution also depends on the structure of the model. If it were possible to make measurements without any uncertainties, would the experimentally obtained values be sufficient to determine the parameters in a model of the experiment? A detailed discussion on this topic is beyond the scope of this book, but we can exemplify the question using the model illustrated in Figure 3.12.

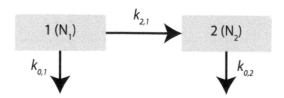

FIGURE 3.12   A simple model illustrating the concept of parameter identifiability.

If several samples are taken at different times from compartment 1, which initially contained a number of atoms, N_0, of a radioactive substance, the observed content, given as the number of atoms, N_1, would be time-dependent, as given in Eqn. 3.18.

$$N_1(t) = N_0 \cdot e^{-(k_{0,1}+k_{2,1})t} \tag{3.18}$$

Sampling from compartment 1 will thus only provide information on the sum of $k_{0,1}$ and $k_{2,1}$, and neither parameter will be uniquely defined. Although the sum is identifiable, the values of the individual parameters are not. The parameter $k_{0,2}$ has no effect on the outcome of the experiment and cannot be determined unless samples are also taken from compartment 2. Programs that use curve-fitting algorithms to estimate parameter values from experimental data may give estimates of both parameters, but since these values are not unique, it is advisable to re-run the curve-fitting algorithm with different initial values. A different outcome will then indicate that the parameters are not identifiable, or, if they are in fact identifiable, that the model is sensitive to the choice of initial values in the curve-fitting algorithm.

3.2 THE ATMOSPHERE

3.2.1 Composition and Circulation Patterns

The atmosphere is composed of a mixture of gases, as well as impurities in the form of particulates that can be of natural origin or the result of human activity. Most of the atmosphere, 99.9% by volume of dry air, consists of nitrogen (78%), oxygen (21%), argon (0.93%) and carbon dioxide (0.039%), together with other minor constituents. The proportions between these gases persist to an altitude of about 100 km, while the amounts of the gases will decrease with altitude because the pressure drops as the strength of the earth's gravitational field is reduced. Above 100 km the proportions of the gases change, and a larger proportion of the lighter elements and molecules are present at these higher altitudes.

The composition of the air results in a molecular mass of 28.97 g mole^{-1} for dry air at standard temperature and pressure (STP), i.e., 0 °C and $1.013 \cdot 10^5$ Pa (760 mm Hg, 1013 mbar). The elements found in the atmosphere originate from the earth's interior and from life on its surface. For example, O_2 is supplied via photosynthesis in plants, and N_2 is a product of metabolism.

The total mass of the atmosphere has been estimated to be $51.5 \cdot 10^{17}$ kg, including about 0.5% by volume water vapour, which is the component of the atmosphere that is the most variable, and which has the greatest influence on the atmosphere's thermodynamic properties. The concentration of water vapour is lowest, about 0.001%, in the colder regions of the atmosphere, and can reach 5% in the tropics. Both water vapour and CO_2 are important *greenhouse gases* that control the temperature of the earth's surface by regulating the balance between energy radiating inwards and outwards through the atmosphere. These gases are also essential for sustaining life on earth, since C, O, and H are the building blocks of living organisms.

The density of dry air at STP is about 1.3 kg m^{-3}. The pressure decreases with increasing altitude, and at a height of 50 km the pressure is only 101.3 Pa (0.76 mm Hg), i.e., about one thousandth of the pressure at sea level. Above an altitude of 7 km the partial pressure of oxygen is too low to sustain life for anything more than short periods. The temperature also varies with altitude and, as will be shown below, the temperature gradient has a significant effect on the fate of airborne radionuclides and other pollutants. Figure 3.13 schematically shows the temperature gradient in the atmosphere and the subdivision of the atmosphere into different layers.

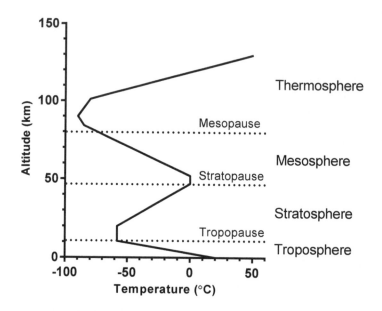

FIGURE 3.13 The variation in temperature with altitude in the atmosphere (here shown schematically) defines the four atmospheric layers: the troposphere, the stratosphere, the mesosphere and the thermosphere. The boundaries between these layers are called the tropopause, the stratopause and the mesopause. However, this is not the only way of characterizing the atmosphere. The troposphere can be divided into the *atmospheric boundary layer* and the *free atmosphere*, where the former is affected by friction with the surface of the earth, and extends from a few hundred metres up to one kilometre in height. The atmosphere can also be divided into the *homosphere* and the *heterosphere*, where the latter refers to the stratification of light and heavy elements and molecules above an altitude of 100 km. The part of the atmosphere above about 80 km is also called the *ionosphere* due to the ionization of atoms and molecules by radiation from the sun and cosmic radiation. The mean free path of ions and electrons in the ionosphere is long enough to prevent recombination.

The part of the atmosphere closest to the earth's surface is called the *troposphere* and contains most of the atmosphere's mass, as well as most of the humidity and suspended particles. The troposphere extends to altitudes between 6 and 18 km, being higher at the equator than at the poles, where its height is usually less than 10 km. In this layer, the temperature normally decreases with altitude due to heating of the earth's surface by the sun, and subsequent convection.

The upper boundary of the troposphere is called the *tropopause*, and above this we find the *stratosphere*, in the lower part of which the temperature remains almost constant with altitude, but then begins to rise, because of absorption of radiation from the sun. Absorption and heating at the earth's surface do not affect the temperature of the stratosphere. The highest temperatures are found at an altitude of around 50 km, mostly due to absorption of ultraviolet radiation from the sun by the ozone layer. The *stratopause* marks the boundary with the *mesosphere*. Neither the mesosphere nor the thermosphere are of interest for the transport of radioactive material.

The air in the lower atmosphere moves as a result of uneven heating by solar radiation. On the local scale this will lead to *sea breezes* and *land breezes*, which are due to convection and advection[3] near the coastline. Because the land has a much lower heat capacity than bodies of water, solar radiation warms up the land rather quickly, while the water temperature remains more or less constant. The ground then heats the air above it, which expands and rises (convection), and is replaced by cooler air flowing from the sea (advection), giving rise to a sea breeze. At sunset, the ground cools relatively rapidly, while the water maintains an almost equal temperature. The warmer air above the water rises and is replaced by cooler air from the land, and a land breeze results (Fig. 3.14).

Monsoons are a form of sea breeze or land breeze on a larger scale. These are common, for example, in the Indian Ocean, where hot regions in Central Asia cause a widespread depression due to rising air. This air is then replaced by colder, moist air from the Indian Ocean, leading to the *Southwest monsoon*, which brings large amounts of precipitation in the summertime. In the wintertime, conditions are reversed: warmer air from the ocean rises and is replaced by colder air from Central Asia, giving rise to the *Northeast monsoon*.

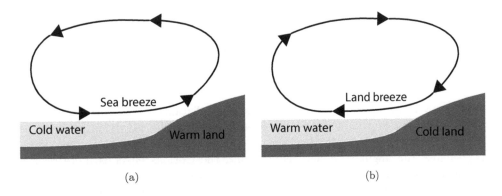

(a)   (b)

FIGURE 3.14   (a) A sea breeze blows towards the shore because of rising warm air above the ground, which is heated by the sun during the daytime. (b) During sunset, a land breeze will blow away from the shore as warmer air rises over the water.

As illustrated above, temperature gradients lead to the transport of air masses in the lower atmosphere. This relocation of air also leads to pressure gradients. For example, ascending air gives rise to a depression. Differences in air pressure tend to even out by a net transport of air from areas of high pressure to areas of low pressure. Because of the rotation of the earth, the wind does not flow perpendicular to lines of equal air pressure, or isobars, but is instead directed along these curves. This is caused by the *Coriolis effect*. In the absence of friction between the air and the earth's surface, the wind would blow parallel to the isobars.

The Coriolis effect can be qualitatively understood by considering a region of low pressure, as shown in Figure 3.15. On a non-rotating earth, the air would flow perpendicular to the isobars in order to fill the area of low pressure and even out the pressure. However, apart from small deviations, the atmosphere exhibits the same angular velocity as the earth itself, and so, in the Northern Hemisphere, a packet of air moving from south to north will have a greater velocity to the east than the air it encounters. This is indicated by the velocity v_{ref} in Figure 3.15, representing the difference in velocity. The final effect of this easterly component is that the wind blows almost parallel to the isobars.

The velocity, v, of the atmosphere as it follows the rotating earth is related to the angular

---

[3]Convection is often used to describe the *vertical* transport of a liquid or a gas due to density differences. Horizontal transport is usually called *advection*. Sea breezes and land breezes are examples of advection.

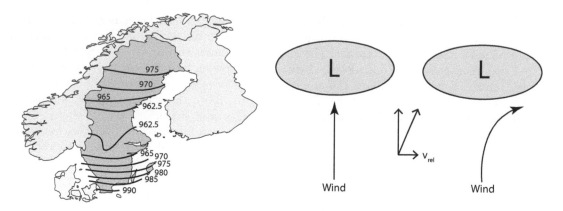

FIGURE 3.15 Isobars indicating the air pressure in mbar during a storm in Sweden in 2005. To the south of the low pressure, westerly winds with velocities over 40 m s^{-1} were encountered. The direction of the wind is depicted schematically in the right-hand part of the figure, assuming that the earth does not rotate (left) and with rotation (right). Data from the Swedish Meteorological and Hydrological Institute.

velocity, ω, by $v = r\omega$, where r is the perpendicular distance to the earth's axis of rotation (see Fig. 3.16). For a given distance travelled by the air, the reduction in r will be larger if the air is further to the north (i.e. the "steps" in Fig. 3.16 become increasingly larger). Thus, because the angular velocity is constant, the Coriolis effect is more prominent at higher latitudes, while the effect at the equator is zero.

Applying the above arguments to a packet of air in the Northern Hemisphere approaching a region of low pressure from the north, we find that the air is deflected to the west due to the Coriolis effect. However, this explanation cannot be applied to air moving to the east or west, which is also affected by the Coriolis force. A stringent treatment, taking into account the physical conservation laws in the rotating system (conservation of energy, linear momentum, angular momentum, and mass) would require quite advanced vector algebra. This is beyond the scope of this text, and the reader is referred to textbooks on classical mechanics. For the sake of a qualitative understanding of the air transport in the longitudinal direction, we can consider a parcel of air, moving to the east in the Northern Hemisphere; Figure 3.17. The eastward velocity of the air parcel adds to the eastward velocity of the earth, giving the air parcel an extra centrifugal force. The centrifugal force is directed outwards in the plane of rotation, parallel to r in Figure 3.17, and can be decomposed into an outward component, A, perpendicular to the earth's surface, and a component, B, in the southwards direction. Thus, air moving eastwards in the Northern Hemisphere is deflected southwards. In the Southern Hemisphere, the deflections are reversed: an eastward velocity gives rise to a northward deflection and a northward velocity gives rise to a westward deflection.

In many situations, for example, when dealing with the release of radionuclides, it is very important to use the same convention when describing the direction of the wind. To avoid confusion, the *wind direction* should always be given as the direction *from* which the wind blows. Thus a northerly wind is one that blows from north to south.

Global atmospheric air flows can be divided into six main wind belts, determined by the interaction between solar heating and the Coriolis effect. The positions of these belts show a seasonal dependence; moving north during the Northern Hemisphere's summer and south during its winter. The wind belts are also roughly related to the climate zones: i.e.

Axis of rotation

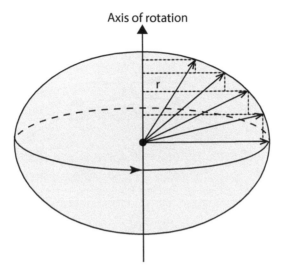

FIGURE 3.16 The perpendicular distance, r, to the earth's axis of rotation will decrease at higher latitudes. This change will be greater for a given increase in latitude at a higher latitude than when close to the equator.

the *tropics, subtropics, temperate zone* and the *frigid zone*. The tropics are located between the *Tropic of Cancer* (about 23° N) and the *Tropic of Capricorn* (about 23° S). Between these latitudes and 38° N and S, we find the subtropics, followed by the temperate zones, which end at the Arctic and Antarctic circles (about 66°). The zones near the poles are called the frigid zones.

Solar heating will cause a horizontal, as well as a vertical, flow of air in the lower atmosphere by driving convection and causing pressure gradients. Heated air at the equator rises, and is replaced by air from a latitude of around 30° in each hemisphere. However, due to the earth's rotation, the air is deflected towards the west in both the Northern and the Southern Hemispheres, see Figure 3.18. These north-easterly and south-easterly winds are

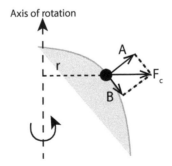

FIGURE 3.17 Air moving eastward in the Northern Hemisphere will be subject to an additional centrifugal force, F_c, which can be decomposed in the components A and B. Component B is responsible for the southward deflection of the air, while component A has been shown to slightly reduce its weight.

called the *Trade Winds* because they were once important for trading vessels powered by sail.

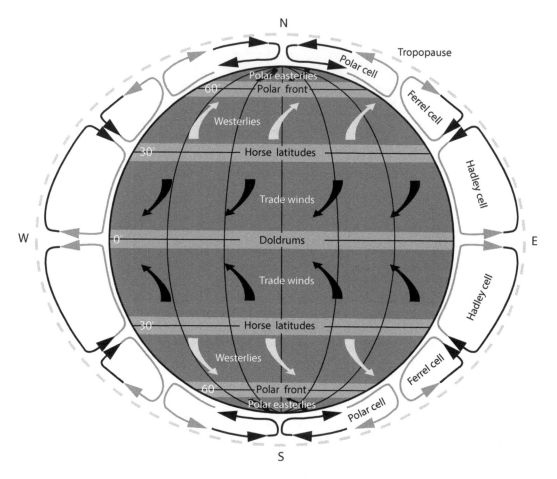

FIGURE 3.18 The figure shows the prevailing wind directions and the corresponding vertical circulation cells in the troposphere. Cold air is depicted by pale grey arrows and warm air by black arrows.

Air that rises at the equator will then flow northwards and southwards, until it starts to sink at about the latitudes of the Tropics of Cancer and Capricorn. This gives rise to a belt of large circulation cells encircling the earth, called *Hadley cells*. The downward vertical motion in Hadley cells is an important pathway for radionuclides in the atmosphere.

However, not all the air that descends in the Hadley cells continues towards the equator. Some of it flows northwards, and some southwards, into the Northern and Southern Hemispheres and, as a result of the Coriolis effect, these flows are deflected eastwards. Such winds, found between latitudes of around 60° and 30°, are called the *Westerlies*. These winds can reach considerable speeds, especially in the Southern Hemisphere, and the zone there that lacks continental land masses which might dissipate some of the wind's energy is referred to as the Roaring Forties.

At about 60° latitude, the air starts to rise again, forming two additional vertical circulation cells called the *Ferrel cell* and the *Polar cell*. The cold air flowing downwards in the Polar cell gives rise to northerly winds that are deflected eastwards—the *Polar easterlies*.

The upward vertical movement of air between the Ferrel and the Polar cells gives rise to a low-pressure region, wandering along the Polar front. This low pressure causes the above-mentioned Westerlies, when air attempting to even out the low pressure is deflected by the Coriolis effect.

The winds at the borders between the wind belts are usually very weak and changeable. At the descent of the Hadley cells, we find the so-called *Horse latitudes*, characterized by dry air and long periods of lull. The corresponding area at the equator is called the *Doldrums*.

The above description of winds and air flow is relevant as long as we consider the lower atmosphere, the troposphere. Close to the tropopause, however, we find much stronger winds, called *jet streams*. These are narrow, meandering, westerly air streams, reaching speeds of around 25–110 m/s, which should be compared with the mean speed of the Trade Winds of 5–6 m/s. The *polar jets* flow at an altitude of around 7–12 km at 30–60° latitude, in both the Northern and the Southern Hemispheres. The *subtropical jets*, which are also present in both hemispheres, are weaker and flow at a higher altitude, about 10–16 km. These jets are located close to a latitude of 30°.

3.2.2 Atmospheric Stability

The temperature gradient in the troposphere affects atmospheric stability and determines how the distribution of a radioactive release develops with time. For example, a continuous airborne release, a *plume*, will essentially retain its shape in a stable atmosphere, while in an unstable atmosphere it will be dispersed.

Atmospheric stability can be understood by considering the movement of a small packet of air. As it rises, it expands because of the decrease in pressure. This expansion lowers the temperature of the air packet, since the energy required to push aside the surrounding air will be taken from the air packet itself. Similarly, the temperature of a falling air packet increases by the same mechanism. If we assume that heat exchange between the air packet and its surroundings is negligible, this can be considered an *adiabatic* process. If the air is dry, it can be shown that the temperature decreases by about 10 °C per km increase in altitude, and this is called the *dry adiabatic lapse rate*. Moist air will have a different lapse rate because of the energy expended in changing the phase of the water. In an elevated moist air packet, the condensation of water vapour will release energy and the lapse rate will be lower.

This condensation of water vapour gives rise to cumulus clouds, characterized by a flat base at the altitude where the vapour starts to condense. The fluffy upper part of the cloud is due to continuing condensation, but the altitude at which all the vapour has condensed is randomly determined by the vertical wind speed, the moisture content and the degree of turbulence. The water droplets forming the cloud actually fall due to gravity, but since the droplets move downwards relative to the surrounding air, the cloud can still persist if the vertical speed of the air is greater than the speed of the droplets. For an observer on the ground, the cloud will then seem to be stationary or even rising.

An example of an adiabatic process is the *Föhn*, a warm wind which blows down a mountain or hillside. The air is very dry and may be considerably warmer than the surrounding air, which may seem surprising since it originates on a cold mountain top. However, as the air moves down the hillside, it encounters an increasing pressure gradient and is adiabatically compressed. The temperature increases as a result of this compression and, eventually, the air becomes warmer than the surrounding air. The air is dry, having lost most of its humidity as it ascended the other side of the mountain.

By using the first law of thermodynamics (conservation of energy), which states that the heat exchanged must be equal to the sum of the internal energy of the air packet and the external work done by (or on) the packet when it expands (or contracts), it can be shown that the dry adiabatic lapse rate, Γ_d, is given by (Marshall and Plumb 2008)

$$\Gamma_d = \frac{g}{c_p}, \tag{3.19}$$

where g is the acceleration due to gravity, and c_p is the specific heat of dry air at constant pressure (1005 J kg^{-1} K^{-1} in dry air at STP). Note that Γ_d is defined as $-dT/dz$ (in units of K m^{-1}), where T (in K) is the temperature and z is the altitude (in m). The *saturated adiabatic lapse rate*, of moist air, Γ_s, can be calculated by taking the latent heat released by condensation into account.

On a sunny day, the earth absorbs energy from the sun and heats the air close to its surface. This may lead to a more rapid temperature decrease than that given by the adiabatic lapse rate. This decrease is described by a *super-adiabatic lapse rate*, and is shown schematically in Figure 3.19. Consider an air packet that is elevated from altitude A_1 to altitude A_2. During elevation, the air packet expands and cools adiabatically in accordance with the adiabatic lapse rate. However, when the packet reaches altitude A_2 it will be warmer than the surrounding air (because the atmosphere will have cooled according to the super-adiabatic lapse rate) and so the packet will continue to rise. Using the same argument, we find that an air packet that descends from altitude A_2 to altitude A_1 will be cooler than the surrounding air, and will therefore continue to fall. This means that vertical air movements are amplified when the atmosphere is super-adiabatic.

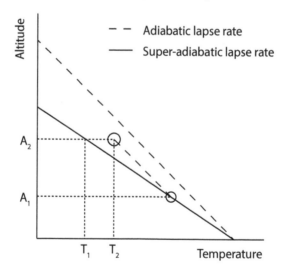

FIGURE 3.19 Illustration of an unstable atmosphere. An air packet that is elevated from altitude A_1 to altitude A_2 will expand and the temperature will decrease in accordance with the adiabatic lapse rate. Since the temperature of the atmosphere is assumed to decrease faster than the adiabatic lapse rate, the temperature of the air packet at altitude A_2, T_2, will be higher than the temperature of the surrounding air, T_1. The air packet will thus continue to rise because of its lower density.

The decrease in temperature with altitude may also be less than the adiabatic lapse rate, i.e., *sub-adiabatic*. The temperature may even increase with increasing altitude, and when the temperature gradient is positive in this way an *inversion* is said to exist. This may be caused by warm air passing over cold air, or by cold air descending to a lower altitude (for example, in a sea breeze). An inversion may also be caused by the earth's surface cooling faster than the atmosphere after sunset. During the night, the earth's surface cools gradually, and may result in a positive temperature gradient, as is schematically depicted in Figure 3.20. An air packet that rises from altitude A_1 to altitude A_2 will expand and cool adiabatically, but this time the packet will be colder than the ambient atmosphere at altitude A_2, and will start to fall. Similarly, an air packet that falls will be warmer than the surrounding atmosphere, and will start to rise. An inversion thus results in very stable conditions in the atmosphere.

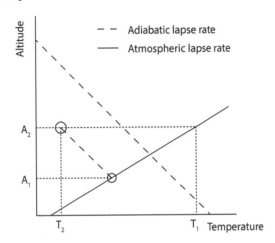

FIGURE 3.20   Illustration of a stable atmosphere (inversion). An air packet that rises from altitude A_1 to altitude A_2 will expand, and its temperature will decrease according to the adiabatic lapse rate. Since the temperature in the atmosphere during an inversion is assumed to increase with altitude, the temperature of the air packet at the higher altitude, T_2, will be lower than the temperature of the surrounding air, T_1. The air packet will thus descend again due to its higher density. This will tend to stabilize the atmosphere.

A pollutant such as a radionuclide that is discharged into the atmosphere is carried by the wind, and will be dispersed by turbulent diffusion in the *boundary layer*, or *mixing layer*, which is the part of the lower atmosphere where the viscosity of the air has a significant impact on air movement. Above the boundary layer, air flows almost parallel to the isobars, while surface drag at lower altitudes forces it to flow more or less perpendicular to the isobars. This surface drag, which can be quantified by the *surface roughness*, will cause a gradual decrease in wind speed, giving rise to a vertical wind speed profile that can be described by a power function. Surface roughness will be further discussed below in connection with dry deposition processes in Section 3.2.4.

The boundary layer is affected by heat transport, as well as the transport of moisture and momentum from the earth's surface, and is characterized by rapid fluctuations in air movement (turbulence). Turbulence causes considerable vertical mixing. The height of the boundary layer, the *mixing height*, is an important parameter when modelling the concentration of radionuclides in a plume and their deposition. However, the height of the boundary layer varies, and is typically some tens to hundreds of metres high during the night. During the daytime, its height can be between a few hundred and one or two thousand metres. In the case of an inversion, the height of the boundary layer coincides with the bottom of the inversion layer.

When radioactive elements are released over an extended period of time, it is very important to consider temporal variations in the stability of the atmosphere. For example, a stable atmosphere during the night-time (no clouds) may develop into neutral or unstable conditions during the daytime, as the sun starts to heat the earth's surface, while at the end of the day, the atmosphere may return to stable conditions. Release will then be affected differently at different times of the day and night. These variations in atmospheric stability are also transferred to higher altitudes, thereby creating a "memory" of the conditions previously prevailing at lower altitudes. Wind blowing from cold water to warmer land (or vice versa) may also cause a change in stability with altitude. These processes may thus create "patterns" of stability in the vertical direction. However, there will always be a stable limit at some altitude that prohibits the release from ascending further. For this reason, it is important to include a "roof altitude" in any model for atmospheric dispersion. Figure 3.21 shows illustrations of a stack plume under conditions of different degrees of atmospheric stability, and various changes in the lapse rate.

In Case a), the atmosphere cools faster with height than the adiabatic lapse rate and gives rise to instabilities, with enhanced vertical oscillations called *looping*. The vertical spread of the plume allows contaminants to reach the ground even close to the point of release, and momentary bursts of high concentration can rapidly be replaced by air free from the pollutant. In Case b), the decrease in temperature with height is smaller than the adiabatic lapse rate, which leads to a more stable atmosphere, and the plume is more constrained; this is called *coning*. However, it is not constrained to the same extent as in Case c), called *fanning*, where the temperature increases with height (an inversion). Fanning occurs during stable conditions and is characterized by very slow vertical dispersion.

Case d) describes the combination of two different temperature profiles: an inversion that extends to the stack height, where it is replaced by a super-adiabatic lapse rate. The opposite situation is depicted in Case e). These two plume types are denoted *lofting* and *fumigation*, respectively. Lofting is a rather advantageous situation, since the inversion prevents the pollutants from reaching the ground. However, lofting may act as a precursor to fumigation. Fumigation also results in rapid mixing of the pollutants that have accumulated at a certain height during a period of stability. This is common after sunrise, when the inversion that prevailed during the night dissipates rapidly due to solar heating of the ground.

The location of the change in lapse rate in relation to the release height is important for the development of lofting and fumigation. For lofting to occur, the change in lapse rate should be located below the release height, otherwise, the release will experience a stable atmosphere and behave like the plume in Case c). For fumigation, the situation is reversed. In this case, the change in lapse rate must occur above the release height. If it does not, the plume will once again behave as in Case c) since the pollutants are released into a stable atmosphere.

Atmospheric stability is usually described by *Pasquill stability classes* (or Pasquill–Gifford stability categories), denoted A–F. Class A corresponds to the most unstable atmosphere, which can be compared with Case a) in Figure 3.21, while class F corresponds to the most stable condition. From A to F, the classes are: extremely unstable, moderately unstable, slightly unstable, neutral, slightly stable and moderately stable. Table 3.2 lists the classes according to solar radiation and wind speed, which is a common method of defining Pasquill classes.

As can be seen from Table 3.2, stable conditions occur only when the incoming solar radiation is low, for example, when the sky is cloudy or during the night; and even then, only in combination with a low wind speed. During the daytime, the stability can at best be neutral, and increasing the wind speed tends to give more stable conditions, regardless of the solar irradiation.

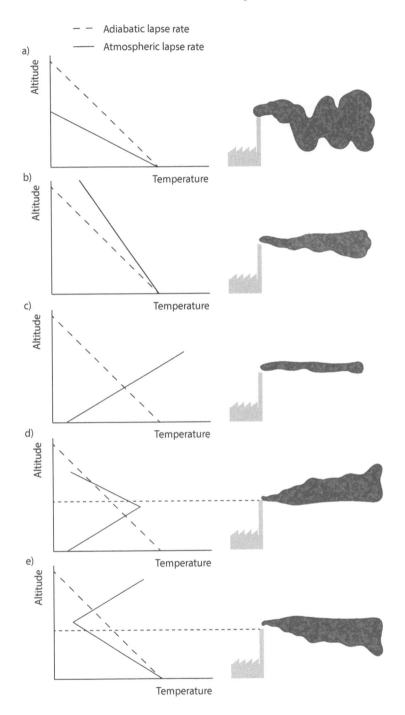

FIGURE 3.21 Appearance of a plume with different temperature gradients in the lower atmosphere. a) Looping (unstable); a super-adiabatic lapse rate will result in an unstable atmosphere, leading to considerable vertical oscillation of the plume. b) Coning (neutral); the rate of temperature decrease is slower (sub-adiabatic) than the adiabatic lapse rate, resulting in a more stable atmosphere. c) Fanning (stable); an increase in temperature with altitude—an inversion—will lead to a very stable atmosphere and constriction of the plume in the vertical direction. d) Lofting (stable below, neutral above); an inversion with a super-adiabatic lapse rate above will enhance the vertical movement of the plume, while there is no movement below the inversion. e) Fumigation (neutral below, stable above); an inversion will prevent the plume from rising. Note the location of the change in lapse rate in relation to the release height in cases d) and e).

186 ■ Environmental Radioactivity and Emergency Preparedness

TABLE 3.2 Pasquill class according to wind speed, u, solar radiation during the daytime, and cloudiness during the night. Solar radiation is characterized as strong, moderate or slight, and cloudiness according to whether more or less than 50% of the sky is obscured. Class G has been added to describe so-called strongly stable conditions.

| Wind speed | Solar radiation | | | Cloudiness | |
u (m s^{-1})	Strong	Moderate	Slight	< 50 %	> 50 %
< 2	A	A–B	B	G	G
2–3	A–B	B	C	E	F
3–4	B	B–C	C	D	E
4–6	C	C–D	D	D	D
> 6	C	D	D	D	D

Source: Data from Slade (1968) and Cooper et al. (2003).

The wind direction and speed can be difficult to estimate since they often vary with altitude and distance from the source. To predict the movement of the plume it is therefore important to have reliable wind data at the altitude of the plume. However, predictions are also complicated by fundamental difficulties in describing turbulence. The variation in wind speed with altitude can be described by a power law

$$u(z) = u_{10} \left(\frac{z}{10} \right)^n,$$
(3.20)

where $u(z)$ is the wind speed at height z m and u_{10} is the wind speed at a height of 10 m. This height is often used as a reference height. The factor n depends on the surface roughness and, to a lesser degree, on the atmospheric stability. Surface roughness is quantified by the *aerodynamic roughness length*, z_0, which can be interpreted as an effective height below which the wind speed profile is affected by the surface. Values of n for a neutral atmosphere are given in Table 3.3. Eqn. 3.20 can thus be used to correct any measurement of the wind speed at the height of the release to the corresponding wind speed at the reference height.

TABLE 3.3 Values used for the calculation of wind speed at source height using Eqn. 3.20 for various surface types, characterized by their aerodynamic roughness length, z_0, and n.

Surface	z_0 (m)	n
Sea	0.0001	0.07
Sandy desert	0.001	0.1
Short grass	0.005	0.13
Open grassland	0.02	0.15
Root crops	0.1	0.2
Agricultural areas	0.3	0.255
Parkland	0.5	0.3
Open suburbia	0.5	0.3
Cities	1.0	0.39
Woodland	1.0	0.39

Source: Data from Clarke (1979).

A useful instrument for predicting wind speed and direction at a given location is the *wind rose*, shown in Figure 3.22. A wind rose can be divided into 8, 16 or 32 sectors, describing the distribution of wind direction and wind speed. The example shown in Figure 3.22 is a 16-sector wind rose. It can be seen from the figure that, at this site, the most probable wind direction is westerly, i.e. *from* the west. The distribution of wind speed can be specified in different intervals, and in this example only four intervals are shown for reasons of simplicity.

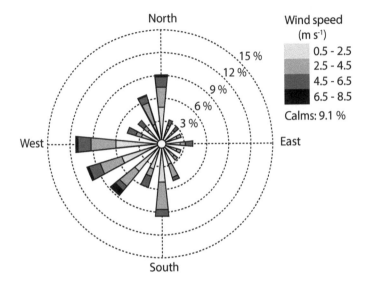

FIGURE 3.22 Illustration of a wind rose showing the frequency of wind speed and wind direction.

When no data are available for wind speed, values can be taken from Table 3.4, which gives typical wind speeds at 10 m height in the various stability categories. This table also gives typical depths of the mixing layer for each stability category.

TABLE 3.4 Typical values of the wind speed at 10 m height, and depth of the mixing layer, for the Pasquill stability categories.

Pasquill class	Wind speed $(m\ s^{-1})$	Depth of mixing layer (m)
A	1	1300
B	2	900
C	5	850
D	5	800
E	3	400
F	2	100
G	1	100

Source: Data from Clarke (1979).

3.2.3 The Gaussian Plume Diffusion Model

A large number of different models of atmospheric dispersion have been developed, with varying degrees of detail in their description of atmospheric phenomena. Discussion of dispersion models in this book will be restricted to the most commonly used model, the *Gaussian plume diffusion model*. For information on other models, and references, the reader is referred to Cooper et al. (2003). The Gaussian plume model assumes that the released pollutant moves with the wind at the mean wind speed, $\overline{u}$. As it moves, diffusion will cause the pollutant to spread out, both horizontally and vertically, forming a plume. The degree of diffusion depends on the atmospheric stability (as described by the Pasquill class), see Figure 3.23.

The degree of dispersion is quantified in the horizontal and vertical directions by the standard deviations, σ_y and σ_z, respectively. This description assumes that diffusion takes place independently in each direction. In addition, diffusion in the direction of the wind (x) is assumed to be much smaller than wind-driven transport of the plume, and is therefore not considered in the model. The standard deviations in y and z are often called the dispersion coefficients.

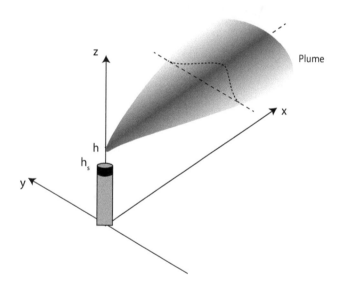

FIGURE 3.23 In the Gaussian plume model, the plume is assumed to travel at the same speed as the wind. The concentration of the pollutant is assumed to be normally distributed in the y and z directions. Note that the model is not valid for very low wind speeds, and at zero wind speed the predicted concentration will be infinitely large.

According to the Gaussian plume model, the concentration, C, of a pollutant at a point (x, y, z) is given by:

$$C(x, y, z) = \frac{Q}{2\pi \cdot \sigma_y \cdot \sigma_z \cdot u_{10}} \cdot e^{-\frac{y^2}{2\sigma_y^2}} \left(e^{-\frac{(z-h)^2}{2\sigma_z^2}} + e^{-\frac{(z+h)^2}{2\sigma_z^2}} \right) \tag{3.21}$$

where Q is the emission rate (or the total amount released), u_{10} is the wind speed 10 m above the ground, σ_y and σ_z are the horizontal and vertical dispersion coefficients, and h is the effective height of the source. The point $(0, 0, 0)$ is a fixed origin on the earth's surface, below the release point. Note that the x-dependence is implicit in the dispersion coefficients, which vary with the distance from the source. If the mean wind velocity is given in m s^{-1} and the standard deviations and height are given in m, the concentration will be given in Bq m^{-3} if Q is given in Bq s^{-1}. If we instead use the total activity released (in Bq), Eqn. 3.21 will give the time-integrated concentration in Bq s m^{-3}; which equals the total number of radioactive disintegrations per m^{-3} at a point (x, y, z) *during the passage of the plume*.

The two exponentials for the dispersion in the z-direction arise from the fact that a plume that spreads vertically will eventually reach the ground ($z = 0$). This is reasonably well modelled by assuming reflection of the plume at the earth's surface, and the resulting dispersion can be considered as a release from a virtual source at a height $-h$ (Clarke 1979). This procedure can also be applied in the case of an inversion. The dispersed pollutant will then be trapped between the ground and the inversion, and the plume can be considered to be reflected at both of these horizontal limits at their specified heights. When the height of the mixing layer, A, is less than the value of the vertical dispersion coefficient, σ_z, the pollutant will be almost uniformly distributed vertically throughout the mixing layer, and Eqn. 3.21 reduces to

$$C = \frac{Q}{\sqrt{2\pi} \cdot \sigma_y \cdot A \cdot u_{10}} \cdot e^{-\frac{y^2}{2\sigma_y^2}}. \tag{3.22}$$

The concentration at ground level ($z = 0$) is given by

$$C = \frac{Q}{\pi \cdot \sigma_y \cdot \sigma_z \cdot u_{10}} \cdot e^{-\frac{y^2}{2\sigma_y^2}} \cdot e^{-\frac{h^2}{2\sigma_z^2}}. \tag{3.23}$$

As is evident from Eqns. 3.21 and 3.22, the concentration is inversely proportional to the wind speed, implying that twice the wind velocity will give half the concentration. This means that the radiation dose will depend on the wind speed, and it is important that this is measured in the case of accidental releases. A fast-moving plume will also give a lower radiation dose because it results in a shorter irradiation time than a slower plume, which spends more time at a given location.

It should be noted that the Gaussian plume model is an idealized model of atmospheric dispersion, and differs from a real plume in many respects. For example, the model assumes that the wind moves at a constant speed in a constant direction. In reality, both the speed and direction can change with time, giving rise to a *meandering* plume. Another problem is that the mathematical representation of the plume using a Gaussian distribution function will overestimate the concentration at large horizontal distances from the centre of the plume. The Gaussian function decreases with distance, but never becomes zero. Therefore, it is recommended that calculations only be made for horizontal distances less than $2\sigma_y$, where the concentration has decreased to about one hundredth of the concentration at the central axis. Over a longer time, the concentration at even greater horizontal distances may be significant because of the effects of changing wind direction. In this case, the Gaussian model with a standard deviation σ_y is no longer a valid description of the transport of the plume.

A modification of the Gaussian plume model for meandering plumes has been proposed by Overcamp (1991), in which the dispersion coefficient σ_y is modified by an empirical parameter that varies with the wind speed. Another approach is described by Slade (1968), in which the fluctuating plume is modelled by distributing plume elements randomly around their mean position. The plume is here treated as a continuous release from disc-shaped plume elements, spreading in the vertical and horizontal directions. The plume elements can be modelled as discs, rather than portions of a sphere, as diffusion in the x-direction is neglected.

The problem of changing wind direction can also be taken into consideration by restricting the calculations to a limited sector. The angle of such a sector is often determined by meteorological data, for example, from a wind rose. The average concentration can be found by calculating the concentration in each of the directions and speeds given by the wind rose, and summing each set of calculations weighted by their frequency of occurrence.

The effective height of the source, h, is given by $h_s + \Delta h$, where Δh is the *plume rise*, i.e. the elevation of the plume above the stack height, h_s (Fig. 3.23). The plume is assumed to rise as pollutants are usually emitted at a temperature above that of the ambient atmosphere. The plume rise is difficult to describe, especially in uncontrolled releases where the emission velocity and stack dimensions may be unknown. Various models have been developed. For example, the plume rise in a neutral atmosphere can be estimated from the *Holland equation* (Abdel-Magid et al. 1996)

$$\Delta h = \frac{v_g}{u}\left[1.5 + 2.68 \cdot 10^{-3} \cdot p \cdot \left(\frac{T_g - T_a}{T_g}\right) \cdot d\right], \qquad (3.24)$$

where v_g is the emission velocity of the gas (m s^{-1}), u is the mean wind velocity at the stack height (m s^{-1}), d is the inner diameter of the stack (m) and p is the atmospheric pressure (mbar). The temperatures (in K) of the emitted gas and the ambient atmosphere are denoted T_g and T_a, respectively.

In order to perform calculations using Eqn. 3.21, we must find numerical values of the dispersion coefficients, σ_y and σ_z. These have been determined experimentally, but can also be calculated from the following empirical expressions

$$\sigma_y = \frac{k_1 x}{\left(1 + \frac{x}{k_2}\right)^{k_3}} \qquad (3.25)$$

$$\sigma_z = \frac{k_4 x}{\left(1 + \frac{x}{k_2}\right)^{k_5}} \qquad (3.26)$$

using the constants k_1–k_5 given in Table 3.5.

TABLE 3.5 The constants used for calculating dispersion coefficients from Eqns. 3.25 and 3.26. Each of the constants should be chosen according to the prevailing Pasquill class.

Pasquill class	k_1	k_2	k_3	k_4	k_5
A	0.250	927	0.189	0.1020	−1.918
B	0.202	370	0.162	0.0962	−0.101
C	0.134	283	0.134	0.0722	0.102
D	0.0787	707	0.135	0.0475	0.465
E	0.0566	1070	0.137	0.0335	0.624
F	0.0370	1170	0.134	0.0220	0.700

Source: Data from Zanetti (1990) and Bualert (2001), presented in Cooper et al. (2003).

Values of σ_y and σ_z calculated from Eqns. 3.25 and 3.26 are shown in Figures 3.24 and 3.25. The dispersion in the horizontal (y) direction with downwind distance is similar for all Pasquill classes, and differs only by a scaling factor. Beyond a few hundreds of metres, the vertical (z) dispersion is strongly dependent on the Pasquill class because of the enhanced vertical movements through the atmosphere discussed above.

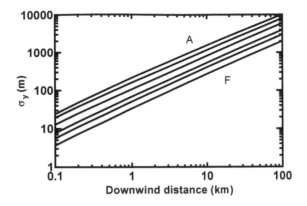

FIGURE 3.24 Dispersion coefficients for horizontal dispersion in the Gaussian plume model, calculated from Eqn. 3.25. The graph shows the variation with distance from the source, and with atmospheric stability, for each of the Pasquill classes A–F.

Having discussed the parameters in the Gaussian plume model, we will now consider some of the model's predictions. We start with the time-integrated concentration (in Bq s m^{-3}) at ground level ($z = 0$), centrally ($y = 0$) along the direction of the plume, following the x-direction. Figure 3.26 shows the time-integrated concentration at various distances from the release point, for the Pasquill categories of atmospheric stability A–F. It is assumed that the activity released is 1 Bq, that the wind speed is 5 m s^{-1}, and that the release originates from a source at ground level ($h = 0$). It is evident that the concentration is highest close to the source for all stability classes. The concentration then decreases, following an approximately straight line on the log–log diagram, to a distance of about 50 km from the source. This approximately power-law decrease is a consequence of how the two dispersion coefficients vary with distance (at least, for stability classes B–E). The concentration thus decreases very rapidly close to the source, but the gradient decreases progressively further away from the source. From a radiation protection point of view, it is therefore more important to evacuate areas close to the source. At larger distances, a much larger area has to be evacuated to achieve a significant decrease in the radiation dose to the population.

Figure 3.27 shows the time-integrated concentration for a source at a height of 100 m. This height can be considered typical of the stack height at nuclear power plants. It is important to realize that the concentration at ground level now reaches its maximum at some distance from the source. This distance varies with the Pasquill stability class, and is greatest for class F, the most stable situation. In this example, the maximum concentration for stability class D is found about 3 km from the source. The distance to the maximum concentration at ground level will increase even more if the source height is increased, as can be seen in Figure 3.28, where the radionuclides were released from a height of 200 m. This can also be the case if the effluent is a warm gas which rises because of its buoyancy. In this case, the distance from the source to maximum concentration at ground level increases to around 10 km, assuming stability class D.

The dose resulting from inhalation in a radioactive cloud will be almost proportional to the distribution of the time-integrated concentration, while the external radiation dose will

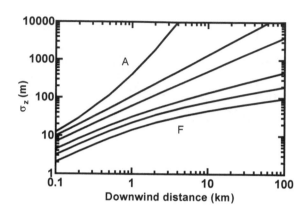

FIGURE 3.25   Dispersion coefficients for vertical dispersion in the Gaussian plume model, calculated from Eqn. 3.26. The graph shows the variation with distance from the source, and atmospheric stability, for each of the Pasquill classes A–F.

follow a somewhat different distribution due to the range of the radiation (a few hundred metres).

Figure 3.29 shows a comparison between data from Figure 3.26 and Figure 3.27, i.e. a release at ground level and a release at a height of 100 m, for Pasquill classes A, D and F. At some distance from the source, the curve describing the dispersion from the source at 100 m height will asymptotically approach the curve describing the dispersion from a source at ground level. Thus, the ground level concentration becomes independent of the height of the source. This is generally the case for each stability class, although the actual distance depends on the atmospheric stability, as is shown in the figure. In the case of real releases, some of the material in the plume will be dry deposited on the ground, and the curve for a release at a height of 100 m will not converge with the curve for release at ground level. However, this may be regarded as a rather small correction.

In summary, if the source is located at ground level, the ground level concentration will have its maximum at the source. If the source is elevated, the maximum will occur at a distance that is determined by the height of the source, the wind speed and the atmospheric stability. In the special case when it can be assumed that $\sigma_y = \sigma_z$, which is true during neutral or slightly unstable conditions, the maximum concentration at ground level, as well as the distance from the source, can be found by differentiating Eqn. 3.23. This gives (Slade 1968, Eisenbud 1987)

$$C_{max} = \frac{2Q}{e\pi u_{10}h^2},$$

(3.27)

where $h^2 = 2\sigma_z^2$ and the wind speed is nonzero.

Some assumptions must be made regarding σ_y and σ_z for the more general case. In Eqn. 3.28 it is assumed that the dispersion coefficients and their derivatives are proportional to each other, i.e. that the vertical and the horizontal spread of the plume are proportional. In this case, the maximum concentration is given by

$$C_{max} = \frac{2Q}{e\pi u_{10}h^2} \cdot \frac{\sigma_z}{\sigma_y}.$$

(3.28)

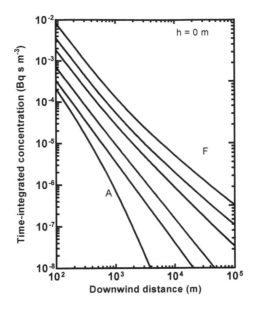

FIGURE 3.26   Time-integrated concentration at ground level, following the central axis of the plume, in the x-direction, for Pasquill categories of atmospheric stability A–F. It is assumed that the activity released from the source at ground level ($h = 0$) is 1 Bq, and the wind speed is 5 m s^{-1}.

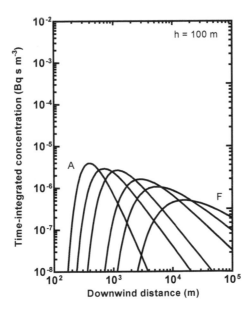

FIGURE 3.27   Time-integrated concentration at ground level, following the central axis of the plume, in the x-direction, for Pasquill categories of atmospheric stability A–F. It is assumed that an activity of 1 Bq is released from a source at a height of 100 m ($h = 100$), and that the wind speed is 5 m s^{-1}.

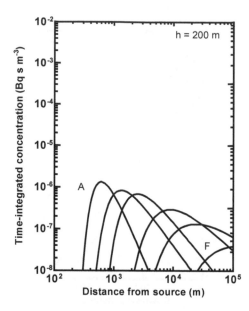

FIGURE 3.28 Time-integrated concentration at ground level, following the central axis of the plume, in the x-direction, for Pasquill categories of atmospheric stability A–F. It is assumed that an activity of 1 Bq is released from the source at a height of 200 m ($h = 200$), and that the wind speed is 5 m s^{-1}.

Again, the maximum concentration occurs when $h^2 = 2\sigma_z^2$. Although Eqns. 3.27 and 3.28 only give the maximum concentration, the distance can be derived from the relation $h^2 = 2\sigma_z^2$. The maximum concentration then occurs when $\sigma_z = h/\sqrt{2}$ and x can be determined from graphs such as Figure 3.25, or calculated from Eqn. 3.26.

Example 3.4 *Calculation of the time-integrated concentration from a radioactive release*

Let us assume that 1 TBq of a radioactive element is released from a stack with a height of 90 m. The plume is assumed to rise 10 m, and the wind speed at the height of the release is 10 m s^{-1}. We wish to calculate the time-integrated concentration at ground level, at $x = 1000$ m and $y = 100$ m, and the maximum concentration and the distance from the source at which it occurs. The atmospheric stability corresponds to Pasquill class D and the environment affected is open grassland.

The time-integrated concentration is calculated from Eqn. 3.23, using values of the dispersion coefficients calculated from Eqns. 3.25 and 3.26 for class D stability, giving values of 69.9 m and 31.5 m, respectively. The value of u_{10}, of 7.1 m s^{-1}, is calculated from Eqn. 3.20, using $n = 0.15$, given in Table 3.4. The effective stack height is increased to 100 m, by the plume rise. Inserting these values into Eqn. 3.23 gives

$$C = \frac{1 \cdot 10^{12}}{\pi \cdot 69.9 \cdot 31.5 \cdot 7.1} \cdot e^{-100^2/(2 \cdot 69.9^2)} \cdot e^{-100^2/(2 \cdot 31.5^2)}. \tag{3.29}$$

which gives the value 47.4 kBq s m^{-3}.

The maximum concentration is found from Eqn. 3.27 as neutral stability conditions are assumed, giving

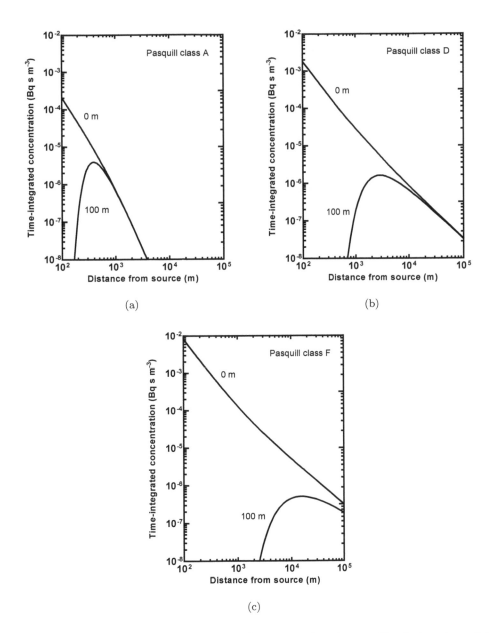

FIGURE 3.29 Comparison of the time-integrated concentration at ground level at various distances from a source at ground level and from a source at an effective height of 100 m, for Pasquill classes A, D and F. Data from Figure 3.26 and Figure 3.27.

$$C_{max} = \frac{2 \cdot 10^{12}}{e \cdot \pi \cdot 7.1 \cdot 100^2},$$ (3.30)

which equals 3.3 MBq s m^{-3}.

The distance at which the concentration is a maximum is found by recalling that $\sigma_z = h/\sqrt{2}$, which gives $\sigma_z = 70.7$ m. From Figure 3.25 we then find that x is about 3 km, in agreement with Figure 3.27.

By comparing with Figure 3.27, we can see that the maximum concentration is about 20 times the concentration 1 km from the source. However, in the example above, the ratio is about 70. Considering the values of σ_y and σ_z at a distance of 3 km we find that there is a discrepancy and we should therefore try to recalculate the concentration using Eqn. 3.28.

At 3 km, σ_z is about 190 m and the ratio between σ_z and σ_y is then 0.4. This means that the calculated maximum concentration will instead be about 1.3 MBq s m^{-3}, or about 27 times the concentration at 1 km. However, it must be kept in mind that these calculations are based on a model that may not accurately reflect the real situation.

The horizontal distribution of a plume at the height of a person inhaling the radionuclides, $z = 2$ m, is shown in Figure 3.30 for a source at a height of 30 m. The figure shows the distribution 500 m from the source in the downwind direction, assuming a wind speed of 10 m s^{-1}. Note that the concentration is now given in Bq m^{-3} because the release is given as a release rate, which in this example is 100 Bq s^{-1}. The curves for stability classes B and D are quite intuitive, as a more stable atmosphere implies a narrower distribution. The curve for stability class F, however, is merely a mathematical effect of the tail of the Gaussian distribution, which approaches zero asymptotically as the distance approaches infinity. In practice, very few radioactive particles will reach the ground at this distance, and the concentration is effectively zero. However, the external radiation dose from the plume may still be significant due to the range of the radiation.

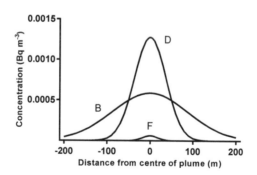

FIGURE 3.30 Variation in concentration along the horizontal (y) direction at a height of 2 m, 500 m from the source for a Gaussian plume during different atmospheric stability conditions: unstable (B), neutral (D) and stable (F). The emission rate is assumed to be 100 Bq s^{-1} at a height of 30 m, and the mean wind speed 10 m s^{-1}.

It is often necessary to estimate the horizontal and vertical distance at which the concentration has decreased to a certain fraction of the value on the plume axis. At ground level, it follows from Eqn. 3.23 that the horizontal distance, y_p, at which the concentration has decreased to p% of the value at the plume axis is

$$y_p = \sqrt{2\sigma_y^2 \cdot ln\frac{100}{p}}$$ (3.31)

and similarly for the vertical distance, z_p,

$$z_p = \sqrt{2\sigma_z^2 \cdot ln\frac{100}{p}}.$$ (3.32)

An example of the dispersion in the horizontal direction, calculated from the above equation, is given in Figure 3.31 for stability classes A–F.

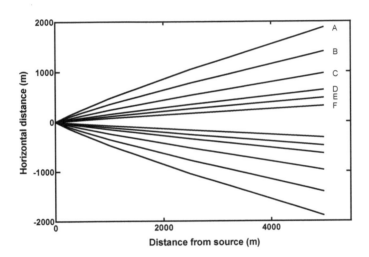

FIGURE 3.31 Isolines, showing where the concentration at ground level has decreased to 10% of its value at the plume axis, for stability classes A–F.

The appearance of the plume at ground level, perpendicular and parallel to the plume axis, is depicted in Figure 3.32 for a ground level release. It is important to realize that the plume will become narrower as the atmospheric stability increases, as indicated by the Pasquill class. Although meandering of the plume will increase the area affected by the pollutant, the plume itself will still be narrow under conditions of high atmospheric stability. The figure also provides an illustration of the important finding that the concentration from a release at ground level will be highest close to the source.

The appearance of plumes from an elevated source is shown in Figure 3.33. It is clear from this figure that the maximum concentration is found at a distance from the source, in contrast to the ground-level source. This distance increases with increasing atmospheric stability, as was also shown in Figures 3.26–3.29.

Deviations from ideal conditions may be caused by buildings and other obstacles in the path of the plume, and these must be taken into account. Figure 3.34 shows how the air is affected when it passes a cylindrical building[4]. The region of distorted flow is typically twice the height of the building and extends downwind for a distance 5–10 times the height of the building (Cooper 1984).

---

[4]This effect can be demonstrated by an efficient method of blowing out a candle. If you place a finger in front of the candle and blow, the turbulence generated behind your finger will efficiently extinguish the flame, reducing the risk of blowing wax onto the table.

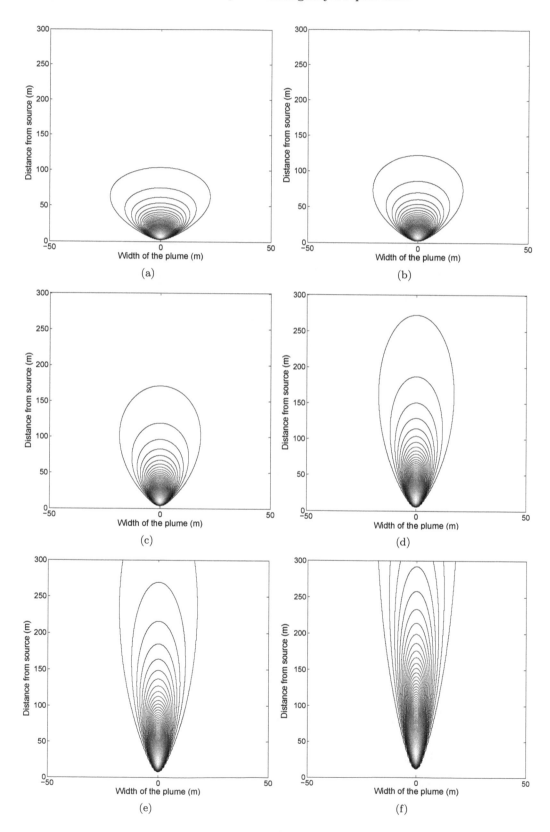

FIGURE 3.32 The appearance of a plume released from a ground-level source. The isolines reveal the concentration gradient at ground level. (a) Pasquill class A. (b) Class B. (c) Class C. (d) Class D. (e) Class E. (f) Class F.

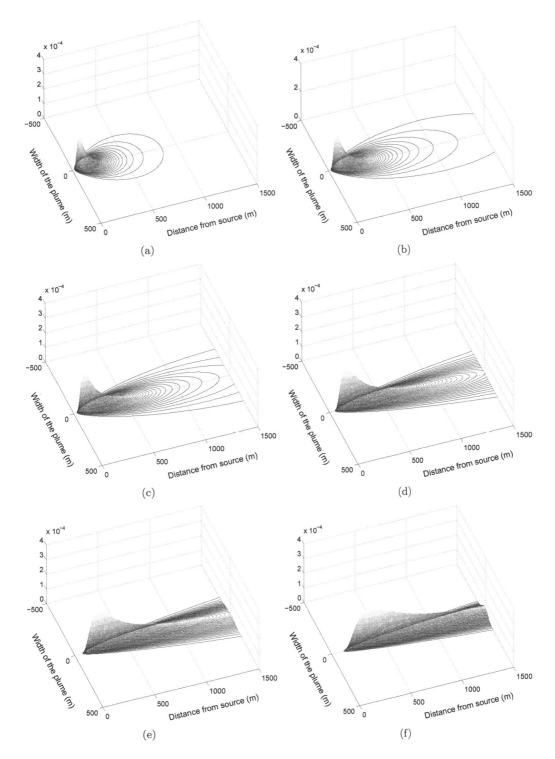

FIGURE 3.33 Variation in the concentration at ground level, of a pollutant released from an elevated source. In this example, an effective height of 10 m was assumed. (a) Pasquill class A. (b) Class B. (c) Class C. (d) Class D. (e) Class E. (f) Class F.

The region in which the air flow is affected can be divided into three zones. The *displacement zone* is the region in which the approaching airflow is displaced, both around and over the building. The *wake* is a turbulent volume of air downwind of the building, and the *cavity* is the volume immediately behind the building where the air flows in a torus-shaped circulating pattern. The wake and the cavity are the result of aerodynamic separation. This occurs when air moving parallel to a surface is forced, by edges or aerodynamic effects, to leave the surface and continue out into the open air flow. The air in the cavity is forced to move downwind by the action of the separated layers, and it must therefore be replaced by a return flow along the ground. Due to the reversal of the air flow close to the surface of the object (see lower part of Fig. 3.34) and at the far end of the cavity, regions of stagnation (low pressure) are formed. Other characteristics of the cavity are low wind speed and high turbulence. Due to the properties of the cavity, particles in the air flow may be trapped, resulting in long residence times within the cavity. The exact behaviour of the air depends on the wind speed and the size of the obstacle, and must be determined by wind tunnel experiments or simulations.

These effects can cause considerable deviations from the model predictions, for example, the reactor building itself will give rise to a wake where the effluents are mixed with the surrounding air. Likewise, a plume approaching an urban area will experience several obstacles, making dispersion modelling in urban areas very difficult. Successful models have been developed, but the discussion of these is beyond the scope of this text, and the reader is referred to Andersson (2009).

The topography of the landscape can also make simple dispersion models, such as the Gaussian plume model, unsuitable. For example, narrow passes or valleys may dramatically change the wind speed as well as its direction. An example of an extreme change in wind direction is when a wind from the north (or south) enters a valley in an east–west direction. The air entering the valley is essentially forced to follow the valley, resulting in perpendicular deflection of the air flow.

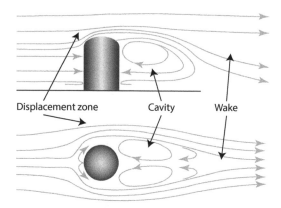

FIGURE 3.34 Schematic view of the streamlines for air approaching a cylindrical obstacle from the left. The lower part of the figure shows the view from above.

3.2.4 Deposition

The environmental impact of an atmospheric release is largely determined by the residence time of the pollutant in the atmosphere. The time that a radionuclide remains airborne will thus affect the geographical area exposed to the release, as well as the radiation dose. If the radionuclides remain airborne for a long time, the release may spread over a larger area, and there may also be a significant contribution to the radiation dose from inhaled radionuclides. If, on the other hand, the radionuclides are deposited on the ground, the affected area may be smaller, but the external radiation dose may increase. In addition, deposited radionuclides may be further transported through the food chain, giving rise to an internal radiation dose from ingestion. This deposited radioactive debris is often denoted *fallout*.

The efficiency of removal of radionuclides from the atmosphere depends on whether deposition is associated with precipitation or not; see Figure 3.35. "Scavenging" of a radioactive plume by precipitation can cause large amounts of material to be deposited, even by low levels of precipitation. It is therefore important to distinguish between *dry deposition* and *wet deposition*, both of which are discussed in this section.

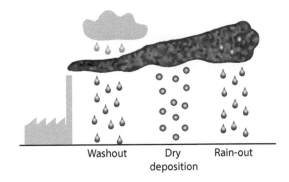

Washout   Dry deposition   Rain-out

FIGURE 3.35 Pollutants in the atmosphere can be removed by dry or wet deposition processes. Wet deposition can be further divided into washout, when the raindrops are formed above the plume, or rain-out if the pollutant aerosols within the plume act as condensation nuclei for the raindrops.

3.2.4.1 Dry Deposition

Dry deposition is quantified by the *deposition velocity*, v_g, which relates the deposited activity to the concentration of the radionuclide in the air. The deposition velocity is defined as the activity deposited per unit area per unit time [Bq m^{-2} s^{-1}] divided by the concentration in the air, C [Bq m^{-3}], at some reference height. If the surface consists of vegetation with a height much less than 1 m, the reference height is usually taken as 1 m. The SI units of v_g are m s^{-1}, but the dry deposition velocity is usually expressed in cm s^{-1}.

The deposition rate, $\dot{D}_d$, can be calculated from Eqn. 3.33 using the dry deposition velocity

$$\dot{D}_d = v_g \cdot C. \tag{3.33}$$

The rate of dry deposition of particles and gases (other than rare gases) depends on several factors, such as the size of the radioactive particles in the plume, the structure of the ground surface, and air movements. As we will see later, large particles fall to the ground rapidly, and this is important in the case of nuclear weapons explosions. The nature of the ground surface will also have a considerable impact on deposition from a plume moving close to the ground. Finally, air movements will transport particles and gases in the plume,

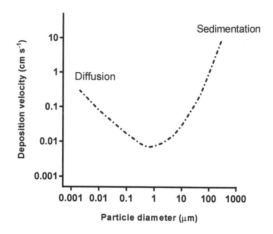

FIGURE 3.36 Dry deposition velocity for a rough surface as a function of particle diameter, showing a minimum for particles in the range 0.1–1 μm. For larger particles the dominating deposition processes are sedimentation and impaction, while smaller particles are mainly deposited by diffusion. Data from Friedlander (2000).

affecting the timescale of the deposition, as well as its location: for example, updrafts may delay deposition at a particular location.

The dependence of deposition velocity on particle size, expressed as the activity median aerodynamic diameter (AMAD), is shown schematically in Figure 3.36. The most important dry deposition mechanisms for large particles, with an AMAD larger than about 1 μm, are *sedimentation* and *impaction*. Sedimentation is simply the gravitational settling of particles, while impaction is an effect of the inertia of the particles in an air stream. When a stream of air flows around an obstacle, the particles' inertia prevents them from following the stream and they collide with the obstacle instead. It can be shown that the maximum speed attained by a large particle, settling under the influence of gravity (sedimentation), is proportional to the square of the radius of the particle. This speed limit is reached when the downward force on the particle due to gravity is balanced by the upward forces of air resistance and buoyancy.

Thus, the total dry deposition velocity is, in general, weakly affected by gravitational settling for particles less than 10 μm in diameter. For particles with aerodynamic diameters in the range 1–10 μm, the gravitational settling velocity, v_s [m s^{-1}] can be calculated as approximately $3.2 \cdot 10^{-5} \cdot D^2$, where the AMAD, D, is given in μm (NRPB 2001).

Particles with an AMAD of less than 1 μm are often deposited by *interception*, in which the particles in an air stream adsorb onto a surface when the distance between the particle and the surface becomes less than half the AMAD. Even smaller particles, much less than 1 μm, are subject to diffusion by *Brownian motion* and to deposition by air turbulence in the vicinity of objects. For particles with diameters less than 0.1 μm the physical diameter is actually of greater importance than the AMAD (NRPB 2001). As can be seen in Figure 3.36, the minimum deposition velocity is found for particles in the range 0.1–1 μm, which includes ordinary household dust. This dust is available for inhalation for a comparably long time, and pollutants in this range will remain airborne even far away from the source.

Both measured and theoretical estimates of the deposition velocity display large variations, which are only partly understood. For example, the dependence on particle size and wind velocity is relatively well understood, while other sources of the variation must be regarded as uncertainties. Modelling of dry deposition velocities is, however, quite complex, and the details of the calculations are beyond the scope of this text. They are presented in detail in NRPB (2001), and the discussion of dry deposition in this section is an overview of this report.

Bulk transfer in the dry deposition process is affected by interactions with the surroundings, which can be modelled as a series of resistances in an electrical circuit. One such interaction is described by the *surface resistance*, R_a, which is a dimensionless parameter that is dependent on the roughness length. (Typical roughness lengths were given in Table 3.4). Some typical values of R_a for various atmospheric stabilities and aerodynamic roughness lengths are given in Table 3.6.

TABLE 3.6 Typical values of the surface resistance at 1 m for different roughness lengths (z_0) and classes of atmospheric stability (A–G).

Pasquill class	Roughness length (m)	
	$z_0 = 0.01$	$z_0 = 0.03$
A	9.89	7.36
B	10.89	8.23
C	11.43	8.69
D	11.51	8.77
E	11.62	8.86
F	12.05	9.21
G	13.67	10.55

Source: Data from NRPB (2001).

For gases and small particles it is necessary to take into account the fact that, immediately before deposition, they are only affected by Brownian motion. This behaviour is described by the *additional sub-layer resistance*, R_b, which is also a dimensionless parameter (defined as the inverse of the so-called Stanton number). The additional sub-layer resistance can be determined for various surfaces, such as urban environments, forests and water.

Using this resistance analogy, we can now write the (normalized) deposition velocity as

$$\frac{1}{V_d} = R_a + R_b, \tag{3.34}$$

where V_d is $v_d/u*$, and $u*$ is the *friction velocity*. The friction velocity is defined as a function of the diffusivity and the change in vertical wind speed, and can be determined from meteorological data.

Table 3.7 gives some typical values of v_s and v_d for particles of various AMADs. The two columns of values for v_d have been calculated with two values of R_b: a best judgement and a conservative estimate. These values are not inconsistent with those shown in Figure 3.36, as the data are derived from different types of surfaces. Deposition velocities for particles of 1, 3 and 10 μm diameter on various urban areas are given in Table 3.8.

For particles, it is assumed that the air concentration at the deposition surface is zero since, according to the model, the particles do not bounce back and they are not liberated by wind action. Deposited particles are thus considered to have been removed from the air. This assumption may not be valid for gases since the concentration at the surface may be nonzero. Dry deposition of gases may, however, be treated similarly to dry deposition of particles by the addition of a third resistance term. In the case of dry deposition on

TABLE 3.7 Gravitational settling velocity, v_d, and deposition velocities, v_d, on meadow grass and low crops, for various particle AMADs. Values are given for $R_a = 10$ and $u* = 0.2$ m s^{-1}. $R_b = 300$ is considered the best judgement, while $R_b = 50$ is regarded as a conservative estimate.

AMAD (μm)	v_s (m s^{-1})	v_d (m s^{-1})	
		$R_b = 300$	$R_b = 50$
1	$3.20 \cdot 10^{-5}$	$6.61 \cdot 10^{-4}$	$3.35 \cdot 10^{-3}$
2	$1.28 \cdot 10^{-4}$	$2.42 \cdot 10^{-3}$	$8.95 \cdot 10^{-3}$
3	$2.88 \cdot 10^{-4}$	$4.76 \cdot 10^{-3}$	$1.30 \cdot 10^{-2}$
4	$5.12 \cdot 10^{-4}$	$7.22 \cdot 10^{-3}$	$1.55 \cdot 10^{-2}$
5	$8.00 \cdot 10^{-4}$	$9.50 \cdot 10^{-3}$	$1.71 \cdot 10^{-2}$
6	$1.15 \cdot 10^{-3}$	$1.15 \cdot 10^{-2}$	$1.81 \cdot 10^{-2}$
7	$1.57 \cdot 10^{-3}$	$1.32 \cdot 10^{-2}$	$1.89 \cdot 10^{-2}$
8	$2.05 \cdot 10^{-3}$	$1.47 \cdot 10^{-2}$	$1.96 \cdot 10^{-2}$
9	$2.59 \cdot 10^{-3}$	$1.59 \cdot 10^{-2}$	$2.02 \cdot 10^{-2}$
10	$3.20 \cdot 10^{-3}$	$1.70 \cdot 10^{-2}$	$2.07 \cdot 10^{-2}$

Source: Data from NRPB (2001).

vegetation, this term represents the transfer of the pollutants from the surface of the plant to its interior parts. We can now write the dry deposition velocity as

$$\frac{1}{v_d} = r_a + r_b + r_s, \tag{3.35}$$

where r_s is the *surface resistance*. The term r_b is called the *sub-layer resistance* and is equal to $R_b/u*$. r_a is the *aerodynamic resistance*, equal to $R_a/u*$ (compare with Eqn. 3.34).

TABLE 3.8 Deposition velocities of particles onto various urban surfaces.

Surface type	v_d (m s^{-1})					
	Best judgement			Conservative estimate		
	1 μm	3 μm	10 μm	1 μm	3 μm	10 μm
Grass	$6.4 \cdot 10^{-4}$	$3.8 \cdot 10^{-3}$	$8.9 \cdot 10^{-3}$	$2.9 \cdot 10^{-3}$	$7.9 \cdot 10^{-3}$	$9.9 \cdot 10^{-3}$
Roofs, paved areas	$6.4 \cdot 10^{-4}$	$3.8 \cdot 10^{-3}$	$8.9 \cdot 10^{-3}$	$2.9 \cdot 10^{-3}$	$7.9 \cdot 10^{-3}$	$9.9 \cdot 10^{-3}$
Walls, windows, doors	$6.7 \cdot 10^{-5}$	$4.8 \cdot 10^{-3}$	$3.5 \cdot 10^{-3}$	$4.9 \cdot 10^{-4}$	$3.2 \cdot 10^{-3}$	$8.5 \cdot 10^{-3}$

Source: Data from NRPB (2001).

The gases of most interest when radioactive elements are released into the environment from a nuclear reactor are the rare gases, and iodine. Since the deposition of rare gases is negligable[5], the following discussion is restricted to iodine. The iodine present in a nuclear reactor can be divided into *volatile* and *non-volatile* forms. The volatile forms consist mainly of methyl iodide (organic iodine), which is loosely deposited on walls and filters, and on the ground. This form is often called *gaseous iodine* and it is assumed that it will behave in a similar way to the rare gases in a release resulting from a reactor accident: i.e., 100%

[5]Rare gases may, however, still contribute to external and internal exposure.

of the gaseous iodine will be released. The non-volatile forms include elemental iodine (I_2) and aerosols of CsI, and have a higher dry deposition velocity than methyl iodide.

It has been shown that the dry deposition velocity of elemental iodine depends on the relative humidity (RH), which is explained by a variation in the surface resistance. For example, at a wind velocity of 5 m s^{-1} and RH of 0.1, the dry deposition velocity of iodine on grass is about 0.003 m s^{-1}, while at a RH of 1.0 it is about 0.014 m s^{-1}. Measured values for methyl iodide are lower, in the approximate range of 10^{-6}–10^{-4} m s^{-1} (NRPB 2001). In urban areas, the dry deposition velocity varies with the composition of the surface; some representative values are given in Table 3.9.

TABLE 3.9 Deposition velocities of elemental iodine onto various urban surfaces.

Surface type	v_d (m s^{-1})	
	Best judgement	Conservative estimate
Grass	$2.6 \cdot 10^{-3}$	$1.0 \cdot 10^{-2}$
Roofs	$3.3 \cdot 10^{-3}$	$1.0 \cdot 10^{-2}$
Paved areas	$4.6 \cdot 10^{-4}$	$2.3 \cdot 10^{-3}$
Walls	$3.0 \cdot 10^{-4}$	$1.5 \cdot 10^{-3}$
Windows and doors	$2.3 \cdot 10^{-4}$	$1.2 \cdot 10^{-3}$

Source: Data from NRPB (2001).

3.2.4.2 Wet Deposition

Wet deposition is far more efficient at removing pollutants from the air, as can be seen from Figure 3.37. Immediately after the plume from the Chernobyl accident reached Sweden on 28 April, 1986, measurements of dry and wet deposition were initiated at several sites. Figure 3.37 shows data from measurements made in Lund in the southern part of the country (Erlandsson and Isaksson 1988), where the first rain fell between 22.00 and 22.34 on 7 May. The amount of precipitation that fell during these 34 minutes was 0.54 mm, which caused an increase in deposited activity of about a factor of 100. A slight increase observed before the onset of the rain was due to dry deposition.

As shown in Figure 3.35, wet deposition can occur either by washout or by rain-out, depending on the origin of the precipitation relative to the pollutants. These two modes are also referred to as *below-cloud scavenging* and *in-cloud scavenging*. We will discuss below-cloud scavenging first.

Below-cloud scavenging is a process in which particles combine with raindrops as they fall and are subsequently deposited on the ground. The process can be modelled by assuming that the intensity of the rainfall is constant over time, that the size of the raindrops does not change with height, and that particles that have combined with raindrops will follow the raindrops all the way to the ground (i.e. the particles will not be re-emitted). If these assumptions are valid, the vertical distribution of particles in the air will not be changed by the removal of particles by scavenging. This means that although the total number of particles in the air is decreased by scavenging, the relative number at various heights remains the same.

The rate of decrease in the mass of airborne pollutants can be described as an exponential decrease in time, determined by the *scavenging coefficient*, $\Lambda(D)$, which has the unit s^{-1}. The concentration, C, of a pollutant in the air at time t is then given by

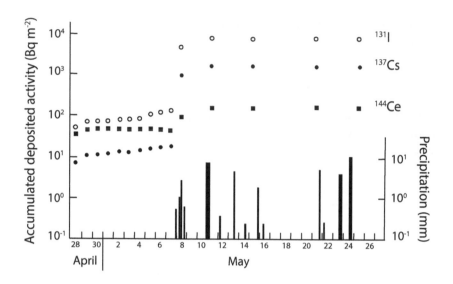

FIGURE 3.37 Total accumulated activity per unit area per unit precipitation in Lund, southern Sweden, after the Chernobyl accident in 1986. The width of the precipitation bars represents the duration of each rain shower. The accumulated activity per unit area increased rapidly when the first precipitation fell at the sampling site. Data from Erlandsson and Isaksson (1988).

$$C(t) = C_0 \cdot e^{-\Lambda \cdot t}, \tag{3.36}$$

where C_0 is the concentration at $t = 0$ (at the onset of precipitation).

The scavenging coefficient is a function of the particle diameter, D, the raindrop radius, the number of raindrops per unit volume, the settling velocity of the raindrops, and the fraction of pollutants removed by the raindrops. If the raindrops are assumed to fall vertically, the wet deposition flux at a location (x, y) can be calculated as

$$W(x, y) = \Lambda \int C(x, y, z) dz, \tag{3.37}$$

where $C(x, y, z)$ is the concentration of the pollutant at height z over the location (x, y), and integration is performed over the whole depth of the plume. If C is given in Bq m^{-3}, the units of W will be Bq m^{-2} s^{-1}.

Wet deposition can also be described by the *washout ratio*, w, defined by

$$w = \frac{W}{J \cdot \bar{C}}, \tag{3.38}$$

where J is the intensity of rainfall (SI units m s^{-1},[6] but this is often given in mm h^{-1}) and $\bar{C}$ is the vertical average of the activity concentration in the air [Bq m^{-3}]. The washout coefficient is thus a dimensionless quantity. The definition given above means that the washout coefficient is the ratio between the deposited activity per unit volume of rainwater and the average activity concentration in the air through which the rain has passed.

---

[6]The intensity, or rate, of rainfall is a measure of the volume of rain that falls onto a surface per unit area per unit time. The units are thus m^3 m^{-2} s^{-1}, or m s^{-1}.

However, the washout ratio is sometimes defined using the concentration in ground level air instead of the vertical average.

If the height of the column of air through which the precipitation passes is denoted b, we find that since $W = \Lambda \cdot \overline{C} \cdot b$,

$$w = \frac{\Lambda \cdot b}{J}. \tag{3.39}$$

The washout coefficient and the scavenging coefficient are thus related, but the two quantities are often used in different situations. Measurements of scavenging coefficients are often related to specific rain events, while measurements of the washout coefficients are often long-term averages.

In order to be consistent with field data, the scavenging coefficient is assumed to be constant in the particle range 0.1–1 μm (AMAD). Values of the scavenging coefficient for various particle sizes are given in Table 3.10.

TABLE 3.10 Values of the scavenging coefficient, Λ, for various particle sizes and a rainfall intensity of 1 mm h^{-1}. The values given should be considered best estimates.

Particle diameter (μm)	Scavenging coefficient (s^{-1})
1	$4.0 \cdot 10^{-5}$
2	$8.0 \cdot 10^{-5}$
3	$1.2 \cdot 10^{-4}$
4	$1.6 \cdot 10^{-4}$
5	$2.0 \cdot 10^{-4}$
6	$2.4 \cdot 10^{-4}$
7	$2.8 \cdot 10^{-4}$
8	$3.2 \cdot 10^{-4}$
9	$3.6 \cdot 10^{-4}$
10	$4.0 \cdot 10^{-4}$

Source: Data from NRPB (2001).

Determination of the scavenging coefficient from in-cloud scavenging requires knowledge of the fraction of the pollutant within the cloud. In addition, the scavenging process is also affected by the fact that cloud droplets are usually much smaller than rain droplets (some tens of μm compared to about 1 mm). A reasonable estimate of the scavenging coefficient lies in the range $3 \cdot 10^{-5}$–$3 \cdot 10^{-4}$ s^{-1}.

Wet deposition of gases is treated differently, since the assumptions made for particles may not always be valid. For example, when gas molecules are transferred to a raindrop, the concentration of the gas in the drop may be in equilibrium with the concentration of the gas in the surrounding air. If the drop falls to a height where the concentration of the gas in the surrounding air is lower, the gas may escape from the drop in order to establish equilibrium with the new conditions. Therefore, the assumption made for particles—that once in the drop, the particle will remain there—may not be valid for gases. The modelling of wet deposition of gases is complicated, and only a few results for iodine will be given here. For a more detailed discussion, the reader is referred to NRPB (2001).

When modelling wet deposition of gases, which may be reversibly attached to raindrops, two limiting cases can be identified for below-cloud scavenging: a thin plume at ground level and a uniform concentration. In a thin plume (some tens of metres), it may be assumed that the time taken for the drops to pass through the plume is short compared to the time taken for the gas to leave the drops. This situation thus resembles scavenging of particles. Under these assumptions it is estimated that the scavenging coefficient $[s^{-1}]$ for elemental iodine is given by

$$\Lambda = 6 \cdot 10^{-5} \frac{J}{J_0}, \tag{3.40}$$

where J_0 is an intensity of 1 mm h^{-1}. This value of the scavenging coefficient can also be considered valid for methyl iodide, due to the large uncertainties involved.

If the concentration in air can be considered uniform, which is more common, an approach similar to dry deposition may be used. In this case, the *wet deposition velocity*, v_w, is defined by

$$v_w = J \cdot p, \tag{3.41}$$

where p is the *partition coefficient*, expressing the ratio between the liquid-phase and gas-phase concentration of the pollutant. NRPB (2001) recommends that $p = 10^4$ is used for elemental iodine if a best estimate is required. If a conservative value is required, $p = 10^5$ is recommended. For methyl iodide, the recommended "best judgement" and conservative estimate are $p = 7$ and $p = 10^4$, respectively.

In addition to dry and wet deposition, a third deposition mechanism called *occult* deposition may be of importance. This process is due to the association of small (submicron) particles with water droplets in fog. The deposition velocity of these particles may then be higher than in dry deposition, by about two orders of magnitude. The droplets may also have a considerable horizontal velocity, and pollutants may then be deposited on vegetation as the wind blows the fog droplets over the landscape.

3.2.4.3 Resuspension

When radionuclides are deposited on the ground, they may be redistributed by various physical as well as biological processes. If the upper layer of the ground is saturated with water, the wet-deposited radionuclides will not be retained but will mostly follow the rain water. This process, denoted *run-off*, thus refers to radionuclides that are relocated as an effect of ongoing wet deposition. If previously deposited radionuclides are relocated by subsequent precipitation, the term *weathering* is usually employed instead. Mechanical processes, such as the passage of traffic on roads, also contribute to weathering.

Another process that will contribute to the relocation of deposited radionuclides is *resuspension*, in which radionuclides on the ground are entrained back into the air by wind and human activities. Resuspension will therefore be of greatest concern in areas characterized by dry, windy conditions. However, this process is significantly less effective than the dry deposition processes described above. Formally, the term resuspension is reserved for the relocation of particles, but deposited gases may also be returned to the atmosphere. In this case, resuspension may be faster than deposition, for example, in the presence of a source of heat. Because of resuspension, radioactive material can be available for inhalation a long time after the deposition event. The process may also redistribute radionuclides to areas previously unaffected by the radioactive deposition, resulting, for example, in the contamination of plants.

The IAEA has listed several factors that may influence resuspension, some of which are given below (IAEA 1992).

- Time since deposition

- Wind speed

- Nature of the surface: vegetation, buildings, etc.

- Surface moisture

- Soil chemistry and texture

- Size distribution of contaminant particles

- Chemical properties of contaminant

- The deposition process: wet or dry, amount of precipitation, etc.

- Mechanical disturbance by traffic, agriculture, etc.

- Depth and method of cultivation

- Intensity and frequency of rain

- Snow cover or freezing of the surface

Resuspension can be described by various parameters (IAEA 1992), but we will restrict the discussion here to the *resuspension factor*, K_s (unit m^{-1}). The resuspension factor is defined as the ratio between the activity concentration in air, C_a, in Bq m^{-3}, and the activity per unit area of the surface, d, in Bq m^{-2}. Thus

$$K_s = \frac{C_a}{d}. \tag{3.42}$$

Although both the numerator and denominator are measurable quantities, some ambiguities may arise from the fact that processes other than resuspension can contribute to the activity concentration in the air. The determination of K is also affected by the measurement method, since the activity concentration in the air is usually determined by filtering the air and analysing the filter with gamma-ray spectrometry. Obviously, the height above the surface at which the air is sampled will influence the resulting activity on the filter. In addition, when defining the activity per unit area of the surface, the activity can be measured in soil samples taken at a specified depth. Ideally, this depth should be equal to the thickness of the ground layer that contributes to resuspension, but the choice of depth is far from straightforward. The chosen depth will also influence measurements if the activity is determined using field gamma-ray spectrometry, since a specific depth distribution must be assumed in order to calculate the activity. In accidental deposition events, it is practical to use the total deposited activity, disregarding redistribution processes such as vertical mixing or weathering (IAEA 1992).

Field experiments and empirical models suggest that K_s is in the range 10^{-6}–10^{-4} m^{-1} soon after deposition. Some models then predict an almost constant value for the first 100–1000 days after deposition, followed by a decline to about 10^{-9} m^{-1} (IAEA 1992).

3.2.5 Dose Calculations from Atmospheric Dispersion

Following a release of radionuclides to the atmosphere, exposure of humans may occur by various pathways, some of which are depicted in Figure 3.38. Airborne radionuclides cause external exposure during the passage of the plume (*cloudshine*) and will also cause internal exposure if inhaled or ingested. After the passage of the plume, there will be no cloudshine, but inhalation of radionuclides may still occur due to resuspended radionuclides. External irradiation also occurs from radionuclides deposited on the ground (*groundshine*). Deposited radionuclides may cause external irradiation for a considerable time, depending on their half-lives and their rate of downward migration into the ground. In a longer time perspective, internal irradiation from ingestion of contaminated food and water may contribute to the radiation dose, to an extent depending on the countermeasures employed. It is important to realize that the radiation dose from each pathway has its own characteristic time and space dependence.

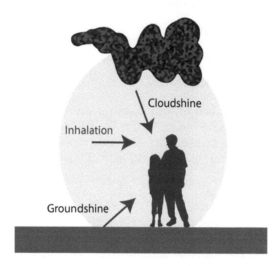

FIGURE 3.38 Three exposure pathways can be identified following an atmospheric release of radionuclides: cloudshine, groundshine and inhalation. Cloudshine will only be relevant during the passage of the plume.

Generally, the effective dose rate from cloudshine, $\dot{E}_{ext}$, is expressed by

$$\dot{E}_{ext}(t) = C(t) \cdot DC_{ext}, \tag{3.43}$$

where $C(t)$ is the activity concentration in the plume [Bq m^{-3}] and DC_{ext} is a dose conversion coefficient [Sv s^{-1} Bq^{-1} m^3], specific to each radionuclide. The activity concentration can be determined from atmospheric dispersion models, such as those discussed in Section 3.2.3, or from measurements. In the case of an accidental atmospheric release of radionuclides, field measurements (*indication*) are used to find the plume limits in order to decide on countermeasures such as evacuation.

The effective dose to an unshielded population during the passage of the plume is found by multiplying the time-integrated concentration [Bq s m^{-3}] by the dose conversion coefficient (see Example 3.5). For an atmospheric release of radionuclides, decisions about countermeasures such as evacuation are based on the effective dose to that population if no action were to be taken. In these cases, the effective dose [Sv] is more relevant than the effective dose *rate* [Sv s^{-1}].

Example 3.5 *Radiation dose from submersion in a radioactive plume*

Following an accident at a nuclear facility, ^{131}I and ^{137}Cs were dispersed in the atmosphere, giving an air concentration of 200 Bq m^{-3} of ^{131}I and 100 Bq m^{-3} of ^{137}Cs. We want to estimate the effective dose to an adult population that is under the plume for 2 hours (7200 s).

The effective dose from cloudshine is found from Eqn. 3.43, and the effective dose from ^{131}I can be found using data from Table 3.11:

$$E_{ext,I}(t) = 200 \cdot 1.69 \cdot 10^{-14} \cdot 7200, \tag{3.44}$$

which equals $2.4 \cdot 10^{-8}$ Sv. The effective dose from ^{137}Cs is given by

$$E_{ext,Cs} = 100 \cdot 2.55 \cdot 10^{-14} \cdot 7200, \tag{3.45}$$

which equals $1.8 \cdot 10^{-8}$ Sv. It is then assumed that ^{137}Cs is in secular equilibrium with it daughter ^{137m}Ba. The total effective dose from submersion in the plume will then be $4.2 \cdot 10^{-8}$ Sv. This dose is so small that no action is required (see Chapter 7.2.1).

The effective dose rate from groundshine can also be estimated using Eqn. 3.43 if $C(t)$ is the activity per unit area [Bq m^{-2}] and the dose conversion coefficient is given in units of Sv s^{-1} Bq^{-1} m^2. The activity per unit area can be obtained from measurements, but it can also be calculated by multiplying the time-integrated concentration [Bq s m^{-3}] by the deposition velocity [m s^{-1}]. The effective dose rate, calculated with Eqn. 3.43, will then refer to the effective dose rate when the plume has passed.

The activity per unit area decreases with time because of radioactive decay, environmental factors such as migration into the ground, and various removal mechanisms. The effective dose during an arbitrary time interval is then found by integrating the effective dose rate for each radionuclide over the time interval, and summing the contributions from all the radionuclides. The activity of each radionuclide in a decay chain can be found using the Bateman equations, Eqn. 1.12, which may be written

$$A_n(t) = A_1(0) \cdot \prod_{j=1}^{n-1} f_{j,j+1} \lambda_j \sum_{j=1}^{n} \left(\frac{e^{-\lambda_j t}}{\prod\limits_{p=1, p \neq j}^{n} (\lambda_p - \lambda_j)} \right), \tag{3.46}$$

where $A_n(t)$ is the activity of radionuclide n at time t, and $f_{j,j+1}$ is the fractional yield in decay from radionuclide j to radionuclide $j+1$. It is assumed that the activity of each progeny is 0 at $t = 0$, i.e. $A_n(0) = 0$ for $n > 1$. For example, if $n = 2$ and the activity of the mother radionuclide at $t = 0$ is $A_1(0)$,

$$A_1 = A_1(0) \cdot e^{-\lambda_1 t} \tag{3.47}$$

and

$$A_2(t) = A_1(0) \frac{\lambda_1}{\lambda_2 - \lambda_1} \left(e^{-\lambda_1 t} - e^{-\lambda_2 t} \right). \tag{3.48}$$

The effective dose from groundshine, assuming an exposure period T, is found by integrating Eqn. 3.46 and multiplying by the dose conversion coefficient, $DC_{gs,i}$, for each radionuclide, n, in the chain, which gives (Eckerman and Ryman 1993)

$$E = A_1(0) \sum_{n=1}^{N} DC_{gs,n} \prod_{j=1}^{n-1} f_{j,j+1} \lambda_j \sum_{j=1}^{n} \left(\frac{1 - e^{-\lambda_j T}}{\lambda_j \prod_{p=1,p\neq j}^{n} (\lambda_p - \lambda_j)} \right). \tag{3.49}$$

Example 3.6 *Radiation dose from groundshine (adapted from Health Canada (1999))*

Following an atmospheric release of ^{95}Zr, the deposited activity per unit area is found to be 1000 Bq m^{-2}. We want to calculate the effective dose from groundshine to an adult population living in the contaminated area for 1 month after deposition. It can be assumed that the dose rate decreases only as a result of radioactive decay.

The effective dose is found by integrating the dose rate from ^{95}Zr ($T_{1/2} = 63.98$ d) and its daughter ^{95}Nb, which decays with $T_{1/2} = 35.15$ d to stable ^{95}Mo. The groundshine dose coefficients are found in Table 3.11 and are $7.04 \cdot 10^{-16}$ Sv s^{-1} Bq^{-1} m^2 and $7.28 \cdot 10^{-16}$ Sv s^{-1} Bq^{-1} m^2 for ^{95}Zr and ^{95}Nb, respectively. The fractional yield, f, for the transformation from ^{95}Zr to ^{95}Nb can be assumed to be 1, neglecting the small fraction of decay that occurs through ^{95m}Nb by $^{95}Zr \rightarrow^{95m}Nb \rightarrow^{95}Nb$.

Given the above decay constants, the half-lives λ, will be $1.25 \cdot 10^{-7}$ s^{-1} and $2.28 \cdot 10^{-7}$ s^{-1} for ^{95}Zr and ^{95}Nb, respectively.

The effective dose from groundshine is given by Eqn. 3.49:

$$E_{gnd}(t) = A_{Zr}(0) \times$$
$$\times \left\{ DC_{Zr} \frac{1 - e^{-\lambda_{Zr} T}}{\lambda_{Zr}} + DC_{Nb} \frac{f \lambda_{Nb}}{(\lambda_{Nb} - \lambda_{Zr})} \left[\frac{1 - e^{-\lambda_{Zr} T}}{\lambda_{Zr}} - \frac{1 - e^{-\lambda_{Nb} T}}{\lambda_{Nb}} \right] \right\}. \tag{3.50}$$

With the values given above, the effective dose in the first 30 d is $1.97 \cdot 10^{-6}$ Sv. This dose is so small that no action is required (see Chapter 7.2.1).

The dose conversion coefficients can be determined from measurements, or Monte Carlo simulations, as described in Section 2.4. The dose conversion coefficients given in Table 3.11 were calculated by assuming a number of monoenergetic sources, distributed in a semi-infinite cloud or on the ground, and computing the distribution of photons impinging on a closed surface surrounding a human phantom. From this distribution, the photon transport and energy deposition in organs and tissues, and thus the organ doses, can be calculated by Monte Carlo methods (Eckerman and Ryman 1993, Eckerman and Legget 1996).

The dose conversion coefficients are calculated for an adult, but doses to individuals will generally vary with their body size. The dose to organs located in the middle of the body will be affected by overlying tissue, which will absorb some of the energy and thus shield the deeper lying organs. Especially for low photon energies, shielding by overlying tissues will therefore be greater in larger individuals, resulting in a lower dose to the deeper lying organs. This effect is less pronounced for high photon energies because of their greater penetration. Doses to the body organs in a small body will thus be higher than in a large body (Health Canada 1999).

The variation in organ dose, and thus effective dose, with body size should be considered when estimating the organ equivalent dose and effective dose to children. It has been suggested that infant organ doses can be up to three times the dose to an adult male when exposed to groundshine or cloudshine with photons below 100 keV (Petoussi et al. 1991). Other studies have shown that the effective dose to infants (below 2 years of age), exposed isotropically to photons above 115 keV, were 20–30% higher than that for an adult (Yamaguchi 1994). For electrons, the differences can be even greater, especially for high energies. Ratios of the effective dose to children and adults have been found to vary between 2 and 20 for electron energies above 600 keV, but only a very small difference (<0.2%) was found for energies below 600 keV (Schultz and Zoetelief 1997).

The committed effective dose rate from inhalation, $\dot{E}_{int}$, can be found from the air concentration by

$$\dot{E}_{int}(t) = C(t) \cdot B \cdot DC_{int}, \tag{3.51}$$

where B is the ventilation (breathing) rate. Recommended values for default ventilation rates have been given by the ICRP and are listed in Table 3.12 (ICRP 1995). The dose coefficients for calculation of the committed effective dose have been discussed in Section 2.8.2, and some examples are given in Table 3.13.

The total inhaled activity [Bq] is found by multiplying the time-integrated activity concentration by the breathing rate. The committed effective dose is then found by multiplying this activity by the dose coefficient (see Example 3.7).

Example 3.7 *Radiation dose from inhalation in a radioactive plume*

We want to estimate the committed effective dose to an adult population under the conditions described in Example 3.5. The committed effective dose from inhalation is found from Eqn. 3.51. The dose coefficients are given in Table 3.11, and the breathing rate for an adult is 22.2 $m^3 d^{-1}$ (Table 3.12). Since 2 h equals 1/12 d, the committed effective doses for ^{131}I and ^{137}Cs are given by

$$E_{int,I} = 200 \cdot 22.2 \cdot 7.4 \cdot 10^{-9} \cdot (1/12) \tag{3.52}$$

and

$$E_{int,Cs} = 100 \cdot 22.2 \cdot 4.6 \cdot 10^{-9} \cdot (1/12), \tag{3.53}$$

which equals $2.7 \cdot 10^{-6}$ Sv and $8.5 \cdot 10^{-7}$ Sv, respectively. The total committed effective dose will then be $3.6 \cdot 10^{-6}$ Sv.

TABLE 3.11 Dose conversion coefficients for cloudshine and groundshine. Contribution from progeny are included in the coefficients for ^{106}Ru, ^{132}Te, ^{137}Cs and ^{144}Ce, assuming secular equilibrium.

Radionuclide	Cloudshine $(Svs^{-1}Bq^{-1}m^3)$	Groundshine $(Svs^{-1}Bq^{-1}m^2)$
^{14}C	$2.60 \cdot 10^{-18}$	$1.27 \cdot 10^{-20}$
^{41}Ar	$6.13 \cdot 10^{-14}$	–
^{60}Co	$1.19 \cdot 10^{-13}$	$2.30 \cdot 10^{-15}$
^{85}Kr	$2.55 \cdot 10^{-16}$	–
^{90}Sr	$9.83 \cdot 10^{-17}$	$1.64 \cdot 10^{-18}$
^{95}Zr	$3.36 \cdot 10^{-14}$	$7.04 \cdot 10^{-16}$
^{95}Nb	$3.49 \cdot 10^{-14}$	$7.28 \cdot 10^{-16}$
^{103}Ru	$2.08 \cdot 10^{-14}$	$4.49 \cdot 10^{-16}$
^{106}Ru	$1.06 \cdot 10^{-14}$	$3.45 \cdot 10^{-16}$
^{125}Sb	$1.87 \cdot 10^{-14}$	$4.09 \cdot 10^{-16}$
^{132}Te	$1.17 \cdot 10^{-13}$	$2.47 \cdot 10^{-15}$
^{125}I	$3.73 \cdot 10^{-16}$	$3.14 \cdot 10^{-17}$
^{131}I	$1.69 \cdot 10^{-14}$	$3.64 \cdot 10^{-16}$
^{135}Xe	$1.11 \cdot 10^{-14}$	–
^{138}Xe	$5.44 \cdot 10^{-14}$	–
^{134}Cs	$7.06 \cdot 10^{-14}$	$1.48 \cdot 10^{-15}$
^{137}Cs	$2.55 \cdot 10^{-14}$	$5.51 \cdot 10^{-16}$
^{140}Ba	$8.07 \cdot 10^{-15}$	$1.90 \cdot 10^{-16}$
^{140}La	$1.11 \cdot 10^{-13}$	$2.16 \cdot 10^{-15}$
^{141}Ce	$3.10 \cdot 10^{-15}$	$6.93 \cdot 10^{-17}$
^{144}Ce	$3.42 \cdot 10^{-15}$	$1.82 \cdot 10^{-16}$
^{192}Ir	$3.61 \cdot 10^{-14}$	$7.77 \cdot 10^{-16}$
^{238}Pu	$3.50 \cdot 10^{-18}$	$6.26 \cdot 10^{-19}$
^{239}Pu	$3.48 \cdot 10^{-18}$	$2.84 \cdot 10^{-19}$
^{240}Pu	$3.42 \cdot 10^{-18}$	$6.01 \cdot 10^{-19}$
^{241}Am	$6.74 \cdot 10^{-16}$	$2.33 \cdot 10^{-17}$
^{252}Cf	$3.63 \cdot 10^{-18}$	$5.24 \cdot 10^{-19}$

Source: Data from Health Canada (1999).

3.2.6 Past Exposure Events and Modelling of Deposition

Debris from a nuclear weapons test can enter the local, regional and global environment, depending on where the explosion takes place. For example, following tests performed on the earth's surface, radioactive debris is deposited as *local fallout* and *regional fallout*, within a radius of a few hundred kilometres and a few thousand kilometres from the test site, respectively (UNSCEAR 2000). *Global fallout* is caused by a proportion of the radioactive debris entering the stratosphere, and this fraction is dependent on the yield of the explosion. The debris from explosions less than about 100 kt tends to remain in the troposphere, whereas stratospheric injection is considerable for yields above around 100 kt. However, for tests performed in the high atmosphere, injection into the stratosphere will be complete, regardless of the yield. The total yield for all nuclear weapons tests is estimated to be 251 Mt, of which 11.6% of the activity was locally and regionally deposited, 6.4% was injected into the troposphere and 57.8% was injected into the stratosphere (UNSCEAR 2000). These

TABLE 3.12 Recommended default breathing rate at various ages.

Age group	Ages	Breathing rate $(\mathrm{m^3 d^{-1}})$
Infant	0–1 y	2.86
1 y	1–2 y	5.16
5 y	2–7 y	8.72
10 y	7–12 y	15.3
15 y	12–17 y	20.1
Adult	>17 y	22.2

Source: Data from ICRP (1995).

TABLE 3.13 Inhalation dose coefficients [Sv Bq^{-1}] for various age groups. The recommended absorption type for each radionuclide is given in brackets.

	Infants	1 Year	5 Years	10 Years	15 Years	Adult
^{14}C (M)	$8.3\cdot10^{-9}$	$6.6\cdot10^{-9}$	$4.0\cdot10^{-9}$	$2.8\cdot10^{-9}$	$2.5\cdot10^{-9}$	$2.0\cdot10^{-9}$
^{60}Co (M)	$4.2\cdot10^{-8}$	$3.4\cdot10^{-8}$	$2.1\cdot10^{-8}$	$1.5\cdot10^{-8}$	$1.2\cdot10^{-8}$	$1.0\cdot10^{-8}$
^{90}Sr (M)	$1.5\cdot10^{-7}$	$1.1\cdot10^{-7}$	$6.5\cdot10^{-8}$	$5.1\cdot10^{-8}$	$5.0\cdot10^{-8}$	$3.6\cdot10^{-8}$
^{131}I (F)	$7.2\cdot10^{-8}$	$7.2\cdot10^{-8}$	$3.7\cdot10^{-8}$	$1.9\cdot10^{-8}$	$1.1\cdot10^{-8}$	$7.4\cdot10^{-9}$
^{137}Cs (F)	$8.8\cdot10^{-9}$	$5.4\cdot10^{-9}$	$3.6\cdot10^{-9}$	$3.7\cdot10^{-9}$	$4.4\cdot10^{-9}$	$4.6\cdot10^{-9}$
^{241}Am (M)	$7.3\cdot10^{-5}$	$6.9\cdot10^{-5}$	$5.1\cdot10^{-5}$	$4.0\cdot10^{-5}$	$4.0\cdot10^{-5}$	$4.2\cdot10^{-5}$

Source: Data from Health Canada (1999), ICRP (2012).

values were estimated for volatile radionuclides such as ^{90}Sr and ^{137}Cs, and additional local and regional deposition of less volatile radionuclides such as ^{95}Zr also occurred.

The dispersion and deposition of radionuclides from atmospheric nuclear weapons tests have been modelled by UNSCEAR in an empirical compartment model, depicted in Figure 3.39 (UNSCEAR 2000). The structure of the model is based on the general behaviour of aerosols in the atmosphere, as discussed in Section 3.2.1. To account for the limited transport of air across the equator and the varying thickness of the troposphere due to the Hadley cell circulation, the model is divided into a polar region and an equatorial region in each hemisphere. The parameter values for the model, based on fitting to empirical fallout measurements, are given in Table 3.14. The parameters given are removal half-lives, and a high value means slow transfer. The most rapid removal occurs during the spring (Q1 in the Northern Hemisphere and Q3 on the Southern Hemisphere).

Global, or large-area, networks of sites for monitoring ^{90}Sr in air were initiated by the former Soviet Union, the United Kingdom, and the United States, to monitor the radioactive fallout from nuclear weapons testing. The network maintained between 1957 and 1983 by the United States Naval Research Laboratory and, later, the Environmental Measurements Laboratory (EML) of the United States Department of Energy, is one example. In addition, several countries have established regional networks.

Monitoring of ^{90}Sr deposition in global networks was performed by EML and the United Kingdom's Harwell Laboratory. These networks are known as the EML and the AERE (Atomic Energy Research Establishment) networks, respectively. The EML network focused on ^{90}Sr, while at a later stage, samples from the AERE network were also analysed for ^{137}Cs. The first results for ^{137}Cs were, however, derived from ^{90}Sr measurements, based on a measured ratio (Pálsson 2012). Fallout data, based on measurements of ^{90}Sr, have been summarized by UNSCEAR and are shown in Figure 3.40. These data have sometimes been regarded as a model, but the compilation is only meant as a description of the global distribution (Pálsson et al. 2013).

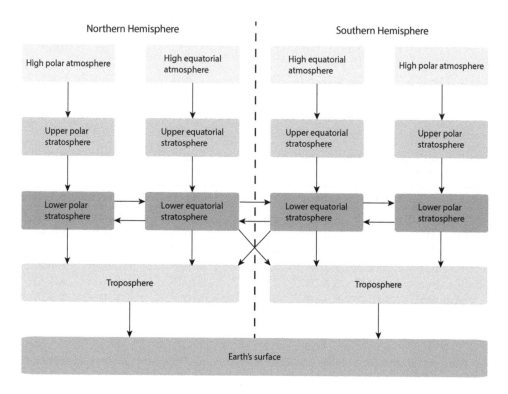

FIGURE 3.39 Compartments and transfers in the UNSCEAR empirical atmospheric model. The polar regions of the high atmosphere and the stratosphere lie between 30° and 90°, whereas the equatorial region is between 0° and 30°. The height of the troposphere is assumed to be 9 km in the polar regions and 17 km in the equatorial regions, while the lower stratosphere is assumed to extend to 17 km in the polar regions and 24 km at the equator. In this model, the upper stratosphere extends to 50 km, in both the polar and equatorial regions. Modified from UNSCEAR (2000).

It has been shown that variations in the deposition density between different sites are closely related to the amount of precipitation at each site (Arnalds et al. 1989, Almgren et al. 2006). The amount of radioactive elements deposited from nuclear weapons fallout, as well as from the Chernobyl accident, can therefore be estimated by relating the measured activity concentration in precipitation at a reference site to the amount of precipitation in other geographical regions. These point estimates may then be used to map the deposition density using various interpolation methods (Wright et al. 1999; Renaud et al. 2003, Almgren et al. 2006).

However, it is important that temporal variations in the activity concentration in precipitation be taken into account. For example, based on measurements made in Western Sweden after the Chernobyl accident, Mattsson and Vesanen showed that the activity concentration in precipitation decreased rapidly during the first few millimetres of rainfall. The activity concentration in samples collected in subsequent rain events was significantly reduced (Mattsson and Vesanen 1988). Geographical variations should also be considered

TABLE 3.14 Parameter values for transfers in the compartment model depicted in Figure 3.39. The values given in the table are removal half-times in months for yearly quarters, determined from fallout measurements. The yearly quarters start with March, for example, Q4 represents December–January–February. An infinite half-time indicates that no transfer occurs during that quarter.

From	To	Q1	Q2	Q3	Q4
Northern Hemisphere					
High polar atm.	Upper polar strat.	24	24	24	24
High equatorial atm.	Upper equatorial strat.	24	24	24	24
Upper polar strat.	Lower polar strat.	6	6	9	9
Upper equatorial strat.	Lower equatorial strat.	6	6	9	9
Lower polar strat.	Troposphere	3	6	12	10
Lower equatorial strat.	Troposphere	8	36	24	12
Lower polar strat.	Lower equatorial strat.	12	12	12	∞
Lower equatorial strat.	Lower polar strat.	12	∞	12	12
Troposphere	Earth's surface	1	1	1	1
Southern Hemisphere					
High equatorial atm.	Upper equatorial strat.	24	24	24	24
High polar atm.	Upper polar strat.	24	24	24	24
Upper equatorial strat.	Lower equatorial strat.	9	9	6	6
Upper polar strat.	Lower polar strat.	9	9	6	6
Lower equatorial strat.	Troposphere	24	12	8	36
Lower polar strat.	Troposphere	12	10	3	6
Lower polar strat.	Lower equatorial strat.	12	∞	12	12
Lower equatorial strat.	Lower polar strat.	12	12	12	∞
Troposphere	Earth's surface	1	1	1	1
Mixing between Northern and Southern Hemispheres					
Lower equatorial strat. N	Lower equatorial strat. S	24	24	24	∞
Lower equatorial strat. S	Lower equatorial strat. N	24	∞	24	24
Lower equatorial strat. N	Troposphere S	∞	24	∞	∞
Lower equatorial strat. S	Troposphere N	∞	∞	∞	24

Source: Data from UNSCEAR (2000).

when relating the deposited activity to activity concentrations in precipitation. If the air activity concentration can be considered reasonably homogeneous, as is the case of atmospheric nuclear weapons tests, a sparse network of reference sites may be used (Isaksson et al. 2000, Almgren et al. 2006). However, a denser network will be required to estimate the deposition when the air activity concentration is highly variable.

Global fallout can thus be estimated by multiplying the decay-corrected activity concentration in precipitation, $C_{R,i}$, at a reference site R during a specified time period, i, by the amount of precipitation, $P_{X,i}$, collected at a site X during the same time period (Pálsson et al. 2006). The total decay-corrected activity in the fallout, D_X, is then found by summing over specified time periods, i.e.,

$$D_X = \sum_i C_{R,i} \cdot P_{X,i}. \tag{3.54}$$

If the amount of precipitation is given in m and the activity concentration in Bq m^{-3},

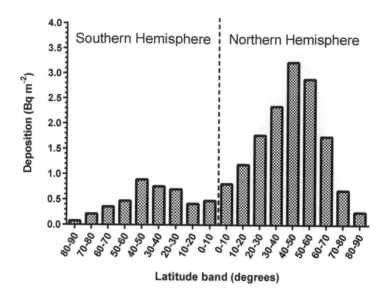

FIGURE 3.40 Distribution of ^{90}Sr fallout, divided into 10° latitude bands. Data from UNSCEAR (2000).

the deposited activity will be given in units of Bq m^{-2}. The choice of time period depends on the frequency of precipitation collection and on the frequency of sampling for activity analysis at the reference site. This kind of sampling may be performed on a quarterly or monthly basis.

Depending on the construction of the precipitation collector at the reference site, dry-deposited radionuclides may be included in the sample used to determine the activity concentration in the precipitation. Since the calculated deposition density is assumed to be proportional to the amount of precipitation, it may overestimate the true deposition density. Temporal variations, discussed above, may also lead to an overestimation, since the activity concentration in the precipitation decreases with time during a sampling period.

Data collected at the reference site, i.e. the amount of precipitation during a collection period and the activity concentration in the precipitation sample, may also be used to calculate the deposition density corresponding to a given amount of precipitation (e.g. Bq m^{-2} per 1000 mm). However, it is important that the time period during which the relationship applies is clearly stated, otherwise, comparisons with data from other sites may be misleading.

The deposition density given by Eqn. 3.54 can be described as the *cumulative deposition density* since the activity concentration in precipitation at the reference site is corrected for decay. This quantity is useful for comparisons with the measured deposition density if both the calculated and measured deposition densities are corrected for decay using a common date. If the deposition density for each time period is not corrected for decay, the calculated quantity will be the *integrated deposition density*, which gives the total activity deposited during the time period in question.

3.3 THE OCEANS

The oceans cover 71% of the earth's surface, or $3.6 \cdot 10^8$ km^2. Almost all of the water on earth (97%) is found in the oceans, which have a total volume of $1.3 \cdot 10^9$ km^3. In comparison, rivers and lakes contain only 0.04% of the total mass of terrestrial water: the remainder being found in ground water, glaciers and, to a small extent, in the atmosphere as vapour, clouds and precipitation. The average depth of the oceans is 3.7 km, and their average temperature is 3.5 °C. This enormous water mass plays an important role in the climate, in part because of its very high heat capacity. Perhaps the most striking example is the Gulf Stream, which makes north western Europe much more temperate than any other region at the same latitude. More than half of the earth's population lives within 100 km of the continental coastlines, which have a total length of 504 000 km. This is about 12 times the circumference of the earth at the equator.

3.3.1 Composition and Circulation Patterns

Typical ocean water contains about 34.5 grams of salt per litre, consisting mainly of chloride and sodium (about 85%). Other constituents of seawater, with an abundance above 1%, are sulphates, magnesium, calcium and potassium. The main source of these salts is weathering of minerals from the earth's crust, and since this process is very slow compared to the turnover of seawater in the oceans, the relative abundances of the salts are almost the same over the whole planet. The amount of dissolved salt in seawater is expressed as the *salinity*, usually in g kg^{-1} or the equivalent: ‰. Salinity can also be defined by the *Practical Salinity Scale* by measuring the conductivity of seawater, which is expressed in the dimensionless units, psu (practical salinity units). The numerical value of the salinity expressed in this way is, however, almost equal to the weight ratio. Thus, a typical salinity for seawater is 34.5 psu or 34.5‰. It is estimated that the oceans contain $5 \cdot 10^{19}$ kg of dissolved solids, and if the salt in the oceans could be removed and spread evenly over the earth's surface, it would form a layer more than 166 m deep.

The mean density of seawater is $1.034 \cdot 10^3$ kg m^{-3}, which is slightly higher than the density of pure water ($0.999 \cdot 10^3$ kg m^{-3}). The density of seawater varies slightly due to differences in both temperature and salinity, but salinity has a greater impact on the observed variation. In contrast to gases, pressure does not significantly affect the density of seawater (except at very great depths) as water is an almost incompressible fluid.

We know that approximately 80% of all life on earth is found under the surface of the oceans. Each year, 70–75 million tons of fish are caught in the oceans, half of which is used for human consumption. Global fish consumption therefore exceeds the consumption of meat, poultry or eggs. The oceans also play an important role in the circulation of carbon and oxygen, since they absorb 30–50% of the carbon dioxide produced by fossil fuels, and produce more than 50% of the oxygen.

The water in the oceans can generally be divided into three horizontal layers, based on density. The density varies with depth due to variations in temperature, salinity, or both. From the surface to a depth of around 300 m, we find the *mixed layer*. At the equator, this layer extends only to a depth of around 100 m, while it is completely absent at latitudes near the poles. The next layer is the *pycnocline*, where the density changes rather rapidly down to between 600 m and 1 km, depending on the latitude. Below this is the *deep layer*. The large differences in density between surface water and deep ocean water effectively prevent vertical currents; the exception being in the polar regions, where there is no pycnocline. Because the pycnocline zone is extremely stable, it acts as a barrier to surface processes.

Thus, changes in salinity and temperature are very small below the pycnocline, but are seasonal in surface waters.

A layer of oceanic water in which the temperature decreases rapidly with increasing depth is called a *thermocline*, and a widespread permanent thermocline exists beneath the relatively warm, well-mixed, surface layer, from depths of around 200 m to around 1000 m, see Figure 3.41. The temperature decreases steadily in this layer, while in deep waters, below the thermocline, the temperature is almost constant. In latitudes characterized by distinct seasons, a seasonal thermocline forms at much shallower depths during the summer as a result of solar heating. It disappears during the winter because there is less solar radiation and surface turbulence increases. Since the density of seawater is governed by temperature and salinity, the thermocline generally coincides with the pycnocline. The middle layer of water in a lake or reservoir during the summer can also be called a thermocline.

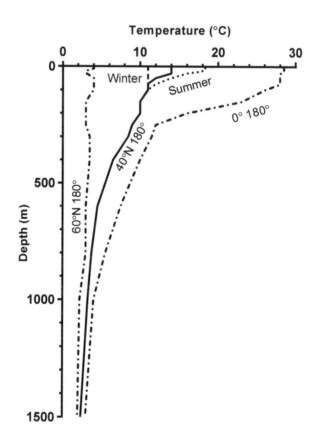

FIGURE 3.41 The variation of temperature with depth at the latitudes 60°N, 40°N and at the equator (0°). All temperature profiles are shown at a longitude of 180° (approximately in the middle of the Pacific Ocean). A seasonal dependence can be seen at 40°N where the thermocline is more prominent in the summertime. This seasonal behaviour is almost absent in the temperature gradient at the equator (the annual average is shown in the figure). There is no thermocline at high latitudes. Data from Locarnini et al. (2010).

The *halocline* is the vertical zone in the oceanic water column in which salinity changes rapidly with depth. It is located below the well mixed, uniformly-saline surface water layer. Particularly well-developed haloclines occur in the Atlantic Ocean, in which the salinity may increase by several parts per thousand from the base of the surface layer to depths of around 1 km. In the North Pacific, at higher latitudes, where solar heating of the surface water is low and rainfall is abundant, the salinity increases rapidly with depth throughout the halocline. Pycnoclines accompany haloclines as the water's density varies directly with its total salt content. Figure 3.42 shows a schematic view of the pycnocline, thermocline and halocline. In this example, the density gradient is determined by the change in salinity and therefore the pycnocline is almost coincident with the halocline.

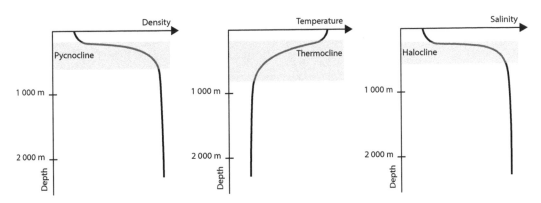

FIGURE 3.42   A schematic view of the gradients of density, temperature and salinity, giving rise to the pycnocline, thermocline and halocline.

The increase in salinity is an important mechanism for the formation of *bottom water* in the ocean basins, which is the densest water found in the oceans. When the surface water close to the poles freezes due to contact with the cold atmosphere, it forms a floating layer of ice. The salt, which cannot be accommodated within the crystal structure of the ice, is excluded from the freezing water, and a layer of cold, saline water is formed below the ice. Because of its high density, this water will sink, and flow along the slope of the ocean bottom towards the equator.

The thermohaline circulation is important for the recycling of nutrients and for the earth's climate. Plants and animals mainly live near the surface of the oceans, since phytoplankton, at the bottom of the food chain, need sunlight for photosynthesis. Since these plants and animals use the nutrients in seawater, the upper layer of water becomes depleted in nutrients if no recycling occurs. However, when organisms die, they sink and decompose. Nutrients are then released into the deep ocean water, which is therefore rather rich in nutrients. These nutrients are then transported to regions of upwelling (see below) by the thermohaline circulation, and so returned to the surface layer.

An ocean current can be described as a more or less permanent, or continuous, directed movement of ocean water in one of the earth's oceans. Ocean currents can extend over thousands of kilometres. The currents are generated by forces acting on the water, such as the earth's rotation, wind, temperature and salinity differences, and the gravitational pull of the moon. Deep ocean currents are driven by density and temperature gradients (buoyancy). The action of wind on the surface of the water gives rise to *wind-driven circulation*, while differences in buoyancy due to changes in temperature and salinity cause a convection called *thermohaline circulation*. The direction and strength of the currents are influenced by the depth contours, the shoreline, and by other currents.

Thermohaline circulation, also known as the ocean conveyor belt, gives rise to the deep density-driven ocean basin currents. These currents, which flow under the surface of the

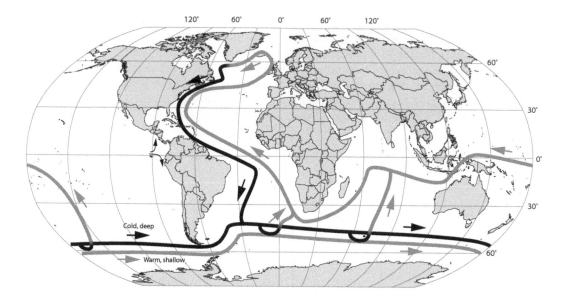

FIGURE 3.43 Global oceanic circulation flows caused by density gradients. The cold, deeper flows are depicted in black, and the warm, shallower flows in grey. The arrows indicate the direction of flow. Heat transport in the Atlantic, with a warm flow across the equator, is quite different from that in the Pacific. On reaching the Arctic, the water cools and sinks, before being transported back to the Southern Hemisphere. This type of overturning cell does not exist in the Pacific.

ocean, and are thus not immediately detectable, are sometimes called submarine rivers. A schematic view of these currents, which in reality are highly complex turbulent flows, is shown in Figure 3.43. These cold, deep currents flow at depths of 2–4 km, while warm, shallow currents flow at depths above 2 km.

The major wind-driven surface currents are shown in Figure 3.44. The deflection and rotation of the currents are results of the Coriolis effect and *Ekman transport* (see below). Many of the ocean's largest currents circulate around warm, high-pressure areas called *gyres* (these include subpolar gyres and subtropical gyres). These currents are very important for the earth's climate, especially for regions lying close to the oceans.

The strong Antarctic subpolar current, which is deep and cold, flows around the Antarctic in a westward direction. Although it is rather slow, it transports a water volume that is about twice that of the Gulf Stream, and it affects other currents in the Southern Hemisphere. For example, the Peru and the Benguela currents receive their water from the Antarctic subpolar current and are therefore cold currents. Since the Arctic is not surrounded by open water, it has no corresponding current, but instead, smaller, cold, currents flow south through the Bering Strait (the Oyashio and California currents) and near Greenland (the Labrador and East Greenland currents).

Warm water is transported to the Arctic Ocean by redirection of the Kuroshio current through the Bering Strait, and redirection of the Gulf Stream by Norway and Spitsbergen. These warm currents give rise to a milder winter climate in the eastern and western parts of North America, the British Isles, Ireland and Scandinavia, and Japan. Higher temperatures promote increased precipitation by increasing the evaporation of water. On the other hand, cold currents such as the Peru and Benguela currents, reduce evaporation, giving rise to the deserts of Peru, Chile and Namibia.

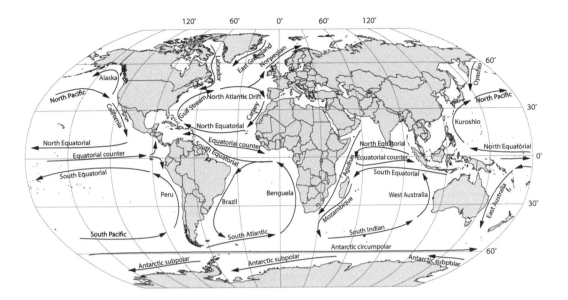

FIGURE 3.44 The main ocean surface currents.

Wind-driven circulation occurs mostly in an upper water layer, extending to a depth of around 100 m, which is called the *Ekman layer*. This circulation is mainly horizontal, in contrast to the vertical circulation caused by the density gradient. Circulation within the Ekman layer can be explained by combining the effects of wind stress on the water surface, the Coriolis force, and the water's internal friction (its dynamic viscosity). Friction between the air (wind) and surface water causes water to move in a direction determined by the wind direction and the Coriolis force, as shown in Figure 3.45 for the Northern Hemisphere. At equilibrium between the driving force from the wind and the Coriolis force, the surface water will flow at 45° to the wind direction.

Wind-driven surface water will drag deeper water with it due to viscosity, although at a slightly lower speed than the surface water. This water is also affected by the Coriolis force, and also moves in a direction different from the wind direction. It will also move in a direction different from the layer above, as shown in Figure 3.45. Applying the same arguments to the deeper water layers, we find that as we move deeper into the oceans the speed of flow decreases, and the direction of the water flow will become more and more deflected from the wind direction. The direction of the water flow describes a spiral, the *Ekman spiral*, as shown in Figure 3.46. The Ekman spiral extends down to the bottom of the Ekman layer, where the water is unaffected by the action of the wind and flows in the opposite direction to it.

The effect of the Ekman spiral is that a net flow of water is directed 90° to the wind direction; to the right in the Northern Hemisphere and to the left in the Southern Hemisphere. In shallower waters, the Ekman spiral does not reach its lowest point and the water therefore flows at a smaller angle to the wind.

Ekman transport has an important effect on the processes known as *upwelling* and *downwelling*, which influence the surface temperature and biological productivity. Upwelling waters may originate below the pycnocline and are therefore colder than the surface waters they replace. Upwelling waters are sometimes confined to the mixed layer, but this depends on the thickness of the warm layer.

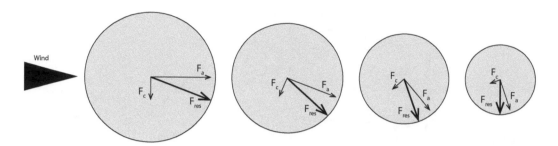

FIGURE 3.45 Surface water will move in a different direction to the wind because of the combined effect of the driving force of the wind and the Coriolis force. The surface water will in turn exert a driving force on the water below it, but this water will move in a direction different from that of the surface water due to the Coriolis force. This process is then repeated in deeper water layers.

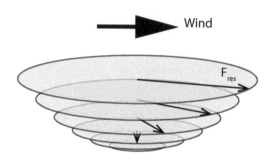

FIGURE 3.46 The water flow below the surface will change direction gradually as the depth increases, due to the effects of drag from upper water layers, the Coriolis force, and dynamic viscosity. At the bottom of the Ekman spiral, the water flows in a direction perpendicular to the wind direction.

An example of coastal upwelling caused by Ekman transport can be found in the Southern Hemisphere, off the west coast of South America. The wind blowing in a northerly direction along the coast causes a net water flow away from the shore due to Ekman transport. The water leaving the coastal areas is then replaced by deeper water, as shown in Figure 3.47.

Upwelling can also be experienced at the beach on a windy day when warm surface water is blown offshore and replaced by colder water from below. Where the thermocline is shallow, the upwelling water is usually rich in dissolved nutrients such as the nitrogen and phosphate compounds required for phytoplankton growth. Nutrients are transported to the surface, where sunlight—also required for phytoplankton growth—is available in the so-called *photic zone*, resulting in rapid growth of phytoplankton populations. Since phytoplankton form the basis of marine food webs, the world's most productive fisheries are located in areas of coastal upwelling (especially in the eastern boundary regions of the subtropical gyres). Approximately half the world's total fish catch comes from these upwelling zones.

In contrast, in zones of coastal downwelling, the surface layer of warm, nutrient-deficient water sinks. Downwelling reduces biological productivity and transports heat, dissolved

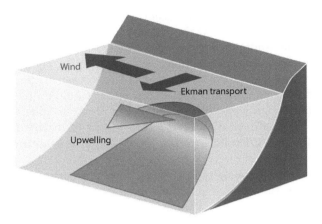

FIGURE 3.47 Wind blowing from south to north gives rise to a net flow of water away from the shore, because of Ekman transport. This water is replaced by deeper, colder water. The process is called upwelling.

materials, and surface water rich in dissolved oxygen to greater depths. This occurs along the west coast of Alaska, in the region of the eastern boundary of the Gulf of Alaska gyre.

3.3.2 Deposition and Transport of Radionuclides

The oceans receive contamination mainly from air pollution (33%) and land-based discharges from rivers and streams (44%). In addition, maritime transportation and dumping contribute 12% and 10%, respectively (FAO 1998). Until 1993, several countries carried out ocean dumping of radioactive waste, despite the disposal of radioactive waste in the oceans having been banned by, for example, the 1972 *London Convention* ("London Convention on the Prevention of Marine Pollution by Dumping of Wastes and Other Matter").

Planned and accidental releases of radionuclides into the environment have occurred from a multiplicity of sources. Given that more than 70% of the earth's surface consists of oceans, it is not surprising that much of the material released in this way now resides in the marine environment. The largest ocean, the Pacific Ocean, has been a major recipient—partly because of its sheer size, but also as a result of local releases. The known sources of anthropogenic radionuclides in the marine environment include global nuclear fallout following atmospheric weapons tests, fallout from the Chernobyl accident, discharges of radionuclides from nuclear installations, dumping of nuclear waste, nuclear submarine accidents, run-off from the sites of nuclear weapons tests, loss of nuclear weapons and radioactive sources, and satellite burnup in the atmosphere. The doses to people from the consumption of seafood are, however, generally much smaller than the doses from food of terrestrial origin. This is because many radionuclides of major concern, such as radiocaesium, are diluted by chemical analogues in salt water. This is, however, not the case in freshwater, where much higher concentrations are generally found in biota. As a result, freshwater fish can constitute a relatively important fraction of the dose to critical groups.

Knowledge of the behaviour of radionuclides in the oceans is important for several reasons. First, the fate of radionuclides must be well understood in order to provide a basis for the assessment of real or perceived adverse consequences for the environment or human health. Second, this accumulated knowledge provides a critical basis for rapid assessment of the impact of future releases, especially accidental ones. Such events include releases from coastal nuclear facilities, nuclear waste disposal sites, and maritime transport of nuclear fuel or wastes. Finally, radionuclides are powerful tracers, providing basic insights into a variety of oceanic processes. For example, plutonium has a high affinity for particulates, and is readily incorporated into several stages of the ocean carbon cycle. Because of the relatively well-defined temporal and spatial inputs of plutonium into the oceans, its movement provides many insights into a large number of processes in the oceanic water column, and in biological and sedimentary systems. Assessment of marine radioactivity in a given region therefore requires knowledge of the source terms and an understanding of oceanic processes.

The abundance of radionuclides in the marine environment varies between geographic regions because both the sources and the subsequent dispersion of the radionuclides are localized, at least to some extent. The main global source of radionuclides in the marine environment is fallout from atmospheric nuclear weapons tests. However, the concentration of ^{137}Cs in some regions, such as the Irish Sea, the Baltic Sea, and the Black Sea, depends on inputs caused by discharges from reprocessing facilities and from the Chernobyl accident. An estimate of inputs of ^{137}Cs to the oceans from global nuclear weapons fallout is shown in Figure 3.48. The Baltic Sea was the marine environment most heavily contaminated by the Chernobyl accident, receiving 4.5 PBq ^{137}Cs, and since then it has formed the main source of ^{137}Cs entering the Atlantic Ocean. The input into the world's oceans from the Chernobyl accident was significantly less than the input from global fallout. It is estimated that about 16 PBq was deposited in fallout, mainly on the North Atlantic and the Arctic Oceans. The reason for this more regional deposition is that debris from the reactor accident at Chernobyl was distributed in the troposphere, while the nuclear weapons tests led to the dispersion of radionuclides in the stratosphere.

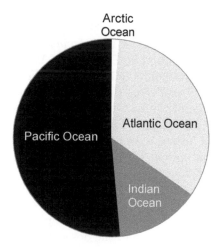

FIGURE 3.48 The total input of ^{137}Cs to the oceans from global fallout is estimated to be around 603 PBq. The Arctic Ocean has received 7.4 PBq (1%), the Atlantic Ocean 201.1 PBq (33%), the Indian Ocean 84.0 PBq (14%) and the Pacific Ocean 310.6 PBq (52%). The estimate for ^{90}Sr is found by dividing the values for ^{137}Cs by 1.6. Data from IAEA (2005).

Figure 3.49 shows inventories for the year 2000 in PBq of ^{90}Sr, ^{137}Cs, and ^{239,240}Pu, in the world's oceans. The figure shows the contributions from various sources in each of four latitude bands: 90° N–30° N, 30° N–0°, 0°–30° S and 30° S–90° S. The distribution of global fallout has been discussed above, in Section 3.2.6, but the contributions designated local fallout are due to nuclear weapons tests performed in the Pacific Ocean. Discharges from nuclear fuel reprocessing plants or dumping of liquid and solid radioactive wastes have been more localized than global fallout, although soluble radionuclides have been transported over long distances by prevailing currents. In order to make reliable measurements and estimates of the input of radionuclides from local sources, the distribution of radionuclides in the world's oceans and seas must be better known.

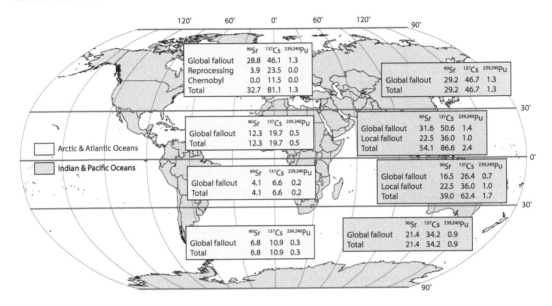

FIGURE 3.49 Inventories in PBq of ^{90}Sr, ^{137}Cs and ^{239,240}Pu in the year 2000 in the Arctic and Atlantic Oceans, and the Indian and Pacific Oceans. Data from IAEA (2005).

The greatest source of radionuclides from nuclear fuel reprocessing has been the Sellafield plant in the UK, which has released radionuclides into the Irish Sea on several occasions. Releases from the French plant, La Hague, have mostly been significantly smaller. The maximum releases of ^{137}Cs, in liquid discharges from Sellafield, occurred between 1975 and 1983, with an annual release of approximately 2–6 PBq (UNSCEAR 2000). Thereafter, releases have decreased, as can be seen in Figure 3.50.

Radiocaesium and ^{99}Tc from Sellafield have been used to trace the flow of water masses from the Irish Sea to the North Sea, further along the Norwegian coast and into the Arctic. Significant amounts of the fission product ^{99}Tc (180 TBq) were released in 1977, and in the years between 1994 and 1997, when 100–200 TBq was released annually (Dowdall et al. 2005). During the last two decades, significant amounts of ^{129}I have also been released (UNSCEAR 2008).

The activity concentration in seawater is generally rather low, on the order of mBq L^{-1}, but radionuclides in marine systems can provide useful information on environmental processes. Measurements of water and algae can provide information on transport routes and sea currents, as well as transport times and dilution factors. Measurements can also be used to determine the residence time of a radionuclide in a sea basin and to obtain sedimentation rates (and, also, to date the sediment). As an example, Figure 3.51 is taken from a study of how radiocaesium is transported from Sellafield into the North Sea and Northern Atlantic. It is possible to determine transit times by comparing the isotopic ratios

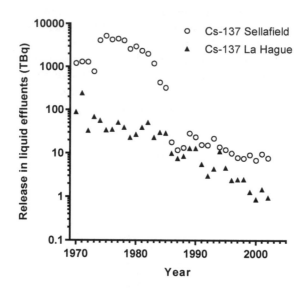

FIGURE 3.50 Annual activity of ^{137}Cs released in liquid effluents from the reprocessing plants Sellafield and La Hague. Data from UNSCEAR (2000) and UNSCEAR (2008).

for ^{134}Cs and ^{137}Cs if the ratio at the time of release is known. Since the half-life of ^{134}Cs is only about 2 years, compared to about 30 years for ^{137}Cs, the isotopic ratio will change with time, as discussed in Example 3.8.

Example 3.8 *Isotopic ratio*

The activity concentrations of ^{134}Cs and ^{137}Cs in seawater due to a release from a nuclear facility to the marine environment, are estimated to be 2.5 mBq L^{-1} and 5.0 mBq L^{-1}, respectively. We want to estimate the isotopic ratio in a sample taken two years after the release.

The isotopic ratio immediately after release is 0.5, and during transport the activity of each of the caesium isotopes will decrease due to decay as

$$A(t) = A_0 \cdot e^{-\lambda t}, \tag{3.55}$$

and the isotopic ratio is then given by

$$\frac{A_{134}(t)}{A_{137}(t)} = \frac{A_{0,134}}{A_{0,137}} \frac{e^{-\lambda_{134}t}}{e^{-\lambda_{137}t}}, \tag{3.56}$$

where λ_{134} and λ_{137} are the decay constants for the two isotopes and the initial isotopic ratio is given by

$$\frac{A_{0,134}}{A_{0,137}}. \tag{3.57}$$

With $T_{1/2} = 2$ y for ^{134}Cs and $T_{1/2} = 30$ y for ^{137}Cs, we obtain $\lambda_{134} = 0.347$ y^{-1} and $\lambda_{137} = 0.023$ y^{-1}. The isotopic ratio will then vary with time as

$$\frac{A_{134}(t)}{A_{137}(t)} = \frac{A_{0,134}}{A_{0,137}} \frac{e^{-0.347 \cdot t}}{e^{-0.023 \cdot t}} = 0.5 \cdot e^{-0.324 \cdot t}. \tag{3.58}$$

Inserting $t = 2$ y yields

$$\frac{A_{134}(t)}{A_{137}(t)} = 0.5 \cdot e^{-0.648}, \tag{3.59}$$

i.e., a value of 0.26.

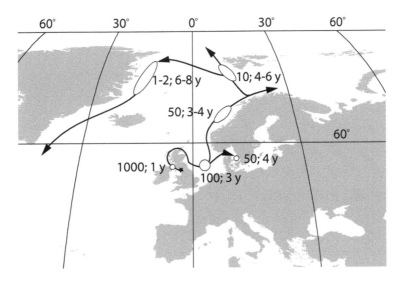

FIGURE 3.51 Route for transport of radionuclides from the Sellafield nuclear fuel reprocessing plant in the North Atlantic. Relative concentrations (^{134}Cs/^{137}Cs) and transit times (in years) are shown. Data from Dahlgaard et al. (1986).

The inhomogeneous input of radionuclides to the oceans leads to a variability in activity concentrations, which is further accentuated by the differences in area between the world's oceans (when considering the activity concentration in surface water). Figure 3.52 shows the activity concentration for ^{90}Sr, ^{137}Cs and ^{239,240}Pu in the surface water of the oceans and seas, corrected for decay to 1 Jan 2000. The location of the oceans and seas included in this figure is depicted in Figure 3.53, which shows the sampling areas used for calculations of the average concentrations. The highest activity concentrations are found in the oceans and seas within, or close to, Europe as the sites of the largest release are found there. These values are shown schematically in Figure 3.54.

It has been estimated by modelling that recent releases of ^{137}Cs from the Fukushima nuclear power plant accident in 2011 will be distributed through the entire Northern Pacific within ten years, resulting in activity concentrations below 1 mBq L^{-1} (UNSCEAR 2013). These concentrations are less than the concentrations in the Northern Pacific before the accident, which were about 2 mBq L^{-1} according to Figure 3.52.

The behaviour of the radionuclides released into the ocean will, to a large extent, depend on their chemical properties, which affect their solubility and affinity for particulates in the water. *Conservative radionuclides*, such as Cs, Sr, C and I, are highly soluble in seawater and will therefore follow the general transport and mixing of the water. The redistribution of these radionuclides thus depends on physical oceanic processes. *Particle-reactive (non-conservative)* radionuclides, for example, isotopes of Pu, Ru and Ce, have a high affinity for the surfaces of particles suspended in water. They are therefore more rapidly removed from the surface water as the particles sink, being transferred into the deep sea and incorporated into sediment.

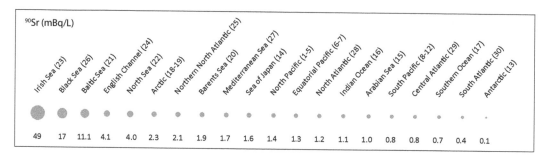

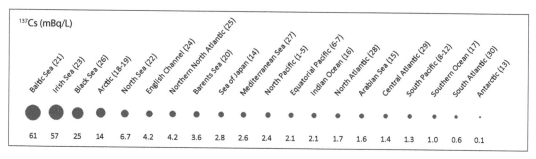

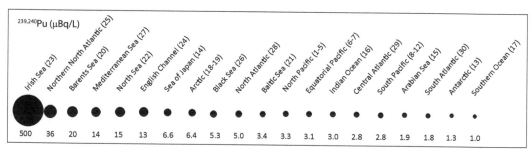

FIGURE 3.52 Activity concentrations of ^{90}Sr, ^{137}Cs and ^{239,240}Pu in surface seawater. The area of the circles is proportional to the activity concentration, given numerically below the circles. Note that activity concentrations for ^{239,240}Pu are given in units of μBq L^{-1}. The numbers given after each location refer to the latitudinal boxes shown in Figure 3.53. Data from IAEA (2005).

However, whether the behaviour of a radionuclide is conservative or particle-reactive also depends on local physical and chemical conditions. For example, ^{137}Cs bound to clay particles in coastal areas can be rather rapidly scavenged, thus acting as a particle-reactive radionuclide. For Pu, the behaviour depends on local chemical conditions in the water since reduced Pu is more particle-reactive than oxidized Pu, which is more soluble (IAEA 2005). The radionuclide may also be incorporated into some other matrix, which will then influence its behaviour in the water.

Examples of the vertical distribution of Cs and Pu are shown in Figure 3.55. In the

Mediterranean Sea, the ratio between the ^{239,240}Pu and the ^{137}Cs activity concentrations increases with depth, down to around 1500 m, indicating that Pu is attached to particles in the water. The different behaviour of a conservative and a particle-reactive radionuclide is clearly seen in the data from French Polynesia (South Pacific), where ^{137}Cs is almost absent at 1000 m depth. The data for Pu from the Enewetak Atoll, in the North Pacific (see Section 1.3.2.1) show a large increase at greater depths compared to the surface water, which is also consistent with the behaviour of a particle-reactive radionuclide.

The data for Pu from the water close to the Enewetak Atoll are derived from measurements using *Inductively Coupled Plasma Mass Spectrometry* (ICP-MS), which enables a distinction to be made between the two plutonium isotopes ^{239}Pu and ^{240}Pu. This discrimination is not possible when using α spectrometry, and data are therefore usually given as the sum of the activity for the two isotopes. These measurement techniques will be further discussed in Chapter 4.

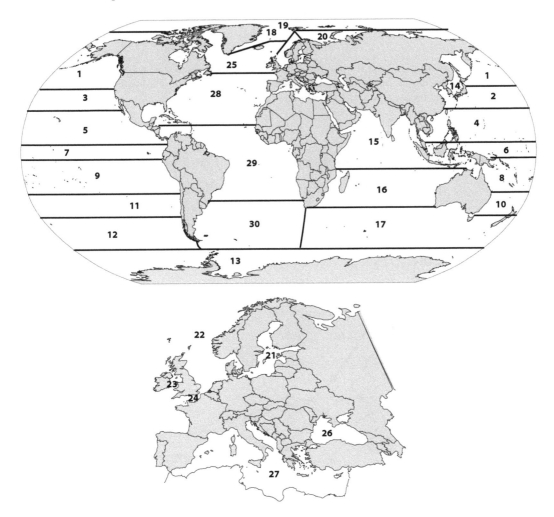

FIGURE 3.53 Latitudinal boxes used to define areas of the world's oceans and seas. Data from IAEA (2005).

The geochemical behaviour of different radionuclides, including particle absorption, re-

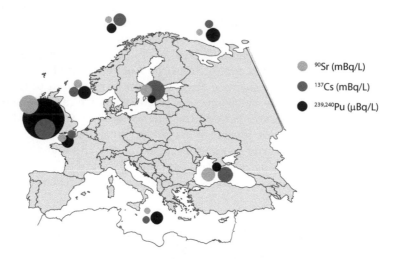

FIGURE 3.54 Schematic illustration of the activity concentration in seas within and around Europe. The area of the circles is proportional to the activity concentration (note that activity concentrations of ^{239,240}Pu are given in units of μBq L^{-1}). Data from IAEA (2005).

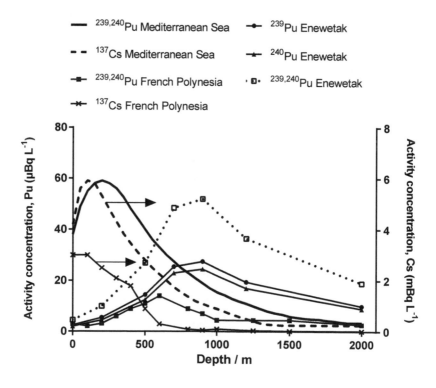

FIGURE 3.55 Vertical distributions of Pu and Cs in seawater in the Pacific Ocean and the Mediterranean. The curves are to guide the eye and do not represent fits to the data points. Data from Fukai et al. (1979), Hamilton (2004) and Povinec (2004).

dissolution, absorption onto plankton, and sedimentation, generally results in a subsurface maximum, as shown in Figure 3.55. These processes are depicted in Figure 3.56. Also, *bioturbation*, the process by which animals or plants rework soil and sediment, can be studied by measurements of radionuclides. Bioturbation increases the exchange of particles between sediment and the water because of the increased volume of the boundary between the two (the *sediment–water interface*).

The relationship between activity concentrations in sediment and water is given by the *distribution coefficient*, K_d, defined by the ICRU (2001) as

$$K_d = \frac{A_m}{A_v}, \tag{3.60}$$

where A_m is the activity concentration in a specified solid phase (e.g. sediment), given in units of Bq kg^{-1}, and A_v is the activity concentration[7] in a specified liquid phase (e.g. water), given in units of Bq L^{-1}. The distribution coefficient is thus given in units of L kg^{-1}. However, K_d may also be defined in terms of the activity concentration per unit *mass* of water. In this case, K_d will be dimensionless (IAEA 2004).

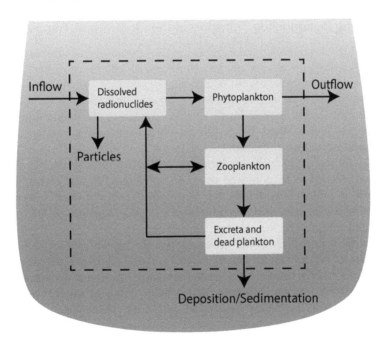

FIGURE 3.56 Radionuclides that enter a body of seawater may be attached to particles or dissolved in the water. The dissolved radionuclides may then be taken up by, or adsorbed onto, phytoplankton, which in turn are eaten by zooplankton. Direct adsorption and uptake are also possible for the zooplankton. Excreta and dead plankton, including, for example, the shells of algae from the phytoplankton, are deposited on the sea floor, but they may also be dissolved and their radionuclides recycled.

The definition of K_d requires an equilibrium between the two phases and that reversible

---

[7] According to the definition given by the ICRU, these activity concentrations are denoted *mass activity density* and *volumetric activity density*, respectively. As mentioned in a previous footnote, since the term activity concentration is widely used, it will be used throughout this book.

exchange takes place between the water and the sediment. This is seldom the case, but the K_d concept can nevertheless be applicable for modelling radionuclide transfer on longer timescales and for homogeneous distributions (for example, when hot particles, i.e. particles with high activity, are not present). Values of K_d have been given by the IAEA (2004) for the open ocean, as well as for the ocean margin. The recommended values of K_d for the open ocean vary over several orders of magnitude, from 1 (H, Na and Cl) to $2 \cdot 10^8$ (Mn and Fe). In the transitional zones between the oceans and the continents (the ocean margin), K_d varies between $3 \cdot 10^{-2}$ (Cl) and $2 \cdot 10^8$ (Fe).

A large distribution coefficient thus implies that a large fraction of the activity is transferred to the sediment, and a smaller amount of the radionuclide will therefore be transported in the water. This is the case for particle-reactive radionuclides such as Pu, where K_d is $1 \cdot 10^5$, both for the open ocean and the ocean margin. Conservative radionuclides, such as Cs, have lower values of K_d: $2 \cdot 10^3$ for the open ocean and $4 \cdot 10^3$ for the ocean margin. The fact that K_d has a finite value indicates that Cs does not remain in dissolved form permanently, and thus that particulate transfer should to be taken into account when high accuracy is needed.

3.3.3 Modelling Radionuclide Transport in the Oceans and Transfer to Biota

Several models have been developed to describe the transport of radionuclides in the oceans. Many of these are hydrodynamic models that predict the three-dimensional flow field by numerical solution of hydrodynamic equations. Depending on the application (studies of nuclear weapons fallout, or regional releases from a nuclear facility, for example), these models can be global, regional or local. In order to study the behaviour of radionuclides, the hydrodynamic model must be combined with a transport model, taking advection, diffusion and scavenging into account.

Numerical solution of the hydrodynamic model requires discretization in space and time. Vertical discretization can be achieved using one of three different approaches, all of which may cause various problems when solving the model. One approach is to define layers that take the bottom topography into account, so that the thickness of the layers varies with location. The thicknesses can also be defined as a constant fraction of the depth, and thus be of varying thickness. In this case, the same number of vertical layers is used at each location, regardless of the water depth. This may, however, cause problems where pressure gradients intersect the layers. The third approach is, therefore, to define layers that coincide with the pressure gradients (or, equivalently, with salinity). Hybrid models have been developed to overcome the specific problems associated with each of these approaches (Harms et al. 2003).

A number of grids have been constructed for horizontal discretization. These grids consist of a mesh covering the modelled area, which extend vertically to the bottom. These volumes are then divided into layers according to the vertical discretization. It is desirable to make the mesh as fine as possible, although a finer mesh requires more computing power. To reduce the computing time, the temporal discretization can be coarser, but increasing the length of the time step in the numerical solutions is limited by the requirement that it should be possible to model the actual flow of the water. Thus gravity waves and horizontal advection velocities place a lower limit on the time step, since the fluid may not travel further than one grid box in that time.

Box models provide an alternative to hydrodynamic models. These are compartment models where the compartments consist of boxes in a horizontal grid, covering the area of interest. The grid size is usually much coarser than that required for hydrodynamic models, and there are no requirements on the shape of the boxes. Since a box model cannot determine water flow, this must be imported into the model from other sources, such as field measurements or hydrodynamic modelling. The box model discussed here, as a representative example, was described by Nielsen et al. (1995). This model includes resuspension of radionuclides from sediment in water, as well as transfer into the sediment by diffusion and bioturbation.

The outlines of the boxes in the North Atlantic and Arctic Sea, which are parts of the larger model, are shown in Figure 3.57. The activity in each box, or compartment, changes with time according to Eqn. 3.15, which can be rewritten as

$$\frac{dA_i}{dt} = -A_i(t) \sum_{\substack{j=1 \\ j\neq i}}^{n} k_{j,i} + \sum_{\substack{j=1 \\ j\neq i}}^{n} k_{i,j} \cdot A_j(t) - k_i A_i + Q_i, \tag{3.61}$$

where A_i and A_j are the activities in box i and j at time t. Radioactive decay is modelled as a loss of material from box i without there being any transfer to another box, and is given by the transfer rate k_i. Q_i is the input from a continuous source into box i.

It is reasonable to assume that the transfer rate $k_{i,j}$ is related to the water flow rate between two boxes, expressed as a volume exchange rate, $R_{i,j}$ [km^3 y^{-1}], and to the volume of the water contained within box i, V_i. In this model, these quantities are related by

$$R_{i,j} = k_{i,j} V_i. \tag{3.62}$$

The radionuclides in the water column can be dissolved in the water or attached to particles (suspended sediment), and partitioning between these two phases is determined by the distribution coefficient, K_d. The fraction of the radionuclide dissolved in the water, F_w, is then given by

$$F_w = \frac{1}{1 + K_d \cdot SSL}, \tag{3.63}$$

where SSL is the suspended sediment load, which has units of kg L^{-1} if K_d is given in L kg^{-1}. Since partitioning is based on K_d, it is assumed that the dissolved radionuclides and those adsorbed onto particles are in equilibrium. This assumption is reasonable because the time steps used in box modelling are usually rather large—days or more (Harms et al. 2003).

The transfer rate from a water compartment, i, to a sediment compartment, j, can be found by assuming a mean water depth, d_i, of the water column, and a mass sedimentation rate SR [kg m^2 y^{-1}]. This gives

$$k_{i,j} = \frac{K_d \cdot SR_i}{d_i(1 + K_d \cdot SSL_i)}. \tag{3.64}$$

The box model described above was later revised to provide a better description of the dispersion of radionuclides in the oceans (Iosjpe et al. 2002). In compartment modelling, it is assumed that the contents of a compartment are homogeneously mixed, and that the mixing takes place instantaneously after input (see Section 3.1). The degree of homogeneous mixing of the activity in each box can be sufficient for dose assessment since in any given time step, integrated concentrations are more relevant than point estimates. The parameters used in the dose assessment (for example, activity concentrations in seafood), are also given as averages over a period of time. However, the assumption that mixing occurs immediately in every box after an input implies that mixing occurs simultaneously over the whole of the

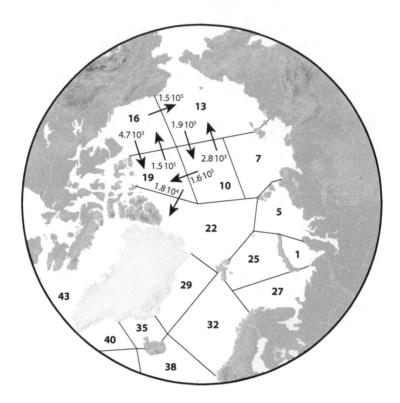

FIGURE 3.57 The regions shown refer to surface water boxes, which have underlying compartments for water and sediment. Examples of volume exchange rates [km³ y⁻¹] are given for boxes 13, 16 and 19 for the flows indicated by the arrows. Modified from Nielsen et al. (1995).

ocean included in the model. This is not realistic, and introduces an additional source of error.

The output of the box model described above is the activity concentration in each box. Provided that a relation can be found between the activity concentration in the water and the concentration in biota, the concentrations given by the model can then be used to estimate uptake from the water by the biota, and thus the effective dose to a population.

The activity in various kinds of biota is usually given as the activity per unit mass, based on either wet or dry weight[8]. The relation between the activity concentration in the water and that found in biota is given by the *concentration factor*, CF, defined by

$$CF = \frac{A_b}{A_w}, \tag{3.65}$$

where A_b and A_w are the activity per unit mass [Bq kg⁻¹] in the organism and in the water, respectively. A concentration factor defined in this way is dimensionless, but CF can also be defined as the ratio of the activity per unit mass in the organism to the activity per unit volume of water, in which case the units are L kg⁻¹. Since the density of seawater is close to unity, the numerical difference between the two definitions is not very large.

---

[8]Strictly speaking, the terms ought to be denoted wet mass and dry mass. However, it is common to use the term weight.

It should be emphasized that CFs are derived assuming equilibrium between radionuclide concentrations in the organism and in the water.

Organisms also take part in the transport of radionuclides in the marine environment. In addition to the physical and chemical processes discussed above, radionuclide distribution may also be affected by transfer in marine food chains and by migration of species. The extent to which these biological processes contribute to the transfer of radionuclides in the oceans is determined by the amount of biomass at a given location (Fowler and Fisher 2004).

Uptake of radionuclides in marine organisms can occur by accumulation from water or from food, depending on the species. However, the relative importance of the various accumulation processes is not reflected in the concentration factors, which only give the relation between the activity per unit mass in the organism and in the water. Furthermore, accumulation in an organism is affected by several external and internal factors. External factors include the exposure time, the physical and chemical form of the radionuclide, the salinity, the temperature, and the presence of other substances able to influence the rate of uptake of a radionuclide. Internal factors are characteristic for a particular species, and include its physiology and feeding habits (Fowler and Fisher 2004).

The concentration factor does not take the various modes of intake into account, since it only relates the activity concentration in an organism to that found in the water. When applying concentration factors in practical situations, it should also be noted that several of them have been determined under laboratory conditions, and may thus deviate from concentration factors in the field (IAEA 2004).

Examples of concentration factors for various organisms are given in Tables 3.15 and 3.16. The variation observed between elements and species can be explained, to some extent, by the various modes of accumulation. Phytoplankton absorb radionuclides rapidly, and equilibrium is reached within minutes or hours. Because they are small, their surface area is large compared to their volume, and concentration factors can be large. Zooplankton take up elements directly from the water and by ingestion. The concentration in an organism is therefore in dynamic equilibrium with its surroundings, and large variations in the concentration factor may be found for a given species. Uptake directly from the water is affected by the behaviour of the radionuclide. Particle-reactive radionuclides tend to show larger concentration factors than radionuclides with more conservative behaviour. This is reflected by the low concentration factors for ^{137}Cs (Tables 3.15 and 3.16).

TABLE 3.15 Concentration factors for macroalgae and plankton.

Element	Macroalgae	Zooplankton	Phytoplankton
Co	$6{\cdot}10^6$	$7{\cdot}10^3$	$2{\cdot}10^3$
Zn	$3{\cdot}10^3$	$1{\cdot}10^5$	$1{\cdot}10^4$
Zr	$3{\cdot}10^3$	$2{\cdot}10^4$	$6{\cdot}10^4$
Ru	$2{\cdot}10^3$	$3{\cdot}10^4$	$2{\cdot}10^5$
Cs	$5{\cdot}10^1$	$4{\cdot}10^1$	$2{\cdot}10^1$
Pb	$1{\cdot}10^3$	$1{\cdot}10^3$	$1{\cdot}10^5$
Pu	$4{\cdot}10^3$	$4{\cdot}10^3$	$2{\cdot}10^5$

Source: Data from IAEA (2004).

One way to model the radionuclide concentration in a marine organism is to use a first-order function of the activity concentration in the water and on particles (Fowler and Fisher 2004). If the concentration of the radionuclide in the organism is denoted C [Bq g^{-1}] and t is the exposure time [d], the change in C can be written

TABLE 3.16   Concentration factors for marine animals.

Element	Fish	Crustaceans	Molluscs*	Cephalopods
Co	$7 \cdot 10^2$	$7 \cdot 10^3$	$2 \cdot 10^4$	$3 \cdot 10^2$
Zn	$1 \cdot 10^3$	$3 \cdot 10^5$	$8 \cdot 10^4$	$6 \cdot 10^4$
Zr	$2 \cdot 10^1$	$2 \cdot 10^2$	$5 \cdot 10^3$	$5 \cdot 10^1$
Ru	$2 \cdot 10^0$	$1 \cdot 10^2$	$5 \cdot 10^2$	$5 \cdot 10^1$
Cs	$1 \cdot 10^2$	$5 \cdot 10^1$	$6 \cdot 10^1$	$9 \cdot 10^0$
Pb	$2 \cdot 10^2$	$9 \cdot 10^4$	$5 \cdot 10^4$	$7 \cdot 10^2$
Pu	$1 \cdot 10^2$	$2 \cdot 10^2$	$3 \cdot 10^3$	$5 \cdot 10^1$

* Except cephalopods
Source: Data from IAEA (2004).

$$\frac{dC}{dt} = (k_u \cdot C_w) + (AE \cdot IR \cdot C_f) - (k_e + g) \cdot C, \tag{3.66}$$

where k_u [L g^{-1} d^{-1}] is the uptake rate from the dissolved phase and C_w [Bq L^{-1}] is the activity concentration in the dissolved phase. The product $k_u \times C_w$ then represents an input to the system, with the units Bq g^{-1} d^{-1}. The corresponding input from ingested particles is given by the product of the assimilation efficiency (fractional uptake) from ingested particles, AE, the ingestion rate of particles, IR [g g^{-1} d^{-1}], and the activity concentration in the ingested particles, C_f [Bq g^{-1}]. The last term in Eqn. 3.66 accounts for the concentration-dependent losses from efflux and growth. These are parameterized by the efflux rate constant k_e [d^{-1}] and the growth rate constant g [d^{-1}], respectively.

If the concentration does not vary with time, $dC/dt = 0$ and the steady-state concentration, C_{ss}, is given by

$$C_{ss} = \frac{(k_u \cdot C_w) + (AE \cdot IR \cdot C_f)}{(k_e + g)}. \tag{3.67}$$

The relation between the concentration in the dissolved phase, C_w, and the total concentration, C_t, is given by Eqn. 3.63, and thus

$$C_w = \left(\frac{1}{1 + K_d \cdot SSL}\right) \cdot C_t. \tag{3.68}$$

The SSL is generally low in oceanic systems ($<$1 mg L^{-1}), and C_w is almost identical to C_t (IAEA 2004).

However, there is no relation similar to Eqn. 3.68 for C_{ss} and C_t, and the steady-state concentration in the organism is therefore not proportional to the total concentration in the water. Physiological parameters and environmental factors may cause the steady-state concentration to vary with time. Taking into account uptake from dissolved and particulate radionuclides, the *bioaccumulation factor* (*BAF*) can be written

$$BAF = \frac{C_{ss}}{C_t} = \left(\frac{k_u}{k_{ew} + g} + \frac{AE \cdot IR \cdot K_d}{k_{ef} + g}\right) \cdot \left(\frac{1}{1 + SSL \cdot K_d}\right), \tag{3.69}$$

where k_{ew} and k_{ef} are the efflux parameters for solute and particulate efflux, respectively.

3.4 FRESHWATER SYSTEMS

3.4.1 Lakes

3.4.1.1 Classification of Lakes

Lakes can be broadly divided into those in which the lake water mixes at least once a year and those where the lake water never mixes completely. The former are termed *holomictic lakes* and the latter *meromictic lakes*. Whether a lake becomes holomictic or meromictic depends on a combination of its average temperature and its depth. However, a meromictic water body may also form because it has very steep sides or a high salinity gradient, creating a layer of dense bottom water. Increased salinity may be caused by factors such as inflow of seawater, or salt from sediment. Stratification has a considerable impact on the transport of nutrients and oxygen within a lake, and thus affects the food web present there.

Lakes that are sufficiently deep for stratification to be possible may be classified according to the Hutchinson–Löffler system, which describes six types of lakes. *Amictic* lakes are meromictic lakes that are always ice-covered. These lakes occur mostly in the polar regions, but also at high altitudes where the average ambient temperature is sufficiently low. *Cold monomictic* lakes are ice-covered a large part of the year, and never warm above 4 °C. The water in these lakes is completely mixed once a year. In *dimictic* lakes, which are ice-covered in the wintertime, complete mixing occurs in spring and autumn, and stable stratification prevails during the summertime, as shown in Figure 3.58. Lakes that are never ice-covered are denoted *warm monomictic, oligomictic* or *polymictic*, depending on the frequency of mixing and stratification. Mixing occurs once a year in warm monomictic lakes, while mixing in oligomictic lakes occurs at irregular intervals longer than a single year. Polymictic lakes mix several times each year.

A revised system for mixing classification was proposed by Lewis (1983). In this system, the amictic, dimictic, warm monomictic and cold monomictic categories are unchanged, but the oligomictic and polymictic categories are replaced. These are replaced by the *continuous* and *discontinuous* cold monomictic, and the *continuous* and *discontinuous* warm polymictic categories, depending on the frequency of stratification and mixing. Using this classification, it becomes possible to determine a lake's type from its latitude and depth, although other factors, such as surface area, can also affect mixing.

According to Lewis (1983), amictic lakes are found between 90° and 75° latitude, regardless of depth. Deeper lakes (around 30 m deep) in the latitude band 75° to 40° tend to be dimictic, while shallower lakes are discontinuous cold polymictic. Even shallower lakes (around 20 m) at these latitudes are classified as continuous cold polymictic lakes. Cold monomictic lakes are found between 75° and 70° latitude. Deep lakes between 40° and 0° latitude are generally warm monomictic, while shallower lakes are discontinuous warm polymictic or continuous warm polymictic. The latitudes given are adjusted according to the elevation of the lakes, which affects their average ambient temperature.

Lakes can also be classified according to their trophic character: *oligotrophic, mesotrophic* or *eutrophic*. In addition, lakes with very high nutrient status (quantified by the *trophic index*) may be considered *hypereutrophic*. Oligotrophic lakes have a low production of organic compounds from carbon dioxide (from the atmosphere, or in the water) and thus have a low nutrient content. Oligotrophic lakes have characteristically clear water, a result of the low production of algae, and high oxygen content. Eutrophic lakes are characterized by high biological productivity, and are thus rich in nutrients. The large quantities of nutrients such as N and P support the growth of plants and algae, which produce oxygen by photosynthesis. Finally, mesotrophic lakes have an intermediate productivity, between oligotrophic and eutrophic lakes.

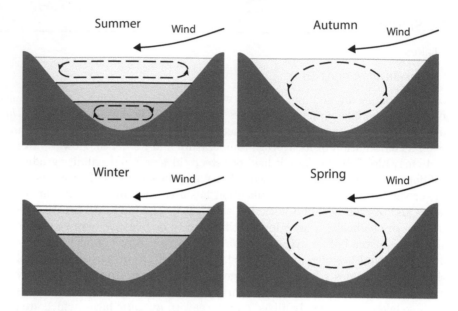

FIGURE 3.58 A dimictic lake, in which complete mixing of the water occurs in spring and autumn due to wind action and an almost homogeneous water temperature. Stratification is caused by summertime heating of the water's surface (the grey shading represents different temperatures). This heating creates a temperature gradient with a top layer, the *epilimnion*, and a thermocline (*metalimnion*), which prevents further mixing with the bottom layer (*hypolimnion*).

3.4.1.2 Transport Processes and Uptake in Biota

Lakes, rivers and seas, can be divided into four different zones: the *littoral*, the *limnetic*, the *profundal* and the *benthic* zones. The definition of the littoral zone is not unique and depends on the type of water body. For example, in seas, the littoral zone extends from the high-tide mark to the continental shelf. In freshwater systems, the littoral zone is the region near the shore, including wetlands. This zone may be more or less vegetated, depending on the condition of the lake, and extends to a depth where sunlight can no longer penetrate to support photosynthesis. The free water body away from the littoral zone constitutes the limnetic zone, where most of the lake's oxygen is produced due to the presence of phytoplankton. This zone therefore includes the surface water down to the depth where sunlight can support photosynthesis. Rooted vegetation is absent in the limnetic zone since the penetration of sunlight into deeper water is insufficient. This deeper zone, consisting of free water, is called the profundal zone, and is often also located below the thermocline. Finally, the benthic zone consists of the bottom sediment.

The most important parameters and processes that affect the flows of radionuclides and biological uptake can be divided into four main categories: *lake morphometry*, *chemical properties*, *physical properties* and *biological properties*. Lake ecosystems are very complex and many of these parameters are correlated to varying degrees. The expected effect of one particular parameter on the biological uptake of a radionuclide may be either enhanced or reduced by other parameters.

Parameters describing lake morphometry include its area and depth, which together define the volume of the water body. Other important parameters are the shape of the lake and the water retention time. The lake's shape can affect the rate of stratification and mixing, as was described above, and the water retention time is affected by the volumetric flow through the inlets and outlets of the lake.

Radionuclides in atmospheric fallout can reach the lake by direct deposition on the lake surface (*primary lake load*) or input from the lake's catchment area (*secondary lake load*). The time lag between the primary lake load and the main part of the secondary lake load can be very short, as was seen in northern Sweden after the Chernobyl accident, where as much as 10% of the ^{137}Cs deposited was rapidly transported to lakes by spring floodwaters (Andersson 1994).

Lake water retention time is an important factor that determines how much of the initial input of a radionuclide is retained by the lake. Lakes with a long retention time retain a comparatively larger part of the initial input. The area and shape of the lake are important because they determine the distribution of radionuclides on the lake bed, and also the distribution in the littoral and profundal zones.

Important chemical properties of the lake water and the sediment include pH and conductivity, as well as the concentrations of K, Ca, Mg, nutrients and organic matter (the ratio between organic and inorganic content is especially important). These properties, in combination with physical factors, such as the settling velocity of particulates, regulate the efficiency with which radionuclides are removed from the water. Studies in Swedish lakes after the Chernobyl accident showed that the chemical properties of the water (pH, K concentration, conductivity, and water hardness) were less important than the chemical properties of the suspended matter for the removal of ^{137}Cs from the lake water (Andersson 1994).

Physical properties that affect the fate of radionuclides in a lake are light conditions, turbidity (haze caused by small suspended particles), stratification, the amount of suspended matter, and the structure and composition of the bottom sediment. A high degree of light penetration promotes growth, increasing the rate and amount of biological uptake. The horizontal and vertical distribution of radionuclides in sediment, and thus the further transport of radionuclides in the lake, is greatly affected by the composition of the sediment. For example, clay-rich sediment binds a large fraction of Cs, which becomes almost immobilized, and is only weakly affected by other environmental changes. Transport of radionuclides bound to carbonates is increased by a reduction in pH, which also decreases the fraction of the radionuclides bound to organic matter. Low levels of oxygen, for example, in stratified lakes or lakes rich in organic matter, may increase the mobility of hydroxides and associated radionuclides. However, oxygen can also increase the binding of radionuclides in the form of less soluble sulphides, reducing their mobility.

Particle-reactive radionuclides (see Section 3.3.2), become attached to particles and may be transferred from the water to sediment via several mechanisms. These include sedimentation of insoluble particles in fallout, adsorption onto inorganic compounds (such as carbonates and clays), sedimentation with humic matter, sedimentation with plankton, and direct adsorption of dissolved radionuclides by sediment.

Relevant biological properties include the relation between the littoral and the profundal zones, biological activity in the water and its sediment, and transport by benthic animals (benthos)[9]. The biological activity of microbes in the sediment will greatly affect the binding and transport of radionuclides, transported to the sediment together with organic material.

The uptake of radionuclides in biota can be described by the concentration factor, CF, (defined in Section 3.3.3) or by the *biota sediment concentration ratio*, CR_{s-b} (IAEA 2010). The concentration factor relates the activity per unit fresh weight of tissue to the activity per litre in the water, taking all possible intake pathways into account. The biota sediment

---

[9]Organisms living in the benthic zone.

concentration ratio is defined as the ratio of the activity concentration in an organism (on fresh weight basis) to that in the sediment.

Table 3.17 gives concentration factors for freshwater fish, and comparison with Table 3.16 shows that the uptake is generally higher than for marine fish. For example, the concentration factor for Cs is 30 times higher in freshwater. This difference can be explained by the abundance of competing elements, such as potassium, in seawater. Although defined as a ratio at equilibrium, the concentration factor will vary with time if the deposition is of short duration. This follows from the time dependencies of the transfer mechanisms, which are determined by a large number of physical, chemical, physiological and ecological processes. Transfer rates, and thus concentration factors, also vary considerably with fish species and size.

TABLE 3.17   Concentration factors [L kg^{-1}] for freshwater fish, given as the geometric mean. Whole-body data are not available for Ru and Pu.

Element	Whole body	Muscle
Co	$4.0 \cdot 10^2$	$7.6 \cdot 10^1$
Zn	$4.7 \cdot 10^3$	$3.4 \cdot 10^3$
Zr	$9.5 \cdot 10^1$	$2.2 \cdot 10^1$
Ru	—	$5.5 \cdot 10^1$
Cs	$3.0 \cdot 10^3$	$2.5 \cdot 10^3$
Pb	$3.7 \cdot 10^2$	$2.5 \cdot 10^1$
Pu	—	$2.1 \cdot 10^4$

Source: Data from IAEA (2004).

3.4.1.3 Modelling of Lake Systems

Models of environmental processes can be used to compare ecosystems, and also to study particular processes in different parts of a single ecosystem. Models can also be used to predict the future concentration of radionuclides in aquatic species such as fish, and to estimate when the concentration will have fallen below the action level for consumption. It is therefore important that these models are based on data that are available to public authorities (Håkansson 1991).

The models that have been developed for studies of radionuclide transfer in lakes vary considerably in complexity, and can be classified into a number of basic groups. Process-oriented models are designed to estimate the transfer and uptake of radionuclides by taking physical, chemical and physiological processes into account. These models are therefore very detailed, but the availability of data and the uncertainty in the parameters describing various processes must be taken into consideration. Statistical regression models have been used for long-term predictions of levels of radioactivity in fish, and to compare the effects of fallout in various types of lake. Compartment models are generally constructed to reflect the trophic structure of the food web in a lake. These models often include parameters describing the intake rates and growth of fish.

The transfer of radionuclides in aquatic systems is governed by several processes, and most models do not attempt to consider all of them separately. Instead, they are often combined into more general parameters, such as bioaccumulation factors and distribution coefficients. These two parameters have been shown to make the largest contributions to modelling errors (Sundblad 1990). However, replacing parameters with process descriptions limits the possibility of reducing modelling errors, as there may be a strong interdependence between parameters in a particular environment. It is also desirable that the application of models in emergency situations is not dependent on a high degree of expert judgement.

A schematic representation of a compartment model for the transfer of radionuclides in an aquatic system is shown in Figure 3.59. Radionuclides entering the lake from the catchment area (including direct deposition on the lake's surface) are either dissolved in the water, forming part of the water table, or attached to particles. Settling particles are then incorporated into the sediment, from which the radionuclides are removed from the system, redistributed by bioturbation, or returned to the water table by resuspension. Dissolved radionuclides are removed from the system by outflow from the lake.

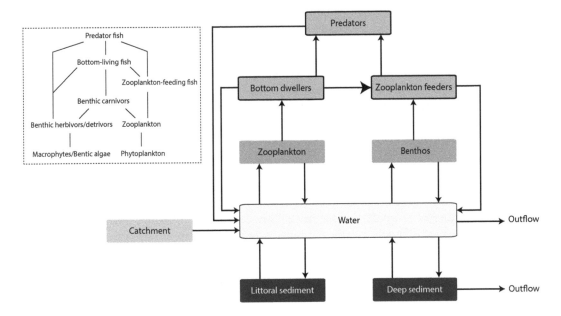

FIGURE 3.59 Compartmental representation of an aquatic food chain. Modified from Broberg and Andersson (1991) and Bergström et al. (1994).

Transfer to aquatic organisms is modelled by compartments representing a simplification of the various trophic levels in the food chain, as shown in Figure 3.59. The activity concentration in phytoplankton is assumed to be proportional to the activity concentration in the water, and these species are therefore included in the water compartment. The benthic algae, herbivores, detritivores, and carnivores are all included in the benthos compartment, while bottom-living fish, fish feeding on zooplankton, and predator fish are represented by individual compartments.

The transfer of radionuclides between species is represented by differential equations, and the activity concentration in a fish species, C [Bq kg^{-1} wet weight], can be given by (Heling 1997)

$$\frac{dC}{dt} = ak_f C_f + bk_w C_w - kC, \tag{3.70}$$

where uptake is assumed to occur both from food and from the water. Uptake from food is determined by C_f, the activity concentration in the food [Bq kg^{-1} wet weight], k_f, the amount of food consumed per unit weight per unit time [kg kg^{-1} d^{-1}], and a, the fraction of food retained in the species. Similarly, uptake from the water is determined by the activity concentration in the water, C_w [Bq kg^{-1}], the intake of water per unit mass per unit time, k_f [m^3 kg^{-1} d^{-1}], and the uptake fraction from water, b. The activity concentration in the water, C_w, can be modelled as a time-dependent parameter. Loss of the radionuclide from the species is assumed to be proportional to the activity concentration in the species, with a transfer rate equal to k [d^{-1}].

If the species is a predator fish, the concentration in its food, C_f, can be modelled by

$$C_f = \sum_{i=1}^{n} C_i P_i \frac{M}{m_i}, \tag{3.71}$$

where C_i is the activity concentration in species i of prey, and the relative preference for that prey is given by P_i, which has a value from 0 to 1. Since the dry weight fraction of different species may vary, this is taken into account by the ratio of the dry weight fraction of the predator, M, to the dry weight fraction of the prey species i, m_i.

Compartment modelling of the activity concentration in lake ecosystems shows that the maximum concentration in different species occurs at different times after the radionuclides are released into the lake. Species at the bottom of the food chain will reach maximum activity concentration sooner than species higher in the food chain. In many cases, the peak activity concentration tends to be significantly higher in predators at the top of the food chain.

The number of compartments in a model depends on how the calculations are to be used. For example, if the activity in several species of fish is to be estimated, each species should be represented by its own compartment. However, in some cases, it may be sufficient to use a smaller number of compartments if the relative uptake of different species in known. For this reason, several different representations of lake ecosystem models have been developed (see e.g. Heling 1997).

3.4.2 Rivers and Estuaries

3.4.2.1 Rivers

Rivers can often be modelled as one-dimensional systems; for example, in the model developed by Schaeffer in 1975, it was assumed that the activity concentration in water decreases exponentially downstream of the point of discharge (Thissen et al. 1999). IAEA (2001) describes a generic model for a river system, discussed below, which was developed to calculate the activity concentration at locations along the river bank. In order to make a conservative estimate of the activity concentration, the parameters are chosen to represent the lowest annual flow rate expected in a 30-year period.

The net freshwater velocity, U [m s^{-1}], can be estimated as

$$U = \frac{q_r}{B \cdot D}, \tag{3.72}$$

where B and D are the river width [m] and depth [m], respectively, corresponding to the 30-year annual flow rate, q_r [m^3 s^{-1}]. The total radionuclide concentration in the water, $C_{w,tot}$ [Bq m^{-3}], at a location on the river bank opposite the point of discharge is given by

$$C_{w,tot} = \frac{Q_i}{q_r} e^{-\lambda_i x/U}, \tag{3.73}$$

where Q_i is the average discharge rate for radionuclide i [Bq s^{-1}], λ is the decay constant for radionuclide i [s^{-1}], and x is the downstream distance from the discharge point [m].

For locations on the same river bank as the discharge point, the water can be considered undiluted at short distances ($< 7D$) from the discharge. For distances at which complete mixing can be assumed, Eqn. 3.73 can be used, however, it is multiplied by a correction factor that takes incomplete lateral mixing into account. In addition, interactions between the sediment and the radionuclides in the water must be considered by the application of a distribution coefficient.

3.4.2.2 Estuaries

Estuaries are semi-enclosed bodies of coastal water. These are connected to the open sea, but estuarine water is often brackish due to dilution by freshwater. The difference in salinity between seawater and freshwater can cause estuaries to be vertically stratified, although turbulence from the tidal current counteracts this by mixing the water. The water in the estuary will then be partially or well-mixed, depending on the strength of the tidal current. The action of the tidal current may vary over the area of the estuary, leading to variations in the degree of vertical mixing in different regions of the estuary. It is important to take the stratification of estuarine water into account when modelling the dispersion of radionuclides in an estuary (Thiessen et al. 1999).

Estuaries also tend to be dynamic systems in terms of the temporal and spatial variation in the pH, salinity, amount of dissolved organic carbon, and turbidity. High levels of dissolved organic carbon may reduce the distribution coefficient, K_d, of Pb and Pu, leading to higher concentrations of these elements in the water. The concentration of Pu can also be increased as Pu attached to particles may be released if it comes into contact with river water with a low pH (IAEA 2004). Equilibrium cannot generally be assumed in estuarine systems, although transitory equilibrium may occur. The non-equilibrium conditions are caused by factors such as tidal currents, but also by the preservation of hot particles, which may occur during deposition or discharges from rivers.

Apart from the special features mentioned above, models for estuaries are similar to those for coastal environments. Close to the banks of the estuary, models for river systems can be used with slight modifications (IAEA 2001).

3.5 THE TERRESTRIAL ENVIRONMENT

3.5.1 Soil Composition and Properties

Natural soils can generally be regarded as consisting of *minerals*, *organic matter*, *air* and *water*, all participating in complicated physical and chemical interactions. The main function of the soil is to provide nutrients and water for plants, as well as support for their roots. A large number of factors are involved in determining the properties of soil that affect the transport of elements in the ground.

The formation of soil can be described as the interplay between four types of processes: *supplies*, *losses*, *transformations* and *translocations* (Foth 1984). These processes are depicted in Figure 3.60. Supplies can, for example, consist of dead plants that rot and become part of the organic matter. Loss mechanisms include the production of carbon dioxide from the carbon in organic matter by microorganisms. An example of a transformation is the transformation of nitrogen from organic to inorganic compounds. Translocations can be caused by water movement or worms. These processes also apply to minerals, which may weather and form new chemical compounds with varying solubility in water, and thus varying transport properties.

Table 3.18 gives the relative abundance of the most common elements in the earth's crust, where it can be seen that oxygen and silicon comprise over 70%. Together with aluminium, they are the most abundant minerals. Minerals are defined as crystalline compounds that occur naturally. Around 4000 different minerals have been identified; examples include halite (NaCl, or rock salt) and quartz (SiO_2). Minerals can be classified into groups according to their chemical composition; the most common groups are silicates, carbonates, sulphates, halides, oxides, sulphides, phosphates, and pure elements (for example, diamond or graphite). The silicates constitute the largest group, with over 1000 known minerals. About 100 minerals with a uranium content exceeding 50% are known, such as uraninite (UO_2), which contains 88.15% U. Rocks are formed from one or several minerals, but only a minor proportion of the known minerals are rock-forming minerals.

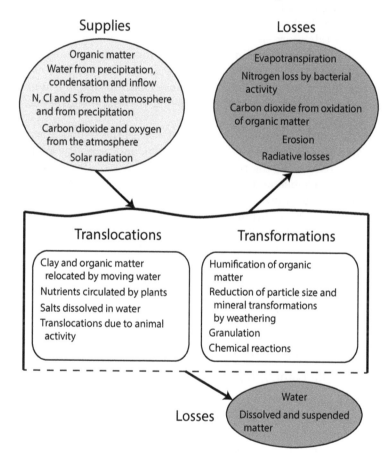

FIGURE 3.60 Schematic depiction of soil-forming processes. Modified from Foth (1984).

The inorganic component of the soil can be divided into soil separates, classified by the size of the particles, which can be determined by sieving. The size distribution of these particles, which are smaller than gravel, is then determined by the relative proportions of *clay*, *silt* and *sand*. Particles with a diameter less than 2 μm are generally classified as clay particles, and particles with a diameter larger than 2 μm, but less than about 50 μm, are generally classified as silt particles. Sand particles are larger than silt particles and are sometimes further divided into fine and coarse sand. The range of particle sizes allocated to each category may vary between countries.

TABLE 3.18 Elements with an abundance >1% in the earth's crust.

Element	Abundance	Element	Abundance
O	46.6	Na	2.8
Si	27.7	K	2.6
Al	8.1	Mg	2.1
Fe	5.0	Others	1.5
Ca	3.6		

Source: Data from Foth (1984).

3.5.1.1 Soil Classification

Soil texture is based on the relative fractions of each size class in the soil. For example, a soil consisting primarily of sand but with a small fraction of clay is called *sandy clay*, while the term *loam* is used to describe soils with approximately equal fractions of clay, silt and sand (e.g. 20% clay, 40% silt and 40% sand). The term loam can also be combined with the three others, for example, a soil containing 30% clay, 60% silt and 10% sand is classified as silty clay loam.

Soils that have similar chemical, physical and biological properties are generally grouped in classes. These properties can give rise to a large number of combinations, and standard classification systems have therefore been developed, starting with the FAO-Unesco *Soil Map of the World* in 1974 (FAO-Unesco 1974). This system has been revised several times, and more modern systems have evolved, for example, the USDA's *Soil Taxonomy*. It is beyond the scope of this book to give a comprehensive description of the various soil classes, but some of them are described briefly below.

Soil differs with depth, defined by soil *horizons*. These are separated, usually horizontal, layers that define a soil profile, as shown in Figure 3.61. The master horizons, denoted O, A, E, B, C and R, can be further separated into subdivisions (denoted, for example, B1 and B2) if distinct layers are present within a horizon. The more detailed characteristics of a horizon are indicated by adding a lowercase letter to the capital letter, which by itself only indicates a general description of the horizon.

The O horizon consists of organic matter that is essentially unaffected by decomposition processes. This horizon is common in forests where the leaf litter provides a source of organic material (however, the litter layer itself is not included in the O horizon). The uppermost horizon in peat bogs is similarly rich in organic material but it is denoted P.

The A horizon, or *topsoil*, contains a large fraction of decomposed or partially decomposed organic matter, and most of the biological activity takes place in this layer. The A horizon may be depleted of soluble elements and matter, such as iron and clay, as a result of the percolation of water. The lower part of the A horizon may then develop into an E horizon, which contains a larger fraction of minerals resistant to leaching. Due to the loss of materials, this horizon has a lighter colour than the A and B horizons. Leached elements and organic material from the E horizon are then accumulated in the B horizon, or the *subsoil*. The C horizon (parent material) is formed by weathering of the underlying bedrock, the R horizon, and is almost unaffected by soil-forming processes.

Soils of natural origin that are rich in organic material are generally referred to as *histosols*. These soils are also referred to as "peat" or "muck", and are formed by the accumulation of organic matter in areas with poor drainage and consequently high water content, which limits the decomposition of the organic material. *Anthrosols* are formed by human activities such as irrigation and cultivation. *Podzols* (spodosols) are common

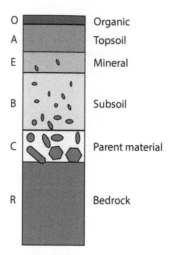

FIGURE 3.61 A generalized soil profile showing the six master horizons.

in coniferous forests and can be formed from a variety of parent materials. Note that soil profiles with the same notation may differ between regions.

In terms of the horizons discussed above, a histosol consists of the O and C horizons, whereas a podzol may consist of the O, E, B and C horizons due to extensive leaching. Another example is a *vertisol*, which has a high clay content (montmorillonite). This soil is subject to extensive mixing due to the high expansion of the clay during the wet season and the cracks formed during the dry season. The resulting soil profile is then OAC, and the B horizon is absent.

3.5.1.2 Chemical and Physical Properties of Soils

The most abundant minerals in soil contain O and Si, and their most stable configuration is reached when a silicon ion (Si^{+4}) is placed between four oxygen ions (O^{-2}) located at the corners of a regular tetrahaedron. This mineral is called quartz and is resistant to weathering. Potassium feldspar is formed by replacing some of the Si ions by Al^{+3} and K^+, resulting in a lower resistance to weathering. Therefore, minerals may be ranked with respect to their resistance to weathering, depending on their atomic composition.

Weathering can generally be described as the decomposition of minerals (or soil and rocks) through reactions with water, biota, or with the atmosphere. However, the transport of soil material by erosion is not considered to be weathering. Weathering occurs by two main processes: *physical weathering* and *chemical weathering*. Physical, or mechanical, weathering is caused by atmospheric conditions that induce thermal stress, and pressure variations due to freezing and melting, water and wind. Chemical weathering changes the composition of the material.

Dissolution, which is a form of chemical weathering, is the result of chemical reactions between minerals and acidic precipitation, although some minerals, such as halite, are soluble in neutral water. Precipitation is always more or less acidic because atmospheric carbon dioxide dissolves in water. Rocks such as limestone, which contain calcium carbonate, are especially susceptible to dissolution. This process can cause widespread formation of cracks and underground caves. In minerals with a high iron or magnesium content, *oxidation* becomes important.

Silicates, which are the most common group of minerals, are mainly weathered by *hydrolysis*, which is caused by hydrogen ions in water. Hydrolysis is enhanced in acidic water due to the higher concentration of hydrogen ions. When hydrogen ions replace other cations in the minerals, the cations are leached from the mineral, and secondary minerals, often clays, are formed. For example, weathering of granite, which can contain a high proportion of the mineral feldspar, produces the clay mineral kaolinite by the reaction

$$2KAlSi_3O_8 + 2\left(H^+ + HCO_3^-\right) + H_2O \rightarrow Al_2Si_2O_5\left(OH\right)_4 + 2K^+ + 2HCO_3^- + 4SiO_2$$

in which feldspar, carbonic acid and water react to form kaolinite, potassium ions, bicarbonate and silicon dioxide.

Clay particles have a considerable impact on the transport of nutrients in the ground, as well as of radionuclides. The clay particles have a large negative surface charge and because their shape is often flat, their surface area is very large in relation to their volume, and a large number of cations can be bound to their surface[10]. However, these cations can be replaced by other cations in the soil water by *cation exchange*. An example of a cation-exchange reaction is shown in Figure 3.62.

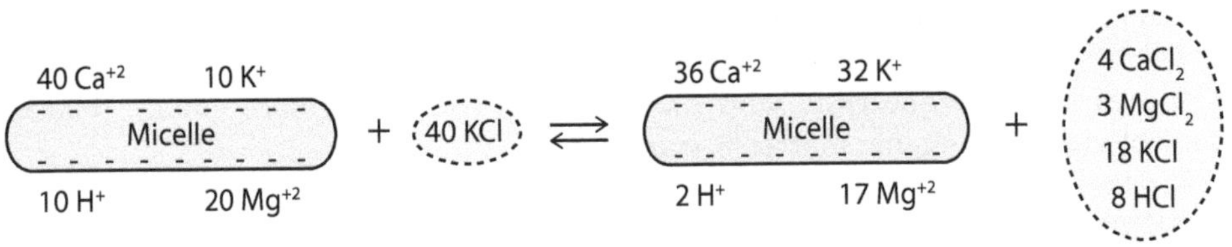

FIGURE 3.62 Example of cation-exchange reactions occurring when KCl is added as fertilizer. Modified from Foth (1984).

In this example, KCl is added as a fertilizer to the soil solution and reacts near the surface of small (colloidal) clay or humus particles (micelles). The negative charges on the micelles are neutralized by cations, which are thus adsorbed onto the surface. The distribution of cations is shown in Figure 3.62, assuming 140 negatively charged sites on a micelle. In reality, there may be thousands of negatively charged sites on each micelle. When KCl is added to the soil, it will form positively charged K ions and negatively charged Cl ions, and the increased number of K ions in solution will lead to exchange with the cations on the micelle. In the case shown here, the result of this exchange is that 22 additional K ions are adsorbed at equilibrium, replacing 4 Ca ions, 3 Mg ions and 8 H ions. These displaced ions form chlorides, which are found in the soil solution.

However, exchangeable cations are not stationary on the surface, but move between exchange sites. They are also *hydrated*, i.e. they attract water molecules, and thus acquire a *hydrated radius* that will affect the strength with which they bind. For example, the ionic radii of Na^+ and K^+ are $0.98 \cdot 10^{-10}$ m and $1.33 \cdot 10^{-10}$ m, respectively, while their hydrated radii are $7.90 \cdot 10^{-10}$ m and $5.32 \cdot 10^{-10}$ m (Foth 1984). The K ion is the most strongly bound to the surface because of its smaller hydrated radius, since the Coulomb force is inversely proportional to the distance squared. The Coulomb force is also proportional to the charge on the ion, and Ca^{+2} is therefore more strongly bound than Na^+. In this case, the hydrated radius is also smaller for the Ca ion, which further contributes to the difference in binding strength.

The ability of a soil to engage in cation exchange (i.e. the number of adsorption sites) is

---

[10]Recent results indicate that iodide (anion) may also interact with clays by forming ion pairs (Miller et al. (2015)).

quantified by its *cation-exchange capacity*, CEC, defined as the number of *milliequivalents*, *mEq*, per 100 g of oven-dried soil. The equivalent mass is calculated by dividing the mass of one mole of the cation by its valency, and is thus the mass of the cation that is chemically equal to 1 g of H. For example, the equivalent mass of Mg (valency +2) is calculated as 24 g /2, or 12 g. One milliequivalent of Mg is therefore 0.012 g, or 12 mg. A soil containing, for example, 14 mEq Mg per 100 g will contain 14·0.012 g, or 0.168 g, of exchangeable Mg per 100 g. Note that the amount of exchangeable Mg can also be calculated directly, without first calculating the mass corresponding to one mEq, by dividing the number of milliequivalents by the valency and multiplying by the molar mass of Mg.

The total CEC for a soil is the sum of the equivalent mass for each of the exchangeable cations it contains. For example, a soil containing Ca, Mg, K, Na and H, with equivalent masses of 14.0, 3.4, 0.5, 0.1 and 9.3 mEq per 100 g will have a CEC equal to 27.3 mEq per 100 g (this value neglects the presence of small amounts of ions of other elements such as Fe and Zn). The total CEC includes both organic matter and mineral soil, where the organic matter generally has the highest cation-exchange capacity. Decomposition of organic matter will produce hydroxyl ions (OH^-) with loosely bound hydrogen ions. These hydrogen ions can be replaced by other positive ions and the organic material therefore acts as a cation exchanger. Silt and sand particles will also have CECs, but these are lower because of their smaller surface-to-volume ratio, and the CEC of a soil is therefore mainly related to the amount of clay and organic matter it contains.

The clay mineral illite is an especially effective adsorbent for caesium because of the internal structure of the mineral. Its structure contains interlayers between sheets of silica, and both the top and bottom surface of a sheet are available for adsorption. In addition, weathering of the mineral creates *frayed edge sites* at the edges of the sheets, which further increase the area available for adsorption.

Since few soil particles carry a positive charge, the exchange of anions is less pronounced than the exchange of cations. However, as plants absorb roughly as many anions as cations, the behaviour of anions is important. Cation and anion exchange are both affected by the pH in the soil, but a change in pH will affect them in opposite ways. The exchange of cations increases with increasing soil pH, while the exchange capacity for anions decreases with increasing pH.

The physical properties of a soil are largely determined by its density and *porosity*. Soil density can be determined in two ways: either as the density of dried and homogenized soil, or as the density *in situ*. The former is called the *particle density*, ρ_p, and is defined as the mass of dry soil per unit volume. The particle density of soil is generally about 2.6 g cm^{-3}. The latter, the *bulk density*, ρ_b, is defined as the dry mass of a sample of soil, including pores and cracks, divided by its volume (before it is dried).

Porosity is expressed as the fraction of the volume of a soil that consists of pores and cavities. If the *total pore volume*, V_p, is defined as the volume of pores and cavities in a soil sample of total volume V, the porosity can be defined as V_p/V. The pore volume can be determined as the difference in mass between a soil sample saturated with water and the dried sample. Since the difference is equal to the amount of water occupying the pores and cavities, the volume of these can be calculated. The total pore volume can also be determined from the bulk and particle densities, since the ratio between these is equal to the fraction of solid matter, R. The total pore volume is then given by $1 - R$.

It can be shown theoretically that spherical particles arranged in a simple cubic lattice will have a porosity of 48%, while a face-centred lattice results in a porosity of 26%. These porosities are independent of the diameter of the spheres. These two lattices are depicted in Figure 3.63.

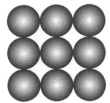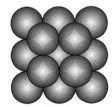

FIGURE 3.63 Spheres packed in a simple cubic lattice (left) will fill about 52% of the total volume, i.e., a porosity of 48%, while spheres packed in a face-centered cubic lattice (right) will fill 74%, giving a porosity of 26%. (The porosity is the fraction of the volume not occupied by the spheres.)

Both the size of the pores and the total pore volume affect the properties of the soil. A small pore size prevents water from flowing away for a considerable time and thereby makes it available for plants. Water in the soil is attracted to the soil particles and is adsorbed onto them, as the water molecule has an electric dipole moment, and because of surface tension. Surface tension gives rise to capillary action, which enables the water to rise in an air-filled pore.

The height of the column of liquid, h, in a narrow tube of radius r is given by

$$h = \frac{2 \cdot \gamma \cdot \cos \alpha}{\rho \cdot g \cdot r}, \tag{3.74}$$

where $\gamma \cdot \cos \alpha$ is the vertical component of the surface tension and α is the angle between the liquid and the wall of the tube. The contact angle between water and mineral soil is approximately zero, meaning that the water wets the surface and does not form droplets. Since the density of water is approximately 10^3 kg m^{-3} and the surface tension is 0.073 N m^{-1}, the height of capillary action in soil can be approximated by

$$h = \frac{0.15}{r}. \tag{3.75}$$

In larger tubes, where capillary action is insignificant, the radius of the tube has a considerable effect on the flow of water through the tube. For example, if the radius of the tube is doubled, the flow increases by 16 times, according to *Poiseuille's law*

$$V = \frac{\pi}{8\eta} \cdot \frac{R^4}{L} \cdot \Delta p \cdot t, \tag{3.76}$$

where V is the volume of a fluid with dynamic viscosity η, flowing through a straight tube with a circular cross section of radius R and length L during the time t, due to a pressure difference Δp between the ends of the tube.

Water that occupies pores and cracks in the ground is called the *groundwater*, and the depth below which all the cavities are filled with water is called the *water table*. At this depth, the pressure of the water is equal to the atmospheric pressure (see below), while below the water table, in the *saturated zone*, the water pressure is higher than the atmospheric pressure. Above the water table, pores are only partly filled with water, and this zone is called the *unsaturated zone*.

The flow of water in the ground is driven by gravity and pressure gradients, which contribute to the *water potential*—the potential energy of the water. Variations in the water potential then act as a driving force, moving water towards regions with a lower water potential. In general terms, the water potential may also include other mechanisms, such as osmosis, but we will restrict this discussion to the gravimetric and the pressure potential.

The gravimetric potential equals the work required to move a unit mass of water against the force of gravity, from a reference level to the level under consideration. This is then equal to the potential energy of the water at this level. The water potential, Ψ, is often given in units of J m^{-3}, but may also be given in m H$_2$O, by dividing by the density of water and the acceleration due to gravity, g. This is analogous to using the units mm Hg in measurements of air pressure. The pressure potential can be defined as the work required to increase the water pressure from ambient atmospheric pressure to the pressure at the level considered.

The movement of water in soil is affected by friction, which may be described in terms of both *laminar* and *turbulent* flow. In laminar flow, the water flows in parallel lines, and the friction is proportional to the speed of the water. However, in turbulent flow, the frictional force is proportional to the square of the speed of the water. The flow, Q [m^3 s^{-1}], of a fluid through a porous medium is given by Darcy's law, which can be written as

$$Q = -K \cdot A \frac{d\Psi}{dx}, \tag{3.77}$$

where A is the cross-sectional area of the ground and K is the *hydraulic conductivity* of the ground (units m s^{-1}), which depends on the size of the pores (r), and the density (ρ) and dynamic viscosity (η) of the water, as well as on the degree of saturation (θ). In terms of these quantities, K is given by

$$K = \frac{1}{8\eta\tau} \sum (\Delta\theta)_i r_i^2, \tag{3.78}$$

where the sum is performed over the distribution of water content and pore size in the ground. The factor τ is the tortuosity, a dimensionless quantity that provides a measure of the number of twists and turns in the water's path as it percolates into the ground. The flow in unsaturated soils will be lower than in saturated soils because capillary forces counteract the flow.

In unsaturated flow, water on the surface of the ground will flow downwards, towards regions that are dryer, under the action of both the gravimetric and the pressure potential. The gravimetric potential will always exert a downward force on the water, and when deeper layers hold less water than the surface, the pressure is reduced in the deeper layers. However, when evaporation occurs, the two potentials act in opposite directions, since in this case, the atmosphere contains less water than the surface.

3.5.2 Transport of Radionuclides in the Ground

Measurements of the vertical distribution of radionuclides in the ground following atmospheric weapons testing (Beck 1966) showed that the vertical distribution of fresh nuclear weapons fallout could be estimated by

$$c(z) = c_0 \cdot e^{-\alpha \cdot z}, \tag{3.79}$$

where $c(z)$ is the activity concentration [Bq kg^{-1}] at depth z and c_0 is the activity concentration at the surface. The parameter α depends on the type of soil, as well as on the time since deposition, and the radionuclide itself, and is an indication of the efficiency of downward transport of the radionuclide. If the depth is given in cm, α will be given in units of cm^{-1}. The *relaxation length* is defined as $1/\alpha$, the depth at which the activity concentration has decreased to e^{-1} (37%) of the activity concentration at the surface, and is about 3 cm for fresh fallout.

An example of a vertical distribution that is rather well described by an exponential decrease with depth is shown in Figure 3.64. The figure shows the activity concentration of ^{137}Cs in a soil core sampled in southern Sweden in 1991 (Isaksson and Erlandsson 1998). The soil core was sliced into layers corresponding to depths of 0–2, 2–4, 4–6, 6–9 and 9–12 cm, and the activity of each slice was determined by gamma-ray spectrometry. Fitting Eqn. 3.79 to the data gave a relaxation length of 5.2 cm.

Deposition of a radionuclide that is confined to the surface (a surface source) will give a relaxation depth of 0 cm and thus $\alpha \to \infty$. This implies that $c(z) = 0$ at all z. This situation can occur with fresh fallout on impermeable ground. On the other hand, if the relaxation length is very large (i.e. $1/\alpha \to \infty$), then $\alpha \to 0$ and $c(z) = c_0$ at all z. This describes a homogeneous distribution with depth, which can often be assumed for naturally occurring radionuclides, such as ^{40}K.

The depth may also be expressed as the *mass depth*, i.e. the mass of unit area of the soil, down to a given depth. The mass depth is given by the product of the depth and the density of the soil, ρ, and is thus given in units of kg m^{-2}. Eqn. 3.79 can then be written

$$c(z) = c_0 \cdot e^{-\frac{\alpha}{\rho}\rho \cdot z}, \tag{3.80}$$

where α/ρ is called the exponential mass activity distribution coefficient.

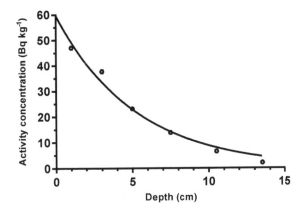

FIGURE 3.64 Activity concentration of ^{137}Cs in a sliced soil core, showing an approximately exponential distribution with depth. The data are plotted at the midpoint of each depth interval and the decay is corrected to 26 April 1986. Data from Isaksson and Erlandsson (1998).

The exponential decrease with depth, as given by Eqn. 3.79 and Eqn. 3.80, does not take into account the fact that both α and ρ may vary with the depth in the soil. It is quite probable that α will be affected by the density, which generally increases with depth. For example, mean values of the density at depths between 0–2 cm, 2–4 cm and 4–7 cm in nine soil samples taken at a grassy site, were approximately 600 kg m^{-3}, 1000 kg m^{-3} and 1100 kg m^{-3}, respectively (Isaksson and Erlandsson 1995). The greatest variation in density was thus found between the organic layer and the topsoil, but other soils may show different patterns in their density variation. Density variations can be taken into account by experimentally determining the mass depth for each layer in a sliced soil core. The mass depth is obtained by dividing the mass of a slice by its cross-sectional area. In this way, the mass depth of the different layers in the soil core can be found.

Although this simple model, predicting an exponential decrease in activity concentration with increasing depth, has proved to be useful in many situations, it fails to describe the fact that the measured depth distribution often has a maximum below the surface, as can be seen in Figure 3.65. This maximum may be situated immediately at the deposition site if the radionuclides are deposited wet. Precipitation then transfers the radionuclides several centimetres into the ground. The maximum below ground can be described by more elaborate models of the type given in Eqn. 3.82 (which will be discussed later in this section).

Other attempts have been made to modify Eqn. 3.79 to provide better fits to experimental data. Mattsson (1975) found that the vertical distribution of ^{137}Cs in lichen (*Cladonia alpestris*) could be described by

$$c(z) = c_0 \cdot e^{-\alpha \cdot z^{0.75}}. \tag{3.81}$$

Isaksson and Erlandsson (1998) found that this model could also be used for soil samples by replacing the value 0.75 with a parameter, p, which was allowed to vary when fitting experimental data.

The models discussed above are time-independent models, describing the vertical distribution at a specific time. Several attempts have been made to model the time dependence of the vertical distribution. Many of these models take physico-chemical processes into account, often expressed as a combination of diffusion and convection. This approach can be summarized by the *convection–diffusion equation*, CDE, given by

$$\frac{dc}{dt} = D \cdot \frac{d^2c}{dz^2} - v \cdot \frac{dc}{dz}, \tag{3.82}$$

where c is the concentration of the substance at depth z. This equation is valid when neither the *diffusion coefficient*, D, nor the *convection velocity*, v, vary with depth.

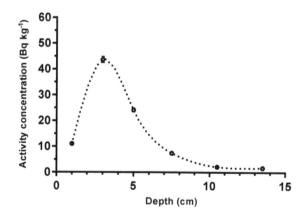

FIGURE 3.65 Activity concentration of ^{137}Cs in a sliced soil core (sampled in 2005), showing a subsurface maximum. The data are plotted at the midpoint of each depth interval and decay corrected to 26 April 1986. The dotted line is not a fit to the data, and is merely shown to guide the eye. Data from Almgren and Isaksson (2006).

Mass transfer by diffusion results from a concentration gradient in the medium. There will be a net transfer of the substance from regions of high concentration to regions of low concentration, driven by the random motion of the particles. This is quantified by the first term in Eqn. 3.82, where the response to a concentration gradient is given by the diffusion coefficient D [m^2 s^{-1}]. The second term in Eqn. 3.82 essentially describes the migration of the substance with water in the ground, which depends on the convection velocity v [m s^{-1}].

The CDE is a partial differential equation of the type also used for calculations of wave motion and heat transfer. In order to solve such a CDE, both the boundary conditions and initial conditions must be specified, and each solution is only applicable under those conditions. One possible boundary condition is, for example, that transport is assumed to take place from the surface down into the ground, and horizontal transfer may be neglected. In this case, the CDE is solved for a semi-infinite volume. An example of an initial condition is that the fallout occurred during a specific period of time, during which a specified amount of the radionuclide was deposited.

Several solutions of the CDE for various boundary and initial conditions can be found in the literature (see, for example, Bossew and Kirchner (2004)). If the deposition is assumed to occur as a pulse event, i.e. all the material is deposited at time $t = 0$ and the duration of the deposition can be regarded as being infinitesimally small, a solution for a semi-infinite volume is given by

$$c(x,t) = J_0 e^{-\lambda \cdot t} \left(\frac{1}{\sqrt{\pi Dt}} e^{-\frac{(x-vt)^2}{(4Dt)}} - \frac{v}{2D} e^{\frac{vx}{D}} \mathrm{erfc} \left(\frac{v}{2}\sqrt{\frac{t}{D}} + \frac{x}{2\sqrt{Dt}} \right) \right), \quad (3.83)$$

where J_0 is the initial deposition density [Bq m^{-2}] and t is the time after deposition. The *complementary error function, erfc*, is defined by

$$\mathrm{erfc}(x) = 1 - \mathrm{erf}(x) = 1 - \frac{2}{\sqrt{\pi}} \int_0^t e^{-t^2} \mathrm{d}t. \quad (3.84)$$

Some assumptions also have to be made concerning the adsorption of the radionuclide during migration. In Eqn. 3.83, a linear *sorption isotherm* was assumed, which means that the adsorption at equilibrium is proportional to the amount of the radionuclide present. Sorption isotherms are often used in these models, and describe the relationship, at equilibrium, between the concentration of a substance on the surface of a medium and the concentration of the same substance in a liquid passing through the medium. Adsorption is assumed to take place at constant temperature, hence the use of the term *isotherm*.

3.5.3 Radionuclide Transfer in Agricultural Ecosystems

After radionuclides have been released into the atmosphere, deposition may take place on plants and trees, or on arable fields and pastures. The radionuclides may subsequently undergo further transfers in the terrestrial ecosystem. Plants can acquire activity as a result of dry or wet deposition on the parts growing above the ground, but there may also be some uptake from the soil via the roots.

3.5.3.1 Transfer to Plants

The uptake, or *interception*, of radionuclides by plants can be quantified in several ways. The least elaborate is to simply divide the activity per unit area retained by the vegetation immediately after deposition, A_i, by the total activity deposited per unit area, A_t. The *interception fraction*, f, is then defined as (IAEA 2010)

$$f = \frac{A_i}{A_t}. \tag{3.85}$$

The interception fraction will generally decrease with the amount of rain since it is closely related to the ability of the plant canopy to store water. When the storage capacity is reached, water will run off the leaves and the activity it contains will not be included in A_i, leading to a decrease in the interception fraction.

Since the interception fraction is dependent on the development of the plant, its value will change as the plant grows. One way to account for plant growth is to normalize the interception fraction to the standing biomass, B, given in units of kg m^{-2}, where the mass referred to is the dry mass. The *mass interception fraction*, f_B, is then defined as (IAEA 2010)

$$f_B = \frac{f}{B}, \tag{3.86}$$

with the units m^2 kg^{-1}.

The valency of the radionuclide will also affect interception since plant surfaces are negatively charged and therefore preferentially attract cations. Anions are less effectively retained. For example, the mass interception fraction found for grass (measured for wet deposition after the Chernobyl accident) was 1.1 for Cs (Cs$^+$), but 0.7 for I (I$^-$) (IAEA 2010). The values will be even higher for cations of higher valency.

Instead of normalizing the interception fraction to the standing biomass, it can be useful to normalize it to the total leaf area. This is quantified by the *leaf area index*, LAI, which is calculated by dividing the total leaf area by the total soil area. The LAI thus gives the fraction of the area available for deposition on plant leaves, and provides a better representation of how the radionuclides in the rain are deposited on the vegetation (IAEA 2009).

The LAI will vary during the growing season. During growth, the LAI will increase, becoming constant when the crop is mature, and falling to zero at harvest. For some crops, the LAI decreases before harvest, as can be seen in Figure 3.66, where the LAI is modelled as a function of the cumulative temperature during the growing season. This approach has been used by Plauborg et al. (2015). An example of the cumulative temperature for two sites, located in different climate zones, is shown in Figure 3.67.

The relation between the interception fraction and the LAI has been modelled for wet deposition by Müller and Pröhl (1993) as

$$f = \min \left[1; \frac{LAI \cdot k \cdot S}{R} \left(1 - e^{\frac{\ln 2}{3kS} \cdot R} \right) \right], \tag{3.87}$$

where S is the storage capability of the plant (in terms of the amount of rain that falls) and R is the amount of precipitation (in mm) in one rain event[11]. The interception fraction thus increases linearly with the LAI, while it is inversely proportional to the amount of precipitation. The storage capability depends on the type of crop, and a value of 0.2 mm is assumed for grass, cereals and corn. All other crops are assumed to have a storage capability of 0.3 mm. The factor k depends on the element under consideration, and is a measure of that element's ability to attach to plant leaves. Anions are assigned a value of 0.5, while

---

[11]If the value of the second item within the square brackets is less than 1, this value is assigned to f. Otherwise, f is given the value 1.

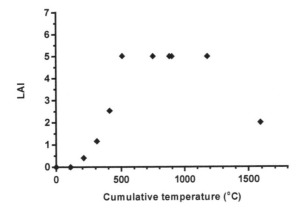

FIGURE 3.66 Calculated leaf area index (LAI) of spring barley during a growing season.

monovalent cations (e.g. Cs) and polyvalent cations (e.g. Sr) are assumed to have $S = 1$ and $S = 2$, respectively (IAEA 2010).

The removal of contamination from plants is called weathering[12] and includes contributions from several processes. For example, intercepted radionuclides on the leaves may be washed off by precipitation or removed by the wind, by abrasion, or by flexing of the leaves. The activity concentration will also decrease as a result of plant growth, but this *growth dilution* does not affect the total activity, as it simply reflects the fact that the area on which the radionuclide is adsorbed increases as the plant grows. The amount of a radionuclide lost from a plant depends on factors such as the chemistry of the element concerned (affecting solubility and adsorption) and the degree to which the element has been internalized within the plant (IAEA 2009).

Weathering is often modelled by the *weathering half-life*, T_w, given by

$$A(t) = A_i \cdot e^{-\lambda_w \cdot t}, \tag{3.88}$$

where $\lambda_w = \ln 2 / T_w$. Weathering half-lives can vary from a few days to around 60 days, depending on the plant and the radionuclide. Data on weathering half-lives can be found in IAEA (2009). The weathering half-life is affected by the growth stage of the plant, and generally decreases as the plant matures.

Translocation of the adsorbed radionuclides to the edible parts of the plant depends on the capability of the radionuclide to enter the interior of the leaves. The leaf surface, the *epidermis*, is covered with a waxy layer called the *cuticle* which prevents water from entering the interior of the leaf (Fig. 3.68). However, if the cuticle contains cracks or other defects, radionuclides may pass through these and enter the leaf. Radionuclides can also enter the leaf, to a lesser extent, through the *stomata* (Bengtsson 2013).

Entrance through the cuticle is determined by several factors, including temperature, light, pH and the valence of the radionuclide. Once the radionuclide has entered the interior of the leaf, it may be translocated within the plant and eventually reach edible parts such as grain, fruit and berries. Since translocation occurs by the diffusion of water and transport in the vascular bundle, the distribution of the radionuclide within the plant will depend

---

[12]This term has been discussed previously, in Section 3.5.1.2.

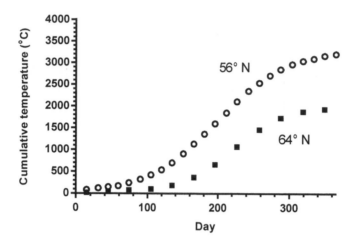

FIGURE 3.67 Cumulative temperature from 1 January at 56° N (in the temperate zone) and 64° N (close to the frigid zone), illustrating the differences in growing conditions between the two climate zones.

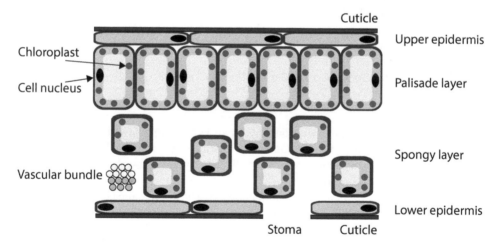

FIGURE 3.68 Schematic cross section of a leaf, showing the three major systems: epidermis, mesophyll tissue, and vascular tissue (depicted by a vascular bundle). Chloroplasts, which contain the chlorophyll used in photosynthesis, are generally absent in epidermal cells, except in the guard cells in the lower epidermis, surrounding the stomata. The upper cell layer in the mesophyll tissue, the palisade layer, consists of cells containing large numbers of chloroplasts, while the cells in the spongy layer, beneath the palisade layer, contain fewer chloroplasts. Water and minerals are transported in the xylem in the vascular bundle, while the phloem (also in the vascular bundle) carries the sugar produced by photosynthesis out of the leaf.

on the physiological status of the plant. The activity concentration in various parts of the plant will thus depend on the growth stage at which deposition takes place.

The activity concentration for any particular part of the plant will depend on uptake from both the leaves and the roots, and on the rate of redistribution within the plant. For example, Mattsson and Erlandsson (1988) found that the activity concentration of ^{137}Cs in spruce needles which developed after deposition was affected by root uptake and by redistribution from older needles. This was also related to the potassium content in the soil; a lack of potassium in the soil resulted in an increase in root uptake, while normal potassium concentrations increased redistribution within the trees. Potassium fertilizers have also been shown to be effective in decreasing root uptake of Cs in barley and grass growing on soils with a low potassium content (Rosén 1991).

The transfer of radionuclides to plants can be modelled by the *transfer factor*, TF, defined as the activity concentration in dry plant tissue (the mass activity density), A_m [Bq kg^{-1}] divided by the activity deposited per unit area (the areal activity density), A_a [Bq m^{-2}] as

$$TF = \frac{A_m}{A_a}. \tag{3.89}$$

The units of the transfer factor are thus m^2 kg^{-1}. The areal activity density is assumed to be the total, depth-integrated, activity underlying the object in which A_m is determined. This transfer factor was developed to represent the transfer from soil to biota, and includes several processes, such as food chain transfer and root uptake, and is often called the *aggregated* transfer factor, or aggregated transfer coefficient[13]. This was previously denoted by the symbols TAG or T_{ag}, but, according to the ICRU (2001), the aggregated transfer coefficient should now be denoted by C_{ag}.

The values of the transfer factors, for example, those given in IAEA (2010), vary by several orders of magnitude, even for the same crop and soil type. This variability can be attributed to six main factors (IAEA 2009).

1. *Deposition properties*, e.g. whether the deposition of a radionuclide occurs as particles, as an aerosol, or in solution. For example, elements that form anions (such as Tc and I) will be more mobile, and may be more available to plants than elements that form cations or stable compounds (such as Cs and Sr).

2. *Time after deposition*: a longer time will usually cause a decrease in uptake due to fixation of radionuclides in the soil matrix.

3. *Soil properties* will affect the uptake due to the processes discussed in Section 3.5.1.2. These properties include texture, organic matter content, and pH.

4. *Crop properties* will affect the transfer factor due to biological variability between plants, for example, with regard to the nutrient requirements of different species. Transfer may also be affected by the length of the growing season and by the part of the plant that is used when determining the transfer factor.

5. *Cultivation practices*, such as fertilization, irrigation and ploughing, are important parameters. Decreased uptake of Cs following the use of potassium fertilizers has already been discussed, but ploughing and liming will also reduce uptake.

---

[13]This should not be confused with the transfer coefficients (transfer rates) between compartments in a compartment model, discussed in Section 3.1.

6. *Radionuclide properties* can generally be described by dividing the radionuclides into five groups, depending on their physicochemical characteristics (IAEA 2009):

   (a) The light, naturally occurring, radionuclides, where the dominating uptake route is foliar (leaf) uptake from water (^{3}H) or carbon dioxide (^{14}C). Uptake of ^{40}K is determined by the nutritional requirements of the plant.

   (b) The heavy, naturally occurring, radionuclides ^{232}Th, ^{238}U, ^{226}Ra, ^{210}Pb, and ^{210}Po. These generally have relatively low transfer factors.

   (c) The transuranic elements, such as Am, Cm, Pu and Np. These may occur in various oxidation states and their soil chemistry is complex, which leads to considerable variation in transfer factors. For example, fruits and grains may have activity concentrations which are up to 1000 times lower than in the vegetative parts of the plant.

   (d) Fission products, such as radionuclides of Sr, Cs, I, Y, Zr, Nb, Ru and Ce, are significantly affected by the soil pH and the organic matter content. Bioavailability and ease of relocation within the plant vary widely within this group.

   (e) Activation products (for example, ^{51}Cr, ^{54}Mn, ^{55}Fe, ^{60}Co and ^{65}Zn). These include elements that are important nutrients for plants. They generally have high mobility in the soil and within plants, and their behaviour is strongly dependent on the pH and the amount of organic matter in the soil.

The variation in the uptake of Cs and Sr within species has been reported by Penrose et al. (2015). Inter-varietal variation, expressed as the ratio of the maximum and minimum activity concentration found for a given species, were reported to range from 1.0 to 6.3 for Cs and from 1.0 to 4.5 for Sr. The uptake in a given crop can therefore vary depending on the properties of the individual plants. This may be used as a remediation strategy (in combination with, for example, fertilization and ploughing): the radiation dose resulting from consumption can be decreased by choosing the variety that exhibits the lowest transfer.

Absalom et al. (2001) developed a model for predicting the transfer of Cs from soil to plants using the characteristics of the soil. The model parameters, used to determine the transfer and bioavailability of Cs, are derived from the soil clay content, the amount of exchangeable potassium, the pH, the concentration of NH_4^+, and the organic matter content. The model parameters were determined from experiments using rye grass and bent grass, but the model also performed reasonably well when tested on uptake data for barley. Examples of some transfer factors calculated by this model are shown in Figure 3.69.

The decrease in the transfer factor with time after deposition is also evident in the field data collected by Eriksson and Rosén (1991) and shown in Figure 3.70. These data were collected from timothy grass grown on various soil types, which also affects the transfer factor. Figure 3.71 shows transfer factors for barley straw and grain. Evidently, most of the Cs is transferred to the straw, which is an important finding when trying to calculate the dose to humans as a result of consumption.

As previously pointed out, the activity concentration in the edible parts of crops will depend on the growing season, and this dependence is quantified by the *seasonality factor*, S, defined as

$$S = \frac{\sqrt{\frac{\sum_{i=1}^{12}(I_m - I_{iap})^2}{12-1}}}{I_m}, \qquad (3.90)$$

where the numerator is the standard deviation of the monthly time-integrated concentration in a sample, e.g. a plant type, observed over 12 months (Aarkrog 1994). Deposition of a radionuclide p, in a given month i, will result in a particular activity concentration in the crop at harvest, while deposition in a different month may result in a different activity concentration. This is modelled by defining the infinite time concentration integral in a specified type of sample (a), I_{iap} [Bq kg^{-1} y]. The seasonality is then given as the relative standard deviation by dividing by the annual mean of the monthly values of I_{iap}, denoted I_m. Seasonality exists if $S > 0$, i.e. if the activity concentration in the crops at harvest depends on the time of deposition.

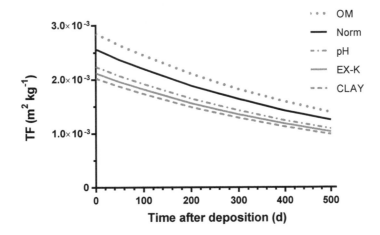

FIGURE 3.69 The transfer factor (TF) decreases with time, and varies with the soil organic matter content (OM), amount of exchangeable potassium (EX-K), amount of clay (CLAY), and pH. The figure shows transfer factors calculated by the Absalom model with default parameter values (Norm), and the transfer factors that result when each parameter value is increased by 40% (while keeping the other parameters fixed at their default values). Default values were pH 6.0, OM 6.0 g g^{-1} soil, EX-K 0.5 cmole kg^{-1} soil and CLAY 15 g g^{-1} soil.

In temperate regions, where crops are cultivated mainly during the warmer seasons, contamination of growing crops will result in larger activity concentration at harvest than contamination in the cold season, when foliar adsorption will be negligible. In regions where crops can be cultivated during most of the year, seasonality will generally be less important. Seasonality thus gives rise to a latitude dependence in the transfer factors, for example, from grass to milk. The degree of seasonality also depends on the relation between foliar and root-mediated uptake. If the root-mediated uptake dominates, seasonality will be low, while the opposite leads to high seasonality.

Bengtsson (2013) found that transfer to the seeds of oilseed rape was at a maximum when contamination occurred close to harvest, and that a smaller fraction of the activity was found in the seeds than in the straw. These experiments were conducted using artificial irrigation with ^{134}Cs and ^{85}Sr, and also showed that a larger amount of Cs was found in the seeds than Sr. The opposite was found for spring wheat, where the transfer processes seemed to favour Sr, which was more abundant than Cs in the grain. Finally, transfer to pasture vegetation was highest when the contamination occurred in the later stages of growth. It it clear that it is important to consider these variations in uptake and relocation within the plant when assessing the radiological impact of a fallout event.

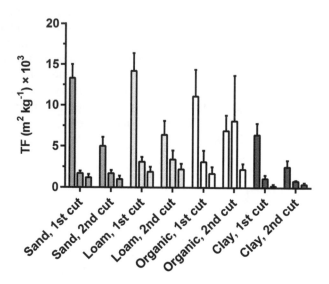

FIGURE 3.70 Transfer factors for ^{137}Cs in timothy grass, sampled in 1986, 1987 and 1988, on various soil types. For most soil types, the transfer factor decreased with time after deposition. In general, the factors were rather low because all the soils were well fertilized. The similarities between the transfer factors for grass grown on loamy soil and organic soil (a fen) are attributed to fertilization and the thickness of the grass cover. Data from Eriksson and Rosén (1991).

3.5.3.2 Transfer to Animals and Animal Products

Transfer to animals and animal products such as milk and meat, is generally described by transfer coefficients. These should not be confused with the rate parameters discussed previously in Section 3.1, and are denoted F_m and F_f, respectively. In milk, the transfer coefficients are defined as the ratio between the activity concentration in the milk [Bq L^{-1}], and the daily intake [Bq d^{-1}], at equilibrium, giving units of d L^{-1} (similarly, d kg^{-1} for meat).

These transfer coefficients do not describe the physiological processes, but simply relate the equilibrium activity concentration of a radionuclide in a given tissue or animal product to the daily dietary intake of the radionuclide. Estimation of the transfer coefficients thus requires that the composition of the animal's diet is known. However, feeding strategies for domestic animals vary between countries, between different parts of a particular country, and with the time of year.

One problem when establishing transfer coefficients is the requirement of equilibrium between the dietary intake of a radionuclide and the activity concentration in the animal. Since intake can be subject to large temporal variations, this assumption may not hold, and the activity concentration may vary over time. However, an approximate equilibrium is often found for milk due to the short time required for the transfer of the radionuclide from grass to milk. An example is shown in Figure 3.72, where the decrease in grass activity is closely correlated to the decrease in milk activity.

As with humans, there are various routes by which internal contamination of an animal may occur: by ingestion, by inhalation, or by absorption through the skin. The most important of these in animals is ingestion of contaminated feed and soil (IAEA 2010). Internal

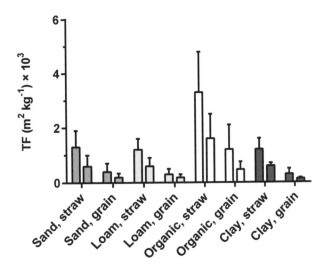

FIGURE 3.71 Transfer factors for ^{137}Cs in barley, sampled in 1986, 1987 and 1988, on various soil types. The transfer factors are lower than for grass, shown in Figure 3.70, and transfer to the grains is generally lower than to the straw. Data from Eriksson and Rosén (1991).

contamination from inhalation is often of minor importance because of the low activity concentration in outdoor air (except if the animals are exposed to a radioactive plume).

Absorption from the human GI tract was discussed in Section 2.8.1.2, and parameter values derived for humans can also be used for monogastric animals. For example, IAEA (2010) recommends gastrointestinal fractional absorption values of 1 for Cs and 0.3 for Sr. The absorption of Cs is also higher than the absorption of Sr in ruminants. However, the fractional absorption of Cs varies with its chemical form. In general, essential elements are absorbed to a greater extent, while elements with high atomic numbers are poorly absorbed, especially if they are not essential, or are not chemical analogues of an essential element.

Radionuclides are distributed within the animal in accordance with the behaviour of their chemical analogue. Thus, radioactive caesium, which is a chemical analogue of potassium, is distributed preferentially in muscle tissue, while radioactive strontium, a chemical analogue of calcium, is preferentially transferred to milk and to the skeleton. Radioactive iodine is taken up by the thyroid, but there is also active uptake by the mammary gland, from where it is transferred to milk (IAEA 2009).

In general, transfer coefficients are affected by the size of the animal. Lower body mass and the associated lower dietary intake rates often lead to a higher transfer coefficient (IAEA 2010). For example, for a given daily intake of a radionuclide, lambs and calves will have a higher activity concentration in their tissues than adult animals. The same holds when comparing small and large animal species. This difference is probably a result of the larger amount of dry matter in larger animals (IAEA 2010).

The transfer of radionuclides from feed to animal products can also be described by the *concentration ratio*, *CR*, which shows less of a difference between species than the transfer coefficient. The concentration ratio is defined as the equilibrium ratio of the activity concentration (fresh weight) in the product to the activity concentration (dry weight) in

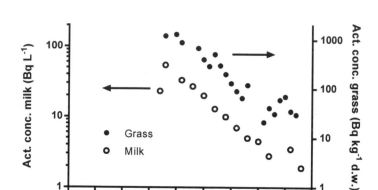

FIGURE 3.72 Activity concentrations in samples collected in western Sweden, shortly after the Chernobyl accident in 1986. The correlation between the activity concentrations in grass and milk indicates rapid transfer from grass to milk. (Note the different scales.). Data from Håkansson et al. (1987).

the feed (IAEA 2010). CR is dimensionless, and does not depend upon the intake, which may be difficult to determine for grazing animals.

TABLE 3.19 Transfer coefficients for caesium to milk from three species vary by more than an order of magnitude, while the variation in concentration ratio is significantly lower. The concentration ratio is calculated by multiplying the transfer coefficient by the daily dry matter intake.

Species	F_m (d L^{-1})	Dry matter intake (kg d^{-1})	CR
Cow	$7.9 \cdot 10^{-3}$	16.1	0.13
Goat	$1.0 \cdot 10^{-1}$	1.3	0.13
Sheep	$5.8 \cdot 10^{-2}$	1.3	0.075

Source: Data from IAEA (2009).

However, if the daily intake is known, the transfer coefficient can be calculated from the concentration ratio, and vice versa. The transfer coefficient is found by dividing the concentration ratio by the daily intake [kg d^{-1}] and the CR is determined by multiplying the transfer coefficient by the daily intake. Table 3.19 gives recommended transfer coefficients for caesium to milk from cows, goats and sheep. Estimated concentration ratios are also given, based on assumed daily dry matter intakes. Values for meat are given in Table 3.20.

The transfer of ^{131}I, ^{134}Cs and ^{137}Cs from grass to cow's milk was studied at three farms in western Sweden after the Chernobyl accident (Wallström 1998). Some of the results are shown in Figure 3.72. The concentration ratios were generally lower than those given in Table 3.19. In 1986, the concentration ratios were 0.041–0.054, 0.043–0.064, and 0.037–0.049 for ^{131}I, ^{134}Cs and ^{137}Cs, respectively. However, measurements in 1987, one year after the fallout was deposited, gave concentration ratios similar to the value of 0.13 in Table 3.19, i.e. 0.10–0.55 and 0.20–0.40 for ^{134}Cs and ^{137}Cs, respectively.

TABLE 3.20 Transfer coefficients for caesium to meat from four species. The concentration ratio is calculated by multiplying the transfer coefficient by the daily dry matter intake.

Species	F_f (d kg^{-1})	Dry matter intake (kg d^{-1})	CR
Beef	$5.0 \cdot 10^{-2}$	7.2	0.36
Lamb	$4.9 \cdot 10^{-1}$	1.1	0.54
Pork	$2.4 \cdot 10^{-1}$	2.4	0.58
Chicken	$1.0 \cdot 10^{1}$	0.07	0.7

Source: Data from IAEA (2009).

3.5.4 Radionuclide Transfer in Natural and Semi-Natural Ecosystems

Forest ecosystems differ from agricultural ecosystems in many respects. For example, deposition of radionuclides over a field in which crops are growing will mainly affect that particular crop, and the food chains in which the crop is included. Deposition over a forest area, however, may affect several types of vegetation, probably belonging to different vegetative strata (the overstorey, the understorey, or the herbaceous layer) and having different susceptibilities to deposition. Also, while the soil in an agricultural system can be regarded as relatively homogeneous, forest soils may be of different soil types and vary in their stratification. Therefore, a high degree of variability is often found in the transfer and redistribution of radionuclides in forest ecosystems.

Various mechanisms may cause trapping and recirculation of radionuclides, and thus affect exposure pathways. Regardless of the nature of the forest (examples include forests intended for timber production, and self-sustaining woodland) the timescale for the turnover of the trees is significantly longer than for agricultural crops. In addition, exposure patterns may vary significantly from year to year, causing variations in the external, as well as the internal dose to the population. For example, favourable climatic conditions for mushrooms will lead to increased activity concentration in some animals, and also an increase in the internal dose to people who consume these mushrooms.

3.5.4.1 Transfer to Vegetation and Forest Products

Trees are also subject to external and internal contamination by radionuclides from fallout, through similar routes as in other plants: interception, translocation and root uptake. In the first 4–5 years after deposition, radionuclides are rapidly redistributed between the trees and the soil. This redistribution includes the removal of radionuclides from the canopy and stems by weathering and leaf litter. Thereafter, changes are rather slow, and the contribution from root uptake will dominate in the trees. This phase is characterized by almost constant soil–tree transfer factors.

Due to the presence of multiple layers in forest soils it is not possible to define a mean rooting zone, and the activity concentration in various forest products is therefore generally calculated using aggregated transfer factors [m^2 kg^{-1}] (see Section 3.5.3.1). This transfer factor relates the activity concentration in the forest product to the areal activity density, and thus integrates over the environmental parameters that affect the activity concentration in forest products. These parameters include physical, chemical, and biological processes in the soil and vegetation.

Interception fractions vary with the character of the deposition (for example, the particle size) and the type of tree, ranging from about 20% for a birch forest to 100% for a tropical rain forest. In a pine forest, the interception fraction may reach 90% (IAEA 2010). These values were measured in studies on global nuclear weapons fallout. Studies performed after the Chernobyl accident showed that coniferous trees were 2–3 times more efficient at trapping the emitted radionuclides than deciduous trees. It has also been found that the interception fraction for coniferous trees was 7–8 times higher than that for meadows and marshy ground (IAEA 2010).

The aggregated transfer factors for both Cs and Sr in trees is on the order of 10^{-3} m^2 kg^{-1}, dry weight, but varies between the different parts of the trees: for example, $5.6 \cdot 10^{-3}$ m^2 kg^{-1} for Cs in foliage and $1.4 \cdot 10^{-3}$ m^2 kg^{-1} in heartwood (Calmon et al. 2009).

The understorey vegetation (shrubs, herbs, mosses and lichens) may become contaminated directly, by interception of atmospheric deposition, or indirectly, by contaminated water reaching the understorey by throughfall and stemflow. The mechanisms of further transfer of radionuclides in these species are similar to those described above for other plants (Section 3.5.3.1). The ability to trap airborne radionuclides makes certain understorey species such as mosses and lichens useful as bioindicators of the extent of an atmospheric release of radionuclides. Other species, for example, ferns, are useful as bioindicators of soil activity concentration because of their relatively high capacity for root uptake. Generally, to qualify as a good bioindicator, a high and predictable uptake capacity is desirable, as well as accessibility.

Mushrooms are classified as saprophytic, symbiotic (mycorrhizal) or parasitic, depending on how they incorporate nutrients. Saprophytic mushrooms live on decomposing material and are thus present in the surface layers of the forest soil. These mushrooms become contaminated by direct deposition, and then by uptake from the soil shortly after deposition. However, their activity concentration will decrease as the radionuclides migrate further downwards into the soil. Symbiotic mushrooms grow on and around tree roots, forming the mycorrhiza. The tree benefits from the high capacity of the mycelium to transfer minerals from the soil to the tree roots. In return, the mushroom receives organic matter produced by the tree via photosynthesis. Finally, parasitic mushrooms live on trees, but instead of having a symbiotic relationship, they damage the tree.

Many symbiotic mushrooms are edible and are important in determining the dose to people. Because of the large area covered by the hyphae, the activity concentration in fruiting bodies can be high, and transfer coefficients for ^{137}Cs in edible mushrooms can vary between 10^{-3} and 10 m^2 kg^{-1}, dry weight (Calmon et al. 2009).

Berries also constitute an important forest product that can contribute to the radiation dose to humans upon consumption. The transfer of radionuclides to berries is much higher than in agricultural products. Aggregated transfer factors vary between 10^{-2} and 10^{-1} m^2 kg^{-1}, dry weight (Calmon et al. 2009).

3.5.4.2 Transfer to Animals and Animal Products

The activity concentration in semi-wild and wild animals depends on the deposition of radionuclides in their habitat and on their feeding habits. Heterogeneous deposition will lead to variations in animals that eat the same types of food. Because the availability of different kinds of plants varies throughout the year, significant seasonal variations are seen for some species such as roe deer and reindeer. These variations are most pronounced for Cs, due to the variability in how it is transferred to plants and mushrooms (Calmon et al. 2009). Figure 3.73 shows the activity concentration of ^{137}Cs [Bq kg^{-1}, fresh weight] in moose from the region around Gävle in Sweden, between 1986 and 1990. This region was highly contaminated by debris from the Chernobyl accident, and the areal activity density exceeded 100 kBq m^{-2}.

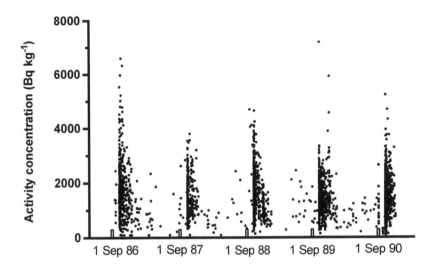

FIGURE 3.73 Activity concentration (fresh weight) in moose from the region around Gävle in Sweden, showing large seasonal variations. The marks on the date axis indicate 1 September in the years 1986–1990. Data from the Swedish Radiation Safety Authority.

The *ecological half-time*[14], T_{ecol}, is often used to model the time evolution of a radionuclide in a particular host, such as a game animal, or humans. It is defined as the time required for the activity to decrease to half its initial value by biological and ecological processes. The ecological half-time thus includes all the processes, except radioactive decay, that influence the change in activity. For humans, these processes include soil transfer processes, transfer from soil to plants, transfer to grazing animals, and transfer to animal products in the human diet. As the radiation dose to the host is affected by any decrease in activity, radioactive decay can be included by defining an *effective ecological half-time*, $T_{ecol,eff}$, given by

$$\frac{\ln 2}{T_{ecol,eff}} = \frac{\ln 2}{T_{ecol}} + \frac{\ln 2}{T_{1/2}}. \tag{3.91}$$

In general, the radiation dose to the general public from eating game is of little concern because average consumption is low. However, some groups, for example, reindeer herders, and hunters generally, consume larger quantities than the general population, and their radiation dose from dietary intakes of game may be significant. Rääf et al. (2006) studied the whole-body activity concentration [Bq kg^{-2}] of ^{137}Cs from the Chernobyl accident in various population subgroups in Sweden. The subgroups receiving the highest committed effective dose were reindeer herders, who received 2–11 mSv, followed by hunters, with 0.4–2.4 mSv. The large variation in the latter subgroup was a consequence of regional variations in deposition density. Farmers received, on average, 1.2 mSv, while rural non-farmers and the urban population were the least affected (0.4–0.9 mSv and 0.03–0.12 mSv, respectively). For comparison, the average annual effective dose (external and internal) to the Swedish population is about 3 mSv (including medical exposure and indoor radon).

---

[14]The ecological half-time is also called the ecological half-life.

Whole-body activity in the urban population was found to exhibit a dual-exponential decrease, characterized by values of $T_{ecol,eff}$ of 1–1.5 y and 20 y for the short- and long-term components, respectively. A dual exponential decrease was also found for farmers and rural non-farmers. However, reindeer herders and hunters exhibited a mono-exponential decrease with values of $T_{ecol,eff}$ of 1.4 y and >5 y, respectively. The data from these measurements are summarized in Figure 3.74.

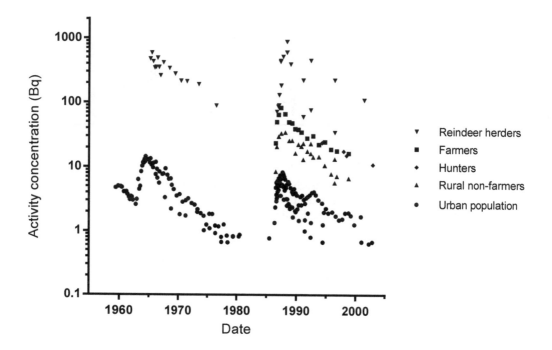

FIGURE 3.74 Data from whole-body measurements of various population subgroups in Sweden. Data from Rääf et al. (2006).

The effective ecological half-time found for reindeer herders for fallout from nuclear weapons tests was 5 y, which is considerably longer than the value obtained for the fallout resulting from the Chernobyl accident. This difference may be attributed to the fact that nuclear weapons fallout was deposited over a much longer period.

The food chain lichen–reindeer–humans has a major impact on the radiation dose to populations in some Arctic regions. Because lichens do not have a root system, they take up nutrients from the surrounding air. If the air is polluted by radionuclides, these will also be retained in the lichens. Lichens are a major part of the diet of reindeer during the winter, and the activity concentration in reindeer meat (venison) reflects the activity concentration found in lichens. This correspondence is shown in Figure 3.75, which shows the activity concentration of ^{137}Cs from atmospheric nuclear weapons fallout between 1962 and 1977 in Finnish lichens and reindeer meat. The whole-body activity in male reindeer herders from one region in Finnish Lapland is also shown for comparison.

Contamination of reindeer can be reduced by feeding them uncontaminated feed, or by adding bentonite, Zeolite or AFCF (ammonium iron (III) hexacyanoferrate) to their feed, which prevents the absorption of Cs by binding it in the GI tract. Feeding with

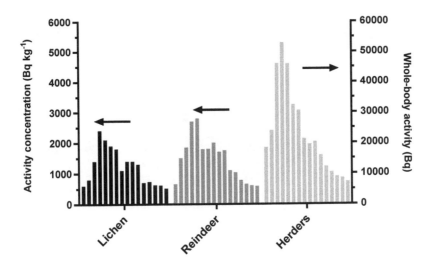

FIGURE 3.75 Activity concentration of ^{137}Cs in lichens (dry weight), in reindeer meat (fresh weight), and whole-body activity in reindeer herders in Finland. The meat samples were collected during the first quarter of the year. Data from Rissanen and Rahola (1990) and Rahola and Suomela (1990).

uncontaminated feed rapidly reduces the activity in reindeer, leading to a biological half-time of around 20 days (Gaare and Staaland 1994). Adding bentonite or Zeolite to the feed can reduce absorption by 85% and 50%, respectively (Åhman et al. 1990).

3.5.5 Radionuclide Transfer in Urban Environments

Deposition in urban environments can be caused by events such as fallout from nuclear weapons tests or accidents at nuclear facilities, but also by intentional actions. Accidents during transport may also result in contamination of the urban environment. In most cases, the radiation dose to inhabitants is governed by persisting γ-radiation from ^{137}Cs in grassy areas or on bare soil, due to its slow downward migration into the ground. However, these kinds of surfaces constitute only a minor part of the total area in cities, and the majority of the inhabitants will therefore be exposed to radiation from other surfaces as well, as shown in Figure 3.76. For example, comparatively high levels of contamination were found on road surfaces and roofs in contaminated areas after the Chernobyl accident. This contamination may affect long-term radiation doses, and it is therefore important to be able to predict how contamination decreases on such surfaces.

Field gamma-ray spectrometry measurements on the streets of the city of Gävle, Sweden, showed that less than 2% of the initially deposited Cs remained 10 y after deposition (Andersson et al. 2002). The decrease with time by weathering could be modelled by a double exponential function

$$C_r(t) = 0.5 \cdot \left(A_1 \cdot e^{-\lambda_1 t} + A_2 \cdot e^{-\lambda_2 t} \right), \tag{3.92}$$

where $C_r(t)$ is the level of contamination relative to a reference grass surface, and A_1 and A_2 are constants equal to 0.7 and 0.3, respectively. The half-times of the short- and

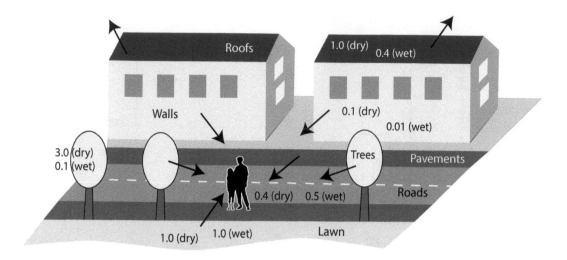

FIGURE 3.76 Examples of sources of external exposure in a contaminated urban environment. The contamination (normalized to that on lawns), resulting from dry and wet deposition, shows that trees and shrubs are most contaminated after dry deposition, while lawns are most contaminated after wet deposition. Data from Roed (1990) and Nisbet et al. (2010)).

long-term decrease, λ_1 and λ_2, are 120 d and 3 y, respectively. The effect of radioactive decay can be considered insignificant on this short timescale.

Contamination of roads is mainly weathered by traffic, and contamination on roofs and walls would be expected to be subject to a lower rate of weathering. Measurements have shown that the level of contamination decreases more slowly on walls than on roads (Andersson et al. 2002). However, the level of contamination can vary considerably, depending on the deposition conditions. For example, wet deposition in combination with an easterly wind may result in a highly contaminated east-facing wall, while west-facing walls will be shielded from deposition. Uneven contamination of walls may also occur through dust or soil adhering to the lower part of the wall.

The removal of radiocaesium from roofs varies considerably and depends on the roof material, as well as on the slope of the roof. Measurements by Andersson et al. (2002) showed that corrugated Eternit (a common asbestos fibre cement panel) and clay tiles retained more of the initial contamination than silicon-treated Eternit. The first two materials also retained most of the contamination over a significant period of time. The results obtained by Andersson et al. are shown in Figure 3.77. If the roof is covered with algae or other growth, a significantly larger fraction of the initial deposition may be retained.

It is assumed that the reason for the greater retention on steeper roofs (45° slope) is that, for at given height of the roof, the path for contaminated water is longer on a steeper roof. Many roofing materials contain the mineral illite, which adsorbs Cs very effectively (see Section 3.5.1.2), and the number of binding sites for ^{137}Cs is increased if the water flows over a longer distance. According to Andersson et al., the retention of ^{137}Cs on roofs can generally be described by a double exponential function, similar to Eqn. 3.92:

$$C(t) = 0.5 \cdot C_0 \cdot e^{\lambda t} \left(e^{-\lambda_1 t} + e^{-\lambda_2 t} \right), \qquad (3.93)$$

where λ is the radioactive decay constant, and λ_1 and λ_2, are 1–4 y and 25–50 y, respectively, depending on the construction of the roof (i.e. the material and firing process).

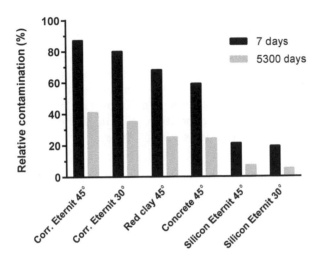

FIGURE 3.77 Retention of ^{137}Cs on corrugated Eternit fibre cement, red clay tiles, concrete tiles and silicon-treated Eternit, for roof slopes of 45° and 30°. The data shown are the contamination levels measured 7 d and 5300 d (about 14.5 y) after deposition, relative to an initial contamination of 760 Bq m^{-2}. Data from Andersson et al. (2002).

The fate of urban contamination can be studied by measurements on sewage sludge, which provides a useful indicator of the transport of radionuclides in the areas served by the plant. The concentration of radionuclides in the sludge depends on the construction of the sewage plant, including factors such as whether run-off water from streets is also fed to the plant. However, if the ratio between the activity in the sludge and that in the outgoing water is reasonably constant for a given radionuclide, the activity in the sludge can be used to estimate the total activity passing through the plant. Figure 3.78 shows measurements made in Lund, Sweden, before and after the Chernobyl accident.

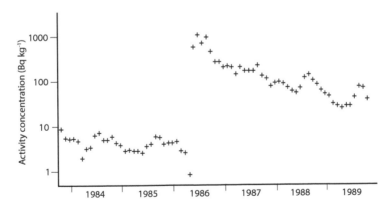

FIGURE 3.78 Activity concentration (dry weight) of ^{137}Cs in sewage sludge from 1983 to 1989, measured in Lund, Sweden. Data from Erlandsson et al. (1990).

The main dose pathways in an urban environment are external exposure from fallout, contamination of the body, the use of contaminated equipment, internal exposure from consumption of contaminated water and food produced locally on contaminated land (e.g. kitchen gardens), and internal exposure from inhalation of contaminated dust (Andersson et al. 2008). The relative importance of each pathway may vary according to the situation, and will also vary with time, as shown in Figure 3.79. The radiation dose from inhalation of radioactive dust indoors is largely affected by the ventilation rate and type of filters in use. The concentration of small particles (0.1–1 μm) may be significant, even with efficient filtration (Nordqvist 2013). Following wet deposition, the level of outdoor contamination will be much higher than for dry deposition, and in this case, the radiation dose from outdoor sources will dominate (Andersson et al. 2008).

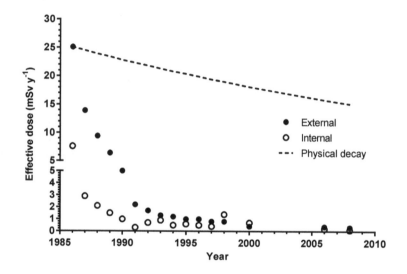

FIGURE 3.79 Average annual effective dose from external and internal exposure from Chernobyl fallout in the Bryansk region, Russia. The decrease of the external component is faster than the decrease due to physical decay, which is explained by weathering and downward migration. Data from Bernhardsson et al. (2011).

Andersson et al. (2008) modelled the external radiation dose from γ-radiation for four types of dwellings: single-storey detached houses, two-storey semi-detached houses, rows of two-storey terraced houses, and multi-storey blocks of flats (5 stories). It was assumed that the contamination was deposited on the walls, roofs, grass and soil surfaces, and on trees. Figure 3.80 shows examples of the results of calculations of the integrated effective dose for the first year after deposition, in μSv, for a contamination of 1 Bq on each type of surface, assuming no measures were taken to reduce the contamination. A number of times after deposition were used, and these results can therefore be used to estimate the radiation dose to inhabitants at varying times after a deposition event.

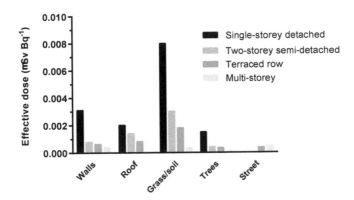

FIGURE 3.80 Integrated effective dose from ^{137}Cs during the first year after deposition on various surfaces, for various kinds of dwellings. The time spent outdoors is assumed to be 15%. Data from Andersson et al. (2008).

3.6 REFERENCES

- Aarkrog, A. 1994. *Direct contamination—seasonality*. In: *Nordic Radioecology: The Transfer of Radionuclides through Nordic Ecosystems to Man*. H. Dahlgaard (Ed.). Elsevier. 149–163.

- Abdel-Magid, I. M., Mohammed, A-W. H. and Rowe, D. R. 1996. *Modeling Methods for Environmental Engineers*. CRC Press.

- Absalom, J. P., Young, S. D., Crout, N. M. J., Sanchez, A., Wright, S. M., Smolders, E., Nisbet, A. F. and Gillett, A. G. 2001. Predicting the transfer of radiocaesium from organic soils to plants using soil characteristics. *Journal of Environmental Radioactivity*, 52, 31–43.

- Åhman, B., Forsberg, S. and Åhman, G. 1990. Zeolite and bentonite as caesium binders in reindeer feed. *Rangifer*, Special Issue No. 3, 73–82.

- Almgren, S., Nilsson, E., Erlandsson, B. and Isaksson, M. 2006. GIS supported calculations of ^{137}Cs deposition in Sweden based on precipitation data. *Science of the Total Environment*, Vol. 368, 804–813.

- Almgren, S. and Isaksson, M. 2006. Vertical migration studies of ^{137}Cs from nuclear weapons fallout and the Chernobyl accident. *Journal of Environmental Radioactivity*, Vol. 91, 90–102.

- Andersson, K. G., Roed, J. and Fogh, C. L. 2002. Weathering of radiocaesium contamination on urban streets, walls and roofs. *Journal of Environmental Radioactivity*, 62, 49–60.

- Andersson, K. G., Ammann, M., Backe, S. and Rosén, K. 2008. *Decision Support Handbook for Recovery of Contaminated Inhabited Areas*. NKS-175, Nordic Nuclear Safety Research.

- Andersson, K. (Ed.). 2009. *Airborne Radioactive Contamination in Inhabited Areas.* Elsevier, Amsterdam, The Netherlands.

- Andersson, T. 1994. Radioecology in freshwater systems: Especially lakes. In *Radioecology. Lectures in Environmental Radioactivity*, ed. E. Holm, 167–179. Singapore: World Scientific.

- Arnalds, O., Cutshall, N. H. and Nielsen, G. A. 1989. Cesium-137 in Montana soils. *Health Physics*, Vol. 57, 955–958.

- Assimakopoulos, P. A., Ioannides, K. G. and Pakou, A. A. 1989. The propagation of the Chernobyl ^{131}I impulse through the air-grass-animal-milk pathway in northwestern Greece. *The Science of the Total Environment.* 85, 295–305.

- Beck, H. L., 1966. Environmental gamma radiation from deposited fission products, 1960-1964. *Health Physics*, Vol.12, 313–322.

- Bengtsson, S. B. 2013. *Interception and Storage of Wet Deposited Radionuclides in Crops. Field Experiments and Modelling.* PhD Thesis, Swedish University of Agricultural Sciences, Uppsala.

- Bernhardsson, C, Zvonova, I., Rääf, C. & Mattsson, S. 2011. Measurements of long-term external and internal radiation exposure of inhabitants of some villages of the Bryansk region of Russia after the Chernobyl accident. *Science of the Total Environment*, 409, 4811–4817.

- Bergström, U., Sundblad, B. and Nordlinder, S. 1994. Models for predicting radiocaesium levels in lake water and fish. In: *Nordic Radioecology: The Transfer of Radionuclides Through Nordic Ecosystems to Man.* H. Dahlgaard (Ed.). Elsevier. 93–104.

- Bossew, P. and Kirchner, G. 2004. Modelling the vertical distribution of radionuclides in soil. Part 1: The convection-dispersion equation revisited. *Journal of Environmental Radioactivity*, 73, 127–150.

- Broberg, A. and Andersson, E. 1991. Distribution and circulation of Cs-137 in lake ecosystems. In: *The Chernobyl Fallout in Sweden.* L. Moberg (Ed). SSI. 151–175.

- Bualert, S. 2001. *Development and Application of an Advanced Gaussian Urban Air Quality Model.* PhD Thesis, University of Hertfordshire, Hatfield, UK.

- Calmon, P., Thiry, Y., Zibold, G., Rantavaara, A. and Fesenko, S. 2009. Transfer parameter values in temperate forest ecosystems: A review. *Journal of Environmental Radioactivity*, 100, 757–766.

- Clarke, R. H. 1979. *A Model for Short and Medium Range Dispersion of Radionuclides Released to the Atmosphere.* NRPB-R91. National Radiation Protection Board, UK.

- Cooper, P. J. 1984. Building Effects in Nuclear Safety Studies: A Review. In: Underwood, B. Y., Cooper, P. J., Holloway, N. J., Kaiser, G. D. and Nixon, W. *Nuclear Science and Technology. Review of Specific Effects in Atmospheric Dispersion Calculations.* Commission of the European Communities. Luxembourg.

- Cooper, J. R., Randle, K. and Sokhi, R. S. 2003. *Radioactive Releases in the Environment: Impact and Assessment.* John Wiley and sons. Chichester, England.

- Crout, N. M. J. and Voigt, G. 1996. Modeling the dynamics of radioiodine in dairy cows. *Journal of Dairy Science.* Vol. 79, No. 2, 254–259.

- Dahlgaard, H., Aarkrog, A., Hallstadius, L., Holm, E. and Rioseco, J. 1986. Radiocaesium transport from the Irish Sea via the North Sea and the Norwegian coastal current to east Greenland. *Rapp. P.-v. Réun. Cons. int. Explor. Mer* 186: 70–79.

- Dowdall, M., Gerland, S., Karcher, M. et al. 2005. *Geostatistical Methods Applied to Sampling Optimisation for the Temporal Monitoring of Technetium-99 in the Arctic Marine Environment.* Strålevern Rapport 2005:19. Norwegian Radiation Protection Authority.

- Eckerman, K. F. and Ryman, J. C. 1993. *External Exposure to Radionuclides in Air, Water, and Soil.* Federal Guidance Report No. 12. EPA-402-R-93-081. U.S. Environmental Protection Agency. Washington, DC. USA.

- Eckerman, K. F. and Leggett, R. W. 1996. *DCFPAK: Dose Coefficient Data File Package for Sandia National Laboratory.* Oak Ridge National Laboratory Report ORNL/TM-13347. Oak Ridge National Laboratory, Oak Ridge, TN. USA.

- Eisenbud, M. 1987. *Environmental Radioactivity. From Natural, Industrial, and Military Sources.* Academic Press. Orlando. USA.

- Eriksson, Å. and Rosén, K. 1991. Transfer of caesium to hay grass and grain crops after Chernobyl. In: *The Chernobyl Fallout in Sweden.* L. Moberg (Ed). SSI. 291–304.

- Erlandsson, B. and Isaksson, M. 1988. Relation between the air activity and the deposition of Chernobyl debris. *Environment International.* Vol. 14, 165–175.

- Erlandsson, B., Isaksson, M., Mattsson, S., Bjurman, B., Eriksson, R. and Vesanen, R. 1990. *Urban Radionuclide Contamination Studied in Sewage Water and Sludge from Lund and Gothenburg.* Report Project SSI P 344.86.

- FAO. 1998. *Integrated Coastal Area Management and Agriculture, Forestry and Fisheries. Part A.* Food and Agriculture Organization of the United Nations. http://www.fao.org/docrep/w8440e/W8440e02.htm, accessed on 18/08/2015.

- FAO-UNESCO. 1974. *FAO-Unesco Soil Map of the World.* Food and Agriculture Organization of the United Nations and United Nations Educational, Scientific and Cultural Organization. UNESCO-Paris.

- Foth, H. D. 1984. *Fundamentals of Soil Science.* John Wiley and sons, New York, USA.

- Fowler, S. W. and Fisher, N. S. 2004. Radionuclides in the biosphere. In: *Marine Radioactivity,* ed. H.D. Livingston, 167–203. Elsevier.

- Friedlander, S. K. 2000. *Smoke, Dust, and Haze. Fundamentals of Aerosol Dynamics.* Oxford University Press, New York, USA.

- Fukai, R., Holm, E. and Ballestra, S. 1979. A note on vertical distribution of plutonium and americium in the Mediterranean Sea. *Oceanologica Acta,* 2, 129–132.

- Gaare, E. and Staaland, H. 1994. Pathways of fallout radiocaesium via reindeer to man. In: *Nordic radioecology: The Transfer of Radionuclides through Nordic Ecosystems to Man.* H. Dahlgaard (Ed.). Elsevier. 303–334.

- Hamilton, T. F. 2004. Linking legacies of the Cold War to arrival of anthropogenic radionuclides in the oceans through the 20th century. In: *Marine Radioactivity*, ed. H.D. Livingston, 23–78. Elsevier.

- Harms, I. H., Karcher, M. J. and Burchard, H. 2003. Modelling radioactivity in the marine environment: The application of hydrodynamic circulation models for simulating oceanic dispersion of radioactivity. In: *Modelling Radioactivity in the Environment*, ed. E.M. Scott, 55–85.

- Health Canada. 1999. *Recommendations on Dose Coefficients for Assessing Doses from Accidental Radionuclide Releases to the Environment*. Health Canada, Health Protection Branch.

- Heling, R. 1997. LAKECO: Modelling the transfer of radionuclides in a lake ecosystem. *Radiation Protection Dosimetry*, Vol. 73, Nos 1–4, 191–194.

- Håkansson, E., Drugge, N., Vesanen, R., Alpsten, M. and Mattsson, S. 1987. *Transfer of ^{134}Cs, ^{137}Cs and ^{131}I from Deposition on Grass to Cow's Milk. A Field Study after the Chernobyl Accident*. Report GU-RADFYS 87:01, Department of Radiation Physics, University of Gothenburg, Sweden.

- Håkansson, L. 1991. Radioactive caesium in fish in Swedish lakes after Chernobyl geographical distributions, trends, models and remedial measures. In: *The Chernobyl Fallout in Sweden*. L. Moberg (Ed). SSI. 239–281.

- IAEA, International Atomic Energy Agency. 1992. *Modelling of Resuspension, Seasonality and Losses during Food Processing*. IAEA-TECDOC-647. International Atomic Energy Agency. Vienna.

- IAEA, International Atomic Energy Agency. 2001. *Generic Models for Use in Assessing the Impact of Discharges of Radioactive Substances to the Environment*. Safety Reports Series No. 19. International Atomic Energy Agency. Vienna.

- IAEA, International Atomic Energy Agency. 2004. *Sediment Distribution Coefficients and Concentration Factors for Biota in the Marine Environment*. Technical Reports Series No. 422. International Atomic Energy Agency. Vienna.

- IAEA, International Atomic Energy Agency. 2005. *Worldwide Marine Radioactivity Studies (WOMARS). Radionuclide Levels in Oceans and Seas*. IAEA-TECDOC-1429. International Atomic Energy Agency. Vienna.

- IAEA, International Atomic Energy Agency. 2009. *Quantification of Radionuclide Transfer in Terrestrial and Freshwater Environments for Radiological Assessments*. IAEA-TECDOC-1616. International Atomic Energy Agency. Vienna.

- IAEA, International Atomic Energy Agency. 2010. *Handbook of Parameter Values for the Prediction of Radionuclide Transfer in Terrestrial and Freshwater Environments*. Technical Reports Series No. 472. International Atomic Energy Agency. Vienna.

- ICRP 1995. *Age-Dependent Doses to Members of the Public from Intake of Radionuclides—Part 4 Inhalation Dose Coefficients*. ICRP Publication 71. Ann. ICRP 25 (3-4).

- ICRP 2012. *Compendium of Dose Coefficients based on ICRP Publication 60*. ICRP Publication 119. Ann. ICRP 41(Suppl.).

- ICRU 2001. *Quantities, Units and Terms in Radioecology.* ICRU Publication 65. ICRU Publications, Bethesda, USA.

- Iosjpe, M., Brown, J. and Strand, P. 2002. Modified approach to modelling radiological consequences from releases into the marine environment. *Journal of Environmental Radioactivity,* 60, 91–103.

- Isaksson, M. and Erlandsson. B. 1995. Experimental determination of the vertical and horizontal distribution of ^{137}Cs in the ground. *Journal of Environmental Radioactivity,* Vol. 27, No. 2, 141–160.

- Isaksson, M. and Erlandsson. B. 1998. Models for the vertical migration of ^{137}Cs in the ground: A field study. *Journal of Environmental Radioactivity,* Vol. 41, No. 2, 163–182.

- Isaksson, M., Erlandsson, B. and Linderson M.-L. 2000. Calculations of the deposition of ^{137}Cs from nuclear bomb tests and from the Chernobyl accident over the province of Skne in the southern part of Sweden based on precipitation. *Journal of Environmental Radioactivity,* Vol. 49, 97–112.

- Jacquez, J. A. 1999. *Modeling with Compartments.* BioMedWare. Ann Arbor. USA.

- Lewis Jr, W. M. 1983. A revised classification of lakes based on mixing. *Canadian Journal of Fisheries and Aquatic Sciences,* 40, 1779–1787.

- Locarnini, R. A., A. V. Mishonov, J. I. Antonov, T. P. Boyer, H. E. Garcia, O. K. Baranova, M. M. Zweng, and D. R. Johnson, 2010. World Ocean Atlas 2009, Volume 1: Temperature. In *NOAA Atlas NESDIS 68,* ed. S. Levitus, U.S. Government Printing Office, Washington, D.C, USA.

- Marshall, J. and Plumb, R. A. 2008. *Atmosphere, Ocean, and Climate Dynamics.* Academic Press, Amsterdam, The Netherlands.

- Mattsson, S. 1975. ^{137}Cs in the reindeer lichen *Cladonia alpestris*; deposition, retention and internal distribution. *Health Physics,* Vol.28, 233–248.

- Mattsson, S. and Erlandsson, B. 1988. Granbarr som bioindikator fr radioaktivt cesium? Proceedings of the Fifth Nordic Radioecology Seminar, August 22-25, Rättvik, Sweden [in Swedish].

- Mattsson, S. and Vesanen, R. 1988. Patterns of Chernobyl fallout in relation to local weather conditions. *Environment International.* Vol. 14, 177–180.

- Miller, A., Kruichak, J., Mills, M. and Wang, Y. 2015. Iodide uptake by negatively charged clay interlayers? *Journal of Environmental Radioactivity,* 147, 108–114.

- Müller, H. and, Pröhl, G. 1993 Ecosys-87: A dynamic-model for assessing radiological consequences of nuclear accidents, *Health Physics,* 64, 232–252.

- Nielsen, S. P., Iosjpe, M. and Strand. 1995. *A preliminary assessment of potential doses to man from radioactive waste dumped in the Arctic Sea.* Strålevern Rapport 1995:8. Norwegian Radiation Protection Authority.

- Nilsson, M. 1981. Mathematical models for simulation of processes of radioecological and biomedical transport of radionuclides, Part III: Application on ^{131}I in the food chain grass-cow-milk following a BWR 1 accident. In *Pathways and Estimated Consequences of Radionuclide Releases from a Nuclear Power Plant*. Dissertation, Radiation Physics Dept., Lund University, Sweden.

- Nisbet, A. F., Brown, J., Cabianca, T., Jones, A. L., Andersson, K. G., Hnninen, R., Ikheimonen, T., Kirchner, G., Bertsch, V. and Heite, M. 2010. *Generic Handbook for Assisting in the Management of Contaminated Inhabited Areas in Europe Following a Radiological Emergency*. EURANOS(CAT1)-TN(09)-03. Health Protection Agency (HPA), UK.

- Nordqvist, M. 2013. *Modeling Protection Coefficients of Buildings during a Release of Radioactive Materials*. MSc Thesis. Uppsala University, Uppsala, Sweden. [In Swedish]

- NRPB. 2001. *Atmospheric Dispersion Modelling Liaison Committee, Annual Report 1997/98. Annex A: Review of Deposition Velocity and Washout Coefficient*. NRPB-R322. National Radiation Protection Board. UK.

- Overcamp, T. J. 1991. Modeling γ absorbed dose due to meandering plumes. *Health Physics*. Vol. 61, No. 1, 111–115.

- Pálsson, S. E., Howard, B. J. and Wright, S. M. 2006. Prediction of spatial variation in global fallout of ^{137}Cs using precipitation. *Science of the Total Environment*, Vol. 367, pp. 745–756.

- Pálsson, S. E. 2012. *Prediction of Global Fallout and Associated Environmental Radioactivity*. PhD thesis. Faculty of Physical Sciences, University of Iceland.

- Pálsson, S. E., Howard, B. J., Bergan, T. D., Paatero, J., Isaksson, M. and Nielsen, S. P. 2013. A simple model to estimate deposition based on a statistical reassessment of global fallout data. *Journal of Environmental Radioactivity*, Vol. 121, 75–86.

- Penrose, B., Beresford, N. A., Broadley, M. R. and Crout, N. M. J. 2015. Inter-varietal variation in caesium and strontium uptake by plants: A meta-analysis. *Journal of Environmental Radioactivity*, 139, 103–117.

- Petoussi, N., Jacob, P., Zankl, M. and Saito, K. 1991. Organ doses for foetuses, babies, children and adults from environmental gamma rays. *Radiation Protection Dosimetry*. 37 (1), 31–41.

- Plauborg, F., Andersen, M. N., Heidmann, T. and Oleson, J. E. 2015. *MARKVAND: An irrigation scheduling system for use under limited irrigation capacity in a temperate humid climate*. http://www.fao.org/3/a-w4367e/w4367e0k.htm, accessed on 27/09/15.

- Povinec, P. P. 2004. Developments in analytical technologies for marine radionuclide studies. In: *Marine Radioactivity*, ed. H.D. Livingston, 237–294. Elsevier.

- Rääf, C. L., Hubbard, L., Falk, R., Ågren, G. and Vesanen, R. 2006. Ecological half-time and effective dose from Chernobyl debris and from nuclear weapons fallout of ^{137}Cs as measured in different Swedish populations. *Health Physics*, Vol. 90, No. 5, 446–458.

- Rahola, T. and Suomela, M. 1990. Radiocesium i renkött—ökad stråldos för samebefolkningen. *Rangifer*, Special Issue No. 4, 38–40 [In Swedish].

- Renaud, P., Pourcelot, L., Métivier, J. M. and Morello, M. 2003. Mapping of Cs-137 deposition over eastern France 16 years after the Chernobyl accident. *Science of the Total Environment*, Vol. 309, 257–264.

- Rissanen, K. and Rahola, T. 1990. Radiocesium in lichens and reindeer after the Chernobyl accident. *Rangifer*, Special Issue No. 3, 55–61.

- Roed, J. 1990. *Deposition and removal of radioactive substances in an urban area*. Nordic Liaison Committee for Atomic Energy, NORD 1990:111.

- Rosén, K. 1991. Effects of potassium fertilization on caesium transfer to grass, barley and vegetables after Chernobyl. In: *The Chernobyl Fallout in Sweden*. L. Moberg (Ed). SSI. 305–322.

- Schultz, F. W. and Zoetelief, J. 1997. Effective dose per unit fluence calculated for adults and a 7-year-old girl in broad antero-posterior beams of monoenergetic electrons of 0.1 to 10 MeV. *Radiation Protection Dosimetry.* 69 (3), 179–186.

- Slade, D. H. (Ed.). 1968. *Meteorology and Atomic Energy 1968*. U. S. Atomic Energy Commission, USA.

- Sundblad, B. 1990. Which processes dominate the uncertainties in the modelling of the transfer of radionuclides in lake ecosystems. In: *BIOMOVS. On the Validity of Environmental Transfer Models. Proceedings. Symposium October 8–10*. Stockholm, Sweden. SSI. 295–306.

- Thiessen, K. M., Thorne, M. C., Maul, P. R., Pröhl, G. and Wheater, H. S. 1999. Modelling radionuclide distribution and transport in the environment. *Environmental Pollution*, 100, 151–177.

- UNSCEAR, United Nations Scientific Committee on the Effects of Atomic Radiation. 2000. *Sources and Effects of Ionizing Radiation, Vol.I: Sources, Annex C*. New York: United Nations.

- UNSCEAR, United Nations Scientific Committee on the Effects of Atomic Radiation. 2008. *Sources and Effects of Ionizing Radiation, Vol.I: Sources of Ionizing Radiation, Annex B*. New York: United Nations.

- UNSCEAR, United Nations Scientific Committee on the Effects of Atomic Radiation. 2013. *Sources, Effects and Risks of Ionizing Radiation, Vol.I: Report to the General Assembly, Scientific Annex A*. New York: United Nations.

- Wallström, E. 1998. *Assessment of Population Radiation Exposure after a Nuclear Reactor Accident. Field Studies in Russia and Sweden after Chernobyl*. PhD Thesis, University of Gothenburg, Sweden.

- Wright, S. M., Howard, B. J., Strand, P., Nylén, T. and Sickel, M. A. K. 1999. Prediction of ^{137}Cs deposition from atmospheric nuclear weapons tests within the Arctic. *Environmental Pollution*, Vol. 104, 131–143.

- Yamaguchi, Y. 1994. Age-dependent effective doses for external photons. *Radiation Protection Dosimetry.* 55 (2), 123–129.

- Zanetti, P. 1990. *Air Pollution Modelling: Theories, Computational Methods and Available Software*. Van Nostrand Reinhold, New York.

3.7   EXERCISES

3.1   A non-radioactive contaminant is assumed to be reasonably well described by the compartment model depicted in Figure 3.4. Assume that the transfer coefficients are 0.8 d^{-1} from soil to the available fraction, 0.2 d^{-1} from soil to the fixed fraction, and 0.5 d^{-1} from the available fraction to pasture. The initial amount deposited on the soil is 100 μg.

    a.   Determine, by solving the differential equations, how much of the contaminant will have reached the pasture after 3 days.

    b.   How much of the contaminant will end up in the pasture at equilibrium?

    c.   Can you find the answer to b without solving the differential equations?

3.2   A scientist is studying the kinetics of a newly developed drug in the human body and has developed a simple compartment model. The model consists of two compartments, 1 and 2, that contain the amounts q_1 and q_2 of the drug after a bolus injection (of very short duration) of an amount, d, into compartment 1 at time $t = 0$. The transfer from compartment 1 to compartment 2 is determined by the transfer coefficient $k_{2,1}$, and the removal of the drug from the system is governed by the transfer coefficients $k_{0,1}$ and $k_{0,2}$. The scientist wishes to determine the transfer coefficients $k_{0,1}$, $k_{2,1}$ and $k_{0,2}$ by measuring the concentration of the drug in a sample taken from compartment 1 at different times after injection. The concentration is given by q_1/V_1 and the volumes of the compartments, V_1 and V_2, are known from previous studies.

    a.   The scientist's model will, unfortunately, not meet expectations and the desired results cannot be obtained. Why?

    b.   What can the scientist do to determine the three transfer coefficients without changing the model?

3.3   The atmospheric inventory of ^{14}C has been estimated to be 140 PBq (see Section 1.1.2.3). Calculate the annual effective dose to a population, assuming that the inventory of ^{14}C is homogeneously distributed in the atmosphere. Only the effective dose from immersion and inhalation needs to be considered.

3.4   A rescue worker is measuring the dose rate during an accidental atmospheric release of radionuclides from a nuclear facility. Between the measurements, the worker is inside a building, but leaves the building four times every hour to make a measurement, which lasts for one minute. It can be assumed that the external dose rate inside the building is 40% of the dose rate outside. It can also be assumed that the activity concentration inside the building is zero, and that no contamination has occurred. The air concentration outside the building is estimated to be 370 Bq m^{-3} of ^{131}I and 100 Bq m^{-3} of ^{137}Cs.

    a.   Calculate the extra effective dose the worker receives by leaving the building for the measurements.

b. Calculate the average committed effective dose from inhalation during 6 h of work, assuming that the worker does not wear a respiratory protection mask.

3.5 The isotopic ratio for ^{134}Cs and ^{137}Cs in a seawater sample is initially 0.5. The isotopic ratio in another seawater sample, taken some distance from the source, is 0.1. Estimate the transit time from the source to the second sampling site.

3.6 Radionuclides in a water column can be dissolved in the water or attached to particulates (suspended sediment material). Partition between these two phases is determined by the distribution coefficient, K_d. Show that the fraction of the radionuclide dissolved in the water, F_w is given by

$$F_w = \frac{1}{1 + K_d \cdot SSL},\qquad \text{(P3.1)}$$

where SSL is the suspended sediment load, given in units of kg L^{-1} when K_d is given in units of L kg^{-1}.

3.7 The transfer rate from a water column, i, to sediment, j, can be found by assuming a mean depth, d_i, of the water column and a mass sedimentation rate SR [kg m^2 y^{-1}]. Show that

$$k_{i,j} = \frac{K_d \cdot SR_i}{d_i \left(1 + K_d \cdot SSL_i\right)}.\qquad \text{(P3.2)}$$

3.8 When offered a glass of water, you are asked if you want a wide straw or four narrow straws. The radius of the wide straw is four times the radius of a narrow straw. Which would you prefer if you want to maximize the flow rate [m^3 s^{-1}], assuming that you can create the same pressure difference in both cases?

3.9 The activity concentration [Bq kg^{-1}] in the ground at a given site decreases exponentially with a relaxation length of 2.5 cm (see Eqn. 3.79). Find the areal activity density [Bq m^{-2}] at the site in terms of S_0, assuming that the density of the soil is 1200 kg m^{-3}.

3.8   FURTHER READING

Benamrane, Y., Wybob, J.-L. and Armand, P. 2013. Chernobyl and Fukushima nuclear accidents: What has changed in the use of atmospheric dispersion modeling? *Journal of Environmental Radioactivity.* 126, 239–252.

Di Stefano III, J. 2013. *Dynamic Systems Biology Modeling and Simulation.* Academic Press. Amsterdam, the Netherlands.

IAEA, International Atomic Energy Agency. 2012. *Environmental modelling for radiation safety (EMRAS).* IAEA-TECDOC-1678. International Atomic Energy Agency. Vienna.

Jacquez, J. A. 1996. *Compartmental Analysis in Biology and Medicine.* BioMedWare. Ann Arbor. USA.

Van der Stricht, E. and Kirchmann, R. (Ed.). 2001. *Radioecology. Radioactivity and Ecosystems.* International Union of Radioecology.

Radiometry

CONTENTS

R ADIOMETRY can be defined as the theory of radiation detection. Radiometry includes the detection of both ionizing and non-ionizing radiation. However, in this chapter/book, only detectors used to measure ionizing radiation will be discussed. When exposed to ionizing radiation, such detectors provide a measurable signal, which may be electronic, resulting from a current of free charges, or optical, resulting from radioluminescence or a chemical reaction caused by free radicals generated by the radiation.

4.1 BASIC STATISTICAL PRINCIPLES OF RADIOMETRY

The most fundamental task in radiation protection is the identification and quantification of radionuclides causing increased levels of radiation. Measurements are carried out to quantify the radiation in terms of e.g. (i) the activity of the source [Bq], (ii) the activity concentration per unit mass of the source [Bq kg^{-1}], or (iii) the activity per unit area of a surface [Bq m^{-2}]. Based on these measurements, the next task is to predict the exposure of individuals to radiation in the affected environment, and to suggest protective measures to reduce this exposure. It is often important to identify and quantify very low levels of radiation, as these may indicate the release or dispersion of other radionuclides that may have serious radiological consequences. Examples are chemically inert fission products such as ^{137}Cs in the water of a nuclear reactor, indicating the onset of fuel damage, or radioactive sodium isotopes such as ^{24}Na in living organisms, indicating exposure to thermal neutrons.

The fundamental processes exploited in radiometry are the radioactive decay of an atom or the nuclear reaction caused by the capture of protons or neutrons, leading to the emission of charged particles (α- and β-particles, conversion electrons) uncharged particles (neutrons) or photons (characteristic x-rays and γ-rays). Radioactive decay is governed by the rules of quantum mechanics, which can be regarded as a set of statistical probabilities of various electromagnetic and nuclear processes. The most important quantity in radiometry is the number of decays that can be expected during a given time interval, t, for a given population of nuclides. Radiometric devices cannot detect the actual nuclides, but are designed to detect the particles emitted when a nuclide decays.

The probability of a specific nucleus decaying during a time interval is constant. In a population of nuclides, the rate of decay per unit time is proportional to the number of nuclides and a constant, λ. If the number of nuclides is large, the number of decays per second, i.e., the activity, A, can be described by the continuous equation

$$A\left(t\right) = A_0 e^{-\lambda t}, \tag{4.1}$$

where A_0 is the activity at the beginning of the observation (at time $t = 0$), t is the elapsed time, and λ is the probability of decay per unit time. The physical half-life, $T_{1/2}$, of the radionuclide is defined as $\ln 2/\lambda$. As the population contains a finite number of radionuclides, the probability of a radionuclide decaying during a finite observation time is thus associated with statistical fluctuations. In a series of repeated observations the mean value and variance of the number of decays recorded per unit time will depend on both the half-life of the radionuclide and the number of radionuclides in the sample.

4.1.1 Statistical Models

The purpose of a measurement is often to count the number of decays recorded in the detector during a specific time interval, or to measure the number of decays recorded successively in a series of observations, referred to here as a series of data $(x_1, x_2, \ldots, x_i, \ldots, x_n)$. Statistical fluctuations will lead to an intrinsic uncertainty in repeated experimental measurements. Therefore, the mean value should always be accompanied by an estimate of the predicted uncertainty or *variance*. The variance is derived from the difference or deviation, ϵ_i, between each data point, x_i, and the mean of the whole data series, $< x >$

$$\epsilon_i = x_i - < x > . \tag{4.2}$$

To take into account the fact that the deviations may be positive or negative, the variance is defined as the square of the mean deviations

$$s^2 = \frac{1}{N} \sum_{i=1}^{N} \left(x_i - < x >\right)^2, \tag{4.3}$$

where N is the number of observations. When N is too small for this approximation to be valid, the *sample variance*, s^2, is often defined as

$$s^2 = \frac{1}{N-1} \sum_{i=1}^{N} (x_i - <x_{sample}>)^2. \tag{4.4}$$

The distribution of the measurements can be described by a function giving the probability of each possible outcome x. Mathematical models are used to simplify the calculation of the probability of each possible outcome. The cumulative distribution function, $P(x)$, is defined as

$$P(x) = \frac{\text{number of outcome in the range } [-\infty, x]}{N}, \tag{4.5}$$

where N is the number of observations or possible outcomes in the whole value range $[-\infty, \infty]$. $P(x)$ thus ranges from 0 at $x = \infty$, and reaches unity as x approaches ∞. The probability density function, $f(x)$, is the derivative of $P(x)$, and represents the relative proportion of outcomes within a narrow range of values $[x, x + \delta x]$. The density function is central in the statistical interpretation of data, as it essentially determines the arithmetic mean, the median and the dispersion (for example, in terms of the standard deviation) of parameter value x.

The arithmetic mean of a parameter; for example when repeatedly casting dice, is given by

$$<x> = \sum_{x=0}^{\infty} x f(x), \tag{4.6}$$

where $f(x)$ is the density function of the probability of detecting a certain number of decays.

For continuous variables the probability function can be also expressed in the form

$$<x> = \int_{x=0}^{\infty} x f(x). \tag{4.7}$$

When the number of observations is small and the outcome can be characterized as a success ($x = 1$) or not ($x = 0$), the *binominal distribution* is used to describe the probability function. The probability of the number of successes, k, in n measurements, $F(k)$, is described by the function

$$P(x) = \frac{n!}{(n-k)!k!} p^x (1-p)^{n-k}, \tag{4.8}$$

where p is the probability of success of each event. When flipping a coin, for example, $p = 0.5$. The most probable result after flipping a coin 10 times is 5 heads and 5 tails, but due to inherent statistical fluctuations, the probability of this outcome, $f(5)$, is less than 25%. Hence, when the number of observations is small, the mean number of successes is not a very accurate approximation of the most probable result, i.e., the true mean. In the case of radioactive decay, the probability of an individual nuclide decaying is described by the binomial distribution above.

In a series of experimental measurements where the number of decays recorded is large ($n > 25$), the variance in the number detected can be described by Poisson statistics. Based on the approximation that $< x >= np$, the *Poisson distribution* can be derived from the binominal distribution, giving

$$P(x) = \frac{< x >^2 e^{-<x>}}{x!},$$ (4.9)

where x is the outcome of a specific observation.

The probability, p, of a single atom decaying during a given time interval is constant. In a sample where the number of atoms of a specific radionuclide is large, and the physical half-life of the radionuclide is long compared to the observation time (i.e., the activity can be considered to be constant), Poisson statistics can be used to describe the statistical fluctuations in the fluence of particles emitted by the sample. Hence, the number of particles or photons detected by a radiation counter (referred to as events) will be proportional to the number of decays in the sample during the observation period.

A Poisson distribution is characterized by the variance being equal to the mean value, $< x >$, in a series measurements. Hence, the variance is directly dependent on the number of events recorded. The square root of the variance, σ, is the familiar standard deviation, which is often used as an estimate of the uncertainty of a measurement. Since $< x >= np$, (p being proportional to the physical half-life of the nuclide, i.e. the number of events per unit time, and n being the observation time), increasing the observation time will reduce the ratio σ/x.

The *Gaussian distribution* is a further simplification of the binominal distribution, which is used when $p \ll 1$, and the number of observations is large enough to assume that $< x >= np$. The Gaussian distribution, also called the normal distribution, is given by the equation

$$P(x) = \frac{1}{\sqrt{2\pi < x >)}} e^{-\frac{(x-<x>)^2}{2<x>}}.$$ (4.10)

The Gaussian distribution is only used when the number of observations is very large, and x can be regarded as a continuous variable, whereas in the Poisson distribution, the sample is still considered to be a finite number of observations. The variance of a Gaussian distribution is still equal to the mean value of the events detected and is thus used as a measure of the inherent statistical fluctuation in the number of events detected. An important characteristic of Gaussian distributions is that a random subset of data from a Gaussian distribution will also be a Gaussian distribution, as long as the number of detected events is sufficiently large to assume that $< x >= np$.

4.1.2 Uncertainties in Radiometry Measurements

Only the intrinsic statistical fluctuations in the number of events detected were described above. However, when the number of detected events is combined with other parameters, such as the mass of the sample containing the radionuclides, to obtain the activity concentration in the sample [Bq kg^{-1}], the uncertainty of each of the parameters will contribute to the uncertainty in the final value.

Uncertainty propagation in Gaussian distributions involves summing the standard deviations of each measured or estimated parameter. If we assume that the experimental result is calculated using a function, F, of the variables a, b, c, ... with standard deviations σ_a, σ_b, σ_c, ..., then the deviation of the resulting value, σ_F, can be calculated using the following expression

$$\sigma_F^2 = \left(\frac{dF}{da}\right)^2 \sigma_a^2 + \left(\frac{dF}{db}\right)^2 \sigma_b^2 + \left(\frac{dF}{dc}\right)^2 \sigma_c^2 + \dots$$ (4.11)

For a more comprehensive and detailed discussion on the propagation of uncertainty in radiation measurements, the reader is referred to ICRU Report 75 (ICRU 2006).

4.1.3   Background Subtraction

To illustrate the stochastic nature of radioactive decay and the detection of the radiation associated with it, we will now consider the detection of a number of decays from a population of radionuclides. The intention is to determine whether the level of radiation is elevated. The events recorded in the detector during a given measuring time, t_m, originate from two sources. The first is background radiation resulting from radionuclides existing in nature, and the second is the radiation from other sources believed to be the cause of the elevated radiation level. The total number of counts, G, during t_m will then be the number of counts from the source, N, plus the number of counts from the background, B_m, i.e. $G = N + B_m$. It is generally assumed that the number of counts due to the natural background and the number of counts due to the radioactive source are both greater than 25, and that the measuring time is less than the physical half-lives of the nuclides involved (i.e., the distribution can be approximated to a Poisson distribution). This is also the case if the detected radionuclides have a very short half-life compared to t_m, but are the daughter products of a mother nuclide with a half-life $T_{1/2} \gg t_m$.

The number of background counts, B_{bkg}, measured during t_{bkg} (which does not necessarily have to be the same as t_m; hence $B_{bkg} \neq B_m$), in the absence of another source of radioactivity, will be

$$B_{bkg} = b_{bkg} \cdot t_{bkg}, \tag{4.12}$$

where b_{bkg} is the background count rate [s^{-1}]. Assuming that the background count rate, b_{bkg}, is constant, the number of counts recorded in the detector during the time t_m is given by the sum of the counts arising from the elevated radioactivity, N, and the background counts, B_m

$$G = N + B_m = N + b_{bkg} \cdot t_{bkg} = N + B_{bkg} \cdot \left(\frac{t_m}{t_{bkg}} \right). \tag{4.13}$$

Thus, the net number of counts is given by

$$N = G - B_m = G - b_{bkg} \cdot t_{m,bkg} = G - B_{bkg} \cdot \left(\frac{t_m}{t_{bkg}} \right). \tag{4.14}$$

According to the relationship between the variance and the number of counts recorded in the detector, the variance will be given by the equation for uncertainty propagation (Eq. 4.11):

$$\sigma_N^2 = \sigma_G^2 + \left(\frac{t_m}{t_{bkg}} \right)^2 \sigma_{bkg}^2 \tag{4.15}$$

Following Poisson statistics, where Var$(x) = <x>$, we then can rewrite Eqn. 4.15 as

$$\sigma_N^2 = G + \left(\frac{t_m}{t_{bkg}} \right)^2 B_{bkg}. \tag{4.16}$$

Thus, for a given measuring time, t_m, the variance in N will be less if t_{bkg} is long compared with t_m. However, the time during which the background is measured, t_{bkg}, is often limited by practical constraints. The relative uncertainty of the net number of counts, in terms of one standard deviation, is then given by

$$\frac{\sigma_N}{N} = \sqrt{\frac{(G+B)}{(G-B)}} \to \frac{1}{\sqrt{N}}; N \to \infty. \tag{4.17}$$

It follows from Eqn. 4.17 that the relative standard deviation will approach $1/\sqrt{N}$ when N becomes large.

Since many applications in radiation protection involve measurements to detect elevated radiation levels in the environment, the detectors must be designed to function when the signal-to-background ratio is low. The detection of small amounts of radiation and the propagation of the uncertainty in these measurements are thus of considerable importance in environmental protection radiometry.

4.2 VARIOUS TYPES OF DETECTORS

4.2.1 General Features of a Radiation Detector

Many types of radiation detectors are available for a broad range of applications. A distinction is made between simple radiometric systems, which essentially measure the total intensity of the incident radiation (count rate meters), and spectrometric systems, which are able to quantify the energy deposited by each photon or particle impinging on the detector. The term spectrometer can be used to describe several types of detector systems, but in this book it refers only to devices used to measure the energy of ionizing particles or photons. In some radiometric applications, detectors can be used to visualize the spatial distribution of radiation intensity.

Another important distinction is made between active and passive detector systems. Active detectors provide a direct signal in response to the radiation detected, for example, Geiger-Müller counters. In passive detector systems, such as personal dosimeters worn by workers in radiological or nuclear facilities, the signal is accumulated over a period of time, and the dose or exposure is then read out using an external device.

In the field of radiation protection, three main detection principles are used for radiometry, based on how the response of the detector is generated: (i) ionization of atoms in the detector material (see Sections 4.2.2 and 4.2.3), (ii) prompt or delayed radioluminescence caused by the excitation of atoms or molecules in the detector material (see Section 4.2.4), and (iii) chemical reactions causing permanent changes in the detector material (see Section 4.2.5). The first category yields a current of free charged particles that can be detected using electronic devices. The second category can also provide an electric signal, by means of light-to-electric-charge converters such as photomultiplier tubes (PMTs) or photodiodes. In the third category the signal is related to the extent of a chemical change, for example, visible or optically measurable darkening of a film. Another detection technique used in environmental monitoring is mass spectrometry which, in contrast to radiometry, involves the detection of atoms of various relevant isotopes (^{14}C, ^{129}I, ^{239}Pu, etc.).

4.2.2 Gaseous Detectors for Ionizing Radiation

It was discovered in the 1890s, that high-energy gamma radiation and high-energy charged particles ionize the medium through which they pass. Gas-filled chambers have been used to detect ionizing radiation by collecting the charge of secondary particles and ionized atoms to provide a signal. A gaseous ionization detector consists of a volume of gas, which may be either enclosed or open. The gas may consist of air or mixtures of several gases. The gas may also be pressurized. The chamber is often cylindrical or spherical. Inside the gas is a thin metal wire, the anode, which is connected to the positive terminal of a voltage supply (Fig. 4.1). The electrodes may also be two parallel plates in the gas. The inner wall of the chamber is connected to the negative terminal. When a high voltage, V_{Bias}, is applied to the terminals, an electric field will form within the chamber. If no external energy is introduced into the gas volume (e.g. in the form of incident ionizing radiation), only noise will be registered due to the thermal excitation of the gas atoms and the quantum fluctuation of the capacitance over the two electrodes. These two noise components are discussed in more detail in Section 4.4.3.

Ionizing radiation passing through the chamber will create a track of positive ions and free electrons. The energy needed to ionize gas molecules is about 30 eV at normal pressure and ambient temperature. If all the energy of a free electron with an initial energy of 100 keV is absorbed by in the detector gas, only 3000 electron-ion pairs will be formed in the gas. Thus, the charge created by a single particle is very small, even when the initial energy is high.

Ions with positive and negative charges in the gas will be attracted by the electric field and migrate to the cathode and anode, respectively, if the electric field is sufficiently high. If the electric field strength is weak, < 100 V cm^{-1}, the charges will not reach the electrodes before being absorbed in the gas (recombination). The charge collected at the electrodes will lead to electric pulses being recorded. Depending on the subsequent electronics, these pulses can either be collected continuously as a current, the amplitude of which represents the average intensity of the incident radiation, or individually, for further analysis of the count rate or energy distribution of the radiation particles.

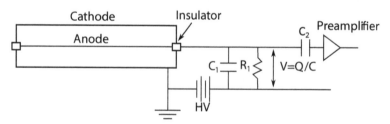

FIGURE 4.1 Schematic illustration of a gaseous detector with a cylindrical cathode and an anode wire.

TABLE 4.1 Ionization energy, w_e, for electrons and α-particles from ^{210}Po, and charge drift velocity, v, at 1 atm for some common detector gases. Data taken from IAEA (1987).

Gas	w_e (eV/ion pair)		μ (cm^2 V^{-1} s^{-1})
	Electrons	**α-particles**	
Air	33.8	35.1	1.3
Ar [1)]	26.4	26.4	1.00
CH$_4$	27.3	29.2	2.26
He	42.3	42.7	10.8

[1)] μ refers to Ar$^+$.

Since the positive charge carriers in gaseous detectors consist of ions rather than electron holes (as is the case in semiconducting materials, see next section), their mobility is much lower than that of electron holes (Table 4.1). This means that the time resolution is much poorer in most gaseous detectors than in semiconductor detectors, due to the longer time required to collect the positive charge carriers at the negative electrode. The rise time of the signal can be shortened by removing the slow ionic charge component using so-called gridded ionization chambers. The mobility of positive and negative ions in air is the same, 1.3 cm^2 V^{-1} s^{-1} (Table 4.1). The mobility of electrons in the media used in gaseous detectors is on the order of 1000 cm^2 V^{-1} s^{-1} at 1 atm and a field strength of 10^4 V m^{-1} (Kaye and Laby 1995).

The detector response is highly dependent on the electric field strength in the gas, as will be discussed below. If the electrodes are two parallel plates, the electric field strength will be uniform throughout the gas volume. However, if the electrodes have a coaxial configuration, i.e. a cylindrical cathode with a thin central wire as the anode, the electric field will not be uniform.

In coaxial-type chambers the electric field decreases radially, and is highest close to the centre. The increasing field strength is given by the expression

$$\varepsilon(r) = \frac{V_{Bias}}{r \ln\left(\frac{b}{a}\right)}, \qquad (4.18)$$

where ε is the electric field strength [V m^{-1}], r is the position along the radial axis [m], HV_{Bias} is the potential between the anode and the cathode [V], a is the radius of the cathode and b is the radius of the anode. A very high electric field strength can be achieved close to the anode using this configuration.

If the electric field strength is sufficiently high, the electrons will be accelerated to kinetic energies high enough to cause secondary ionization in the gas (Fig. 4.2). The secondary electrons thus created will in turn accelerate towards the anode and cause more ionization. The charge reaching the anode can be a factor of 10^6 greater than the charge created by primary ionization in the gas.

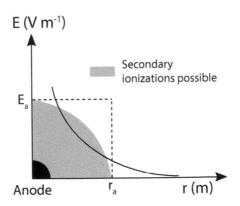

FIGURE 4.2 The solid line shows how the electric field strength decreases with distance from the anode inside a cylindrical ionization chamber. The shadowed area represents the volume around the anode where the electric field strength is sufficient for secondary ionizations.

Gaseous detectors can be divided into three categories: (i) ionization chambers, (ii) proportional counters, and (iii) Geiger-Müller (GM) counters. The difference between them is the number of charges collected after each primary ionization event. The three categories are described below.

4.2.2.1 Ionization Chambers

In an ionization chamber, the electric field is lower than that causing secondary ionization, usually around 100 V. This means that the electrode design can be cylindrical, or parallel plates.

Ideally, all the electrons created by primary ionization in the gas or in the chamber wall should be collected by the anode. However, electrons may recombine with ions before reaching the anode. Since there is no secondary ionization occur, the pulse after each event in the detector is very weak, and the detection limit is often too high for many applications, especially in environmental radiometry. To increase the efficiency, the gas may be pressurized or replaced by a liquid. However, in such chambers, the walls cannot be made thin enough for charged particles to enter the detector.

In an ionization chamber, the charge collected by the anode is proportional to the number of electrons created by primary ionization in the gas. If the total mass of the gas is known, the absorbed dose can be calculated. Since ionization chambers can fulfil the requirements of the Bragg–Gray theorem (Section 2.2.2), they are often used as dosimeters in photon or particle beams.

4.2.2.2 Proportional Counters

Proportional counters are gaseous detectors in which the electric field strength close to the anode is sufficient to cause secondary ionization. This means that each primary ionization event in the gas will release an avalanche of events due to secondary ionization. The voltage bias is highly dependent on the detector design, but is typically on the order of 100–1000 V. As suggested by the name, the magnitude of the electron multiplication is proportional to the voltage between the cathode and the anode. The charge collected after each primary event in a proportional detector will be on the order of 100 times higher than in an ionization chamber, and the detection limit of the absorbed dose in the volume is thus much lower.

Proportional counters can be used to detect the energy deposited by a single particle, and if corrections are made for recombination, they can be used for spectrometry (as discussed in Sections 4.5 and 4.6). In environmental radiometry and radiation protection, proportional chambers are used as stationary contamination control detectors.

4.2.2.3 GM Counters

When the electric field strength is higher than in proportional counters, the discharge avalanches will be more violent, causing the emission of ultraviolet (UV) radiation. These UV photons penetrate the gas and may cause further ionization in the gas or in the cathode, thereby generating new avalanches. The continuous discharges will cease when the concentration of cations close to the anode is too high to sustain a sufficiently high electric field strength to enable further secondary ionizations. If the fluence of incident particles is too high, the detector may not have time to recover between each pulse, since the heavy cations must reach the cathode before new avalanches can be started.

Since several avalanches are started after each primary event in the detector, the charge collected will not be proportional to the initial charge created by primary ionization. Detectors operating in this configuration are called Geiger-Müller (GM) counters. In contrast to ion chambers and proportional counters, the pulses from a GM counter are often large enough to be recorded without amplification. GM counters are therefore suitable in many applications in radiometry using portable detector systems.

4.2.2.4 Role of Quenching Gas

Since binding energy is released when an electron leaves the cathode and neutralizes the ion, extra electrons may be emitted from the cathode into the gas and may start a new avalanche. The probability of this is higher if the binding energy of the gas atoms is high. To prevent this, a quenching gas is often added to the gas inside the chamber (especially in the case of GM counters). Gas-filled detectors are often filled with argon (Ar), a rare gas with a high binding energy. Gases with heavier atoms, such as methane, are used as quenching gases, as these will be easily ionized by argon ions in the gas. The positively charged quenching gas molecules will migrate to the cathode, but will not release sufficient energy to cause new avalanches when absorbing electrons from the cathode.

4.2.3 Solid-State Detectors for Ionizing Radiation

Semiconductors have found wide use since the 1970s due to their specific property of being conductive only under certain conditions. This property is useful when processing electric pulses and currents. Semiconductors have also become widespread for the detection of radiation, since semiconducting materials can be made as capacitors that only detect a pulse if incident radiation causes ionization within the semiconductor. Hence, it is possible to design an electric circuit that records individual ionization events. In terms of radiometry, semiconductors can be considered as the solid-state equivalent of the ionization chamber, both of which detect radiation by means of the secondary charge produced within the detector volume.

TABLE 4.2 Band gap, E_g, mean energy expended per ionization (electron–hole pair), ϵ, and charge drift velocity, v, for some common semiconductor materials.

Medium	Atomic number	E_g (eV)	ϵ (eV/e–h pair)	v (cm^2 V^{-1} s^{-1}) [1]	
				Holes	Electrons
Si	14	1.1	3.62	$\leq$ 450	$\leq$ 1400 [2]
Ge (High purity)	32	0.67	2.96	$\leq$ 1900	$\leq$ 3900 [3]
CdZnTe	50	1.4 [4]	4.6	30–80	800–1000
Diamond	6	5.5	13	1200	1800

[1] Depending on the concentration of dopants.
[2] For commercial Li-doped Si detectors at 300 K.
[3] Example of commercial HPGe detectors at 77 K.
[4] Depending on energy.

Semiconductors are normally based on pure crystals of silicon (Si) or germanium (Ge). Both these elements belong to the carbon group in the periodic table, i.e. they have 4 electrons in the outer shell. Compounds such as gallium nitride may also have semiconducting properties. The atoms of a semiconductor material in its intrinsic form are arranged in a crystal structure, and the electrons will have quantum energy levels given by the Schrödinger equation for periodic potential, where each atom in the lattice consists of a potential well. A simplified expression is given in Eqn. 4.19 and illustrated in Figure 4.3.

$$\psi(x) = e^{ikx} u(x) \tag{4.19}$$

The solution of the wave function is a periodic function related to the stoichiometry of the atoms in the crystal lattice of the semiconductor. There are a number of possible quantum energy levels for the atomic electrons. The number of electrons populating a given state at any time is governed by the temperature of the crystal according, for example, to the so-called Stefan–Boltzmann law (Kittel 2005).

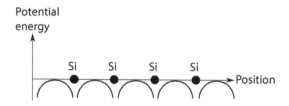

FIGURE 4.3 Schematic representation of the electromagnetic potential $(V(r))$ as a function of the spatial location in a crystal lattice of Si. The potential wells are located in phase with the positions of the atom nuclei.

Some of these energy levels exceed the valence energy level of the atoms in the lattice. These energy levels are referred to as belonging to the conduction band (Fig. 4.4). If an electron in a state belonging to the valence band attains enough energy to be excited into a quantum state in the conduction band, then the electron will no longer be bound to a specific atom, and can move freely within the lattice. The electron will leave a hole in the atomic shell, which may be filled by an electron from an adjacent atom. Thus, the "hole" can be considered a free positively charged particle in the lattice. If an external electric field is applied over the semiconductor, a free electron such as this is interpreted macroscopically as the conduction of charge. The energy difference between the uppermost state in the valence band and the lowest possible energy state in the conduction band is referred to as the band gap (see Table 4.2 for characteristics of common semiconducting materials).

This width of the band gap depends on the fundamental crystal lattice and the temperature of the crystal. Due to the inherent thermal energy of the electrons, a fraction of the electrons in the lattice will always be excited to quantum energies corresponding to unbound or conduction band states. At equilibrium, the number of electrons in the conduction band is equal to the number of electron holes in the valence band. Again, if an external electric field is applied over the volume of the lattice, no current would flow apart from that due to thermally induced excitations. If, however, external energy is introduced into the system, e.g. photons or ionizing particles, the electrons near the upper edge of the valence band can attain sufficient energy to move up into the conduction band. The band gap is the minimum amount of energy needed for an excitation. This corresponds to the excitation of p-electrons in the valence band. However, the mean energy absorbed per electron–hole pair produced is higher than this due to the higher binding energy of the electrons in the inner shells.

In addition to energy levels and the band gap, the existence of impurities is very important for the properties of a semiconductor as a radiation detector. No material is completely chemically pure. Even at an impurity level of 1 in 10^{15} atoms, the disarrangement of the lattice structure will introduce new energy states near the band gap. Impurities may be inherent in many metals and compounds, but in semiconductor physics the influence of impurities can be controlled and exploited by intentionally introducing other elements into the lattice in a process called doping. Doping is achieved by drifting and diffusion of certain atoms into the bulk of the intrinsic semiconductor material to introduce energy levels that are desirable for a given application.

Two types of impurities can be introduced into semiconductors: donors and acceptors. Donors are generally elements from group 15 in the periodic table, i.e. elements that have 5 electrons in their outer shell. Lithium (Li) in group 1 has only 1 electron in its outer

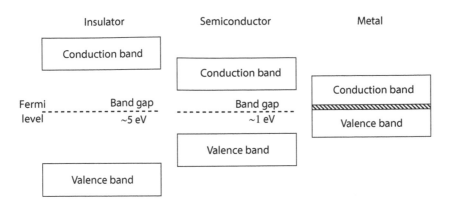

FIGURE 4.4 Illustration of the conduction and valence bands of insulators, semiconductors and metals.

shell but can also act as a donor. Donors are impurities that create energy levels close to the conduction band, and therefore provide the conduction band with electrons. A donor concentration of about 1 ppm in Si, which has an intrinsic band gap of 1.1 eV at 300 K, will result in an equilibrium electron concentration that is more than 10^{13} times higher than the hole concentration. A donor-doped semiconductor material is referred to as an n-type material due to its greatly enhanced electron concentration in the conduction band.

Acceptors are elements from group 13 of the periodic table, i.e. elements that have 3 electrons in the outer shell. Acceptor impurities will create energy states close to the valence band. A thermally excited electron from an atom in the conduction band may be caught in an energy level resulting from an acceptor impurity. The electron will be trapped and immobilized. However, an electron hole will be created that behaves almost exactly like a positively charged particle in the conduction band. In analogy with the donor-doped semiconductor, the concentration of holes in the conduction band will be governed by the concentration of the acceptor impurity in the material. An acceptor-doped semiconductor material is thus referred to as a p-type material.

When an n-doped material and a p-doped material are brought together, a so-called, p-n junction will be formed (Fig. 4.5). The free electrons on the n-side and free holes on the p-side will migrate to the junction, where the electrons and holes will recombine. The region around the junction will thus be almost completely free from free charge carriers. As the electrons from the n-side have migrated to the junction between the semiconductors, the n-doped material will have a net positive charge, and an intrinsic electric field, or contact potential, V_{int}, will form between the two sides. These p-n junctions are used to obtain several desirable electrical properties in semiconductor components, and they are referred to as *diodes*.

In an electric circuit, a diode conducts electric current from the p-side to the n-side, i.e. the n-side is connected to the negative terminal and the p-side to the positive terminal. The external potential across the contacts of the diode, V_{Bias}, will continuously feed the n-side with electrons that can migrate towards the p-side. The diode will not conduct any current if the bias is reversed. The excess negative charges on the n-side will be attracted by the anode, and will not be able to recombine with holes from the p-side. Thus, a larger depletion region, with no free charges, will form.

To create a fully operational radiation detector requires one more reconfiguration of the semiconducting material. A p-n junction or diode, without any external field, suffers

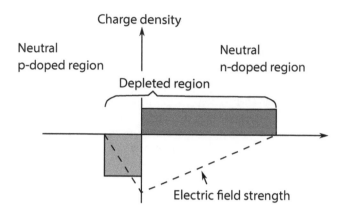

FIGURE 4.5 The charge distribution and associated intrinsic electric field over a p-n junction.

from two main complications: the mobility of the induced charge is too slow to allow time resolution of the events caused by the incoming radiation, and the volume of the p-n junction cannot be made large enough for practically useful counting efficiency. However, if a reverse external bias is applied over the junction, the total potential will be the sum of the inherent contact potential, V_{int}, and the reverse potential (V_{Bias}) (Fig. 4.6). This means that the electrons on the n-side will migrate to the anode instead of migrating over the junction to the p-side.

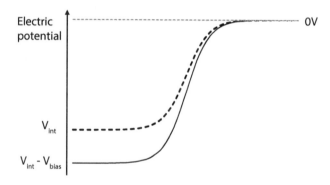

FIGURE 4.6 Electric field over the detector: without external bias (dotted line) and with an external reverse bias (solid line), which increases the electric potential over the diode.

It should be remembered that these charges were both the minority carriers on their respective sides of the diode. This means that the thermally induced excitation creating the charge on either side will be relatively small provided that the electric field applied across the junction is moderate (0.1–10 kV cm^{-1}). This provides a configuration in which stray and thermal noise are relatively low. The p-side of the semiconductor can be extended, thus increasing the volume of the depletion region. This provides a detector with enhanced charge mobility, yet with a low noise level, and a higher counting efficiency for externally induced ionization events, due to the large volume of the depletion region. The relationship between the depth, d, of the depletion region and the bias voltage, V_{Bias}, is given by

$$d = 2 \cdot \varepsilon \cdot V_{Bias} \cdot \mu_{charge} \cdot \rho_{dop}, \tag{4.20}$$

where ε is the dielectric constant of the semiconductor, μ_{charge} is the mobility of the majority charge carrier (electrons in the n-doped part of the detector), and ρ_{dop} is the resistivity of the doped detector (if the p-part is dominant this refers to the resistivity over this part of the junction). This relationship can be rewritten so that the depth of the depletion is expressed as a function of the high voltage bias, V_{Bias} [V], and the dopant concentration [m^{-3}]:

$$d \sim \sqrt{V_{Bias} \cdot \rho_{dop}}. \qquad (4.21)$$

At higher electric fields, however, the reverse leakage current may become important. In many cases, there is thus an upper practical limit on the voltage that can be applied across such diodes.

Three main geometries are used for semiconducting radiation detectors: (i) the thin planar configuration, (ii) the bulk planar configuration and (iii) the coaxial configuration. The geometrical configuration determines the time resolution and the energy resolution of the detector. The first category is intended for charged particle detection, and these detectors can be made so thin that they record the energy loss per unit normal path length of the impinging charged particle, dE/dx. This means that only some of the energy of the charged particles is deposited in detector. In the bulk planar configuration, the surface of a bulk semiconductor can be doped to create a depletion region extending deep enough into the crystal that all the kinetic energy of the charged particles is deposited. This allows full spectroscopic information to be obtained from the pulses generated by not only short-range particle radiation but also low-energy photons. This planar bulk geometry is useful for the detection of low-energy γ-rays and x-rays in the range 10–100 keV. A small bulk volume, or a thinner depletion layer, provides a means of reducing the background from higher energy photons.

Detectors for higher energy photons must have a larger active detector volume than detectors for charged particles or low-energy photons. In detectors with large volumes, where the contacts are far apart, long charge collection time can be a problem. It is therefore preferable to collect the faster moving negative charge carriers in a centrally positioned anode. The coaxial configuration of semiconducting diodes (Fig. 4.7) is thus common in gamma-ray spectrometry systems for energies between 100 and 3000 keV. In practice, most coaxial detectors are closed-ended, giving rise to a varying electric field, E.

Bulk coaxial detectors can either have a depletion volume of predominantly p-doped or n-doped material, for example, p-type and n-type high-purity (HP)Ge detectors. Both types have advantages and disadvantages. In p-type coaxial detectors, the negative bias is at the centre of the coaxial crystal, consisting of a p+ contact. The n+ contact is on the surface of the crystal, as shown in Figure 4.7. However, n-type contacts cannot be made thinner than about 0.6 mm, as this leads to a dead layer that prevents low-energy radiation from reaching the radiation-sensitive depletion region of the detector. On the other hand, an n-type detector, which consists predominantly of n-doped Ge, has its p+ contact on the outer coaxial surface. This type of contact can be made much thinner, thus allowing lower energy radiation to enter the sensitive volume of the detector. The drawback, however, is that charge collection by the p+ contact attracting the electron holes is slower, resulting in a system with a charge collection time that depends on the location of the event in the sensitive volume, causing a reduction in energy resolution.

Semiconductor materials intended for spectrometry are used as ionization detectors in pulse mode. Ideally, there should be no free charge in a semiconductor detector in the absence of radiation. As the probability, $P(T)$, of thermally induced electron–ion pair production in a crystal material is given approximately by Eqn. 4.22, it is evident why semicon-

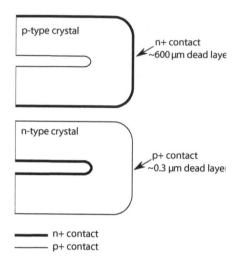

FIGURE 4.7   Illustration of the electrical contacts of a p-type and an n-type HPGe crystal.

ductors are often cooled to the temperature of liquid nitrogen (77 K) to reduce the thermally induced charge and subsequent noise when the external field is applied. The band gap, E_g, is not significantly affected by the decreased temperature, but the overall probability of thermal charge induction is considerably decreased, for example, for HPGe.

$$P(T) = \text{const} \cdot T^{3/2} \cdot e^{-E_g/2kT} \tag{4.22}$$

The energy required for pair production is also relatively independent of radiation quality in semiconductor spectrometers. Charge collection per unit field strength is also more rapid than in gaseous ionization detectors, which have a mobility of 1000 cm^2 V^{-1} s^{-1}. The quantity *scattering time* refers to the time that the charge is accelerated by the external electric field, before being scattered by an atomic electron. The product of the mobility and the scattering time thus represents the distance over which a given electric field will accelerate the charge in the semiconductor. This quantity is many orders of magnitude higher for Si and Ge than for other semiconductors, thus these materials are suitable for high-count-rate systems, with rapid charge collection.

For comparison, the pair production energy, or average ionization energy, for gaseous ionization detectors is about 30 eV (Table 4.2). For a given type of gas, the values for fast electrons and α-particles differ slightly. The mobility of charge pairs in gas is 10–100 times lower than in solid Si and HPGe crystals, which leads to a somewhat poorer time resolution in gaseous detectors. The low density of gas also leads to poorer counting efficiency of both light charged particles and non-charged particles.

There are alternative semiconductor materials to Si and Ge. An interesting material is cadmium zinc telluride (CdZnTe). The material has an effective atomic number of 50, which means that high counting efficiency can be obtained for small crystal volumes. The band gap of CdZnTe is broader ($\sim$1.5 eV) than in Ge (0.7 eV), and the detector does not need cooling. Since the band gap is broader, the pair production energy is also higher. This in turn, leads to less charge being generated per unit energy deposited in CdZnTe than in Si and HPGe, and thus to lower energy resolution. Improvements in crystal quality have led to CdZnTe detectors with a resolution of better than 1% at 662 keV.

Another semiconducting material that can be used at room temperature is mercuric iodide (Hg_2I_2). This material has a high effective atomic number but the charge mobility is poor. Thallium bromide (TlBr) has a high effective atomic number, and high energy resolution has been achieved, but the material must be cooled to about $-20\ °C$.

Materials with a lower effective atomic number, such as crystalline carbon (diamond), are less efficient as photon detectors. However, such materials may be suitable for dosimetry as their atomic number matches that of tissue.

4.2.4 Luminescent Detectors

Luminescent materials emit visible light in the form of photons in the wavelength region (1–5 eV; 250–1200 nm), after being irradiated with ionizing radiation. When ionizing radiation or secondary radiation interacts with a luminescent material, some of its energy can be imparted to the electrons in atomic or molecular states of the material, leading to excitation to higher energy levels. When these states decay back to their original level, energy is released in the form of light (luminescence).

This de-excitation may be immediate (prompt emission), but if transition to the lower energy level is not permitted by quantum mechanics, or if an electron is unable to recombine with an ion, de-excitation may be delayed. Prompt emission is called *fluorescence* and delayed emission is called *phosphorescence*. Luminescence can therefore be prompt or delayed, enabling either a direct measurement of the dose rate, or a measurement of the cumulated radiation dose.

Luminescent materials were among the first used to detect ionizing radiation. Screens made of barium platinocyanide ($BaPt(CN)_4 \cdot 4H_2O$) or calcium tungstate ($CaWO_4$) were already in use for the detection of ionizing radiation at the end of the 19th century. Fluorescent screens were used in diagnostic x-ray imaging well into the latter half of the 20th century, before they were replaced by photographic film, and later by digital sensors. However, luminescent detectors are still used in several dosimetry applications. Both organic and inorganic materials may have the property of luminescence, and they are referred to as scintillators.

4.2.4.1 *Organic Luminescent Detectors*

Organic scintillators are usually made of aromatic carbon compounds. Since these compounds consist mainly of carbon and hydrogen, their atomic composition is close to that of tissue. The carbon atoms in these molecules are organized in benzene rings (6 carbon atoms) of various conformations. Each carbon atom is bound to two adjacent carbon atoms and one hydrogen atom. The orbitals of the bound electrons (σ-bonds) all lie in the same plane, i.e. the molecule is planar. However, one p-electron at each carbon atom will not be bound and its orbit is orthogonal to the molecular plane. This electron is called the p_z-electron. These six p_z-electrons may form a π-orbital delocalized from the rest of the molecule. Excitation and de-excitation of this orbital may cause the emission of light.

The Schrödinger equation can be solved to describe the different energy states of the orbital. The wave function of the orbital must be continuous and periodical. The simplest function that fulfils these requirements is the sine wave. Since the sine wave can have several nodes and still be continuous and periodical, a quantum number, q, the ring quantum number, has been defined, which may take any integer value. Thus, the wave function can be written in three forms:

$$\psi_0 = \left(\frac{1}{l}\right)^{\frac{1}{2}} \tag{4.23}$$

$$\psi_{q_1} = \left(\frac{1}{l}\right)^{\frac{1}{2}} \sin\left(\frac{2\pi q x}{l}\right) \tag{4.24}$$

and

$$\psi_{q_2} = \left(\frac{1}{l}\right)^{\frac{1}{2}} \cos\left(\frac{2\pi q x}{l}\right), \tag{4.25}$$

where l is the perimeter of the orbital, q is the ring quantum number and x is a spatial point on the orbital. The Schrödinger equation describing such a rigid plane rotor is

$$-E\psi = \frac{\hbar^2}{2m_0} \frac{d^2}{d^2 x} \psi, \tag{4.26}$$

where E is the energy of the state, m_0 is the electron mass and $\hbar$ is the reduced Planck's constant $(h/2\pi)$. Solving the equation gives the energy states

$$E = \frac{\hbar^2 q^2}{2m_0 l^2}. \tag{4.27}$$

Thus, the energy of each state is given by the ring quantum number q.

However, when the orbital is excited, the angular momentum (spin) may change. When the total spin of the system is 0, i.e. the electrons have opposite half-integer spin, the state is called a singlet state. If the spin is changed by excitation, i.e. the electrons have the same half-integer spin, the total spin becomes +1 or -1, and the state is called a triplet state. Triplet states have a slightly lower energy than the corresponding singlet states, as shown in Figure 4.8.

The lowest energy level corresponds to $S = 0$. When radiation energy is absorbed, the electrons may make the transition to a higher energy level. After excitation, the electrons will de-excite to the lowest energy state. In some transitions, usually S_1 to S_0, a photon is emitted. The intensity of the photon emission decays exponentially after excitation. The time constant of prompt fluorescence is on the order of nanoseconds. However, transitions from the lowest triplet state, T_0, to the lowest energy state, S_0, are forbidden by the laws of quantum mechanics. This leads to phosphorescence with a much longer time constant, of several μs. This so-called afterglow can be a problem in some applications. Another path of decay from a triplet state is by crossing, and this is called delayed fluorescence. Since the light is emitted during the transition from S_1 to S_0 and not from T_1 to S_0, the light emitted after an inter-system crossing will have the same wavelength as the prompt fluorescent light.

Excitation of a molecule can also make it vibrate. This vibration adds further levels of freedom in which energy can be absorbed. Since absorbed energy can lead to excitation of the orbitals and vibration, the absorption spectrum is broadened, and energy is absorbed at the discrete energies corresponding to the difference between different singlet and triplet states (Fig. 4.9). For a more thorough explanation of this process we recommend Birks (1964).

Organic materials have several desirable properties for the detection of ionizing radiation. Since they consist mostly of carbon and hydrogen atoms, the cross section for photon interactions is very similar to that in water or soft tissue, which is useful for dosimetry. Many luminescent organic materials can be made into various shapes, or mixed with liquids and plastics. However, their atomic composition also means that the counting efficiency is low compared to detectors made from high-Z materials, and a large detector volume is needed to compensate for the low detection efficiency.

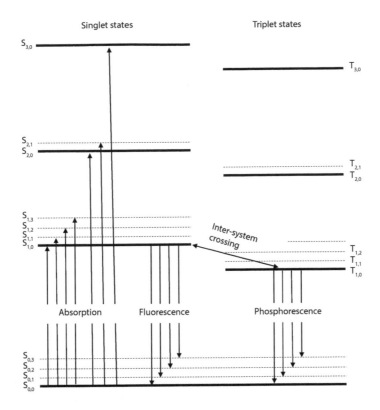

FIGURE 4.8   Molecular singlet and triplet states of a luminescent crystal or solution showing possible paths of de-excitation generating prompt (fluorescence) and delayed (phosphorescence) luminescence.

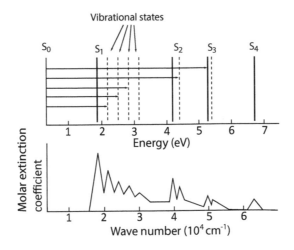

FIGURE 4.9   Example of an absorption spectrum (anthracene in cyclohexane).

Since the energy emitted during a transition is the same as that needed to excite another molecule, the probability of re-absorption of the emitted light is significant, and scintillators are thus not transparent to their own emission light. To remedy this problem, a wavelength-shifting molecule is added to the detector. This molecule absorbs the fluorescence photon and emits a new photon with a longer wavelength (lower energy). The energy of the new photon is not sufficient to excite the molecules of the scintillator and can thus be transmitted through the material without being absorbed. However, self-absorption will limit the size of the detector, even with a wavelength shifter.

4.2.4.2 *Inorganic Luminescent Detectors*

Inorganic materials may also emit light after being irradiated. Crystals of alkali metal halides, where the atoms are strictly arranged in a lattice, are often used. Electrons in the valence band can be excited into energy states determined by the overall lattice structure. These excited energy states can have half-lives ranging from a few nanoseconds to several years. Luminescent materials with relatively short-lived phosphorescent excited states are suitable for spectrometry, as the individual events can be processed on-line. Although organic scintillators generally have shorter decay times than inorganic scintillators, such as crystals of thallium-activated sodium iodide (NaI(Tl)), the higher effective atomic number of many inorganic crystals increases the probability of photons interacting with the detector material. This makes them suitable for either total γ counting or for gamma-ray spectrometry (Section 4.6).

Since de-excitation from the conduction band to the valence band releases the same amount of energy as is needed for re-excitation, re-absorption of the luminescence photons is also a problem in inorganic scintillators. This is remediated by adding small amounts of an impurity. Impurities create recombination centres in the lattice, where the energy gap is smaller. If an electron passes from the conduction to the valence band at a recombination centre, the emitted photon will have a longer wavelength, and hence a lower energy. Thus, the light emitted at recombination centres will not be reabsorbed by the crystal.

The following properties, described by the parameters given in Table 4.3, are desirable in prompt scintillators used for spectrometry.

- The kinetic energy of the charged particles should be efficiently converted into scintillation light photons.

- The number of scintillation photons emitted must be proportional to the amount of kinetic energy absorbed from the charged particles in the detector.

- The crystal material must be transparent to the scintillation light generated in order to enable efficient light collection by the photocathode of the light collection unit (PMT or photodiode).

- Prompt emission of fluorescence is necessary in spectrometers and fast count rate meters.

- When performing measurements at high dose rates, the background arising from phosphorescence or delayed fluorescence must be low. The refractive index of the crystal must be similar to that of glass or other optical conductors to enable efficient optical coupling with PMTs.

- The composition of the crystal material should be such that it can be grown in bulk sizes to be useful as a gamma photon detector.

The scintillation light efficiency is defined as the amount of luminescence produced in a scintillating medium per unit energy loss, dL/dE, of a charged particle, and can be expressed as in Eqn. 4.29 (Koba et al. 2011)

$$\frac{dL}{dE} = \frac{a}{1 + b \cdot \left(\frac{dE}{dx}\right) + c \cdot \left(\frac{dE}{dx}\right)^{-1}}, \qquad (4.28)$$

where L is the luminescent energy (number of light photons multiplied by their average energy) [J], E is the kinetic energy of the charged particle, and dE/dx is the kinetic energy loss per unit path length. dL/dE is thus dimensionless and varies typically from a few per cent up to 12 per cent for the most efficient scintillator crystals.

Bulk inorganic NaI(Tl) crystals were developed in the 1940s (Hofstadter 1949) and exhibit many of the desirable properties listed above. NaI(Tl) detectors were the standard detectors in gamma-ray spectrometry and γ-ray intensity measurements for several decades. However, other scintillating organic and inorganic solid crystals emerged in the 1950s and 1960s, many of which could be grown with the addition of so-called crystal growth modifiers. The choice of detector depends on the application and is a trade-off between robustness, sensitivity, and time resolution (short afterglow in terms of phosphorescence). If the phosphorescence is slow, too sensitive a crystal combined with a light-to-charge converter (either a photomultiplier tube or a photodiode; see Section 4.4.6), will lead to the detector being overloaded by events that cannot be resolved at relatively moderate γ fluences. Therefore, even high-performance scintillators such as NaI(Tl) detectors will be inappropriate in high-intensity gamma radiation fields.

TABLE 4.3   Properties of inorganic and organic scintillators used for γ-ray detection.

	Density	Effective Z	Principal decay constant	Pulse 10–90% rise time	Photon yield
	(g cm^{-3})		(μs)	(μs)	(MeV^{-1})
NaI(Tl)	3.67	50	0.23	0.5	38 000
CsI(Tl)	4.51	54	1.0	4	52 000
BGO	7.13	83	0.30	0.8	8200
LBr$_3$(Ce)	5.29	47	0.026	0.0007	61 000
			to	to	
			0.035	0.0010	
NE102A [1]	1.03	4.5	0.0024	0.0006	10 000

[1] Fast plastic scintillator.
Source: Data from Knoll (2010), Cherepy et al. (2011) and Wiener et al. (2013).

The luminescence yield is dependent on the temperature, and for inorganic scintillators varies with the material. However, since the signal from many scintillators is amplified using PMTs (although light amplification by Si diodes has become more common), the overall temperature dependence of the detector system will be dictated by the much stronger temperature dependence of the dynode amplification of the PMT (Fig. 4.10). Various kinds of grease or fluids with a refractive index of n≈1.5 are used to couple scintillators and PMTs to maximize light transfer to the PMT. The photocathode of the PMT must also be designed to match the emission spectrum of the luminescence photons of the scintillator to obtain the maximum electron yield in the cathode for a given amount of luminescence, as illustrated schematically in Figure 4.11.

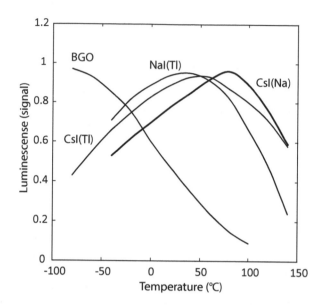

FIGURE 4.10 Temperature dependence of the luminescence yield in some common scintillators.

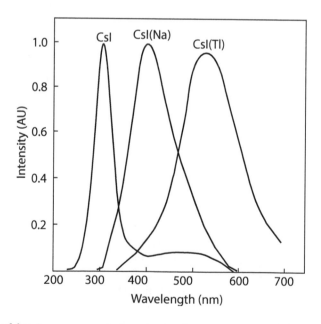

FIGURE 4.11 Typical luminescence emission spectra from some commonly used scintillators.

4.2.4.3 Integrating Luminescent Detectors

When the intention is to obtain information on the accumulated dose over a period of time, it is necessary to use integrating detectors made of materials with extremely long-lived phosphorescence states (in contrast to the use of prompt scintillators when analysing radiation on-line). These luminescent detectors are often used as prospective dosimeters to monitor occupational exposure (e.g. staff at nuclear power plants or medical radiology departments), for environmental surveys, or for retrospective dose assessments. The charge collected in these long-lived energy states, which are sometimes denoted "electron traps" and "electron hole traps" can then be transformed into luminescence by means of external excitation. The external excitation can either be induced by thermal energy, causing thermoluminescence (TL), or by optical illumination, causing optically stimulated luminescence (OSL). Figure 4.12 illustrates the energy levels involved in an integrating luminescent material. Thus, in contrast to prompt luminescent crystal materials, integrating luminescent crystals are designed to have long-lived or stable electron and electron hole traps that can be emptied of their charge by means of either TL or OSL. Such luminescent detectors are vital for retrospective dosimetry, when past radiation exposure is investigated long after the exposure event (as discussed in more detail in Section 7.3.3).

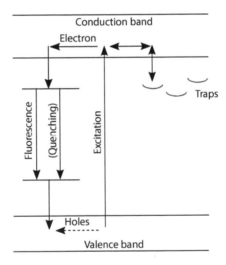

FIGURE 4.12 Illustration of the mechanism of delayed luminescence in an integrating luminescent detector material, induced by an external energy.

The use of thermoluminescent crystals, such as lithium fluoride (LiF), for radiation dosimetry was explored by e.g. Daniels et al. (1953), not long after the development of the NaI crystal in the late 1940s. It exhibits several desirable properties for applications in radiation protection, namely: (i) linear or supra-linear dose–response over a wide range of radiation exposures (Table 4.4), (ii) high sensitivity to γ-rays, with detection limits down to a few tenths of a μGy, (iii) energy-independent counting efficiency (i.e., dose response) over a large range of electron and γ-ray energies, which is essential when estimating the absorbed dose at a point in an irradiation field, and (iv) high stability of the latent luminescence, due to long-lived energy states, making detector read-out possible many years after exposure.

Upon increasing the temperature of the crystal from room temperature to 200–250 °C (below the crystal melting point) at a constant rate (so-called linear temperature ramping), the electrons trapped in the various electron traps in the crystal will attain sufficient energy

TABLE 4.4 Applications of integrating luminescent detectors in personal dosimetry, and their typical dose ranges.

Application	Dose range (Gy)	Uncertainty, 1 σ (%)
Personal dosimetry	$10^{-5} - 5 \cdot 10^{-1}$	-30 to $+50$
Environmental radiology	$10^{-6} - 10^{-2}$	± 30
Radiotherapy	$10^{-1} - 10^{2}$	± 3.5
Diagnostic x-ray examinations	$10^{-6} - 10$	± 3.5
Occupational exposure, industrial	$10^{1} - 10^{6}$	± 30

to become excited into the conduction band, and will then de-excite to the valence band (Fig. 4.12) at a recombination centre. As mentioned above, luminescence in the form of photons in the visible energy range will then be generated. If the heated crystal is positioned close to a small PMT, the intensity of the luminescence can be recorded. A plot of the luminescence as a function of the temperature is called the glow curve. The glow curve is characteristic of a given type of detector material, and for a given dose from a specific radiation quality, e.g. 1 Gy of ^{60}Co photons (1173.2 and 1332.5 keV), and can be used to compare different crystals and crystal activators (Fig. 4.13, right figure). This means that the optimal read-out temperature varies depending on the crystal material. The peak in the glow curve represents the energy depth of the electron traps, as the thermally induced energy is just sufficient to excite the electrons in a given trap into the conduction band (Fig. 4.13)

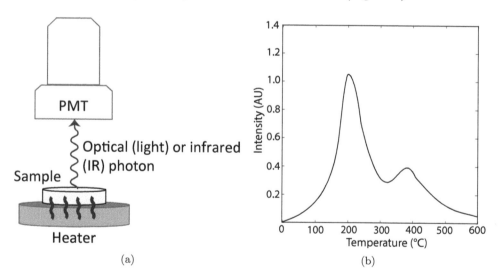

(a) (b)

FIGURE 4.13   (a) Illustration of the read-out of a thermoluminescent dosimeter. (b) Typical glow curve for NaCl (ramping rate $\approx 2\ ^\circ\text{C s}^{-1}$).

As mentioned above, it is important that the half-life of the electron traps is long in an integrating luminescent material, otherwise the population of the traps may decay, leading to so-called fading. It is also desirable that the crystal exhibits a simple, distinct glow curve, indicating at least one well-defined electron trap. The crystal must be able to withstand some degree of mechanical stress without the electron traps being affected. For crystals whose luminescence can be retrieved by thermal stimulation (as in TL), the latent signal in the material should be unaffected by exposure to visible light. For optically stimulated luminescent crystals, exposure to external light during irradiation measurements will affect the latent signal, and they must therefore be sealed against light. A typical irradiation and signal read-out protocol for TL with crystals that can withstand very high temperatures (up to 600 °C) is given below.

1. *Calibration of individual dosimeters*: Highly homogeneous replicas of luminescent dosimeter material can be achieved in terms of doping concentrations. However, they still vary in terms of their sensitivity (signal per unit absorbed dose), the dependence of sensitivity on energy, reproducibility in the signal per unit radiation exposure, and non-irradiation-associated background. Therefore, each dosimeter requires individual calibration to ensure precision within 20% at low doses (<100 mGy). The dosimeters are exposed to a precise absorbed dose (within a few per cent) from a well-known source of radiation. The dosimeter is then read out by ramping the temperature to 200–250 °C , while the luminescence signal generated by the de-excitation of the emptied electron traps is recorded, giving the sensitivity of the dosimeter.

2. *Annealing*: In order to completely empty all the electron traps, the dosimeter is heated to about 500 °C in an oven with precise temperature regulation. It should be noted that many naturally occurring materials that are sensitive to optical stimulation cannot withstand such a high temperature.

3. *Cooling and repetition of steps 1–2*: The dosimeter is cooled and then steps 1–2 may be repeated until the sensitivity stabilizes.

4. *Irradiation/exposure*: After steps 1–3, the dosimeter is ready for use as a personal dosimeter, for example, in a radiology department or in the field where workers are operating in an area contaminated by radioactive fallout.

5. *Preheating/pre-annealing*: When the dosimeter is retrieved for read-out, it is often preheated to empty low-energy electron traps that do not contribute to the more predictable and reproducible luminescence signal from the deeper electron traps.

6. *Read-out*: The dosimeter is then heated further to the optimal read-out temperature and the luminescence is recorded by a photosensitive device such as a PMT or a photodiode.

7. *Calculation of absorbed dose*: The absorbed dose is calculated using the individual sensitivity calibration factor obtained in Step 1.

Although the thermoluminescent and optically stimulated luminescent detectors used today have excellent dosimetric qualities, the reproducibility of the doses obtained, especially at low radiation exposures, is still not optimal. One reason for this is that the increase in signal per unit radiation dose in many of these materials increases slightly as a function of the accumulated dose, even after being read out and emptied. This means that previous exposures and read-out will cause small changes in the electron traps that sensitize the crystal, leading to a somewhat higher signal in subsequent read-outs. This appears to reach equilibrium after an initial increase in sensitivity. Carefully designed read-out protocols must therefore be employed, and it is important that the protocol is followed once a batch of dosimeters has been calibrated in order to ensure reproducibility. Another factor affecting reproducibility of the results is incomplete emptying of the electron traps from the previous irradiation and read-out cycle. This can arise if the external thermal or optical stimulation of the crystal is not sufficiently long to empty very deeply located electron traps.

At the beginning of the 2000s, carbon-doped aluminium oxide (Al_2O_3:C) dosimeters, using optical stimulation for read-out instead of thermal excitation of LiF , were developed and became commercially available (Bøtter-Jensen et al. 1997). OSL devices have partly replaced TL devices in personal dosimetry, mainly due to their higher sensitivity at low radiation doses. Al_2O_3:C can also be read out using thermal stimulation in a similar fashion to LiF. Furthermore, at the beginning of the 1990s, it was discovered that naturally occurring materials such as quartz and feldspar also exhibited luminescent properties that enabled retrospective dosimetry for doses as low as 10 mGy (Bötter-Jensen et al. 2000). Other naturally occurring or manmade materials, commonly found in the environment and in society, were found to exhibit integrating luminescent properties. These materials include household salt (NaCl), with detection limits as low as 0.1 mGy, electrical components in mobile phones (the ceramic parts of capacitors and resistors as well as glass and display components), dental repair materials, and desiccants, among others. Much research is currently being conducted on both prospective dosimetry and retrospective dosimetry using various common materials as dosimeters.

4.2.5 Chemical Detectors

The detector systems described so far were based on the detection of the free charged particles or the luminescence emitted as a result of the interaction of radiation with the detector material. However, ionization and the excitation of atoms and molecules not only leads to the release of charge or the induction of luminescence, but can also cause chemical reactions in the detector material. If these chemical reactions are predictable and reproducible, they can be related to the absorbed radiation dose and utilized to assess the dose from ionizing radiation, provided the chemical change can easily be quantified.

A commonly used kind of detector utilizing a radiation-induced chemical reaction is traditional radiographic film, often consisting of grains of silver bromide (AgBr) salt contained in a thin layer of emulsion between the protective coating and the transparent base material (Fig. 4.14). The interaction of photons or charged particles with the emulsion causes an electron to be released from the Br ion into the conduction band. As in most crystal lattices, impurities or delocalized atoms create electron and electron-hole traps that capture the excited electron hole, which recombines with an interstitial Ag ion, to produce metallic Ag. A cluster of metallic Ag is thus created around these sites of irregularities or impurities when the film is exposed to ionizing radiation. A so-called latent image is created, where areas that have been exposed to more radiation have a higher concentration of metallic Ag clusters. This latent image is then developed by exposing the emulsion to a chemical that reduces the grains to Ag metal (photographic developing). The temperature and the concentration of the developer must be carefully controlled. The image is then fixed by removing the remaining Ag using a fixing agent such as sodium thiosulphate.

The sensitivity of a photographic film can be modulated in several ways: by varying the impurity concentration in the AgBr in the emulsion layer; by varying the thickness of the emulsion layer; or by adding a screen containing fluorescent material to convert the impinging γ-rays, x-rays or high-energy electrons more efficiently into light photons in the energy range 1–3 eV, which then precipitate the photo-emulation of the AgBr. In the middle of the 20th century, x-ray films were often combined with fluorescent detector materials (intensifying screens) to enhance their sensitivity to photons with energies in the range 10 to 140 keV, so-called screen-film systems. The intensifying screens often consisted of $CaWO_4$

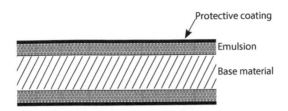

FIGURE 4.14 Configuration of a traditional radiosensitive film for radiographic imaging. The double emulsion layers are positioned above and below a flexible base material, and covered with a protective layer to prevent mechanical and optical contact with the emulsion layer.

or of rare earth metals. Note that the fluorescent materials used in the intensifying screens are somewhat ambiguously referred to as "phosphors" within the field of radiology.

The optical density of the developed film as a function of radiation exposure (S to D curves) is characteristic for a given film or film-screen system. The films can be adjusted to suit the anticipated range of fluence of photons in various exposure set-ups, such as in diagnostic radiographic examinations or in industrial radiography. Most film detectors had been replaced by digital computed radiography by the end of the 1990s, but some are still used within emergency preparedness for first responders.

Photosensitive films have also been used extensively for personal dosimetry in environmental radiology and radiation protection applications, although they have now largely been replaced by TL or OSL devices. However, radiochromic films containing a photosensitive emulsion that can be observed without processing or read-out equipment (process-less film) are still in use, in such diverse applications as radiotherapy and emergency preparedness. Instead of the chemical conversion of Ag ions to metallic Ag, the polymerization of diacetylene is used (Fig. 4.15). The change in colour as a function of radiation exposure makes such films useful for first responders who need a swift visual indication of the accumulated exposure to ionizing radiation.

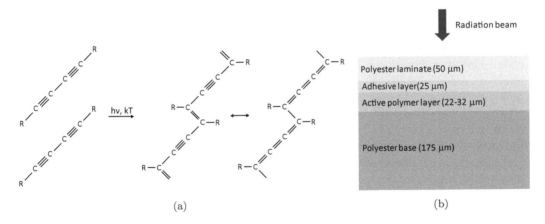

(a) (b)

FIGURE 4.15 (a) Polymerization of monomers caused by irradiation. (b) Film structure in second-generation radiochromic film used, for example, for dose verification of diagnostic and therapeutic radiation beams. These films have also found use in environmental radiology as they change colour depending on the absorbed dose.

In the 1920s, a dosimetric method based on the oxidation of ferrous ions (Fe(II)) to ferric ions (Fe(III)) was developed (Fricke and Morse 1927). When the ions are oxidized, the colour of the solution changes, and the dose can be quantified by measuring the optical density. In 1984, Gore at al. (1984) suggested that ferrous sulphate could be mixed into a gel solution, although dosimeters based on chemicals that change colour after irradiation mixed in a gel solution had already been suggested by Day and Stein in 1950 (Day and Stein 1950). A 3D dose distribution can be obtained by quantifying the concentration of ferric ions formed in the gel after irradiation. When the iron is oxidized, the magnetic properties of the gel change and the dose distribution can be read out using magnetic resonance tomography (MRT). Since the gel solution consists mostly of water, the effective atomic number is close to that of soft tissue.

An alternative to Fricke gel dosimeters is polymer gel dosimetry based on the polymerization of monomers. These monomer-based dosimeters have a higher sensitivity and the polymers formed are more stable in the gel structure than iron ions, which tend to diffuse in the gel. Polymer gel dosimeters can be read out after irradiation either by quantifying the spin–spin relaxation time using MRT or by measuring the optical density. Neither Fricke gels nor polymer gel dosimetry are common in radiation protection, but they are worth mentioning as they provide a unique possibility for 3D dosimetry.

4.2.6 Mass Spectrometry

Mass spectrometry is a technique used to determine the amount of a certain radionuclide without involving its decay products. The technique is based on the analysis of the mass-to-charge ratio of ionized elements. In the first part of the mass spectrometer the sample is vaporized and ionized (upper left in Fig. 4.16). When the electrons have been removed from the atoms, the nuclei are accelerated in a linear accelerator (upper right in Fig. 4.16) towards an electromagnet (lower left in Fig. 4.16). The ions are deflected in the electromagnetic field. The deflected ions are then registered by a detector on the other side of the electromagnet. Since the defection angle will depend on the mass-to-charge ratio, (m/q), of the nucleus, the intensity of ions in different parts of the beam at the detector will give a measure of the abundance of specific isotopes in the sample. An example of a typical spectrum is shown in the lower right of Figure 4.16.

A number of mass spectrometry techniques are in use today. Accelerator mass spectrometry (AMS) and inductively coupled plasma mass spectrometry (ICP-MS) are those most often used in environmental measurements of radionuclides. In AMS the elements to be analysed are accelerated to very high kinetic energies (several MeV) compared with other mass spectrometry techniques. This provides better resolution in the mass-to-charge ratio, especially when there are many other similar elements in the sample. AMS is especially useful for quantifying low-Z radionuclides, as the technique allows suppression of interfering/perturbing atomic isobars, such as [10]Be, [26]Al and [36]Cl. AMS is also used for quantifying ^{14}C in various environmental and biological samples. Hence, AMS is an important tool for monitoring long-lived neutron activation products from operational nuclear power plants.

There are several kinds of ICP-MS: quadrupole ICP-MS, high-resolution ICP-MS, and multi-collimator ICP-MS. In ICP-MS, the sample is ionized by creating a plasma by argon gas flowing through concentric channels surrounded by radiofrequency coils. As the argon gas passes through the channels, a fraction of the argon in ionic form is oscillated at a high frequency providing a column of gaseous ions that act as a plasma. This configuration is referred to as a torch. The temperature in the torch during analysis is about 6000 K.

The sample to be analysed must be introduced into this plasma flow in the form of an aerosol. This can be achieved in several ways, such as nebulization (a tech-

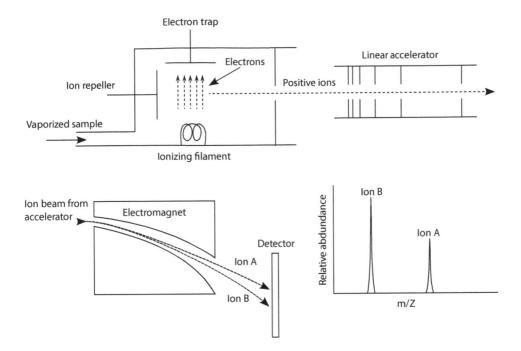

FIGURE 4.16 Schematic view of the various stages of a mass spectrometer.

nique used, for example, in asthma medication to create small droplets from a liquid) or laser ablation. These aerosols are then injected into the plasma torch and are subsequently disintegrated into atoms and then into ions at the end of the torch. Nebulized aerosols in plasma usually leads to positive ions, which makes it difficult to analyse electronegative elements such as the halogens I and F with this technique.

After ionization, the ions are injected into the high-vacuum mass analyser. In the vacuum chamber, the beam of positive ions is focused by means of electro-dipole lenses or quadruple magnets that guide the ions and separate them from unwanted neutral particles that have leaked into the vacuum tube from the ionization chamber. The field strength and geometry of the electromagnets can be adjusted to change the trajectories depending on their m/q ratio. This ratio determines the trajectory of the ion beam in the magnetic field. The efficiency of transmission of the ions from the ionization chamber by the mass analyser into the detector can vary depending on the method used to focus the beam.

The detector unit of the mass spectrometer may consist of dynodes, in a similar fashion as in PMTs. When the focussed incoming ion beam hits the initial dynode, a cascade of photoelectrons are created that are accelerated to the following dynodes by an applied bias voltage. The accelerated electrons create a charge pulse similar to that in a sensor (e.g. a charge-coupled device), which is recorded in a data memory and can be readily retrieved by a computer.

The advantages of mass spectrometry over radiometry in the identification and quantification of radionuclides are that the analysis is often more rapid, and the quality of the results is improved compared with α spectrometry for many α-emitting radionuclides (see Section 4.5). Mass spectrometry of ratios between various uranium isotopes, such as $^{238}U/^{235}U$, is now often used in applications such as dating studies. In contrast to α spectrometry, the

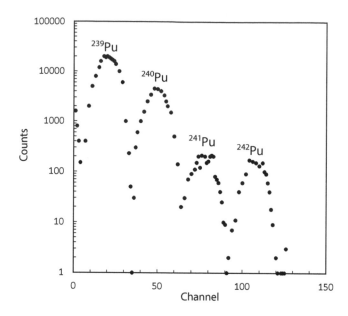

FIGURE 4.17 Mass spectrum obtained from a sediment sample from the Irish Sea, in which plutonium isotopes have been radio-chemically separated. The spectrum illustrates how clearly the peaks of ^{239}Pu and ^{240}Pu can be separated and quantified.

two plutonium isotopes ^{239}Pu and ^{240}Pu, which are commonly found in global fallout from atmospheric nuclear weapons tests, can be separated and quantified using mass spectrometry. Furthermore, the sensitivity is higher and the detection limit in terms of radionuclide concentration is lower for mass spectrometry than for a spectrometry for α-emitting radionuclides with longer half-lives. The half-life at which the sensitivity of mass spectrometry exceeds that of α spectrometry is about 400 years (Fig. 4.18). Table 4.5 presents various mass spectrometry techniques used for the accurate detection of uranium isotopes.

However, mass spectrometry has a number of drawbacks compared with α spectrometry: (i) isobaric interference between ^{99}Ru and ^{99}Tc (^{99}Tc is a very important radionuclide within the field of environmental radioactivity), (ii) polyatomic interference of, for example, ^{197}Au and ^{40}Ar (giving a mass number of $197 + 40 = 237$) and the often analysed actinide ^{237}Np, (iii) peak tailing (although also occurring in α spectrometry), (iv) isotopic fractionation, (v) beam instability leading to signal loss, (vi) the sample matrix can interfere with the spectrometry of the atoms, and (vii) unavailability of appropriate background samples, or so-called "method blanks", which are useful to verify the presence of signal background or perturbation in the detector.

It is important to analyse the actinide plutonium and its various isotopes as it is a tracer for many environmental and ecological processes. Mass spectrometry has found use in Pu detection and can differentiate between the radionuclides ^{239}Pu and ^{240}Pu (see Fig. 4.17). These isotopes cannot be resolved by α spectrometry as their energy peaks overlap. The variation in time of the fallout can be used for dating of continuous and single dispersion events, such as tracing sea currents, etc. Various types of ICP-MS provide precision down to 0.1% in the determination of the $^{239/240}$Pu ratio, whereas AMS has a precision of 1% at best.

Other isotopes relevant in environmental radioactivity that are preferably analysed by mass spectrometry are ^{99}Tc and ^{129}I, which are fission products emanating, for example,

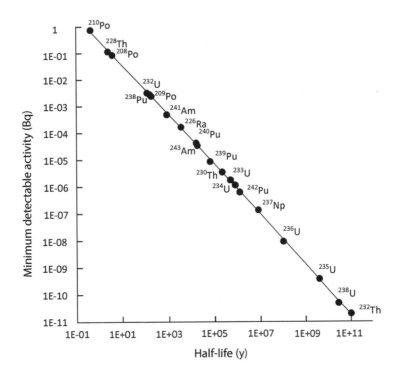

FIGURE 4.18 Minimum detectable activity [Bq] in a 1-mL solution, assuming a 5-pBq limit, as a function of half-life of a number of high-Z radionuclides using a typical quadrupole mass spectrometer.

from releases from reprocessing plants or from nuclear weapons fallout. AMS has been used to determine low-level concentrations of ^{99}Tc, but the lack of stable Tc isotopes is a problem since a stable isotope is required as an internal yield monitor and for instrument normalization. However, detection limits of ^{99}Tc using AMS are almost 50 times lower than with radiometric methods such as liquid scintillation counting (see Section 4.5.3). For the transuranic radionuclides with a physical half-life longer than 5000 years, AMS is the most sensitive method for low-level quantification in samples, and much lower concentrations can be determined than with traditional α spectrometry.

4.3   BASIC CHARACTERISTICS OF A RADIATION DETECTOR

The choice of the type and material of a detector depends on the specific application. Most radiation detectors are designed for specific applications in terms of energy range, radiation quality, and time and energy resolution. A compromise must often be made between sensitivity and detector size, or between energy resolution and robustness in field applications. It is often necessary to estimate the absorbed dose in water or in soft tissue. However, the response of the detector depends on both the energy of the radiation and the detector material. It is therefore often necessary to make a compromise between the material and the efficiency of the detector. Before choosing a detector, it is necessary to consider the fundamental parameters characterizing the detector's performance. The presentation of detector parameters in the sections below is not exhaustive, but is relevant for applications within environmental radioactivity and radiation protection.

TABLE 4.5 Mass spectrometric techniques used for the quantification of uranium isotopes.

Method	Amount of U required for detection	Amount of U required for isotope ratios	Isotopes reported	Typical accuracy (%)	Typical precision (%)
High- resolution γ-ray spectrometry	10 μg	1 mg	^{235}U ^{238}U	10	10
α-particle spectrometry	10 ng	10 μg	^{234}U ^{238}U	10	5
Quadrupole ICP-MS	5 pg	1 μg	^{235}U ^{236}U ^{238}U	2	5
High-resolution ICP-MS	50 fg	5 μg	^{234}U ^{235}U ^{236}U ^{238}U	1–8	0.1–1
Thermal ionizing mass spectrometry	1 fg	1 ng	^{234}U ^{235}U ^{236}U ^{238}U	0.1–2	0.1–0.2
Secondary-ion mass spectrometry	5 pg	5 ng	^{235}U ^{238}U	1–5	10
Multi-collimator ICP-MS	5–50 fg	1 pg	^{234}U ^{235}U ^{236}U ^{238}U	0.1–0.2	0.1–0.3

4.3.1 Spatial Resolution

The detector response provides a measure of the mean absorbed dose in the active detector volume. The spatial resolution of a detector is thus limited by the size of the detector. Radiation detectors can be designed such that the variation in intensity of the radiation over the cross section of the detector can be recorded and visualized. An example of such an instrument is the gamma camera used in nuclear medicine, where a matrix of detectors is used to quantify the activity concentration in a patient (Fig. 4.19). The spatial resolution of a detector is defined as the minimum distance at which two physical objects can be discerned, and is related to the line spread of a detector system when imaging a sharp line.

High spatial resolution is vital in radiographic imaging as this determines the overall image quality, and may thus affect the ability of the radiologist to make a reliable diagnosis. Imaging systems with pixel elements smaller than 100×100 μm^2 have been achieved. In medical imaging applications high resolution is required to visualize fine anatomical structures with dimensions of 0.1 mm or less, or for the localization of radioactive metal fragments and splinters in wounds or the respiratory tract (Ören et al. 2014).

High spatial resolution (on a scale of 0.1 to 10 cm) is generally not so important in environmental radiology and radiation protection. Spatially resolved detection is mainly important when characterizing the presence and variability of local clusters of radioactive deposits, so-called *hot particles*, for example, from the atmospheric fallout mentioned in Section 1.3.2.1. Measurements of contaminated soil or dust on a shielded detector with an

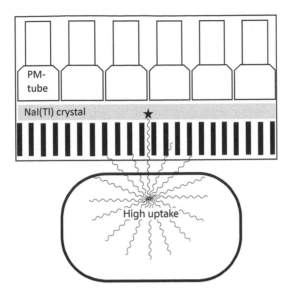

FIGURE 4.19 The principle of the visualization of the location of γ-emitting radionuclides in a body organ using a luminescent detector connected to an array of PMTs.

appropriate cross section ($>10\times10$ cm^2) can reveal the presence of β-emitting hot particles (Bernhardsson et al. 2011).

A gamma camera can be used to localize and quantify activity in the body following the ingestion or inhalation of radioactive materials or gases. The efficiency and resolution of clinical gamma cameras are usually sufficient for detecting such radioactive materials (Fig. 4.20). Spatially resolved measurements of the radiation intensity can also be obtained by repeated measurements at different locations. The data obtained from such measurements can then be visualized using graphics software to provide maps of the radiation intensity over a confined surface or a geographical area. This technique is referred to as mobile radiometry, and is discussed further in Section 7.3.6.

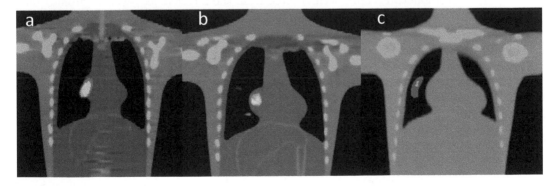

FIGURE 4.20 Images (matrix size 64×64) obtained from a combined gamma camera tomography (SPECT) and computed tomography (CT) system using a medium-energy collimator and point sources of (a) 160 kBq ^{137}Cs, (b) 800 kBq ^{60}Co, and (c) 2.3 MBq ^{90}Sr/^{90}Y, positioned in the right lung of an anthropogenic phantom. Modified from Hansson and Rääf (2011).

4.3.2 Energy Resolution

Detectors that can resolve the energy distribution of the incident particles or photons, referred to as spectrometers, play a central part in radiation protection and environmental radiology. The output from a spectrometer is a pulse height distribution, commonly referred to as a spectrum. The energy resolution of a detection system ultimately depends on the number of information-carrying quanta produced per unit of radiation energy deposited in the detector volume.

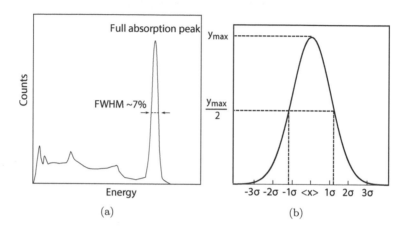

FIGURE 4.21 Typical pulse height distribution (a) recorded using a gamma detector (NaI(Tl) crystal), showing the Gaussian-shaped energy peak (b) with its centroid at the energy $< x >$ and a full width at half maximum (FWHM) of 2.35 σ.

Several kinds of spectrometric systems are available for the identification and quantification of α- and γ-emitting radionuclides. β-particles, on the other hand, are not emitted at distinct energies, making β spectrometry less appropriate for radionuclide identification. However, β spectrometry is used in some applications where a single or only a few types of β-emitting radionuclides have been isolated from the interference resulting from α and γ emitters.

The energy resolution of a spectrometer is commonly expressed as the full width at half the maximum height (FWHM) of the peak in a pulse height distribution (Fig. 4.21). The energy resolution is governed by the quantum fluctuations in the number of charge carriers collected by the signal amplification device of the detector:

$$\frac{\delta E}{E} \sim \frac{\sqrt{Var\,(N_e)}}{< N_e >}, \tag{4.29}$$

where E is the energy of the particle reaching the detector, and N_e is the number of charge carriers generated.

An energy resolution better than 10% is required to resolve the gamma emitters, ^{40}K, ^{238}U and ^{232}Th, especially in geological surveys, where NaI(Tl) crystals are traditionally used. Much higher resolution is required when analysing unknown samples, or when measuring fallout from recent nuclear weapons testing. Many modern γ-ray spectrometry systems have a resolution of 0.1–0.2%. An energy resolution of 0.5–1% is usually sufficient to distinguish between various γ peaks.

4.3.3   Time Resolution

The signal from radiation detectors is often amplified in one of two ways: (i) current mode, in which the incoming pulses are transformed into a current, whose amplitude represents the intensity of the incident energy fluence carried by the radiation particles, or (ii) pulse mode, in which the energy deposited by each radiation particle is individually amplified and displayed (Fig. 4.22). The amplification of individual pulses is often necessary and is generally more demanding than amplification in current mode, and is therefore described in greater detail here.

Detectors intended for spectrometry or for measuring the temporal variation in radiation intensity must have a short detection and pulse processing time, to avoid consecutive events not being recorded separately. The time resolution of a detector system depends on (i) the detector material, as the initial interaction between the radiation and detector material determines the time required to generate a pulse in the detector, (ii) the time required to collect the information carriers generated by the initial interaction between the radiation and the detector material, and (iii) the time required for the electronics to process and amplify the signal (see Section 4.4).

Due to the time required by the electronics to amplify and correctly store the individual pulses (described in Section 4.4), there will be a fraction of the time when the detector cannot handle incoming pulses. For a measurement that has taken a certain time, referred to as *real time*, the sum of all the time fractions when the detector system was locked, or unable to handle incoming events, is referred to as *dead time*. The difference between real time and dead time is referred to as the *live time*, which is the actual time the detector system (including the detector material, the electronics, and the display functions) has been open for detection and not busy with pulse processing.

It is important to distinguish between the pulse processing time and the integration time, i.e. the time taken for the system to display the result. In many detector systems used for γ dose rate monitoring, the value read-out is the mean of several consecutive pulse rate measurements. The integration time may be pre-set by the manufacturer or can be adjusted by the operator. Averaging over time reduces the statistical fluctuations and thus gives a more reliable signal. However, it is important to consider the integration time so that the detector is allowed to stabilize before the value is read out.

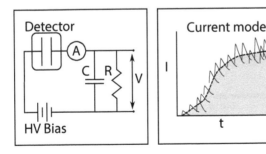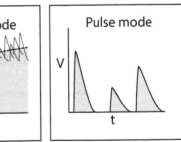

FIGURE 4.22   Left: Schematic illustration of an ionization detector. Middle: In current mode the signal is the current of free charges created in the detector. Right: In pulse mode each pulse is amplified and presented separately.

In passive detectors, chemical reactions are initiated by free radicals generated by the interaction of the radiation with the detector material. Since these chemical processes may continue for hours or days after irradiation, detectors based on the polymerization of monomers (e.g. radiosensitive films or gels), or the oxidization of ions (Fricke solutions), are usually read out the day after irradiation.

4.3.4 Sensitivity and Counting Efficiency

The term sensitivity is often used to characterize detector systems in terms of how rapidly and distinctly (in terms of signal intensity) they react to subtle changes in the radiation level. A more sensitive detector can detect a smaller change in the radiation level during a specific measuring time than a less sensitive detector. The sensitivity of a detection system can be defined as the proportionality between the detected signal and the fluence of the incoming radiation particles. Sensitivity can also be defined as the inverse of the counting efficiency, as sensitivity can be defined as the ratio between the fluence of incident radiation particles and the instrument reading. Generally, the larger the active volume of detector or the higher the density of the material, the higher the counting efficiency. For hand-held count rate meters, for example, their size is often a trade-off between the increase in counting efficiency with volume on the one hand, and the increased weight and reduced portability on the other hand. The time resolution may also be insufficient to cope with high-intensity radiation fields, leading to dead time losses.

In spectrometry, where the aim is to accurately quantify the particle fluence or the activity of a source, it is more relevant to use the definition of counting efficiency than sensitivity. In charged-particle spectrometry, which involves the measurement of radiation from sources inside the detector, the detection efficiency can be defined as the fraction of the emitted α- or β-particles that deposit all their energy in the detector. So-called 4π geometry can often be obtained in liquid scintillators where the radioactive source has been dissolved in the detector solution, giving a counting efficiency approaching 100%. However, a small fraction of the short-range particle emission can occur close to the wall enclosing the solution, and can escape or be only partially detected.

In α spectrometry using semiconductor detectors, the source is often deposited on a disc-shaped backing, either as a single point or homogeneously distributed over the surface of the disc. If the area of the detector surface is sufficient, a 2π geometry can be achieved, meaning that almost 50% of the emitted particles will be recorded in the detector. Figure 4.23 illustrates these cases.

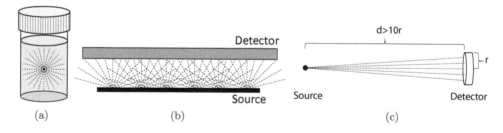

FIGURE 4.23 (a) 4π geometry with the sample dissolved in a liquid scintillator cocktail. (b) 2π geometry with a sample deposited on a surface parallel to a planar detector. (c) Point source geometry where the sample is at a distance at least ten times greater than the radius of the detector.

For γ-ray spectrometers, where the activity of a source is often measured outside the detector, the counting efficiency can be defined as the intrinsic efficiency, ε_i. This quantity is a measure of the number of pulses originating from a γ source recorded by the detector system, N_γ, per unit fluence rate from a γ-emitting source outside the detector which is sufficiently far away that the photon flux can be approximated as being normally incident on the detector surface (configuration (c) in Figure 4.23). This quantity will be independent of the sample geometry, and is unique for each detector configuration. The measured intrinsic efficiency, $\varepsilon_i\left(E_\gamma\right)$ for this source–detector geometry is given by

$$\varepsilon_i\left(E_\gamma\right) = \frac{\left(\frac{N_\gamma}{t_m}\right)}{\left(\frac{A_{source}\cdot n_\gamma(E_\gamma)}{4\pi\cdot d^2}\right)\cdot\varphi\cdot e^{-\left(\frac{\mu}{\rho}\right)_{E_\gamma,air}\cdot\rho_{air}\cdot d}}, \tag{4.30}$$

where

- A_{source} is the activity of the source,

- $n_\gamma\left(E_\gamma\right)$ is the branching ratio of the emitter (e.g. 0.85 for the 661.6 keV photons of ^{137}Cs),

- d is the source-to-detector distance (defined as the distance between the source and the centre of the sensitive volume of the detector),

- μ is the linear attenuation coefficient in air for photons of energy E_γ,

- N_γ is the net number of counts detected in the full energy peak of the measured photons of energy E_γ during the measurement time t_m, and

- φ is the cross-sectional surface of the detector, πr^2, where r is the detector radius.

Various definitions of detection efficiencies used in spectrometry are discussed further in Section 4.6.4.

4.3.5 Energy Dependence

The counting efficiency of the detector depends on the energy (E_γ), and may be decisive when choosing the appropriate detector in a given application. Spectrometric systems often exhibit well-defined relationships between the energy of the incident radiation and the detection efficiency. The energy dependence is also important for non-spectrometric devices, especially if the detector signal is converted into quantities related to the absorbed dose. It is thus necessary to know the energy interval over which a γ dosimeter for personal dose monitoring is suitable.

The dependence of counting efficiency on energy is determined by three main factors. The first is the thickness of the detector casing or shield, which may stop or attenuate low-energy components of the radiation, preventing them from entering the sensitive volume of the device (e.g. Fig. 4.24). The second factor is the atomic composition of the detector material, which determines the kind of interactions that will take place between the incident radiation and the detector material. Finally, the mass (or size) of the detector affects the relationship between counting efficiency and radiation particle energy. Depending on the atomic composition of the detector material, some incident radiation may have insufficient cross section for interaction with the detector if the active volume is small (e.g. high-energy γ photons). This can be compensated for to some extent by using systems with large detectors, but this is not possible in all applications, as this will affect both the time resolution and portability, which are necessary properties for hand-held instruments intended for field measurements.

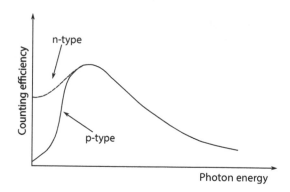

FIGURE 4.24 Counting efficiency vs. photon energy for HPGe gamma detectors with a thin (n-type) and a thick outer contact (p-type) (described in Section 4.2.3). A thin entrance window into the detector allows the transmission and subsequent detection of low-energy γ photons.

4.3.6 Signal-to-Noise or Signal-to-Background Ratio

Most radiation environments contain components of ionizing radiation emanating from cosmic radiation, and the natural background arising from primordial radionuclides, such as ^{40}K and radionuclides from the ^{238}U series and the ^{232}Th series. These, together with other components, will contribute to the detector signal, despite the fact that they are not related to the reason for the measurements. The following factors will contribute to the background counts in the detector.

- Electronic and thermal noise due to pulses resulting from spontaneous excitation of charge carriers in the detector material due to thermal motion.

- Stochastic noise resulting from quantum uncertainty in the detected pulses. This quantity is intimately linked with the uncertainty in the detector reading, but may not always be considered as noise per se. For example, a variation in the detected signal between successive measurements can be observed even in case of a constant flux of radiation particles. This variation is caused by the stochastic fluctuations in the number of quanta that undergo interactions in the detector volume.

- The contribution from background radiation: for example, from ^{40}K and radon daughter products present in the laboratory or in the field.

- Interference in the detector electronics (in addition to thermal and detector noise): for example, microphonic noise (i.e. the transformation of mechanical vibrations into unwanted electrical signals leading to noise), which contributes to the spectra recorded on-line during vehicle-borne surveys.

- Internal and instrument-inherent interference. Background radiation in the materials used in the instruments, damage and defects in the detector crystal, and time-varying interference such as shifts in the gain of the detector electronics (so-called drift), may cause stray losses or contributions within a specific energy interval.

- Scattered or distorted pulses arising from the radiation to be detected, such as pulses from Compton scattering in the crystal, incompletely absorbed photons, or electrons, etc.

It is thus important to consider all the possible background contributions before carrying out quantitative radiation measurements in any environment.

4.3.7 Response to Various Types of Radiation Particles

Most detectors are designed for the measurement of a specific radiation quality, such as α, β, γ or neutron detection. These types of radiation particles have widely different interaction cross sections with the materials of various atomic compositions in a detector. Detector materials must therefore be chosen bearing in mind the radiation particles of interest. However, in many applications it is necessary or desirable to monitor more than one kind of radiation simultaneously, and the detector system must then be able to distinguish the signals from each radiation quality. The ability to discriminate and distinguish pulses from a detector system requires that the atomic composition provides a non-zero cross section for all of the radiation particles to be detected, and that the detector response to the initial interactions of the radiation is dependent on the linear energy transfer (LET), which is related to the stopping power, as described in Section 2.1.1.

The isotope ^{7}Li can be added to TL dosimeters made of ^{6}LiF (LiF) so that the material interacts not only with high-energy γ photons, but also with thermal neutrons. Another example is the response of liquid scintillator detectors when both α- and β-emitting radionuclides are present in the detector solution. An α-particle produces luminescence for a longer period than an event resulting from the interaction of a β-particle, when the same amount of energy is deposited. This difference in the time pattern of the luminescence can be utilized in an electronic pulse shaping process to provide simultaneous on-line spectra of both α- and β-particle emission (see Section 4.5.3).

Combining two or more different detector materials provides another means of measuring several types of radiation particles simultaneously. An example of this is hand-held probes for simultaneous surveying of gross α- and gross β-particle detection using a combination of a thin inorganic scintillator (e.g. Ag-doped zinc sulphide (ZnS(Ag)), which is sensitive to α-particle detection, and a plastic scintillator that responds to β-particles. Optical filtration of the combined luminescence can enable pulse discrimination of luminescence generated by α-particles and β-particles. Such a combination of solid luminescent detectors for simultaneous detection of more than one radiation quality is referred to as a *phoswich detector*.

Some detector materials that interact with more than one type of radiation quality can be geometrically configured such that the sensitivity to a particular radiation quality is enhanced while that to other radiation particles is reduced, for example, ^{3}He-gas-filled proportional counters. Due to the low density of the ^{3}He gas, they have a low probability of interaction with high-energy γ photons, whereas the cross section for neutron capture is relatively high. These will be discussed briefly in Section 4.7.2.

4.3.8 Tissue Equivalence

Instruments designed to measure quantities related to the effective dose, such as the personal dose equivalent (see Section 2.5), must be composed of materials that have a similar atomic composition to that of human soft tissue (effective atomic number ~ 7.8). Detectors filled with air ($Z_{eff} = 7.6$), or other materials with similar effective atomic numbers can be used as simple dosimeters to measure exposures to γ photons with energies ranging from 60 keV to 3 MeV.

The air kerma rate (defined in Section 2.2) is the radiation energy deposited at a point in space located in a volume element of air at normal temperature and pressure. It is the product of the photon fluence rate, $d\Phi/dt$, photon energy, E_γ, and linear energy transfer coefficient in air, $(\mu_{tr}/\rho)_{E_\gamma,air}$, for that particular photon energy.

$$\frac{dK}{dt} = \frac{d\Phi}{dt} \cdot E_\gamma \cdot \left(\frac{\mu_{tr}}{\rho}\right)_{E_\gamma,air} \tag{4.31}$$

This corresponds to an absorbed dose rate of

$$\frac{dD}{dt} = \frac{dK}{dt}(1-g). \tag{4.32}$$

The term g is the fraction of the released radiation energy that is not deposited locally around the point considered (see Section 2.2.1. In this context, "local" means approximately the mean free path of an electron with an energy of 20 keV.

For an exposure to a fluence of different photon energies ranging from 0 to E_{max}, the dose increment from the fluence of photons $d\Phi$ within the energy range $[E_\gamma, E_\gamma - dE_\gamma]$ can be expressed as in Eqn. 4.33,

$$dD = d\Phi(E_\gamma) \cdot \left(\frac{\mu}{\rho}\right)_{E_\gamma,air} \cdot (1 - g_{E_\gamma,air}) = \left(\frac{d\Phi}{E_\gamma}\right) \cdot dE_\gamma \cdot \left(\frac{\mu_{tr}}{\rho}\right) \cdot (1 - g_{E_\gamma,air}) \tag{4.33}$$

and the absorbed dose is given by

$$D_{air} = \int_0^{E_{max}} \left(\frac{d\Phi}{dE_\gamma}\right) \cdot \left(\frac{\mu_{en}}{\rho}\right)_{E_\gamma,air} \cdot dE_\gamma. \tag{4.34}$$

It is evident from Eqn. 4.34 that a detector intended for measurements of the air kerma or the absorbed dose in air must have a well-established detector response over a wide range of photon energies. The more the energy fluence ($d\Phi/dE_\gamma$) differs from that used to calibrate the detector (i.e. a reference beam of irradiation), the greater the risk of inaccurate measurements of the absorbed dose. Ionization chambers can be designed to have a relatively energy-independent response, and are used to measure the absorbed dose when high accuracy is required, e.g. in radiation therapy. However, caution should be exercised when using solid detectors, which often have high atomic numbers ($Z > 14$), for personal dosimeters to detect γ-radiation. These are usually calibrated so as to give an equivalent dose in tissue for γ-rays of a particular energy (e.g. for 661.6-keV photons from the decay of ^{137}Cs). If the person wearing the dosimeter is exposed to γ-rays of considerably lower energy than that for which the dosimeter is calibrated, the equivalent dose will be overestimated, while it will be underestimated if the person is exposed to γ-rays of a higher energy.

In many practical applications, a rough estimate of the kerma rate resulting from γ-rays is made from very simple instruments such as GM counters. This means that the detector must be able to translate a pulse rate (representing the interaction of γ photons in the detector volume) into a corresponding kerma rate contribution. Low-energy photons may not have sufficient energy to penetrate the walls of the detector, and will thus not contribute to the signal. To avoid this problem, the walls can be modified to include thin metal windows. These detectors are called energy-compensated detectors.

The energy-dependent filtering factor, $F(E_\gamma)$, initially increases with increasing γ energy, but above 100 keV it decreases roughly exponentially, as shown in Figure 4.25. The device can be made relatively independent of γ energy by using a thinner wall or a different wall material (by combining the filtering factor with the product $E_\gamma \cdot (\mu_{tr}/\rho)_{E_\gamma,air}$. This will give a compensated instrument in which the response or signal is proportional to the absorbed dose deposited in the detector for photon energies from 100 eV to 3 MeV.

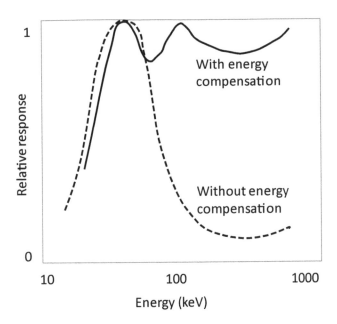

FIGURE 4.25 Schematic illustration of the energy response of an uncompensated GM tube and an energy-compensated GM tube.

4.3.9 Robustness

A detector must be able to perform reliably in the environment in which it is used, especially in outdoor environments. Most detector systems rely on amplification of the pulses generated by radiation (the exception being GM counters), meaning that the robustness of the electronics is also important.

Humidity and moisture will affect detectors based on crystals, such as Na(Tl) detectors, and these must therefore be protected as the crystals are hygroscopic. Temperature can affect the luminescent properties of a material. This may be a problem in light-to-charge converters, such as those used in PMTs. In extreme environments such as outer space (bearing in mind that this high-intensity radiation environment causes the radiation protection of astronauts to be of great concern), the melting point and condensation point of materials used in detectors must be considered. Mechanical effects must also be taken into consideration, especially in hand-held devices used in the field, as the NaI and CsI crystals used in crystalline detectors may crack if exposed to mechanical strain. The electronics, for example, in high-resolution γ-ray spectrometers, is susceptible to microphonic disturbances.

In the aftermath of the 9/11 terrorist attacks in the USA in 2001, rapid development of portable detection devices intended for mobile measurements began. This led to new generations of ruggedized portable detector systems that can withstand the outdoor environmental conditions associated with many homeland security applications. Many of these ruggedized detector systems consist of detector materials that have traditionally only been used under laboratory conditions, but the mechanic strength has been improved, and the electronics have been made much more robust (see, e.g. Upp et al., 2005).

4.3.10   Ageing

The response of a detector system may change over time due to ageing of some of the components, or to radiation-induced damage. In solid-state detectors, guard rings made of elastic material may lose their elasticity, leading to leaks and loss of vacuum. Moisture may also build up in the cavity between the casing and the sensitive part of the detector, causing deposits on the detector surface. Prolonged exposure of, for example, HPGe crystals to neutrons may change the stoichiometry of the crystal lattice, causing so-called Frenkel defects, which will reduce the energy resolution. Repeated periods at room temperature, instead of being cooled to 77 K, may lead to a slow increase in the dead-layer of the semiconductor (see Section 4.2.3), which in turn has negative effects on the counting efficiency for photons with low energies (>100 keV).

The proximity of the detector to radionuclides leaking into the environment and volatile decay products from radionuclides in the sample, may also lead to a gradual increase in the contamination of the detector. In scintillation detectors, the waveguide connecting the detector and the light sensor may degrade with time. Inorganic scintillators are often hygroscopic, and will therefore be sensitive to temperature change and mechanical stress. In liquid luminescent detectors, light-induced chemical reactions may increase the so-called chemical and colour quenching of detected signals. In gaseous detectors, the failure of guard rings may lead to air leaking into the chamber, reducing the concentration of the quenching gas, and eventually reducing the counting efficiency of the detector.

The effects mentioned above will lead to deterioration in the performance of the detector, and a long-term quality control programme is necessary to monitor and, if possible, to remedy these effects.

4.4   ELECTRONIC PROCESSING OF DETECTOR PULSES AND SIGNALS

4.4.1   General Introduction

Most radiation detectors consist of some kind of detector material, connected to an external voltage supply or bias, and electronics to process the signal. For detectors based on ionization of the detector material, the detector volume can be regarded as a capacitor with a capacitance, C, over which no current will flow unless there in an external energy input. When an incoming radiation particle interacts with an atom in the detector material, charge is generated (e.g. electron-ion pairs) and is collected by the electrodes under the influence of the electric field created by the high-voltage bias (V_{Bias}). As mentioned in Sections 4.2.1 and 4.2.2, the time required for charge collection depends on the mobility of the charges, which in turn is a function of the atomic composition of the detector.

A resistor, R_f, in series with the electrodes will then make it possible to register any changes in voltage caused by the charges generated in the detector volume. The electric pulses resulting from the deposition of energy in the detector are illustrated in Figure 4.26. The amplitude and shape of the pulse contain information that must be processed by the electronics connected to the detector. The amplitude of the pulse is usually only a few mV, and can be from a few nanoseconds long in a fast detector system, up to several hundred μs in a slow detector system. Due to the low amplitude of the pulse, it must be amplified before being presented on the display of the detector system. However, the noise inherent in the electronics will delay or reshape the pulse. The main challenge in designing a signal processing chain is thus to maintain the information contained in the original pulse.

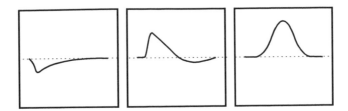

FIGURE 4.26 Left: Pulse shape before preamplifier. Centre: pulse shape after preamplifier. Right: pulse shape after amplification and shaping.

4.4.2 Preamplification

The first step in pulse processing is preamplification. The charge collected at the electrodes is first transformed into a voltage pulse by an operational amplifier. There are two types of preamplifiers: (i) *resistive feedback* and (ii) *transistor reset* preamplifier.

In the resistive feedback preamplifier, the pulses are collected in a so-called RC circuit, which consists of a resistor through which the charge passes in parallel with a capacitor where the charge is collected. The resulting preamplifier output is illustrated in Figure 4.26. The product of the load resistance R_L and the capacitance C_L, is referred to as the time constant, τ, of the circuit, and gives the decay time of the collected pulse. The greater the value of τ the slower the voltage of the pulse will decay through the preamplifier. The time constant must be matched to both the duration of charge collection and the pulse frequency. The charge mobility of the detector material is therefore an important property of a detector.

A long time constant may result in pile-up of consecutive pulses, as can be seen in Figure 4.27. Pulse pile-up may result in increased dead time if most of the pulses are rejected, or if the voltage over the capacitor constantly exceeds the dynamic range ("lock-up"). Pile-up may also result in the addition of pulses if the pulses cannot be separated in time. To prevent pile-up, the resistance must be low enough to allow the capacitor to discharge between each pulse.

Transistor reset preamplifiers consist of a capacitor only. Hence, the charge collected in the capacitor will only increase upon the arrival of a new pulse (upper left in Fig. 4.27). When the sum of the voltage of the pulses reaches a pre-set threshold, the circuit will be reset. Thus, transistor reset preamplifiers will not suffer from pile-up at high pulse rates. However, dead time will be introduced each time the amplifier is reset. The number of pulses that can be recorded between each reset depends on the energy deposited by each event. Hence, the dead time will be longer when detecting particles or photons with high energy.

The shapes of the pulses are dependent on the detector material and the design of the detector. When the load resistance, R_L, and load capacitance, C_L, are appropriately adjusted, the shape of the preamplified pulse will have a characteristic rise time, t_i, that is close to, or related to, the duration of charge collection (as illustrated in the centre of Fig. 4.26), and a decay time given by the time constant $\tau = 1/RC$.

4.4.3 Noise Propagation and Pulse Shaping

It is important that the information carried in the rise time and amplitude of the pulse is not lost during shaping. The electronic noise generated by thermal effects in the detector and in the preamplifier circuit will propagate throughout the signal processing system. The higher

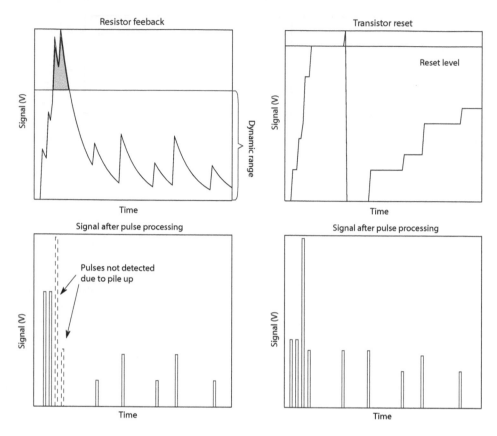

FIGURE 4.27 The outputs from a resistive feedback preamplifier (upper left frame) and transistor reset preamplifier (upper right frame). The lower frames show the respective signal after pulse processing.

the fraction of spurious variations in the pulse train, the more distorted the information will become when the signal is processed.

The total noise in the detected signal, N_{signal}, is made up of contributions from three main kinds of noise (Eqn. 4.35): the noise in the original detector signal (N_q) due to the variation in the number of quantum charge carriers produced in the detector volume, the serial noise (N_s) due to the fluctuation in the applied bias voltage and capacitance over the detector volume, which in turn causes fluctuations in the number of charges collected, and the parallel noise (N_p) due to the predominantly thermally induced electronic noise that is accumulated during pulse collection.

$$N_{signal} \sim \sqrt{N_q^2 + N_s^2 + N_p^2} \qquad (4.35)$$

Pulse shaping is carried out in order to reduce the pulse length while the information of the maximum amplitude is preserved. The first step in the pulse shaping process is to pass the preamplified pulse through a differentiating CR circuit, also referred to as a high-pass filter. This attenuates the low-frequency variations of the incoming pulse and thereby also the low-frequency noise. The differentiated pulse has a time-decay represented by the time-constant, $\tau = 1/CR$, of the CR circuit. In the next step, this differentiated pulse is then directly coupled to an integrating circuit (an RC circuit) that integrates the whole differentiated pulse. If the integrating RC circuit has the same time-constant as the CR circuit, the resulting pulse is a signal with a rise time of about 1.2τ (i.e. the time it takes for the pulse to rise from 10% to 90% of the maximum amplitude) and a pulse decay extending about 6τ (Fig. 4.28).

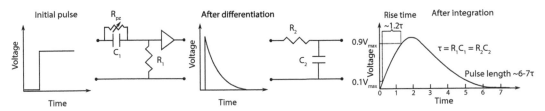

FIGURE 4.28 Pulse shaping by means of a CR (differentiator) circuit and RC (integrator) circuits coupled in succession. The pulse shaping time constant can often be set by the operator of the device to obtain the minimum total noise. The resistor R_{pz} acts as a so-called pole-zero adjuster, which will be described below.

The shaping time is set to minimize the total noise given by Eqn. 4.35 (see Fig. 4.29. The longer the shaping time, the lower the noise contribution from the original detector signal (i.e., the serial noise, N_s). However, the thermal noise of the RC–CR circuits will increase with increasing shaping time, thereby increasing the parallel noise (N_p).

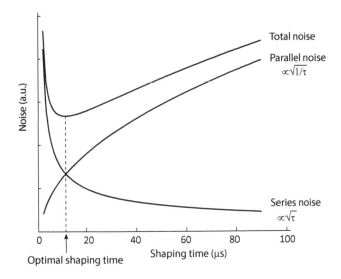

FIGURE 4.29 Noise contributions as a function of the shaping time constant of the circuit presented in Figure 4.28.

4.4.4 Pole-Zero and Baseline Restoration

If only successive CR–RC coupling is employed in the pulse shaping chain, this will result in so-called undershoot in the unipolar pulse shape due to the way in which the operational amplifier transforms an exponential signal. It can be shown mathematically that a purely mono-exponential output cannot be obtained from CR–RC coupling. An additional resistor is therefore connected to the input capacitor to adjust the input pulse shape so as to obtain a more unipolar pulse shape. This resistor (R_{pz}) acts as a so-called pole-zero adjuster (Fig. 4.28).

The consecutive coupling of capacitive circuits means that when measuring the current and time, the time-averaged value of the current must be zero as capacitors do not allow the flow of direct current. This means that the tail of the pulse resulting from the pulse shaping electronics must decay to a fluctuating baseline. Older electronic circuits employed bipolar signal shapes instead of monopolar, but relatively noise-free compensatory processes are now available, which rapidly restore the shaped pulses in the pulse train back to the baseline (*baseline restoration*).

4.4.5 High-Voltage Bias Supplies

The bias provided by a high-voltage (HV) supply is generally applied to the detector via the preamplifier. An a.c.-coupled arrangement can be used, with a coupling capacitor between the preamplifier and the detector. Alternatively, the high voltage can be applied independently of the preamplifier, providing the preamplifier with less noisy detector signals. Ripples in the bias can add serial noise to the signal, and filters are thus used to attenuate potential drifts and fluctuations (N_s). Applying too high a bias voltage to a semiconductor detector or an ionization chamber may cause electric field breakdown close to the charge-collecting electrodes, and may damage the detector. HPGe crystals must be cooled while the bias is applied, otherwise the large current caused by thermal excitations will damage the electronics. Modern HPGe detector systems have automatic bias shutdown.

In the case of scintillation detectors, the HV bias is connected directly to the PMTs (Fig. 4.30), and the stability of the HV supply is thus more important in scintillation detectors than in ionization detectors. Users of spectroscopic instruments must be aware of the fact that the energy resolution is directly affected by the bias for scintillators employing PMTs, whereas it primarily determines the counting efficiency of ionization detectors. HV stability is also important in proportional counters in the mid-range of bias (about 100 to 500 V).

4.4.6 Photomultipliers

Traditional PMTs consist of a photocathode, a vacuum tube in which electron multiplier dynodes are positioned, and an anode at the opposite end (Fig. 4.30). The size of PMTs ranges from 1 to 50 cm in length and diameter. The electron multiplier dynodes consist of metals with low threshold energy for secondary emission of electrons, such as beryllium oxide (BeO) or manganese oxide (MgO). The function of the PMT is twofold: (i) to convert incident photons in the visible or UV range into photoelectrons and (ii) to amplify the photocathode electrons to produce a charge pulse. The latter is accomplished by accelerating the photocathode electrons by the electric field resulting from the high-voltage bias. As they are accelerated through the vacuum tube towards the anode, they collide with the dynodes, which are at increasing voltage along the acceleration path. Upon each collision the beam of initial photocathode electrons will be multiplied by a factor of 10 or more, until a charge pulse of up to several mV is obtained and transferred further in the pulse chain (in analogy with the charge pulse in ionization detectors as described above).

Today, digital and semi-digital alternatives to PMTs are available in the form of semiconducting diodes, allowing the signal to be digitalized at an earlier stage of the pulse shaping process. This means that large and often fragile, vacuum tubes can be replaced by arrays of semiconductor sensors for the detection of luminescence. Multiplication can be achieved in the diode using a high-bias voltage, in a similar way to that in a PMT. So-called avalanche photodiodes are designed with a photocathode consisting of an exit layer, over which the

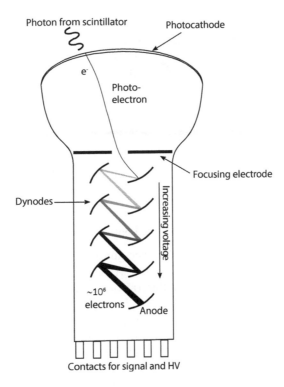

Photon from scintillator

Photocathode

e⁻

Photo-electron

Focusing electrode

Dynodes

Increasing voltage

~10⁶ electrons

Anode

Contacts for signal and HV

FIGURE 4.30 Schematic illustration of a photomultiplier tube with 8 dynodes, which successively amplify the signal. The high-voltage bias is applied between the anode and the photocathode.

bias is applied, multiplying the charge created in the entrance photocathode, which consists of SiO_2. Avalanche photodiodes can have several different designs. In the reach-through design, for example, light enters through a thin n-doped layer, as illustrated in Figure 4.31, and interacts somewhere in a p-doped region behind the thin n-doped layer. The p-doped layer constitutes most of the diode thickness. The avalanche effect can be compared with a GM tube (i.e. the sensitive part of the GM counter) since secondary ionization may occur as the electrons migrate towards the n-doped layer. The multiplication factor, i.e. the number of secondary particles (electrons or ions) created per ionization event, is on the same order as in a conventional PMT.

Avalanche photodiodes have both advantages and disadvantages compared to conventional PMTs. Due to the high efficiency in translating incident photons into photocathode electrons, these diodes can be made very small (less than 1 mm²), which may be an advantage when designing detectors for radiological imaging where high spatial resolution is required. Furthermore, they are more robust than PMTs. Nevertheless, PMTs have to be used in applications where large bulk scintillators are required, as diodes cannot be made as large as conventional PMTs. Semiconductor sensors operating without an external bias are also available. Such detectors can be very simple, since only a small electric current is required for signal processing.

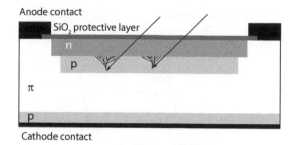

FIGURE 4.31 An avalanche photodiode with an entrance window consisting of a layer of SiO_2. Arrows indicate impinging γ-rays.

4.4.7 Analogue-to-Digital Converters and Multichannel Analysers

The final stage in pulse processing involves the analysis of the pulse height and digitization of the data in an *analogue-to-digital converter* (ADC), and subsequent sorting of the pulses in a *multi-channel analyser* (MCA). An illustration of an ADC as a number of *single channel analysers* (SCAs) coupled in parallel is shown in Figure 4.32. Each SCA is set individually to match a gradually increasing pulse height (normally between 0 and 12 V). The pulse height interval; $[H_{LLD}, H_{ULD}]$ is divided into n increments. The lower-level and upper-level pulse height discriminators, LLD and ULD, are set so as to remove low-energy noise that does not contain any important information, and to remove detector pulses from events with energies too high to be relevant for the measurement. Each pulse interval thus corresponds to $[H_m, H_{m+\Delta H}]$, where $\Delta H = (H_{ULD} - H_{LLD})/n$. As the pulse passes through the array of parallel coupled SCAs, the SCA with the pulse height discriminator setting covering the pulse height will provide a signal indicating a true event. The LLD and ULD, as well as the amplification gain (often set by the operator), must be adjusted so that pulses of interest, i.e. those resulting from the radiation being studied, fall within the energy window.

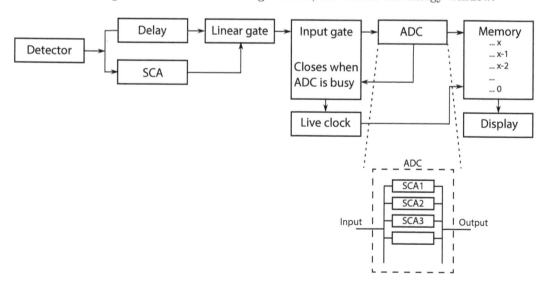

FIGURE 4.32 Illustration of the principle of analogue-to-digital conversion of electric pulses in a spectrometric detector system.

The task of the ADC is to linearly convert and pass on the incoming pulses to the MCA in the shortest possible time. There are several types of ADCs, the simplest of which compare the incoming pulse height with discrete reference energy values. The pulse processing time of these ADCs depends on the shape of the incoming pulse. However, an ADC with a processing time that is independent of the pulse height is often desirable, as this will result in almost constant processing time for all incoming pulses. These ADCs, in which a constant number of steps is used to generate the digitized pulse, are referred to as successive approximation ADCs. Flash-type ADCs, on the other hand, use as many comparators as the number of channels. In such a device, the incoming analogue pulses can be processed into the correct channels much faster. Modern ADCs have a processing time of about 20–30 μs per incoming pulse.

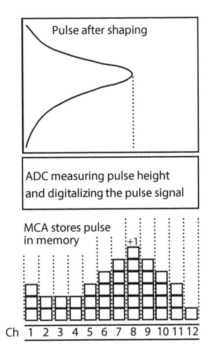

FIGURE 4.33 Illustration of the principle of an ADC and an MCA.

In spectrometric detector systems, the pulse height distribution is analysed during signal acquisition to allow the energy distribution of the incident radiation particles to be shown on-line. This is done by a MCA connected to the ADC (see Fig. 4.33). The number of channels in the MCA ranges from 256 in low-resolution spectrometers (such as NaI(Tl) detectors for gamma-ray spectrometry or organic liquid scintillation counting (LSC) detectors for β-particle spectrometry), to 8192 (or even 16k) in high-resolution spectrometry systems such as HPGe detectors for gamma-ray spectrometry or Si diodes for α spectrometry. However, one drawback of using many channels is that the stochastic noise in the detected quanta (given approximately by the Poisson distribution, as mentioned in Section 4.1), becomes significant. The optimal number of channels is generally considered to be 4 channels per FWHM of the intrinsic energy resolution of the detector system. For gamma-ray spectrometry the FWHM varies with photon energy, so the FWHM value in the mid-energy interval of interest, e.g. 1332.5 keV (corresponding to the gamma decay of ^{60}Co), is often used.

The energy range in α spectrometry is often narrower than in gamma-ray spectrometry as >90% of the α-particle energies of interest are in the range 4 to 6 MeV. It is, therefore, generally sufficient to connect one MCA to two or more detectors. Pulses that originate from α-particles with energies below 4 MeV are supressed by setting the LLD accordingly. The unipolar output signals should have a typical shaping time of a few μs (as in high-resolution gamma-ray spectrometry), and a d.c. level that matches the ADC input level within $\pm 0.1\%$ ($\sim$10 mV/10 V). The MCA is then subdivided into groups of channels, each representing one of the serially connected detectors, and the subdivided regions of channels are then individually visualized using appropriate software.

4.4.8 Digital Signal Processing

Digitization of the signal as early as possible in the signal processing chain is desirable as dead time and noise can be reduced more efficiently than when using analogue amplification. However, a very fast ADC is needed to prevent loss of information. Since the sampling rate of a flash ADC can be as high as 5000 MHz, the data must be analysed on-line and cannot be written to a disk for later analysis.

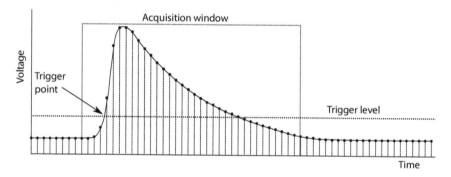

FIGURE 4.34 Pulse integration of a preamplifier signal output with an acquisition window.

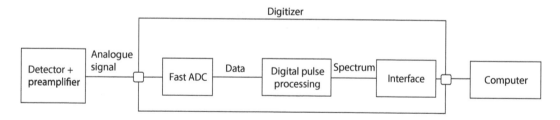

FIGURE 4.35 Simplified figure showing the various components in a digitizer.

The ADC continuously sends the digitized data to the pulse processor, which processes them. If the amplitude of the signal exceeds a certain trigger level, the pulse processor integrates the incoming data and updates the stored spectra, as illustrated in Figure 4.34, where the preamplifier signal is digitized instead of being pulse shaped with RC–CR coupling. Although digital systems may be more complex, they have several advantages over analogue systems. Provided that the data can be saved rapidly after each pulse, the dead time will be shorter. The components can usually be fitted inside a single digitizer unit, as shown in Figure 4.35, and the variables in the pulse shaping algorithms can easily be changed. The performance of digital components is generally more uniform and reliable, and they are not affected by changes in temperature, which is often a problem in analogue systems.

Most of the electronic devices described in this section were bulky and were installed in rack systems. However, developments in electronics have allowed miniaturization of all these components, and many of them can often be manufactured in a single compact unit, called a detector interface module, which is connected to the computer system.

4.5 SPECTROMETRY OF CHARGED PARTICLE RADIATION

4.5.1 General Introduction

Charged particles have a much shorter range in matter than uncharged particles such as x-rays, γ-rays and neutrons. The mean range of α-particles emitted from ^{238}U or ^{232}Th daughters is only 3–4 cm in air, and only 0.1–0.5 mm in most solid crystals. The range of β-particles of energies that are typical in radiation protection and emergency preparedness (0.1–2.5 MeV) is longer, up to 1 m in air. However, the protective encapsulation of a β-particle detector must be less than a few mm to ensure that a high proportion of the radiation is detected. For the efficient detection of α- and β-particles, the samples must often be concentrated and dissolved in liquids before being deposited on sample discs (Fig. 4.36). Extensive sample preparation, referred to as radiochemistry, is thus often needed in the fields of radiation protection and environmental radioactivity. (This is discussed in more detail in Section 5.3.)

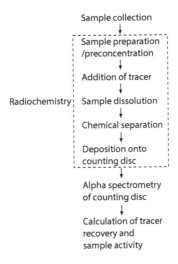

FIGURE 4.36 Processing of samples for charged particle spectrometry.

Full energy absorption of the emitted α-particles can be achieved using liquid scintillators or bulk gaseous ionization chambers. If the α-emitting radionuclide can be obtained in a gaseous state, the gas could be mixed with a counting gas in a gas-flow proportional counter. Thin scintillators, such as polycrystalline ZnS, can be used for total α-counting, but not for α spectrometry. In α spectrometry, the ideal source configuration when the source is outside the detector is a single atomic layer of the α-emitting radionuclide. Even after laborious radiochemical processing, α-particle detection requires a system with as thin a barrier or entrance window between the sample and the detector as possible. Counting efficiency can be improved considerably by using a so-called 4π measurement geometry, which refers to the situation when the source to be measured is completely surrounded by a detector material, and when all emitted radiation particles from the source will reach the detector. Such a geometry is achievable when using liquid scintillator systems or some proportional gas flow counters. However, these systems are not practical for many types of samples and external detectors are more commonly used. An example of an external counting system is shown in Figure 4.37.

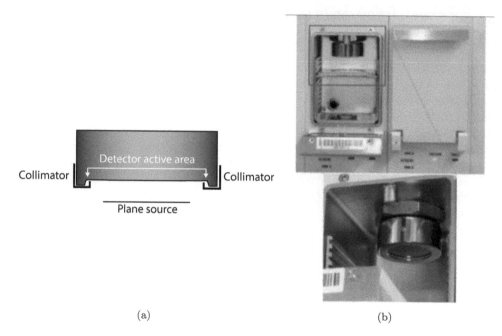

(a) (b)

FIGURE 4.37 (a): The source geometry often used for the detection of α-particles in a vacuum chamber enabling a barrier-free pathway between the α-emitting source and the detector entrance window. (b): Photograph of a Si diode inserted into a vacuum chamber for the measurement of thin α-particle sources.

4.5.2 α Spectrometry Using Solid Semiconductor Detectors

Detectors used for charged particle spectrometry are designed and optimized for inelastic collisions between the incident particles and the atoms in the detector material. The equation for the collisional stopping power (Eqn. 2.2 in Section 2.1.1) can be used to predict the energy transferred from the incoming particle to the atoms and atomic nucleus in the detector material. The stopping power has two components: the radiative component, where energy is transformed into bremsstrahlung, and the collision component, where energy is transferred to charged particles by soft or hard collisions. The collisional stopping power, which determines the number of charges produced in a gaseous detector or a scintillator, is proportional to the Z/A ratio of the detector material, whereas the radiative loss is proportional to Z^2 (Eqn. 2.4). This means that a fraction of the energy of the incoming charged particle will not be detected in a small semiconductor crystal. For this reason, Si ($Z = 14$) is preferable to Ge ($Z = 32$) in charged particle spectrometry. Si has therefore become the

FIGURE 4.38 Sample disc (consisting of e.g. nickel or silver) often used for α spectrometry in vacuum chamber systems.

standard detector material for α spectrometry, whereas Ge is mainly used in high-resolution gamma-ray spectrometry.

Semiconducting charged particle detectors are made by two main methods: ion implantation in a Si material (e.g. PIPS® detectors), or the surface-barrier technique (Fig. 4.39). Using ion implantation, the barrier can be made very thin, which improves the energy resolution in spectrometry. This configuration will also be less sensitive to electronic noise. Surface-barrier detectors, on the other hand, can be made very thin, and are suitable as so-called transmission detectors, where the energy dependence of the energy loss per unit path length (dE/dx as a function of E) provides a measure of the kinetic energy of the charged particle. Such transmission detectors are often $\sim$10 μm in thickness, with an entrance window of about 0.05 μm, allowing minimum energy loss of the α-particles impinging on the sensitive volume. Solid-state detectors must be shielded from background radiation. An aluminium coating is often used, which will add an additional barrier.

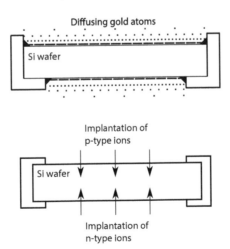

FIGURE 4.39 Two common Si detector designs. Above: a cross section of a surface-barrier detector where a thin layer of gold is diffused onto a Si wafer forming a metal–semiconductor junction. Below: an ion-implanted detector where the p-n junction is formed by accelerating doping ions into a Si wafer.

The most commonly used detectors for α-spectrometers today consist of a thin Si-based diode, with its depletion region close to the surface (Fig. 4.39). The depletion region is achieved in principle by a surface barrier, created by a p-n junction, or by ion implantation in a semiconducting material. Recall that the depletion region is a volume in which there is no free charge, although there is an electric field that can attract any charge released in the region in the event of an interaction between incident radiation and the detector. A depletion region with a depth >15 μm will completely absorb a 5-MeV α-particle. However, such a small distance, d, between the contacts would create excessive capacitance, C, given by Eqn. 4.36:

$$C = \frac{a \cdot A}{d} = \frac{a \cdot A}{b \cdot \sqrt{\rho \cdot V_{Bias}}}, \tag{4.36}$$

where a and b are constants related to the product between the dielectric constant and the relative permittivity, $\varepsilon_0 \cdot \varepsilon_r$ [F m^{-1}] of the diode, ρ is the resistivity per unit length [Ω m^{-1}] of the semiconducting material, and V_{bias} [V] is the bias voltage applied over the semiconductor. The depth of the depletion region, d, in Eqn. 4.36 is thus proportional to the square root of the bias voltage (compare also to Eqn. 4.21). As a high capacitance leads to higher serial noise (N_s in Eqn. 4.35), it is better to increase the depletion depth to at least 100 μm. If the voltage is further increased, however, the thermally induced leakage current in the reverse direction will increase instead. Figure 4.40 gives a schematic illustration of the relationship between the location of the alpha peak and the bias voltage (left) and that of the energy resolution and bias voltage (right).

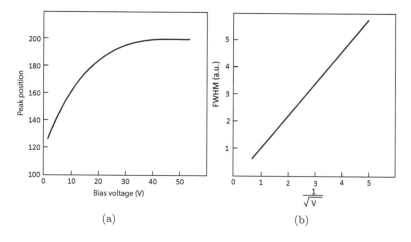

(a) (b)

FIGURE 4.40 Left: The positon of alpha peaks will asymptotically reach a limit value as the bias voltage is increased due to the related increase in the depth of the depletion region. Right: The relationship between bias voltage [V] and energy resolution, in terms of the FWHM [keV] of a Si-based thin ($<$500 μm) diode detector used for charged particle spectrometry.

Another reason why Si has become a standard detector material for charged particle spectrometry is the small band gap, 1.12 eV at room temperature, which results in a low average charge pair creation energy of $\sim$3 eV (Table 4.2). This means that for a given amount of energy deposited by the charged particle, more charge is generated than in many other materials, leading to a lower relative quantum uncertainty and higher energy resolution of the system.

In contrast to spectrometry of uncharged particles, α and β spectrometry is based on full energy absorption of all incident particles (with the exception of very thin transmission detectors). The counting efficiency in charged particle spectrometry is therefore mainly dependent on the geometric configuration of the sample and the detector geometry, and is independent of the energy of the particles impinging on the detector. The energy resolution will, however, depend on how the sample is geometrically positioned in relation to the detector. Energy calibration in α spectrometry is therefore more difficult than in gamma-ray spectrometry, since the energy of the incident particles reaching the detector depends on both the position of the source relative to the detector, and on how the sample material is distributed in the sample disc.

Energy straggling, i.e., the process in which a charged particle loses kinetic energy through interactions with material outside the sensitive volume of the detector, will arise in all external detector systems (in contrast to liquid scintillator systems). Figure 4.41 illustrates how the peaks recorded by a Si detector are shifted towards lower energies due to the presence of air between the α source and the entrance window of the detector. In addition to the contact, a dead layer greater than 0.5 μm can be expected in ion-implanted Si systems currently used for α spectrometry, causing a straggling of several 100 keV before the α-particles reach the sensitive volume. Straggled particles contributing to pulses in the α-spectrometer will arise from obliquely impinging α-particles, as well as from α-particles incident normally on the detector surface close to the collimator, where charge collection is expected to be somewhat less efficient.

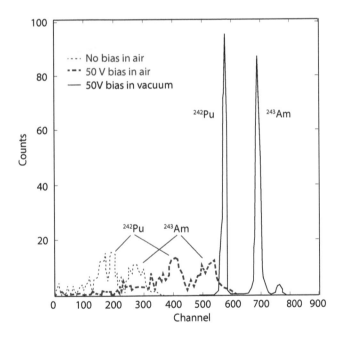

FIGURE 4.41 Pulse height distribution from a sample of ^{242}Pu-^{243}Am (^{242}Pu, $E_\alpha = 4901$ and 4856 keV; ^{243}Am, $E_\alpha = 5275$ keV and 5233 keV) electrodeposited on a stainless steel disc, measured with an ion-implanted planar Si detector (ISD) with an active area of 300 mm^2. Laboratory measurements with an ISD in air (with and without bias) and in vacuum with bias.

In a well-designed calibration source for α spectrometry, for example, ^{239}Pu deposited on an aluminium disc, the radioactive atoms should be distributed over a depth of less than a few μm. Standard calibration sources with deposits as thin as 6 μm on the sample discs are commercially available. The calibration of a specific source–detector setup with such a source will be accurate as long as it is used to measure a sample with a very similar source depth distribution. Most α-particle-emitting sources used for calibration or fabricated by radiochemical separation are electrodeposited.

The energy resolution for full energy absorption in α-particle detectors varies from 10 keV for small detectors (<10 mm^2) to 50 keV for detectors with larger surfaces (300 mm^2). A larger fraction of straggled particles reaches the detector as the detector surface increases. This also means that a larger fraction of the detected particles will deposit energies less than the maximum energy of the α-peak, creating an asymmetric full energy peak elongated towards lower energies (Fig. 4.42). The particular shape of the tail in an α-spectrum depends on several factors: (i) the energy difference between two full energy peaks, (ii) the source–detector geometry (the number of α-particles incident on the periphery of the spectrometer), (iii) the distance between the sample and the detector entrance window, (iv) the way in which the detector is collimated, and finally, (v) the thickness of the sample (i.e., how deep the α-emitting radionuclide is in the sample disc).

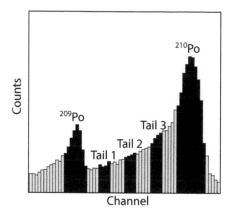

FIGURE 4.42 Examples of low-energy tailing in an α-particle spectrum. Three ROIs in the tail of the ^{210}Po peak (Tail 1, Tail 2 and Tail 3) are used to correct for the tail interfering with the peak from ^{209}Po.

The counting efficiency of an external α-counting setup can be calculated analytically or by means of Monte Carlo simulations, and can be determined experimentally using a standard α-emitting source, preferably in the form of a thin point source mounted on a sample disc. Using an α source covering the whole disc may introduce uncertainties related to the lack of homogeneity of the distribution of the radionuclide over the source. In many cases, there is no need to determine the counting efficiency as radiochemical tracers are often used to calculate the activity in the original sample (see Section 5.3). However, when evaluating various radiochemical separation methods, for example, it is necessary to determine the radiochemical yield, and the absolute detection efficiency must therefore be determined. Depending on the radiochemical process used to extract the element of interest, the actual α-emitting source may either be a small spot ($\sim$1 mm diameter) deposited at the geometrical centre of the disc (see Fig. 4.38), or dispersed over the whole disc (e.g. using electrodeposition of ^{210}Po.). In the latter case, the inhomogeneity in the deposition of the atoms on the sample disc is compensated for by assuming that the sample radionuclide and yield tracer radionuclide are similarly dispersed over the disc.

Yield determinants, or tracers, are thus used in α spectrometry to relate the number of counts from an α-emitting radionuclide in a sample deposited on a sample disc, to the known activity of a similar radionuclide (often of the same element as that being measured) added to the original sample. Consider a sample disc containing two α-emitting radionuclides of the same element, with activities A_1 [Bq] and A_2 [Bq], the first of which is the one to be determined and the second is the yield determinant. The number of counts from each radionuclide, obtained during a certain measuring time, will be N_1 and N_2 (Eqns. 4.37 and 4.38). In Si detectors with depletion depths sufficient for full energy absorption, the counting efficiency for the two sources, ε_1 and ε_2, will be the same. We wish to know the activity of the α emitter in the original sample, A_{sample}. We do not know A_1 or A_2, but we do know that A_1 is the product of the radiochemical yield, R, and A_{sample}. We also know that A_2 is the product of R and the known amount of activity added as a tracer, A_{tracer} (Eqns. 4.39 and 4.40).

$$N_1 = \varepsilon_1 \cdot t_m \cdot A_1 \tag{4.37}$$

$$N_2 = \varepsilon_2 \cdot t_m \cdot A_2 \tag{4.38}$$

where

$$A_1 = R \cdot A_{sample} \tag{4.39}$$

$$A_2 = R \cdot A_{tracer}. \tag{4.40}$$

As the radiochemical yield, R, is assumed to be identical for two isotopes of the same element, the following expression can be written:

$$\frac{N_1}{N_2} = \frac{\varepsilon_1 \cdot t_m \cdot A_1}{\varepsilon_2 \cdot t_m \cdot A_2} = \frac{R \cdot A_{sample}}{R \cdot A_{tracer}}, \tag{4.41}$$

which gives

$$A_{sample} = A_{tracer} \frac{N_1}{N_2}. \tag{4.42}$$

Hence, the ratio of the net counts in the two α peaks (N_1/N_2) allows the activity of the α emitter in the sample, A_{sample}, to be deduced without knowing the exact counting efficiency or the chemical yield of the radiochemical process used to extract the sample.

The leakage current through the detector consists of thermal electrons generated in the diode, together with stray currents generated in the contacts. These spurious pulses must be minimized to enable detection of the pulses generated by α-particle interactions in the sensitive volume of the detector. The sides of the p-n junction cause the diode to act as a capacitor, and the capacitance is a function of the cross-sectional area of the diode and the depletion depth, d, of the diode (Eqn. 4.36). The preamplifier connected to the diode/detector is itself a high-quality RC circuit, whose noise level is often given by the manufacturer in terms of the FWHM without load. The unloaded FWHM is typically 2 keV. However, when the detector is connected to the preamplifier, additional thermal noise, N_p, and noise from fluctuations in the capacitance, N_s, lead to a much broader FWHM (>10 keV). To limit N_s, the depletion depth is extended far beyond the range required for the detection of α-particles by increasing the bias over the diode.

The most important factor in performing high-quality α spectrometry is sample preparation. The radionuclides of interest must be efficiently separated from the rest of the sample matrix. The deposition of the radionuclide onto the sample disc is also important. As in the manufacturing of a standard source, a single atomic layer should be homogeneously distributed over the surface of the disc. In reality, clusters of atoms will form, making deposition slightly inhomogeneous, giving rise to variable straggling and degradation of the energy resolution. Even without straggling, it may be difficult to resolve several α-emitting isotopes, due to the relatively poor energy resolution in α detection compared with that in high-resolution gamma-ray spectrometry.

One problem encountered when measuring samples in a vacuum chamber is that the recoil atoms from nuclear α-particle decay may have energies (50–80 keV) such that they may be deposited on the entrance window or other parts of the detector. Furthermore, volatile elements, such as thorium and polonium, may evaporate from the sample and be deposited on the detector surface (see Fig. 4.43). It is therefore recommended that the distance between the detector and the sample is increased when measuring volatile elements. Moreover, the background in the detector should be determined regularly to monitor any increase in the background due to recoil contamination and volatile escape. Apart from increasing the distance, other methods of ensuring correct analysis include placing a Mylar foil over the sample disc, decreasing the vacuum (from below 10–50 mTorr to a few hundred mTorr), or the application of a negative potential to the sample disc. However, all these measures lead to degradation of the energy resolution.

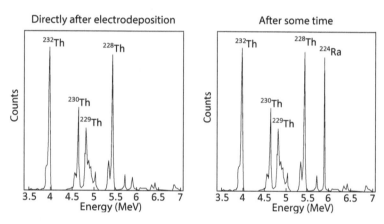

FIGURE 4.43 An example of a typical background in an α spectrometry cavity with a history of Th and U sample measurements.

4.5.3 Charged Particle Spectrometry and Total Beta Counting Using Liquid Scintillators

α spectrometry is preferably performed using semiconducting detectors, whereas β emitters are more commonly detected using liquid scintillation counting, LSC. The β-particle emission from nuclear decay is by nature more difficult to interpret from a pulse height distribution due to the continuous energy distribution of the β-particles emitted from most β^- and β^+ emitters. As a rule of thumb, the average energy emitted is approximately one third of the maximum energy emitted, $E_{\beta,Max}$. However, some β-emitting radionuclides, such as ^{99}Tc, emit conversion electrons with relatively distinct energy distributions that can be used for quantitative determination of β. As is the case in α spectrometry, sample processing is often required for the extraction of β-emitting elements such as ^{3}H, ^{14}C, and ^{32}P, to increase the specificity and sensitivity of liquid scintillation measurements.

The principal use of LSC within radiation protection and emergency preparedness is in the screening of many biological samples, especially in cases where they contain both β- and α-emitting radionuclides. Liquid scintillation is suitable for isotopes emitting β-particles in the low- and mid-energy range (50–200 keV). Liquid scintillation is the main method of detection for volatile β-only emitters such as ^{3}H. Sophisticated computational techniques are used to extract spectral information from LSC and to separate different β-emitting radionuclides, but these will not be considered here.

A transparent vial, containing a mixture of the sample solution, a solvent and a scintillating organic material in liquid form (a *fluor*), is placed in a measuring chamber. The mixture of the solvent and the scintillating organic material is often referred to as a *cocktail*. At least one, but usually two PMTs positioned opposite each other and protected from external light, are mounted in the cavity close to the vial, and record the luminescence or Cherenkov radiation (see Section 2.1.1) produced. In theory, counting efficiencies up to 100% can be achieved. The sample to be measured may be in liquid form (e.g. urine, blood, tap water), or transformed into liquid form by means of radiochemical processing (see Section 5.3). The vial may be transparent plastic or glass. Liquid scintillation detector systems are usually equipped with automated sample changers so that large batches of samples can be analysed. Figure 4.44 shows a typical liquid scintillation system.

The measuring chamber with the PMTs is surrounded by shielding (often lead) to suppress background radiation, referred to as *passive shielding*. *Active background suppression* is obtained by connecting the two PMTs to a coincidence circuit requiring simultaneous detection (within about 20 ns), so that only events arising from a charged-particle interaction in the sample mixture are recorded. Other events, for example, from the interaction of external γ-radiation (background), can thus be discriminated from real events arising from the emission of α- and β-particles.

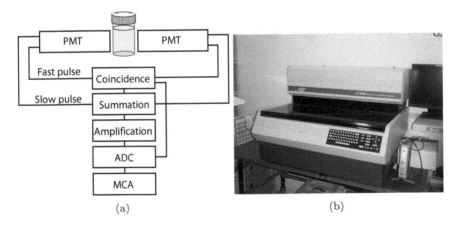

(a) (b)

FIGURE 4.44 (a): Illustration of a liquid scintillation counter with its preamplifier and pulse processing chain. (b): A liquid scintillation counter with an automatic sample exchanger, containing vials of samples to be measured. Photograph by Kristina Stenstrm.

PMT-based liquid scintillation systems can be used for two types of radiation measurements: (i) luminescence generated by the excitation of molecular electrons in a cocktail of organic scintillator in a vial between two oppositely positioned PMTs, and (ii) Cherenkov light resulting from the decay of charged particles, in a setup similar to that with a vial containing an organic scintillator. The latter method cannot be used directly for spectrometry.

An event is generated by the disintegration of radioactive isotopes in the cocktail. Most cocktails will completely absorb the energy of α- and β-particles, provided they have a kinetic energy above $\sim$4 keV. Particles with this kinetic energy will cause ionization and excitation of molecules in the solvent. When the excited molecules return to the ground state, photons in the ultraviolet wavelength region are emitted. The UV photons will be absorbed by fluorescent molecules and re-emitted at a longer wavelength. The luminescence from the fluorescent molecules will decay with a time constant of $\sim$10 ns or less. However, decay with a longer decay constant is also present, leading to delayed phosphorescence similar to that in inorganic scintillators.

There may also be sources of background luminescence. Light in the visible region may penetrate the vial. Hence shielding from external light is necessary, not only for the protection of the PMTs, but also to prevent spurious pulses being generated in the cocktail and included in the signal. Light-induced chemiluminescence in the fluor of the cocktail may last many minutes after exposure, and samples should therefore be kept in the dark for 30–60 minutes, depending on the fluor and solvent used, before being run in the liquid scintillation system. There may also be some thermally induced excitation of the fluor, in analogy with the thermal noise in a diode at room temperature. Furthermore, radioactive contamination and impurities may be present in the cocktail. The vials may also attract radon daughters from the ambient air due to static electricity, which will further contribute to the background.

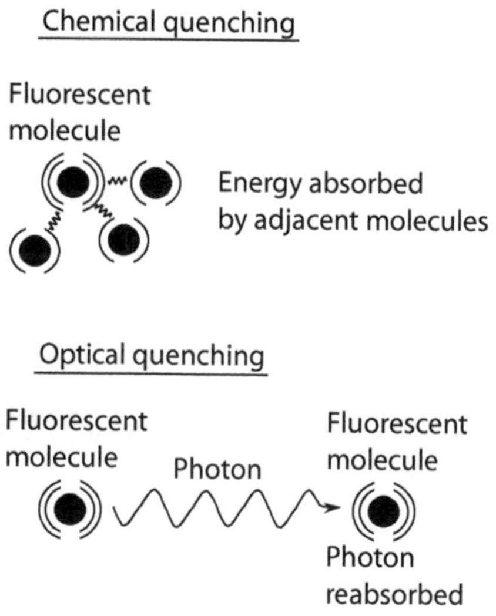

FIGURE 4.45 Examples of chemical and optical quenching.

Vials containing a mixture of distilled water and the cocktail are used as blanks to obtain an estimate of the background associated with the system. Tests with blanks should be performed regularly in routine quality assessment of the liquid scintillation system. The blanks must not contain any of the radionuclides of interest (e.g. ^{3}H, ^{14}C or ^{210}Pb), and should match the samples to be analysed in terms of the cocktail and the degree of chemical and optical quenching, described below (see also Fig. 4.45). The background pulse height distribution can then be subtracted from that obtained from the samples, to obtain the net counts originating from the β and α emitters in the sample. Thermal noise in the PMTs may also lead to spurious signals. Most PMTs exhibit thermal noise, which is sometimes referred to as dark current. The photocathode efficiency of a PMT depends on the fluorescence photon energy, and this must be taken into account when calculating the detection efficiency of the liquid scintillation system for a given cocktail–sample combination.

Quenching is the term applied to any process by which the fluorescence intensity of a substance decreases (Fig. 4.45). The main form of quenching is *chemical quenching*, which takes place when the constituents of the solution interact chemically with the fluor. This can make some of the molecular energy states unavailable, thereby decreasing the amount of fluorescence per unit energy deposited. *Colour quenching*, or *optical quenching*, on the other hand, arises from the absorption of the fluorescence photons from the fluor by atoms in the solvent or sample, for example, the absorption and re-emission of fluorescence at a different wavelength by the yellow pigmentation in urine samples. The amount of fluorescence can also be significantly decreased by the absorption of light by opaque substances in the sample. The main effect of quenching is a decrease in the total amount of fluorescence and luminescence recorded by the PMT, resulting in events being interpreted as having a lower energy, causing a shift of the α- or β-pulse height distribution towards lower energies. Complete absorption of certain luminescence energies can lead to a reduction in the number of counts in the pulse height distribution (see Fig. 4.46).

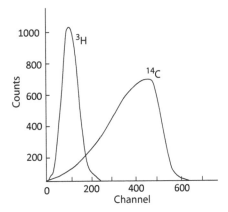

FIGURE 4.46 Typical pulse height distributions of β-particles measured with liquid scintillation. Note the effects of quenching that cuts off the contribution from low energy beta emission by the source.

The effects of quenching can be accounted for by using an internal standard. A known amount of a standard solution is added to the cocktail in the vial, and the measurement is repeated, assuming that the counting efficiency for the solution with the added radionuclides from the standard solution is the same as for the original sample in the vial. Corrections can also be made for quenching for a given type of cocktail by means of quench-indicating parameters that relate the number of pulses recorded to the number of true events (Eqn. 4.43):

$$DPM = CPM/QIP, \tag{4.43}$$

where DPM is the number of α- or β-particle depositions in the sample (this term is used in LSC measurements rather than activity concentration), CPM is the number of pulses recorded in the pulse height distribution, and QIP denotes the quench-indicating parameter used to correct the CPM to give the DPM of the radionuclide. The QIP for an individual LSC sample can be determined from its pulse height distribution using the ratio between the count rates in various energy windows (ROIs) (Fig. 4.47). These ratios are then related to the QIP using a function obtained from a calibration measurement using the pulse height distribution acquired from a number of α- or β-emitting sources with known activity using a similar cocktail.

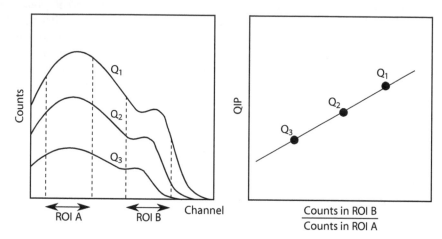

FIGURE 4.47 Left: Pulse height distribution obtained from a β-emitting sample with various levels of quenching in the cocktail. The ratio of the number of counts in ROIs A and B will depend on the QIP. Right: A plot of QIP vs. the ratio between the numbers of counts in the two ROIs.

The advantage of liquid scintillation systems is that they can be used to measure and differentiate between full energy events resulting from α-particles and lighter charged particles simultaneously. The phosphorescence produced by α- and β-particles differs in many organic scintillators; the α-particle-induced phosphorescence having a longer lifetime (Fig. 4.48: left). A pulse shape analyser (PSA) can be used to extract the temporal pattern of the phosphorescence recorded by the PMT, and this can then be used to distinguish between α- and β-particles. The PSA is connected to a pulse discriminator that sorts the recorded events into either an α- or a β-counting window.

In addition to the difference in the temporal pattern of the phosphorescence, there is also a difference in the total amount of luminescence (both fluorescence and phosphorescence) per unit energy deposited by α- and β-particles. For an α- and β-event depositing the same amount of energy, the event from the α-particle will be recorded at a significantly lower energy in the MCA than the event from the β-particle (Fig. 4.48: right).

When charged particles travel faster than the speed of light in a transparent dielectric medium with a refractive index greater than unity, Cherenkov radiation is produced (Mietelski et al. 2005). Cherenkov light has the same wavelength range as the luminescence generated by many organic scintillators, making it possible to use both methods for radionuclide detection in the same LSC-PMT system. For a solution with a given refractive index, n, the threshold energy [keV] of β-particles for the generation of Cherenkov radiation is given by Eqn. 4.44

$$E_{th} = m_0 \cdot c^2 \cdot \left(\frac{1}{\sqrt{1 - \frac{1}{n^2}}} - 1 \right), \tag{4.44}$$

where m_0 is electron rest mass and c is the speed of light. It follows that for an aqueous solution with a refractive index of $n = 1.333$, the threshold energy for Cherenkov radiation

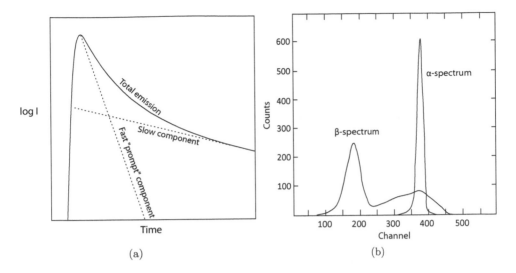

FIGURE 4.48 Luminescence decay curves consisting of a slow (phosphorescent) and fast (fluorescent) component. The proportions of these two components differ between α- and β-particle-induced fluorescence. Right: Pulse height distribution recorded from a sample containing the β emitters ^{210}Pb, ^{210}Bi, and the α emitter ^{210}Po.

is $E_{th} = 512$ keV. This threshold can be lowered by adding acids or alkaloids with a higher refractive index to the sample solution.

The use of Cherenkov counting in liquid scintillation counting is especially advantageous for isotopes emitting β-particles with energies above 500 keV, as sample processing is cheaper and more convenient. Instead of complex radiochemical preparation and dissolution in an organic scintillator, the sample is simply diluted in an acid with a suitable refractive index to adjust the threshold energy according to the anticipated spectrum of the β emitter. The organic waste produced is relatively easy to dispose of, and it exhibits no chemical quenching. However, the total counting efficiency is lower, and the spectral information on the β-particles emitted by the sample is lost. Cherenkov counting is used within radiation protection and emergency preparedness, for example, when determining the presence of high-energy β emitters such as ^{90}Y in fresh vegetation and food samples. Other β-counting devices that require less handling of solvents are gas flow counters, often operating in the same voltage range as GM counters (>500 V). GM counters using a gas flow of Ar and methane (CH_4) can be used to quantify the activity of low-energy β emitters, provided the sample contains only the radionuclide of interest.

4.6 GAMMA-RAY SPECTROMETRY

4.6.1 Interactions between Uncharged Particles and the Detector and Surrounding Materials

The longer range of x-rays and γ-rays in matter means that they are detected using different principles from those used to detect short-range charged particles. These detection devices can be designed with a greater variety of measuring geometries than for charged particles. Stationary gamma-ray spectrometry has a similar source–detector setup to that used for α spectrometry, including a shield surrounding a cavity where the sample containing the radionuclide is placed during pulse acquisition. However, the differentiation of a true event from perturbing events is more complex than for charged particle interactions. Incomplete absorption in the detector material, as well as interactions in the materials surrounding the detector, will introduce background that must be suppressed or corrected for before assessing the pulse height distribution.

Let us consider a detector material with an atomic composition and of a size that is appropriate for detecting photons in the energy range 30–3000 keV. When a mono-energetic γ-ray emitter, such as a ^{137}Cs source, is placed in front of the detector, the pulse height distribution should ideally reflect the flux of the γ-rays emitted by the source (Fig. 4.49). However, a number of events can occur in the γ-detector, as described below.

4.6.1.1 Complete Absorption of the Incident γ Energy

The incident γ photon passes through the wall and other potential barriers between the source and the sensitive volume of the detector, and then interacts with the atoms in the sensitive volume by one of two mechanisms. The first is the photoelectric effect, resulting in the release of a characteristic x-ray photon from one of the atoms in the detector (Fig. 4.50), which is in turn completely absorbed by successive Compton scattering or photoelectric energy absorption. The second mechanism is successive Compton scattering, in which all the electrons released and the scattered photons are completely absorbed within the sensitive volume of the detector. In both cases the charge created represents the energy of the incoming γ photon, albeit with some interference from detector noise (see Section 4.4.3). The MCA will then allocate the signal to a channel representing the energy of the full energy (or full absorption) peak of the incoming γ photon (Fig. 4.49).

4.6.1.2 Incomplete Absorption of the Incident γ Energy

In this case, the fate of the incident photon is the same as described above, with the exception that one of the secondary particles released is not completely absorbed in the detector. Secondary particles can cause so-called escape events in the detector through three mechanisms. Characteristic x-rays can be produced by the photoelectric effect, which can escape without being detected in the sensitive volume of the detector (see Table 4.6). If the incoming photon has an energy higher than twice the rest mass of an electron (> 1022 keV), either one or both annihilation photons generated by pair production can also escape the detector volume. Finally, the atomic electron released by Compton scattering may have sufficient kinetic energy to escape the detector volume. The third is especially probable if scattering occurs very close to the boundary of the sensitive volume of the detector. Each of the events described above will lead to a pulse in the spectrum with a lower energy than the incident γ photon. When annihilation photons or characteristic x-rays escape detection, distinct peaks may be visible in detector systems with high energy resolution.

If a ^{60}Co source is placed in front of the detector, two primary peaks will be seen in the pulse height distribution, with energies of 1173.2 and 1332.5 keV. However, it is likely that two distinct peaks will also be observed at $1173.2 - 511 = 662.2$ keV and $1332.5 - 511 = 821.5$ keV. If the source is replaced with a ^{137}Cs source, a characteristic x-ray escape peak will be observed at an energy of $661.6 - E(K_\alpha(Z))$. For a HPGe detector with $Z = 32$, the value of K_{α_1} is 9.9 keV, thus a small escape peak can sometimes be discerned at ~652 keV.

For events involving the escape of charged secondary particles, the pulses will be distributed more randomly since the energy lost will depend not only on the average energy of the Compton scattered electrons, but also on the location of the interaction point. For a large detector, where the size of the detector is much greater than the mean free path of the photons, there will be virtually no escape peaks.

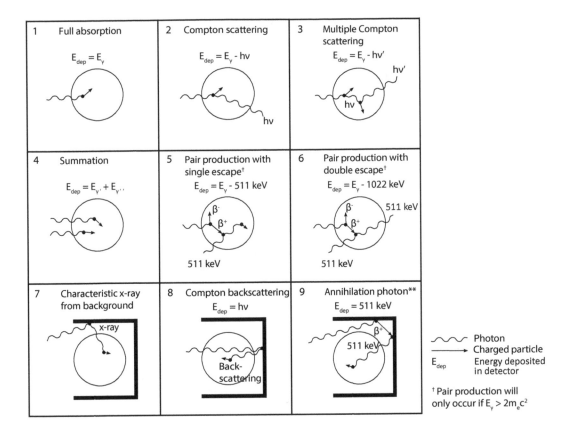

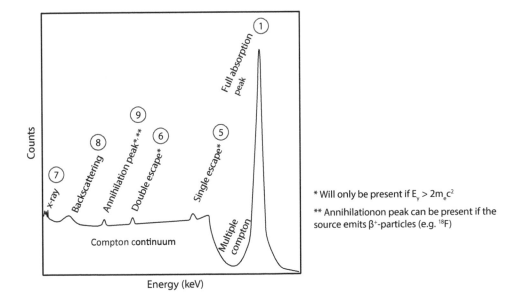

FIGURE 4.49 Top: Illustration of 9 possible events leading to a pulse being registered in a γ-ray spectrometer. Bottom: Typical pulse height distribution from a monoenergetic photon source.

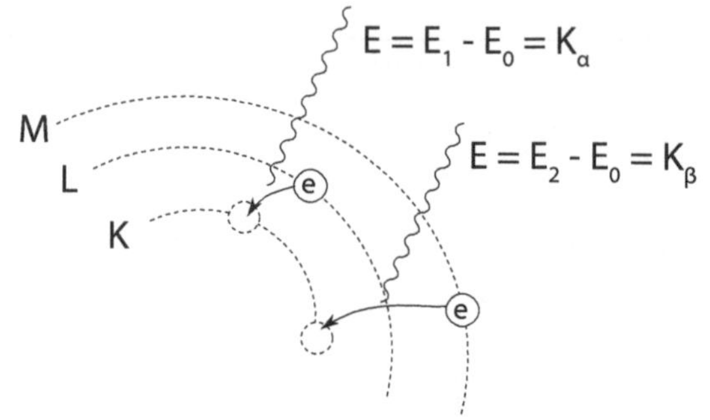

FIGURE 4.50 Schematic illustration of the electron orbitals of an atom and a characteristic x-ray defined by de-excitation to the inner K shell.

TABLE 4.6 The characteristic x-ray energies (K_α and K_β lines) [keV] of elements relevant in gamma-ray spectrometry as a function of the atomic composition of the detector material and the shielding.

Element	Z	K_{α_1}	K_{α_2}	K_{β_1}
Na	11	1.0409	1.040 98	1.0711
Si	14	1.739 98	1.739 38	1.835 94
Cu	29	8.047 78	8.027 83	8.905 29
Ge	32	9.886 42	9.855 32	10.9821
Sn	50	25.2713	25.0440	28.4860
I	53	28.6120	28.3172	32.2947
Pb	82	74.9694	72.8042	84.936
U	92	98.439	94.665	111.300

Source: Data from Bearden (1967).

4.6.1.3 Incomplete Energy Deposition of the Incident Photons

In this kind of event, the incident photon from the γ source undergoes Compton scattering in the detector, but the scattered photon does not interact with the detector volume. The scattered photon is said to have escaped from the sensitive volume of the detector, which means that only a fraction of the incident photon energy is deposited in the detector volume. Since the energy distribution of Compton scattered photons is continuous, a continuous background will arise. A measure of the proportion of events originating from Compton scattered photons is given by the so-called peak-to-Compton (PTC) ratio, Eqn. 4.45:

$$PTC = \frac{N_{peak}}{R_{Compton}} \tag{4.45}$$

where N_{peak} is the net number of counts in the full energy peak. The full energy peak is defined over an energy range covering at least 99% of the full energy peak (see next section on energy resolution). $R_{Compton}$ denotes the number of counts in the energy interval between the lower discriminator level (LLD), which is the lowest pulse height accepted by the MCA (see Section 4.4.7), and the energy of the so-called *Compton edge*. The Compton edge represents the maximum energy that can be deposited in a Compton event at a scattering angle of 180°. The PTC ratio of a spectrometric detector will depend on the effective atomic number of the detector, as well as the detector volume. For a given crystal size, NaI(Tl) detectors have a higher value of PTC than HPGe semiconductor detectors due to their higher effective atomic number.

4.6.1.4 Backscattering of Photons Interacting with Surrounding Materials

Gamma photons emitted by the source may be backscattered and hit the shielding or detector casing. If this photon undergoes photoelectric absorption, releasing a characteristic x-ray photon, the x-ray photon may reach the detector and be detected and interpreted as a true event with the energy of the characteristic x-ray. This is often the case when using lead shielding, whose high-energy characteristic K lines (74–85 keV; see Table 4.6) have a high probability of penetrating the detector casing and interacting with the detector. A similar event may occur if the emitted photon interacts in the surrounding material by Compton scattering, resulting in a distribution of scattered γ photons that can reach the detector volume. If the photon undergoes pair production, with the emission of annihilation photons, these secondary photons may also reach the detector volume and be detected as annihilation peaks at 511 keV, adding to the background.

4.6.1.5 Interaction of Background Radiation in the Detector

All the mechanisms described above will contribute to the background (Table 4.7). The most common background peak is that at 511 keV, but other background peaks arise from the short-lived daughters of the ^{238}U and ^{232}Th series, especially from radon and downwards in the decay chain. Moreover, primordial nuclides, such as ^{40}K, may also reach the detector volume, even with relatively thick shielding of 5–10 cm lead. Therefore, it is important to quantify the background accurately, and to keep a log of any fluctuations or systematic changes in the various background peaks over time. Radon daughters and cosmogenic ^{7}Be are known to vary in intensity on a daily basis due to the dynamics of the ambient air and the generation of ^{7}Be in the upper atmosphere due to solar protons.

4.6.1.6 Coincidences and Other Artefacts

The higher the γ-ray emission rate of the source, the higher the probability that two events will occur within the time window set by the rise time in the preamplifier (Fig. 4.27). In this case, the amplifier will interpret this as a signal that is either the sum of the incoming events or the sum of one event plus a fraction of the other event. The former will lead to a distinct peak and the latter to a continuous tail. There are two types of coincidence summing: *true coincidence* summing and *random coincidence* summing.

True coincidence summing (TCS) occurs when a γ source emits two or more photons in a very short time interval ($< 10^{-15}$ s). For example, ^{134}Cs decays with the highest probability to various short-lived energy states at more than 1 MeV above ground level. The half-lives of the intermediate states of ^{134}Cs are much too short to be resolved in the detector. This means that during the complete decay of a ^{134}Cs nuclide, the two main γ photons (604 and 795 keV), as well as other γ photons will have a high probability of reaching the detector simultaneously for short source-to-detector distances. If the energy of these γ photons is completely deposited in the detector, they will be recorded as a single event with an energy of, for example, $604 + 795 = 1400$ keV.

Other radionuclides often encountered in applied radiation protection with similar properties are ^{125}I, ^{133}Ba, ^{152}Eu, ^{154}Eu and ^{155}Eu. The probability of two photons reaching the detector volume simultaneously increases with the solid angle subtended by the source at

TABLE 4.7 Some x-ray and γ energies in the range 30–3000 keV often found in the background in stationary gamma-ray spectrometry systems. Count rates refer to a HPGe detector with a relative efficiency of 55% shielded by $\sim$8 cm lead. Data on branching ratios are taken from ENSDF Decay Data.

Photon energy (keV)	Radionuclide	Branching ratio[1]	Net count rate in full energy peak (10^{-3} cps)
45.54	^{210}Pb	0.042	2.66
63.29	^{234}Th	0.048	5.47
92.38	^{234}Th	0.028	11.9
143.8	^{235}U	0.110	1.93
185.71	^{235}U	0.572	8.25
186.2	^{226}Ra	0.036	*2)
238.63	^{212}Pb	0.433	5.93
295.17	^{214}Pb	0.193	2.51
338.32	^{228}Ac	0.1127	1.20
351.74	^{214}Pb	0.376	4.95
583.19	^{208}Tl	0.845	2.59
608.86	^{214}Bi	0.461	1.15
911.20	^{228}Ac	0.258	2.54
968.97	^{228}Ac	0.158	1.62
1001.03	^{234m}Pa	0.008 37	0.760
1021.87	^{208}Tl + ann.[3]	2.226	18.1
1120.3	^{214}Bi	0.151	1.49
1460.83	^{40}K	0.110	20.95
1764.5	^{214}Pb	0.154	3.22
2103.53	^{208}Tl SE[4]		1.20
2204.2	^{214}Bi	0.0508	1.18
2614.53	^{208}Tl	0.99	10.3

[1] Gamma yield per decay.
[2] Net count rate above refers to the combined peak 185.71 & 186.2 keV.
[3] ^{208}Tl (510.87) + Annihilation (511.0). Branching ratio is 0.226 + $\sim$2.
[4] Single escape.

the detector, assuming that the photons are emitted isotropically by the source. The net effect will be the loss of two pulses at their respective energies, and a pulse with an energy of the sum of the two γ photons, leading to a so-called coincidence peak. TCS is thus inherently associated with the type of radionuclide(s) in the γ source.

The second type of summation, random coincidence summing (RCS), is the direct result of the high intensity of incoming γ photons, which increases the probability of two events not being completely separated by the preamplifier. The net effect is the loss of an event of the true energy, and a coincidence pulse with an energy of $E_\gamma + k \cdot E_\gamma$, where k depends on the time between the two incoming photons and the rise time of the preamplifier.

In both TCS and RCS, pulse pile-up can be reduced by increasing the source–detector distance. However, it is important to note that for a given source–detector geometry TCS will *not depend on the activity* of the γ-ray source, which is a common misconception. The principal effects of TCS are illustrated in Figure 4.51.

The loss of pulses in the full energy peak, due to coincidence, can be compensated to

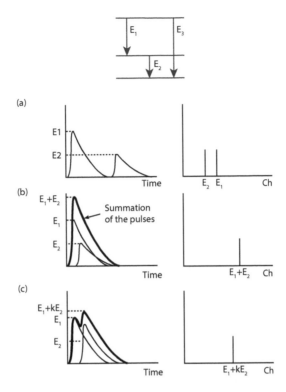

FIGURE 4.51 Top: Principal decay scheme of a γ-ray-emitting radionuclide with two photon energies, E_1 and E_2, respectively. (a) Preamplifier signal and the resulting pulse height distribution from two events, caused by the successive decay of the γ-ray-emitting radionuclide, being registered separately. (b) Preamplifier signal and the resulting pulse height distribution when the two events are both completely registered as one during the pulse rise time, causing a distinct peak located at E_3 (E_1+E_2). (c) Preamplifier signal and the resulting pulse height distribution when the second event is partially causing a count being registered in the interval between E_1 and E_1+E_2.

some degree by applying a correction factor. For the peak corresponding to energy E_1 in Figure 4.51, this factor is given by

$$C_1 = \frac{1}{1 - P_{12} \cdot \varepsilon_{T_2}}, \tag{4.46}$$

where P_{12} is the probability that a photon with energy E_2 is emitted simultaneously with a photon of energy E_1, and ε_{T_2} is the total efficiency (see Eqn. 4.52) for energy E_2. Similarly, the loss of pulses in the full energy peak corresponding to energy E_2 is given by

$$C_2 = \frac{1}{1 - P_{21} \cdot \varepsilon_{T_1}}, \tag{4.47}$$

where P_{21} is the probability that a photon with energy E_1 is emitted simultaneously with a photon of energy E_2, and ε_{T_1} is the total efficiency for energy E_1. The increased number of pulses in the full energy peak corresponding to the sum of the two photons, E_3, is compensated by the factor

$$C_3 = \frac{1}{1 + \frac{n_{\gamma_1}}{n_{\gamma_3}} \frac{\varepsilon_1 \cdot \varepsilon_2}{\varepsilon_3} P_{12}}, \tag{4.48}$$

where n_{γ_1} and n_{γ_3} are the intensities for γ-emission (number of photons per decay) in the transitions with energies E_1 and E_3, respectively, and ε denotes the full energy peak efficiency.

Of all these cases of events, it is mainly the one described in 4.6.1.1 that will lead to the full energy event that is of primary interest of the operator. Various configurations of the detector volume and shape, the shielding, amplifier settings and the ambient air masses must be considered in order to find ways to suppress undesired contributions from the other kinds of events described above. However, in uncharged particle spectrometry there will be significant contributions from all these kinds of events, both in the laboratory and the outdoor above-ground environments. The probability of each kind of event depends on the size of the γ-detector (Fig. 4.52); for example, the larger the detector, the higher the probability of a full energy absorption event, and the less likely are the remaining events such as single and double escape peaks, or Compton escapes. The PTC (as defined in Eqn. 4.45) will increase roughly as the square root of the detector volume. However, backscattering cannot be avoided as this is due to interactions in materials close to the detector volume.

Large detectors are, however, not always practical, as the collection of the charge (ionization detectors) or luminescence (scintillators) to obtain the full signal will take a longer time and thus affect the time resolution. Slower charge collection will also lead to more noise and poorer energy resolution of the full energy peaks.

For x-ray spectrometry in the energy range 10–100 keV, large detectors may only enhance the background contribution. The sensitive volume is normally configured as a cylinder with an axis length equivalent to 1–3 mean free paths of the relevant photon energies in the detector material. This will reduce the probability of events from higher energy photons that would otherwise contribute to the background. Si detectors are therefore preferable to HPGe detectors for low-photon-energy spectrometry, such as x-ray fluorescence in connection with, for example, the detection of heavy metals in internal organs.

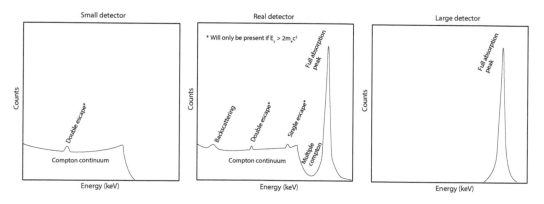

FIGURE 4.52 Illustration of the pulse height distribution in a very small and very large detector, compared to a "real" detector.

If two equally large crystals of atomic number Z_1 and Z_2 are placed in front of a monoenergetic γ source, the one with the lower effective atomic number will exhibit a higher proportion of events in which only part of the energy is deposited, mainly due to Compton scattering. This continuous Compton background is therefore more pronounced in semiconducting crystals such as Si and Ge than in NaI(Tl) detectors. The semiconductor CdZnTe, with a high effective atomic number of 50, has a higher peak-to-Compton ratio and could be an alternative to HPGe in some applications.

The continuous Compton background can be suppressed by using coincidence detector configurations, also called Compton-suppressed systems. This technique is used for both low-resolution γ-ray detectors such as NaI(Tl) detectors, and high-resolution Si or HPGe systems, and involves an additional detector positioned close to the main detector (Fig. 4.53). The detectors are coupled in such way that the pulses from the preamplifiers of both detectors are recorded simultaneously. When only part of the energy of a photon is absorbed by the main detector, the scattered photon or annihilation photons will have a high probability of interacting in the other detector, referred to as the shielding detector. Such an event, involving, for example, Compton scattering in the main detector, and Compton scattering or full energy photon absorption in the shielding detector, will result in two pulses being collected almost simultaneously by the preamplifiers of the two detectors. These pulses are then compared in a so-called time-to-amplitude-converter (TAC), which sends a gating signal to the MCA of the principal detector (D1), suppressing all events that arrive within a certain time delay in the TAC. The design of Compton-suppressed systems is described in more detail by Knoll (2010).

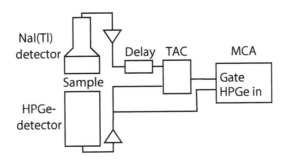

FIGURE 4.53 A schematic view of a slow anticoincidence system for Compton suppression, with an output of logic pulses created by the amplifiers of the two detectors. These pulses are compared in a time-to-amplitude-converter (TAC), which sends a gating signal to the MCA of the principal detector (D1), suppressing all events that arrive within a certain time delay in the TAC.

In most stationary γ-ray spectrometry systems, the detector is surrounded by a shielding material (Section 4.6.1.5). Intuitively it may appear to be best to use as much shielding around the detector as possible to suppress the background radiation. Using shielding materials with a high atomic number, such as lead ($Z = 82$), will suppress the natural background by several orders of magnitude. However, fast neutrons (kinetic energy >1 MeV) from cosmic radiation interact with lead by reactions such as Pb(n,n). This will lead to a build-up of thermal neutron flux, and if the thickness of the shielding is increased above ~15 cm, no gain will be obtained in the suppression of the γ-ray background. Characteristic x-rays from high-Z shielding material can perturb or interfere with the measurements when detecting radionuclides such as [109]Cd (88.04 keV), or characteristic x-rays in the energy range 10–100 keV from other materials. This may not be too serious a problem in gamma-ray spectrometry, but in x-ray fluorescence spectrometry, it is vital that such contributions are suppressed.

To reduce the fluence of characteristic x-rays from lead and neutron-induced γ-rays, another shielding material with a lower Z, e.g. copper ($Z = 29$), can be used on the inner walls of the shield. In many low-activity counting applications the design of the shielding is often an elaborate trade-off between cost and background contributions.

4.6.2   Energy Resolution of Gamma-Ray Spectrometers

The ability of a detector system to resolve the energy distribution of a photon radiation field is a central feature of gamma-ray spectrometry. In contrast to charged particle spectrometry, where radiochemical procedures must be used to extract the radionuclides of interest, gamma-ray spectrometry is usually performed on samples without any major chemical processing. The greater the number of anticipated γ-ray-emitting radionuclides present in the source, the higher the demand on the energy resolution (in terms of FWHM) of the γ-ray detector.

For γ-sensitive scintillators, the FWHM has traditionally been expressed as a percentage of the full absorption energy, whereas for semiconductors and ionization chambers, the FWHM is usually expressed in terms of energy [keV]. The energy resolution is essentially determined by the quantum uncertainty in the number of charge carriers generated per unit energy deposited by a radiation particle per event (1–3 eV for Si and HPGe detectors, 30 eV for gaseous ionization detectors, and 100–300 eV or above for inorganic scintillators such as NaI(Tl); Tables 4.2 and 4.3). However, other factors also determine the FWHM. Statistical fluctuations in the number of charge carriers resulting from a detected event will influence the FWHM. This component can be modified by the electric field across the diode in semiconductor detectors, or the voltage across the PMTs in scintillators.

Electronic noise will significantly affect the FWHM in semiconductor detectors. Serial noise (N_s in Eqn. 4.35) arising from fluctuations in the capacitance over the diode, decreases with increasing pulse collection time, while parallel noise (N_p in Eqn. 4.35) arising from thermal noise in the capacitor and preamplifier, increases with increasing shaping time. It is important to note that electronic noise may also arise from microphonics and mechanical vibration. This is the cause of the long delay in the breakthrough of electrically cooled high-resolution systems, but the energy resolution of modern systems is very close to that of liquid-nitrogen-cooled systems (e.g. 2.2 keV vs. 1.81 keV at 1332.5 keV for a HPGe crystal with a relative efficiency of 50%). Mechanical vibration and microphonics also contribute to noise when high-resolution systems are used in vehicles, leading to significantly poorer resolution. In PMT-based spectrometers, noise arises mainly from the stray currents generated in the PMT by the thermal excitation of electrons in the dynode material. The quantum noise, on the other hand, arises from the stochastic nature of the number of electrons amplified per dynode, which will increase with increasing high voltage. Therefore, the value of V_{Bias} across the PMT must be set so as to minimize the sum of the stray current and the quantum noise contributions.

Electric field gradients in the sensitive volume of the spectrometer give rise to a positional dependence of the charge collection time at the electrodes. This in turn leads to fluctuations in the number of charges collected per unit energy deposited in the detector, leading to broadening of the range of pulse height from events with the same energy deposition. Electronic drift in the amplifier during pulse acquisition caused, for example, by temperature changes in the PMT, will also result in variation in the number of charge carriers being produced per unit energy deposited, over time, in turn causing broadening of the full energy peak.

Variation in the interaction history of the detected event leading to a full energy pulse may also contribute to broadening of the FWHM. As an example, we can consider two photons that deposit their full energies at two separate times. The first photon may have been subject to several Compton scattering events, whereas the second photon may have undergone a photoelectric interaction in which all its energy was deposited. This phenomenon is more pronounced at photon energies below 100 keV.

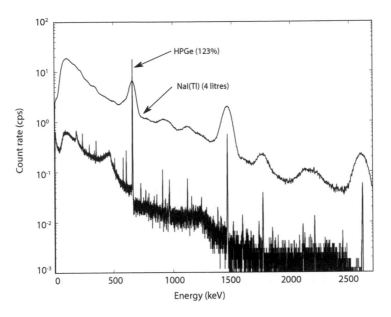

FIGURE 4.54 Pulse height distribution recorded by two different gamma-ray spectrometry systems at the same distance from a ^{137}Cs source.

The energy resolution of a scintillation or semiconductor γ detector can be described as a function of photon energy for a given bias voltage by the following relation:

$$FWHM = \sqrt{(a + b \cdot E_\gamma)}, \qquad (4.49)$$

where a and b are arbitrary constants characteristic of the individual detector. The square root dependency arises from the quantum uncertainty in the number of charge carriers produced per event (Eqn. 4.35). An event in which an energy of 1000 keV is deposited will give rise to twice the number of charge carriers as one in which 500 keV is deposited. However, the relative stochastic uncertainty (which for charged particles follows the Poisson law; Eqn. 4.9) in the 1000-keV event will only be improved by a factor of $\sqrt{2}$ compared with the 500-keV event.

The number of charge carriers is also dependent on the relationship between the rate at which charge is collected and the pulse acquisition time. For an ionization chamber with a given shaping time, the energy resolution will be dependent on the electric field strength over the sensitive volume, as this determines the velocity of the charge carriers created along the path of deposition. This effect is exemplified in Figure 4.40, where the energy resolution of a Si detector is plotted against the voltage applied across it. Typical values of FWHM for HPGe crystals are ~1.7 keV for small crystals (<50% relative efficiency), and up to about 2.2 keV for large crystals (>100% relative efficiency).

The resolution of scintillators is defined by two factors. The first is the *intrinsic* resolution, before amplification by the PMT, and the second is the resolution following amplification by the dynodes. The intrinsic resolution of NaI(Tl) crystals at 662 keV is about 6.6%

(e.g. Menge et al. 2007). The pulses are then amplified, and the higher the bias voltage over the PMT, the lower the relative stochastic variation in dynode amplification per dynode. However, increasing the bias over the PMT reduces the linearity between the energy deposited and the amplitude of the output pulse of the PMT. The optimum bias voltage must thus be determined for each combination of scintillator and PMT, and is usually set such that the FWHM is minimized for a given photon energy.

4.6.3   Time Resolution of Gamma-Ray Spectrometers

The ability of a spectrometer to resolve incoming events and correctly represent them in an MCA is referred to as the *time resolution* of the system, and is composed of three main stages.

1. Electromagnetic and nuclear interactions between the incoming photon and the detector material, which typically occur during a period of 1 to 1000 ns).

2. The response generated by the detector, either in terms of secondary charged particles that migrate towards the electrode, or luminescence that is emitted during the decay of the singlet states (see Section 4.2.4), which have typical time constants of 10 ns to 10 μs.

3. Shaping of the pulse and subsequent processing by the ADC, which is the most time-demanding step, typically requiring 10 to 20 μs.

For ionization detectors, such as bulk HPGe crystals, the time resolution (and indirectly also the energy resolution) can be improved if as many charge carriers as possible are collected as rapidly as possible. This means that the anode must be geometrically configured such that, on average, all the electrons created by an impinging photon will experience the same electric field over the whole of the sensitive volume. Therefore, the coaxial geometry, where the anode is the central electrode, is the most common geometry in high-resolution gamma-ray spectrometry at energies above 100 keV.

The drawback of this configuration is that the anode on the outer surface of the crystal is physically much thicker (500–600 μm) than the p-doped cathode ($\sim$ 0.3 μm), (as illustrated in Fig. 4.7). This configuration is referred to as a *p-type* detector. This means that the dead layer will absorb photons with lower energies (especially below 60 keV), preventing them from reaching the sensitive volume of the detector. Detectors where the electrodes are reversed, i.e. the thick n-doped anode is in the middle of the detector and the thin p-doped cathode is on the outside of the crystal, are referred to as *n-type* detectors. Since the outer dead layer is thinner, n-type detectors are more sensitive to low-energy photons. However, the charge collection time will be somewhat longer, and depends on the location of the interaction. Hence, the energy resolution of n-type detectors is poorer than that of p-type detectors. n-type HPGe detectors have no benefits over p-type detectors for spectrometry above about 100 keV.

Crystals of CdZnTe, which have a high effective atomic number, have significantly lower charge mobility than HPGe crystals (Table 4.2). Although CdZnTe detectors can be operated at room temperature, the poor time resolution of larger crystals has essentially confined their application to environments with high-intensity gamma radiation where small crystals can be used.

Scintillators with luminescence half-lives shorter than 10 ns are considered fast (such as the organic scintillators used in LSC), while others with longer half-lives are considered to be slow. Generally, organic materials, especially plastic scintillators doped with certain fluorescent materials, can be made extremely fast with afterglow half-lives of a few nanoseconds. These detectors are therefore widely used for rapid monitoring of tissue equivalent doses and for rapid detection of elevated radiation, for example, at international boarders to detect radioactive material. However, their low luminescence yield per unit of absorbed radiation energy results in very poor inherent energy resolution. On the other hand, the short luminescence afterglow, in combination with the possibility of making very large bulk crystals of plastic (several hundreds of dm^3), make them useful in some particular applications of gamma-ray spectrometry. The combination of large volume and high time resolution leads to high counting efficiency, which is useful for in vivo whole-body counting of γ-ray-emitting nuclides such as ^{40}K.

Among the slow inorganic scintillators are the commonly used NaI(Tl) detectors, which until recently excelled among bulk crystal scintillators in terms of energy resolution. NaI(Tl) has an afterglow time constant of 230 ns, and therefore has limited time resolution when large crystals (4–50 litres) are used in mobile gamma-ray spectrometry, or when the detector is used in high-intensity gamma radiation fields. A new scintillator material, lanthanum bromide doped with cerium, LaBr$_3$ (Ce), was introduced onto the market at the beginning of the millennium. This material has a much shorter afterglow, 26 ns, which is comparable to that of fast scintillators. LaBr$_3$(Cr) also has a significantly higher luminescence yield per unit energy deposited, leading to a factor of 2 lower FWHM than for NaI(Tl). Figure 4.55 shows a plot of the count rate vs. fluence rate for NaI(Tl) and LaBr$_3$(Ce)[1] detectors.

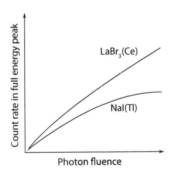

FIGURE 4.55 Schematic view of the full energy count rate from a γ-ray-emitting source recorded by two scintillator systems using the same signal processor (including ADC).

4.6.4 Detection Efficiency and Detection Limits of Gamma-Ray Spectrometers

The energy dependence of a spectrometer refers to the variation in the detection efficiency of gamma quanta with their energy. Defining the efficiency as the number of correctly recorded counts per incident photon with a given energy means that this quantity will depend on three principal factors: (i) the size of the sensitive volume of the detector, (ii) the atomic composition of the detector material (Z_{eff}), and (iii) the barrier thickness between the sensitive volume and the emitted fluence (as illustrated in Fig. 4.24).

---

[1]The drawback of LaBr$_3$(Ce) compared with NaI(Tl) is the inherent activity of ^{138}La ($T_{1/2} = 102 \cdot 10^9$ years), whose full energy peaks at 788 keV and 1435 keV interfere with that of ^{40}K at 1460.8 keV, for example. It is thus unlikely that LaBr$_3$(Ce) will replace NaI(Tl) for geological or environmental surveys of natural background radiation.

The relationship between the atomic composition and the capacity to detect a full energy event has been described in previous sections. In this section we introduce a number of definitions of counting efficiency used within gamma-ray spectrometry. In high-resolution gamma-ray spectrometry there are essentially four parameters that define the capacity of a detector to translate the incoming flux of photons from a γ-ray-emitting source of specific source activity, A_{source} [Bq], into detectable pulses in the MCA (Gilmore and Hemingway 1995).

1. The *intrinsic efficiency*, ε_i, defined in Eqn. 4.30, is the net number of counts in a narrow energy interval around the full energy peak of a *normally* incident beam of mono-energetic gamma photons, divided by the flux of normally incident photons of the same energy. This parameter can be used to compare the performance of various crystals as it is independent of the measurement geometry. However, other efficiency parameters are more useful when comparing different crystal sizes and source–detector geometries. In practice, the intrinsic efficiency is determined using a point source positioned along the central axis of the detector at a distance, d, which is at least 10 times the diameter of the front surface of the detector. The point source may contain several γ-ray-emitting radionuclides, but should not contain gamma-ray emitters that exhibit TCS, such as ^{133}Cs or ^{133}Ba. All γ-ray energies emitted by the radionuclides in the source should be separated by at least 3 times the FWHM of the detector.

2. The *relative efficiency* of HPGe crystals, ε_{rel}, is defined as the net number of counts in the full energy peak at 1332.5 keV acquired using a ^{60}Co source positioned 25 cm away from the detector surface, divided by the net number of counts in the corresponding full energy peak in a 7.54 cm × 7.54 cm (diameter) NaI(Tl) detector using the same source (Eqn. 4.50). This parameter has become a standard measure of the efficiency of HPGe crystals. This is useful for comparing the performance of crystals of different sizes, but can sometimes be misleading, as HPGe crystals with relative efficiencies up to 250% are now available. The fact that using an efficiency parameter that can be more than 100% can be confusing since this quantity refers to a reference detector of a very limited size. Furthermore, the relative efficiency is *not* proportional to the counting efficiency when the source is very close to the detector. For example, the counting efficiency of a HPGe crystal with a relative efficiency of 100% when detecting ^{60}Co photons in a 60-mL source close to the crystal is *less than* twice that of a HPGe crystal with 50% relative efficiency. Also, when comparing the performance of a HPGe crystal with that of a NaI(Tl) crystal, e.g. for mobile gamma radiation surveys, this parameter may be misleading, as a 250% HPGe detector only has about 3% of the counting efficiency of a 4-litre NaI(Tl).

$$\varepsilon_{rel} = \frac{N_{\gamma,1332.5}/A_{source}}{N_{\gamma,NaI,1332.5}/A_{source}} \qquad (4.50)$$

3. The *absolute efficiency*, ε_{abs}, is the most commonly used parameter for stationary gamma-ray spectrometry, and relates the number of net counts in a given full energy peak to the emission rate of gamma photons of the same energy from a sample using a certain geometry (Eqn. 4.51). For a given crystal and source–detector geometry, this parameter is dependent on the photon energy. The parameter ε_{abs} is experimentally determined using a calibration source of a specified volume, density and atomic composition. As in the case of the point calibration source used for the determination of ε_i, the calibration source can contain several gamma-ray-emitting radionuclides,

provided that the energies of the emitted photons are separated by at least 3 times the FWHM.

$$\varepsilon_{abs}\left(E_{\gamma}\right) = \frac{\left(\frac{N_{\gamma}}{t_m}\right)}{\left(A_{cal} \cdot n_{\gamma}\right)} \qquad (4.51)$$

4. The *total efficiency* is a measure of the total response over the whole energy spectrum and can be used to calculate correction factors for the effects of TCS. It is determined in the same way as ϵ_{abs}, but in this case the calibration source must contain only one, preferably a mono-energetic, γ-ray emitter. The pulses arising from both completely deposited and incompletely or coincident events (Section 4.6.1) are related to the emission rate of a mono-energetic gamma source. The number of pulses attributed to the source over the whole energy window of the gamma-ray spectrometer, e.g. 30 to 3000 keV, is obtained by dividing the number of counts acquired with the source present, G_{tot}, by the measuring time, t_m, and then subtracting the background counts, G_{bkg}, recorded during an acquisition time of t_{bkg} (Eqn. 4.52).

$$\varepsilon_{tot}\left(E_{\gamma}\right) = \frac{\left(\left(\frac{G_{tot}}{t_m}\right) - \left(\frac{G_{bkg}}{t_{bkg}}\right)\right)}{\left(A_{cal} \cdot n_{\gamma}\right)} \qquad (4.52)$$

4.6.5   Calibration and Quantitative Assessment in Gamma-Ray Spectrometry

To enable accurate quantification of the radionuclides in a γ-ray-emitting source using a gamma-ray spectrometry system, it is necessary to determine the optimal detector amplification settings and to perform calibration measurements. The gain and fine gain of the amplifier must be set so that the MCA can discriminate photons in the energy range of interest (30–3000 keV). For high-resolution systems it is often necessary to select a shaping time and adjust the pole-zero (R_{pz} in Fig. 4.28) before performing measurements. Energy calibration can be performed by performing measurements on a well-known γ-ray-emitting source. For regular monitoring of the energy calibration, the accuracy of the source activity, A_{source}, is of little concern. The spectrometer software will usually identify the centroid of the full energy peak in terms of channel number. The position of the centroid, ch_0, can be calculated using Eqn. 4.53.

$$ch_0 = \frac{\sum_i^{i+n} ch_i \cdot N_i}{\sum_i^{i+n} N_i} \qquad (4.53)$$

For many high-resolution systems, a second-order polynomial function of the channel number, ch, is sufficient for calibration with an accuracy of 0.1% (Eqn. 4.54):

$$E_{\gamma} = a + b \cdot ch + c \cdot ch^2. \qquad (4.54)$$

Modern MCAs are usually highly linear, and the quadratic component is often negligible compared with the linear component (i.e. b ≫ c). For energy calibration, about 10 000 pulses (sometimes less) are necessary in each full energy peak to obtain an acceptable level of stochastic uncertainty. However, in some γ-ray spectrometric systems the software performs energy calibration simultaneously with the FWHM calibration, which means that not only the centroid, but also the width of the peak (in terms of the FWHM) is required. This quantity is calculated by a somewhat complex formula using the counts in several channel intervals (Gilmore and Hemingway 1995). From the propagation of uncertainty it is then clear that the FWHM is much more sensitive to stochastic noise than is the centroid. Therefore, at least 50 000 counts are recommended in each full energy peak when the FWHM is also to be calibrated.

Electronic drift between calibration and measurements may lead to serious inaccuracies when performing quantitative assessments of the acquired pulse height distribution. This is especially the case for measurements lasting longer than 1 week, where a slow drift may shift the amplification so that the peaks are either smeared, or shifted towards higher or lower energies. Regular calibration is recommended to detect and correct for such drifts.

Modern stationary high-resolution gamma-ray spectrometry systems need calibrating less often than previously, due to the improved stability of the amplifier and other electronics. Quality assurance must include at least a quarterly check of the linear amplification and energy calibration, preferably more frequently. The stability of newly purchased low-resolution gamma-ray spectrometry systems using PMT-based scintillators is often adequate for use in *in vivo* counting, but regular checks must still be performed if not daily, then on a weekly basis, to check for drifts in the amplification. For mobile systems intended for outdoor environmental surveys, such as *in situ* or vehicle-borne gamma-ray spectrometry, it is important that the accuracy of the calibration is checked by visual inspection in real time, or that an alert algorithm is implemented which detects electronic drift, for example, by monitoring the centroid of an ever-present peak such as that at 1460.78 keV from ^{40}K.

In both low- and high-resolution gamma-ray spectrometry it is sometimes necessary to adjust the amplification of the detector to restore the relationship obtained during energy calibration (Eqn. 4.54). However, it is preferable to maintain a constant relationship between E_g and ch_i over a long period of time, rather than recalibrating the energy response. Small adjustments to the linear gain are often sufficient to correct for any shifts that may occur.

The absolute efficiency, ε_{abs}, as well as, in many cases, the intrinsic efficiency, ϵ_i, are determined using a standard sample with a reference source whose content of γ-ray-emitting radionuclides is determined within at least $\pm 1.5 - 2\%$ *standard uncertainty* (here defined as one standard deviation, 1σ of the estimated value, divided by the estimated value). Standards with activity contents traceable to so-called *secondary standard laboratories*, such as the National Institute of Standards and Technology (NIST), are commercially available from a number of manufacturers. Observed values of ε_{abs} are plotted as a function of photon energy, to which a mathematical function is fitted. An exponential function, using a high-order polynomial of E_γ ($n > 7$) as an exponent, is the one most commonly used for high-resolution gamma-ray spectrometry systems (Eqn. 4.55).

$$\varepsilon_{abs}\left(E_\gamma\right) = a_0 \cdot exp\left(a_1 \cdot E_\gamma + a_2 \cdot E_\gamma^0 + a_3 \cdot E_\gamma^{-1} + a_4 \cdot E_\gamma^{-2} + a_5 \cdot E_\gamma^{-3}...\right) \qquad (4.55)$$

The uncertainty in the standard used for the calibration must be combined with the uncertainty in the curve fit of ε_{abs} to obtain an estimate of the relative uncertainty in ε_{abs}, $\Delta\varepsilon_{abs}$. Carefully performed calibrations will at best give a $\Delta\varepsilon_{abs}$ of 2–3% relative uncertainty. Having established this mathematical expression for ε_{abs}, it is then possible to carry out quantitative γ measurements. Based on a pulse height distribution acquired during t_m the activity at a reference time, t_{ref}, of a certain radionuclide in a source positioned in the given source-to-detector geometry can be estimated (Eqn. 4.56):

$$A\left(t_{ref}\right) = \frac{\left(\frac{\left(\frac{N_\gamma - b \cdot t_m}{t_m}\right)}{\varepsilon_{abs}}\right) \cdot \frac{1}{n_\gamma} \cdot T\left(t_{ref}, t_m, t_{acq}\right)}{m_{sample}}, \qquad (4.56)$$

where b is the background pulse count rate, given by

$$b = \frac{B_\gamma}{t_{bkg}}, \tag{4.57}$$

where B_γ is the number of background counts in the same energy window obtained by measuring the background using a blank sample for a time t_{bkg}. The blank sample can consist of the beaker normally used for measurements, filled with distilled water or sugar, which have a low natural γ activity. The background count rate in the energy window of the full energy peak of interest is often referred to as the *peaked background*, and is assumed to be relatively constant over time. However, this assumption must be confirmed regularly, as part of the overall quality assurance of the gamma-ray spectrometry setup.

All samples intended for quantitative gamma-ray spectrometry must be associated with a specific reference time, t_{ref}, in order to determine the activity. This is especially important when quantifying short-lived γ-ray emitters. The operator often defines this reference time based on prior knowledge of the sample to be measured, e.g. the sample collection date. The time function, T, corrects for the physical decay of the radionuclide between the reference time and the date of pulse height distribution acquisition:

$$T\left(t_{ref}, t_m, t_{acq}\right) = e^{\left(\frac{\left(t_{acq} + \frac{t_m}{2}\right) - t_{ref}}{T_{1/2}}\right)}, \tag{4.58}$$

where t_m is the measuring time. To compensate for the physical decay during acquisition, the time function can be modified by adding half of the measuring time to t_{acq} when subtracting the reference time.

If the uncertainties in t_m, t_{acq}, t_{ref} and t_{bkg} are set to zero, the uncertainty in N_{peak} can be expressed as

$$N_{peak} = N_\gamma - \left(\frac{B_\gamma}{t_{bkg}}\right) \cdot t_m \rightarrow \Delta N_{peak} = \sqrt{\Delta N_\gamma^2 + \Delta B_\gamma^2 \cdot \left(\frac{t_m}{t_{bkg}}\right)^2} = \sqrt{N_\gamma + B_\gamma \cdot \left(\frac{t_m}{t_{bkg}}\right)^2}. \tag{4.59}$$

The uncertainty in A at $t = t_{ref}$ can then be expressed as

$$\Delta A\left(t_{ref}\right) = \sqrt{\left(\frac{\Delta N_{peak}}{N_{peak}}\right)^2 + \left(\frac{\Delta \varepsilon_{abs}}{\varepsilon_{abs}}\right)^2 + \left(\frac{\Delta m_{sample}}{m_{sample}}\right)^2}. \tag{4.60}$$

In this equation ε_{abs} is assumed to have a relative uncertainty determined by the relative uncertainty of the curve fit. The uncertainty in the standard calibration source, typically 1.5–3% (1 SD), must also be considered. This can be done by considering a correction factor, k_{cal}, that has the value of unity, but a standard uncertainty, $\Delta k_{cal} = 1.5 - 3\%$. In some cases, it is necessary to introduce additional correction factors, to take into account factors such as differences in attenuation properties between the measured sample and the standard calibration source, k_{att}, or a correction factor for TCS of certain full energy peaks, k_{TCS}, etc. Each of these correction factors is associated with a standard uncertainty, and these must be propagated in quadrature (Eqn. 4.61).

$$\Delta A\left(t_{ref}\right) = \sqrt{\left(\frac{\Delta N_{peak}}{N_{peak}}\right)^2 + \left(\frac{\Delta \varepsilon_{abs}}{\varepsilon_{abs}}\right)^2 + \left(\frac{\Delta k_{cal}}{k_{cal}}\right)^2 + \left(\frac{\Delta k_{att}}{k_{att}}\right)^2 + \left(\frac{\Delta k_{TCS}}{k_{TCS}}\right)^2 + \ldots + \left(\frac{\Delta m_{sample}}{m_{sample}}\right)^2} \tag{4.61}$$

If all the corrections apart from the uncertainty in the calibration standard used to obtain the efficiency curve for ε_{abs}, are disregarded, the *total relative uncertainty* in the activity of a sample, $A(t_{ref})$, determined from a pulse height distribution containing >1000 net counts in the full energy peak (giving a relative stochastic uncertainty of about 3%), will be at least 5%, assuming both $\Delta\varepsilon_{abs}$ and Δk_{cal} to be 2–3%.

4.6.6 Analysis of γ Pulse Height Distributions

A number of functions and parameters should be included in the software for γ-ray spectrometric evaluation. In addition to recording and storing the pulse height distribution, the reference time, t_{ref}, sample weight, m_{sample}, energy calibration, efficiency calibration (e.g. ε_{abs} for a given source–detector geometry) and a nuclide library, containing data on half-lives and branching ratios of the radionuclides of interest should also be included.

The software should also include algorithms for separating and localizing peaks and their respective centroids, and for determining the background and net counts in these peaks. An algorithm for calculating the FWHM of the localized peaks must also be available. These algorithms are usually provided with the basic version of the spectrometry software. However, it is advisable to consult the manufacturers manual, which describes how some of these parameters are estimated by the algorithms, as ambiguous results may arise, especially in parts of the pulse height distribution with many full energy peaks. Figure 4.56 gives a schematic view of how the background counts and net counts are calculated from a pulse height distribution of a full energy peak.

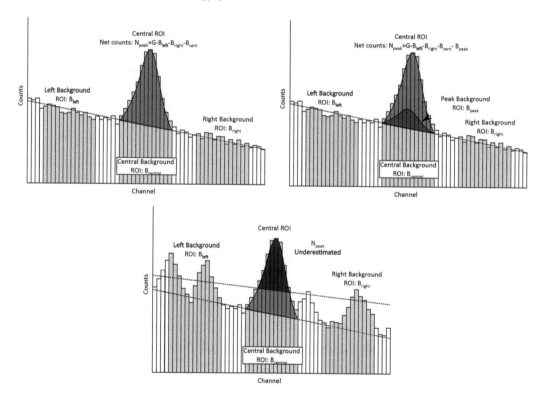

FIGURE 4.56 Top left: Background correction of a full energy peak using ROIs on each side of the peak. Top right: Corresponding background correction in the presence of a peaked background, B_{peak}. Bottom: Example of how the background correction can be disturbed by nearby peaks.

The software should also have routines for performing energy and efficiency calibration with options for various mathematical fits (e.g. Eqns. 4.54 and 4.55) and storage of energy calibration parameters traceable to the date of creation. Figure 4.57 shows examples of the graphical representation of energy and efficiency calibration.

Sophisticated routines are available for automated calibration of gamma-ray spectrom-

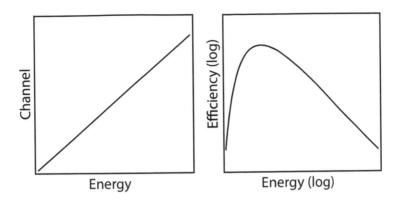

FIGURE 4.57 Graphical representation of the energy calibration curve (left) and the absolute efficiency calibration curve (right). Data from the calibration curves can be entered in the gamma-ray spectrometry software.

etry systems and the analysis of the recorded pulse height distribution. These may include algorithms for the deconvolution of superimposed photon peaks, nuclide libraries that can be used for manual or automated analysis, and a nuclide library editor for customizing and updating nuclide data. Deconvolution procedures for the separation of full energy γ-ray peaks may be complicated when several peaks overlap. Libraries containing γ-ray data on radionuclides that are likely to be found in the samples are often used. However, the more comprehensive the library, the greater the risk of erroneous identification. The software should therefore provide a means of customizing dedicated nuclide libraries for various measuring applications to ensure high accuracy in identifying radionuclides, while identifying as many nuclides as possible (Fig. 4.58).

Finally, gamma spectrometric analysis software should also include (i) automated analysis, including identification and quantification of full energy peaks attributable to suspected γ-ray emitters, (ii) a means of generating reports on counting statistics in manually defined ROIs, (iii) a feature for providing exhaustive reports on the radionuclide content of γ-ray emitters in the sample, including a list where unidentified γ-ray peaks are compared and matched with an exhaustive master library (also referred to as a suspect library), and (iv) various algorithms for corrections such as density corrections. Figure 4.59 gives an overview of the main steps in the automated assessment of a γ pulse height distribution.

4.6.7 Detection Limits

It is important to estimate the uncertainty in the radiometric quantities measured if they are to be used as the basis for decision making in emergency preparedness. In spectrometry and in other radiometric measurements, each value must be accompanied by the uncertainty or confidence level associated with the value. The relative uncertainties in measurements made in low-level radiation condition close to the background will be high, and there is a risk that erroneous conclusions can be drawn based on the observations.

Measurement errors can be divided in two main categories, depending on how the data are presented. As an example, we consider the hypothesis, H_0, where there is no elevated level of radioactivity in the sample. However, the measured signal at a given instance appears to be high compared with the expected background, due to statistical fluctuations, leading to the conclusion that the level of radioactivity is elevated. This can be categorized as a *type*

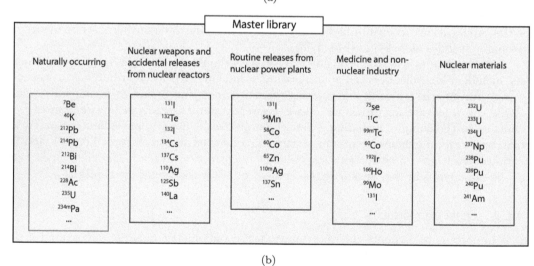

(a)

(b)

FIGURE 4.58 Top: Excerpt of radionuclide library as shown in a γ-ray spectrometry software. Bottom: Principle of using situation-based customized nuclide libraries.

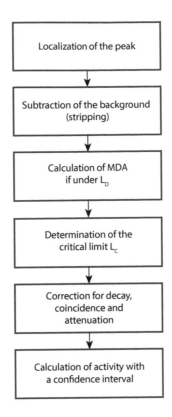

FIGURE 4.59 The various steps involved in gamma-ray spectrometry of a pulse height distribution with the aim of identifying and quantifying the γ-ray-emitting nuclides present.

I error: i.e. H_0 is rejected. Indirectly, this means that we have the hypothesis H_1, where there is indeed an elevated level of radioactivity in the sample. In simple terms, a type I error means that an effect has been detected which is not present, and the finding given is a so-called *false positive* result.

If, on the other hand, the value measured appears to be too low in relation to the background (also due to stochastic fluctuations in the signal), the opposite can occur, leading to the conclusion that the activity is not elevated. This is classified as a *type II error*, i.e. the rejection of a true positive result, and thus failure to identify an effect that is actually present, i.e., a so-called *false negative* observation. Figure 4.60 illustrates the relationship between the hypothesis (H_0 and H_1) of a model and the outcome of a measurement.

Let us now assume that a background spectrometric measurement is carried out yielding the number of pulses, B_1, during a measuring time, t_m in a well-defined energy interval (ROI). The following examples refer to the number of counts in that ROI. When the measurement is repeated, using the same time for convenience, the number of pulses recorded is B_2. If the measurement is repeated n times, a distribution of the number of background pulses will be obtained, which can be visualized in a histogram (an idealized histogram is shown in the upper part of Fig. 4.61). From this distribution, an expectation value, $< B >$, and a standard deviation, s_B, can be extracted. In most cases, the average background is identical to the expectation value, and the standard uncertainty of the mean is given by $s_{<B>} = s_B \sqrt{n}$.

We now assume that we are conducting a measurement on a blank sample (i.e. a sample

		State	
		H_0	H_1
Decision	H_0	Correct acceptance	Type II error (β)
	H_1	Type I error (α)	Correct rejection

FIGURE 4.60   Illustration of type I and type II errors.

with no added radioactivity). The gross number of counts is denoted G_1. The number of net counts is thus $N_1 = G_1 - <B>$. The measurement on the blank sample is repeated m times, and a histogram is plotted where the average value of the net counts, denoted $<N>$, approaches zero as m increases. The standard deviation of the mean number of net counts, $s_{<N>}$, will be somewhat greater than $s_{<B>}$, but the values will converge as m approaches infinity (Fig. 4.61).

A very important parameter within charged particle and gamma-ray spectrometry can be extracted from this distribution. A limit can be defined, above which it can be claimed that the level of radiation detected is significantly above the expected background. The *critical limit*, L_C, is commonly defined as the level above which there is only a 5% probability that the recorded pulses originate from the background only. Assuming that the Poisson distribution can be applied in spectrometry, this 95% confidence level represents $L_C = 1.645 \cdot s_N$. In the case of a well-established background (large m), this quantity is equal to $s_B \cdot \sqrt{1 + 1/m} = s_B$, and hence, in most cases, $L_C = 1.645 \cdot s_B$. This means that there is, at most, a 5% probability that a type I error will be made when the number of counts recorded from the sample is L_C or higher. L_C is used to test whether there is any significant elevation in the level of radioactivity following spectrometric measurements.

We will now consider measurements on a sample that contains a certain amount of radioactivity, A_{source} [Bq], giving rise to G counts during pulse acquisition. The gross number of counts recorded during t_m is G_1. The net number of counts, N_1, is $G_1 - <B>$, and the associated stochastic uncertainty is given by $s^2(N_1) = s^2(G_1) + s^2(<B>)$. The measurement is repeated m times to obtain a distribution of counts, as in the previous measurements, giving $<N>$ as the expectation value and a standard deviation of $s_N = s_G/\sqrt{m}$. This distribution will be broader than that of the distribution of G due to the background.

The *detection limit*, L_D, is often used to define the minimum number of counts indicating a significant increase in radioactivity above the background. The level of significance in the value is usually the 95% confidence level, i.e. there is only a 5% probability of committing a type I error (false positive), or a type II error (false negative) (as shown in Fig. 4.61). For a particular detector setup, the pulse distribution of the net counts for this elevated radioactivity will be exactly centred around $L_C + 1.645 \cdot s_N = 1.645 \cdot s_B + 1.645 \cdot s_N$. For Poisson distributed pulse counts close to the background level, it can be assumed that $s_B \approx s_N$. This means that L_D (95% confidence level) is approximately $3.27 \cdot s_B$, which is often simplified to $3 \cdot s_B$. In other words, this expression corresponds to three standard deviations of the measured background, which is a very commonly used definition of the detection limit for radiometric systems when the background is well known (see e.g. ICRU 1972).

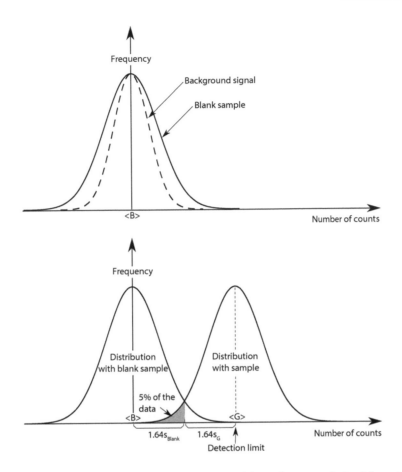

FIGURE 4.61 Illustration of type I and type II errors and how these are derived from the idealized histograms of observed background counts with and without a sample blank (above), and how the histogram from the blank sample compares with the histogram from observations of a source with an activity A_{source} which, on average, gives rise to a number of detected pulses at the detection limit, L_D (see below). The shaded area is the probability of a type II error, which will depend on the separation between $<B>$ and $<G>$.

The detection limit is used to characterize the ability of a given system to detect elevated radioactivity during a given measuring time. There are other definitions of this quantity and related quantities (see e.g. Calmet et al., 2008), which in practice are relatively close in low-activity situations. These quantities are, nevertheless, useful when comparing the ability of various detector setups to detect significant amounts of radioactivity, both in stationary and mobile systems. For high-resolution gamma-ray spectrometry, where the activity of a sample can be expressed according to Eqn. 4.56, the corresponding detection limit can be derived by inserting the number of net counts, N_g, into the equation

$$N_g - b \cdot t_m = 3.27 \cdot s_B. \qquad (4.62)$$

We then have a situation in which the number of counts in the particular ROI equals the detection limit, L_D. If the background pulses registered in the same ROI are Poisson distributed, $s_B = \sqrt{B}$, the background obtained during t_m is then given by

$$s_B = \sqrt{B} = \sqrt{b \cdot t_m} = \sqrt{\frac{B_g}{t_{bkg}} \cdot t_m}.$$ (4.63)

Inserting Eqns. 4.62 and 4.63 into Eqn. 4.56, and expressing B_g as $b \cdot t_m$ (as the background count rate is assumed to be constant for a particular detector setup), gives the *minimum detectable activity*, MDA:

$$MDA\,(t_{ref}, t_m) = \frac{\left\{ \left(\frac{3.27 \cdot \sqrt{\frac{b \cdot t_m}{t_{bkg}}}}{\varepsilon_{abs}} \right) \cdot \frac{1}{n_\gamma} \cdot T\,(t_{ref}, t_m, t_{acq}) \right\}}{m_{sample}}.$$ (4.64)

From Eqn. 4.64 it can be seen that the MDA for a certain spectrometer setup depends on the square root of the ratio between the measuring time and the background measurement time. It is important to consider this when deciding the acquisition time for a series of samples in order to obtain the lowest MDA for each sample. Finally, one must bear in mind how MDA is related to reducing the risk of both type I and type II errors when performing a *qualitative assessment* of the acquired pulse height distribution. The aim of the assessment is to both determine whether the presence of a given radionuclide can be established or not based on the measurement, and to quantify the elevated radioactivity, (e.g. A_{source} in a sample collected for gamma-ray spectrometry) with a certain confidence level.

A quantification limit, L_Q, can be defined for use in radiometry based on a specific confidence level. Since the sum of the number of counts recorded in the spectrometer from both the background and the source itself can be considered Poisson distributed, the relative uncertainty in N_{peak} (Eqn. 4.59) is roughly proportional to the gross number of counts (especially if $b_{bkg} \gg N_{peak}/t_m$). In both charged particle and gamma-ray spectrometry, the operator usually adjusts the t_{acq} to obtain, if possible, at least 1000 net pulses in a given full energy peak in order to obtain a quantitative estimate that has a relative standard uncertainty of roughly 3%. Further discussions on detection and quantification limits can be found in various references (Currie 1968, DeGeer 2004). Detailed descriptions of definitions and concepts used for detectability in radiometry can also be found in e.g. ISO 11929:2010 (ISO 2010).

In some cases, for example, in exemption measurements of radioactive waste, very small amounts of a certain radionuclide may be indicative of other much more prevalent radionuclides that are difficult to detect by spectrometry. The MDA of these indicator radionuclides must be established in order to determine whether the content of a radionuclide in the sample is less than the intervention level and can thus be exempted. The software used must therefore provide algorithms that can compare the low count rates at the energy of a suspected γ-ray with a pre-set value of L_D, and determine the probability of whether the nuclide has been detected or not. However, the more sensitive the search algorithm is in identifying various peaks (meaning that L_C is set at a lower value, e.g. at the 90% cumulative probability level instead of 95%; Fig. 4.61), the higher the risk of a type I error, since algorithms that are too sensitive may interpret spurious clusters of elevated pulse counts as peaks.

Furthermore, in some applications of gamma-ray spectrometry, such as the analysis of water samples from reactors at nuclear power plants, in which many hundreds of full energy γ peaks must be separated, high sensitivity alone is not sufficient for the algorithm to adequately detect the gammas present. The algorithm may assign an elevated value of the activity with a certain level of confidence, but identify it incorrectly, leading to a type II error. The *specificity* of a gamma-ray spectrometry system depends on three factors:

1. the energy resolution of the system,

2. the accuracy in the FWHM modelling of the energy dependence at this energy resolution (see Eqn. 4.49), and

3. the γ-ray energies in the nuclide library used by the algorithm.

If calibration of the FWHM is erroneous, as may be the case if it is based on a measurement with too few counts in the peaks, the peak search algorithm may fail to properly locate and fit the observed γ peaks in the pulse height distribution. In view of the risks of committing either type I or type II errors in gamma-ray spectrometry, automated analysis by system software should be used with caution. If a laboratory receives samples that deviate in terms of density, volume or atomic composition from the standard calibration sources used, or if the sample is of unknown origin, it is advisable for the automated assessment to be checked by an experienced gamma-ray spectrometry operator.

4.7 NEUTRON DETECTORS

4.7.1 Neutron Interactions in Matter

Very few instruments are designed to directly measure the energy distribution of a neutron flux. Due to the complexity of the interactions between neutrons and stable elements, there is no single type of detector element or material that is sensitive over the broad energy range of neutron energies, in the way that high-resolution gamma-ray spectrometry can be used for the whole range of photon energies.

In detector physics and dosimetry, neutrons are categorized according to their kinetic energy. The terminology used varies depending on the field of physics and the application. In general, the term *fast neutrons* refers to neutrons with a kinetic energy above 1 MeV. A number of terms are used for neutrons in the range from room temperature (0.025 eV) up to 1 MeV. Neutrons with energies below about 0.1 eV are referred to as *thermal neutrons*, while those in the range of 0.1 eV–1 MeV are called *epithermal neutrons*. A further division within this range is sometimes made, where neutrons with energies between 0.1 eV and 1 keV are referred to as *slow neutrons*, and those in the energy range 1 keV to 1 MeV are referred to as *intermediate neutrons*. Thermal neutrons generally have a much higher cross section for interactions with matter, such as neutron capture with subsequent fission of the nuclei, than fast and epithermal neutrons (Fig. 4.62).

Two kinds of interaction between neutrons and matter are used in neutron detection as outlined below.

- *Elastic scattering* between the neutron and atomic nuclei leads to the dissipation of energy from the recoiling nuclei to atomic electrons, which leads to ionization or luminescence in analogy with photon and charged particle interactions with atomic electrons. In practice, only light atomic nuclei such as H and He have sufficient recoil energy for ionization to take place.

- *Inelastic interactions* with nuclei involve some sort of nuclear reaction such as neutron capture followed either by prompt γ-ray release and subsequent spallation or fission, or the emission of protons or α-particles. These charged particles can cause ionization in the detector material, while the prompt γ-rays (with energies in the range of 5–10 MeV) can be detected in a secondary detector system. The prompt γ flux can then be related to the incident neutron flux in the primary detector.

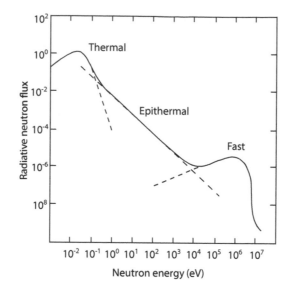

FIGURE 4.62 Example of a neutron spectrum consisting of a thermal component (<0.1 eV), epithermal neutrons (0.1 eV–1 MeV) and fast neutrons (>1 MeV).

Both these kinds of interactions occur mainly at thermal neutron energies, but the cross section for inelastic interactions can be 100 to 1000 times higher at thermal energies than at higher neutron energies. In order to improve the detection efficiency for neutrons with a wide range of energies, neutron detectors often include some kind of moderating material to reduce the kinetic energy of fast neutrons to thermal energies. The aforementioned recoil effects of light nuclei are used for the detection of fast and epithermal neutrons, since the cross section for elastic scattering of neutrons by H or He as a function of neutron kinetic energy is similar to the neutron spectrum obtained from ^{235}U fission (Fig. 4.63). Solid plastic scintillators or liquid scintillators are most common, but hydrogen can be used in the gaseous form. Gaseous ionization or proportional counters can be used for the detection of epithermal and fast neutrons, but it is difficult to discriminate between gamma radiation and neutrons as the cross section for corresponding γ-ray energies is comparable to that for neutrons.

The energy of the neutrons is not completely transferred to the nuclei in either of the two types of neutron interactions described above. Therefore, it is difficult to record a full energy absorption event even in a large bulk detector. This means that complete information on the initial neutron energy cannot be obtained in the same way as for x-rays and γ-rays. Furthermore, the use of a moderator in many neutron detectors to improve the neutron counting efficiency means that information on the initial neutron energy is lost.

In the energy range 20 keV to 2 MeV, neutrons have a relatively continuous and well-defined probability of interaction with ^{3}He(n,p)^{3}H, which can be exploited using 3He-gas-filled proportional counters, which detect the energy of the emitted protons. Such a device must, however, be shielded from thermal neutrons, to prevent the detection of events arising from the thermal ^{3}He(n,p)^{3}H reaction, which is three orders of magnitude more probable. Materials including B or Cd are useful since they have high cross sections for thermal neutron capture. For this reason, they are used in control rods in nuclear power plants. A combination of thin layers of materials containing elements with high cross sections for

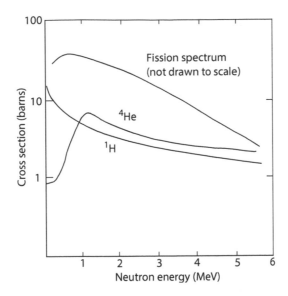

FIGURE 4.63 Scattering cross section as a function of neutron energy for ^{1}H and ^{4}He.

a specific neutron–nuclear reaction, leading to the emission of charged particles or γ-rays, could, in theory, be used to characterize the energy distribution of a neutron flux.

4.7.2 Design of Neutron Detectors

When designing a neutron detector one must also consider the ever-present γ flux, especially when monitoring samples from spent nuclear fuel as most of the neutron detectors are also, to a various degree, sensitive to γ photons (see Table 4.8). The challenge is therefore to efficiently detect and discriminate between the pulses arising from neutrons and from γ-rays. The following factors, related to the relative counting efficiencies for γ and neutron interactions, must be considered when designing a neutron detector.

- Physical (passive) shielding of γ-rays: Shielding should be chosen so as to enhance the possibility of discriminating γ-rays from neutrons. Pb ($Z = 82$) effectively absorbs the gamma photon flux, without significantly stopping fast ($>$1 MeV) neutrons.

- Choice of detector material: Some materials exhibit considerable differences between the cross sections for thermal neutron absorption and γ-ray interaction. Examples of such materials are He- and B-based materials. Detector systems can be physically shielded from γ-rays using, for example, Al, without significantly affecting the incoming neutron flux, hence maintaining a high ratio of the cross sections for n and γ-ray interactions.

- Choice of detector dimensions: The secondary particles generated by the interactions of neutrons and photons differ. Most of the energy imparted by thermal neutrons is transferred to heavy charged particles, whereas photon interactions produce mainly electrons. The volume and pressure in gaseous ionization detectors can be chosen so that the dimensions match the effective path length of the heavy charged particles, while having only a small effect on secondary electrons. One drawback of gaseous detectors is their relatively slow response time, compared with liquid and solid organic scintillators.

- Signal collection in the detector and subsequent amplification: Differentiation between γ-rays and neutrons can be greatly enhanced using pulse shape analysis when one or both of the uncharged particle events are collected rapidly in the detector. Liquid and solid organic scintillators are best suited for this.

For gaseous ionization detectors, proportional counters operating in pulse mode are usually preferred, although little energy information can be retrieved from the pulses. The gas should be polyatomic to afford a more rapid and well-defined avalanche. The significant cross section for Compton scattering of gamma photons will be a problem since this interaction generates free electrons leading to ionization similar to that resulting from recoil nuclei and charged particles from neutron interactions. Another drawback of gas-filled detectors is that they are cumbersome, especially if a constant supply of gas is needed requiring gas-flow techniques.

TABLE 4.8 Examples of neutron detectors and their approximate detection efficiency.

Detector type	Size	Neutron active material	Incident neutron energy	Neutron detection efficiency[1] (%)	γ-ray sensitivity (mGy s^{-1})[2]
Plastic scintillator	5 cm	^{1}H	1 MeV	78	$2.8 \cdot 10^{-5}$
Liquid scintillator	5 cm	^{1}H	1 MeV	78	$2.8 \cdot 10^{-4}$
Loaded scintillator	1 mm	^{6}Li	thermal	50	$2.8 \cdot 10^{-3}$
Hornyak button	1 mm	^{1}H	1 MeV	1	$2.8 \cdot 10^{-3}$
Methane (7 atm)	5 cm diam.	^{1}H	1 MeV	1	$2.8 \cdot 10^{-3}$
^{4}He (18 atm)	5 cm diam.	^{4}He	1 MeV	1	$2.8 \cdot 10^{-3}$
^{3}He (4 atm), Ar (2 atm)	2.5 cm diam.	^{3}He	thermal	77	$2.8 \cdot 10^{-3}$
^{3}He (4 atm), CO$_2$ (5%)	2.5 cm diam.	^{3}He	thermal	77	$2.8 \cdot 10^{-2}$
BF$_3$ (0.66 atm)	5 cm diam.	^{10}B	thermal	29	$2.8 \cdot 10^{-2}$
BF$_3$ (1.18 atm)	5 cm diam.	^{10}B	thermal	46	$2.8 \cdot 10^{-2}$
^{10}B-lined chamber	0.2 mg cm^{-2}	^{10}B	thermal	10	2.8
Fission chamber	2.0 mg cm^{-2}	^{235}U	thermal	0.5	$2.8 \cdot 10^3$

[1] Interaction probability of neutrons impinging at right angles.
[2] Approximate upper limit for usable output signals.
Source: Data from Crane and Baker (2005).

In emergency preparedness and homeland security applications there is a need for portable, low-maintenance instruments for mobile field searches of orphan neutron-emitting sources. Organic and solid scintillators have traditionally been used for the detection of epithermal and fast neutrons. These must either be hand-held or be mountable in vehicles for orphan source searching. Solid-state detectors are compact, but they will be heavier if a moderator, in the form of plastic material rich in hydrogen, is included. Compact neutron detectors also have the disadvantage that the higher the density of the detector material, the more likely the simultaneous registration of a neutron and gamma event in the detector.

Another important aspect when performing radiometry in an environment where both γ-rays and neutrons are present is that the neutron flux may interact with the detector materials in γ-ray spectrometers based on HPGe and especially CdZnTe, causing nuclear reactions that change the energy states in the semiconductor. These changes are often irreversible, and the effect is cumulative, leading to degradation of the detector performance over time. This irreversible effect must therefore be considered when installing γ spectrometric survey systems in nuclear power plants.

4.7.3 Application of Neutron Detection in Radiation Protection

There are a number of applications within environmental radioactivity in which neutron detection is used. Within radiation protection, the instrumentation for the detection of neutrons is primarily intended for qualitative measurements, that is, measurements carried out to determine whether neutrons are present or not. Examples of the kind of detectors used are large gas-filled ^{3}He chambers exploiting the elastic scattering of thermal neutrons from He atoms, in vehicle-borne monitoring of elevated γ-ray and neutron levels in the outdoor environment, or for searches of orphan sources within confined areas, e.g. by customs officials. Apart from its relatively high cross section for fast neutrons and low cross section for γ-rays, ^{3}He is attractive as it is non-toxic and chemically inert. ^{3}He tubes configured as proportional counters are commonly enclosed in a moderating material to further increase the cross section for lower-energy neutrons, making them suitable for the detection of ^{239}Pu.

The combination of LiF tablets containing the more abundant isotope ^{7}Li and tablets containing the isotope ^{6}Li is common in personal dosimetry at nuclear power plants. The ^{6}Li isotope has a significant cross section for neutrons allowing the ^{6}Li (n, ^{3}H)He reaction over a relatively wide energy range (Fig. 4.64), and a combination of tablets of the two isotopes provides personal dosimeters that can measure the dose contribution from neutrons in a mixed neutron and γ field.

Solid-state semiconductors containing an element with a high cross section for thermal neutron capture, such as Cd in the compound CdZnTe, are used for the detection of thermal neutrons in combination with hand-held instrumentation in searches for orphan sources.

Neutron activation analysis involves detection methods that measure the presence of neutrons indirectly by detecting prompt or delayed γ radiation resulting from neutron reactions. The detection of neutron-activated atoms in tissue is used in retrospective dosimetry to determine the neutron flux and their air kerma contribution, and to convert this into an absorbed dose and an effective dose. An example of this is gamma-ray spectrometry of ^{24}Na in whole-body measurements of astronauts based on the ^{23}Na(n,γ)^{24}Na reaction in the sodium contained in the body, e.g. in physiological NaCl. β-particle counting of neutron-activated sulphur (^{32}S(n,p)^{32}P) in the hair of workers exposed to neutrons in accidents involving thermal or epithermal neutrons has also been reported.

At the beginning of this millennium the increase in the global demand for neutron detection (precipitated by increased awareness of homeland security after 2001) led to an increase in the demand for ^{3}He. The demand for ^{3}He in US alone is currently estimated to be 60 m^3 per year (Kouzes 2009), and there is now a shortage of this element.

In addition to emergency preparedness and homeland security, other applications of neutron detection include research, for example, neutron spallation as a powerful tool for imaging low-Z structures in various elements. Alternative materials for neutron detection must now be considered, such as scintillators combined with moderating materials.

In summary, it is relatively easy to construct neutron detectors with a high counting efficiency for fast neutrons, that can discriminate against thermal neutrons. However, the detection of neutron fluxes with a wide range of energies is more difficult. The cross sections for neutron capture and other reactions are highly irregular at low energies, and many materials that are highly efficient for the detection of thermal neutrons also have a high cross

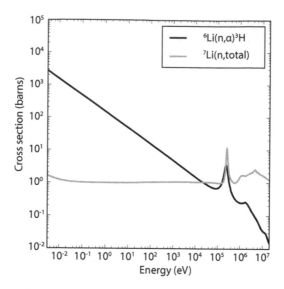

FIGURE 4.64 Cross section as a function of neutron energy for ^{6}Li and ^{7}Li.

section for γ-ray detection, resulting in poor γ-ray discrimination. No universal detector has yet been developed for neutron spectrometry, as in α-particle and gamma-ray spectrometry, and quantitative neutron detection for emergency preparedness is still a challenge.

The challenge in designing detector systems that can simultaneously detect several types of radiation particles is no more evident than for radiation monitoring in galactic applications, e.g. in Marsal expeditions. To prepare manned expeditions and voyages in space, e.g. to Mars, which may become incerasingly relevant in the near future, composite detector systems that can detect both light and heavy charged aprticles as well as uncharged particles such as g and neutrons have been tested in spacecrafts. An example of such a system is the Radiation Assessment Detector (RAD) made for the Mars Science Laboratory, descibed by Hassler et al. (2012). The system consists of a combination of three types of detectors: solid state Si-diodes, and two types of scintillators (inorganic CsI(Tl) and plastic scintillator). The detector combination can fully detect protons and α-particles in the energy range 3–100 MeV, β-particles and electrons from about 0.2 to 10 MeV, and neutrons ranging from 3–100 MeV. Furthermore, the system can detect the very important galactic flux of heavy charged particles, ranging from Litium ($Z = 3$) to iron ($Z = 26$), in the similar energy range (3–100 MeV).

4.8   REFERENCES

- Bearden, J. A. 1967. X-ray wavelengths. *Review of Modern Physics*, 39(1); 78–.

- Bernhardsson, C. 2011. *Radiation Exposure of Human Populations in Villages in Russia and Belarus Affected by Fallout from the Chernobyl Reactor: Measurements Using Optically Stimulated Luminescence in NaCl, Tl-Dosemeters and Portable Survey Instruments*. ISBN 978-91-86871-11-6, Lund University, Faculty of Medicine Doctoral Dissertation Series 2011:62, 10 June 2011.

- Birks, J. B. 1964. *The Theory and Practice of Scintillation Counting*. Pergamon Press.

- Bøtter-Jensen, L., Agersnap Larsen, N., Markey, B. G. and Mc-Keever, S. W. S.. 1997. Al_2O_3:C as a sensitive OSL dosemeter for rapid assessment of environmental photon dose rates. *Radiation Measurements*. 27, 295–298.

- Bøtter-Jensen, L., Solongo, S., Murray, A. S., Jungner, H. 2000. Using the OSL single-aliquot regenerative-dose protocol with quartz extracted from building materials in retrospective dosimetry. *Radiation Measurements*, 32(5-6), 841–845.

- Calmet, D., Herranz, M. and Idocta, R. 2008. Characteristic limits in radioactivity measurements: From Currie's definition to the international standard ISO-11929 publication. *J Radioanalytical and Nucl Chemistry*, 276(2), 299–304.

- Cherpy et al. Website: https://sc-programs.llnl.gov/pdf/hep/cherepy_DOE_110210.pdf. Accessed in 2015.

- Crane, T. W., Baker and M. P. 2005. Neutron Detectors Website: http://www.lanl.gov/orgs/n/n1/panda/00326408.pdf; Los Alamos National Laboratory, Accessed in 2016.

- Currie, L. A. 1968: Limits for qualitative detection and quantitative determination-application to radiochemistry. *Anal. Chem.*, 40, 586–593.

- Daniels, F. Boyd, C. A. and Saunders, D. F. 1953. Thermoluminescence as a research tool. *Science*. 117, 343–349.

- Day, M. J. and Stein, G. 1950. Chemical effects of ionizing radiation in some gels. *Nature*, 166, 146–147.

- De Geer, L. E. 2004. Currie detection limits in gamma-ray spectrometry, *Applied Radiation and Isotopes*, 61, 151–160.

- Fricke, M. and Morse, S. 1927. The chemical action of roentgen rays on dilute ferrous sulphate solutions as a measure of radiation dose. *Am J Roentgenol Radium Therapy Nucl Medicine*, 18, 430–42.

- Gilmore, G. and Hemingway, J. D. 1995. *Practical gamma-ray spectrometry. 1st ed..* John Wiley & Sons, Chichester, UK.

- Gore, J. C., Kang, Y. S. and Schulz, R. J. 1984. Measurement of radiation dose distributions by nuclear magnetic resonance (NMR) imaging. *Phys Med Biol.*, 29, 1189–97.

- Hansson, M. and Rääf, C. 2011. Visualization and quantification of lung content of radionuclides associated with nuclear and radiological emergencies. *Radiat Prot Dosimetry*, 145(4), 341–350.

- Hassler, D. M., Zeitlin, C., Wimmer-Schweingruber, R. F., Böttcher, S., Martin, C., Andrews, J., Böhm, E., Brinza, D. E., Bullock, M. A., Burmeister, S., Ehresmann, B., Epperly, M., Grinspoon, D., Köhler, J., Kortmann, O., Neal, K., Peterson, J., Posner, A., Rafkin, S., Seimetz, L., Smith, K. D., Tyler, Y., Weigle, G., Reitz, G., Cucinotta, F. A. 2012. The Radiation Assessment Detector (RAD) Investigation. *Space Sci Rev*, 170(1), 503–558.

- Hofstader, R. 1949. The detection of gamma-rays with thallium-activated sodium iodide crystals. *Phys. Rev.*, 75, 796.

- IAEA, International Atomic Energy Agency. 1987. *Absorbed Dose Determination in Photon and Electron Beams: An International Code of Practice.* Technical reports Series No. 277. International Atomic Energy Agency. Vienna.

- ICRU, International Commission on Radiation Units and Measurements. 1972. *Measurement of Low-Level Radioacitivty.* ICRU Publication 22. ICRU publications, Bethesda, USA.

- ICRU International Commission on Radiation Units and Measurements. 2006. *Sampling of Radionuclides in the Environment.* ICRU Publication 75. ICRU Publications, Bethesda, USA.

- ISO, International Organization for Standardization. 2010. *Determination of the Characteristic Limits (Decision Threshold, Detection Limit and Limits of the Confidence Interval) for Measurements of Ionizing Radiation: Fundamentals and Application.* ISO 11929:2010.

- Kaye and Laby. *Tables of Physical & Chemical Constants* (16th edition 1995). 2.1.4 Hygrometry. Kaye & Laby Online. Version 1.0. http://www.kayelaby.npl.co.uk/. Accessed in 2016.

- Kittel, C. 2005. *Introduction to Solid State Physics*, 8th ed. ISBN-13:978-0471415268. John Wiley & Sons, Inc., USA.

- Knoll, G. F. 2010. *Radiation Detection and Measurements.* 4th rev ed. John Wiley & Sons, USA.

- Koba, Y., Iwamoto, H., Kiyohara, K., Nagasaki, T., Wakabayashi, G., Uozumi, Y. and Matsufuji, N. 2011. Scintillation efficiency of inorganic scintillators for intermediate-energy charged particles. *Progress in NUCLEAR SCIENCE and TECHNOLOGY.* Vol. 1, 218–221.

- Kouzes, R. T. 2009. *The* ^3He *Supply Problem.* Pacific Northwest National Laboratory. doi:10.2172/956899. http://www.pnl.gov/main/publications/external/technical_reports/PNNL-18388.pdf. Accessed 2016.

- Menge, P. R., Gauter, G., Iltis, A., Rozasa, C. and Solovyev, V. 2007. Performance of large lanthanum bromide scintillators. *Nucl. Instruments and Methods in Physics Research Sec. A: Accelerators, Spectrometers, Detectors and Associated Equipment,* 579(1), 6–10.

- Mietelski, J. Gaca, P. and Bielec, J. 2005. Optimization of counting conditions for Cerenkov radiation by LSC. In: Chalupnik, S., Schönhofer, F. and Noakes, J. (Eds.). LSC 2005, *Advances in Liquid Scintillation Spectrometry. Proceedings of the 2005 International Liquid Scintillation Conference,* Katowice, Poland.

- Ören, Ü., Mattsson, S. and Rääf, C. 2014. Detection of radioactive fragments in patients after radiological or nuclear emergencies using computed tomography and digital radiography. *J. of Radiological Protection,* 34(1), 231–247.

- Upp, D. L., Keyser, R. M. and Twomey, T. R. 2005. New cooling methods for HPGE detectors and associated electronics. *J. Radioanalytical and Nuclear Chemistry,* 264(1), 121–126.

- Wiener, R. I., Surti, S. and Karp, J. S. 2013. DOI determination of rise time discrimination in single-ended readout for TOF PET imaging. *IEEE Trans Nucl Sci.*, 60(3):1478–1486.

4.9 EXERCISES

4.1 What are the time delays in the various steps from the collection of charge pulses by the ionization detector, through preamplification and pulse shaping to the recording of pulse height in a MCA? How is this time delay related to the total amplification dead time?

4.2 How large a problem is dead time in digital amplifiers for spectrometry?

4.3 Discuss four factors that affect the practical time resolution of many detector systems.

4.4 Discuss how flash ADCs compare with traditional ADCs.

4.5 Discuss the advantages and disadvantages of using inorganic scintillators and organic scintillators for the detection of γ-rays with energies between 30 and 3000 keV.

4.6 Sketch the design of a neutron-sensitive detector that can be used as a coarse neutron spectrometer.

4.7 The data given in Table 4.9 are the channel number and number of counts in that channel for a high-resolution gamma-ray spectrometry detector. Use the data to deduce the following: (i) the net number of counts, (ii) the location (channel number) of the centroid and (iii) the FWHM of the peak. The pulse height distribution was obtained from a sample of an alloy containing tin (Sn, $Z = 50$) that had been exposed to neutrons. Given an energy calibration of 0.54187 keV ch^{-1}, suggest the identity of the radionuclide present.

TABLE 4.9 Pulse height distribution recorded in a γ-ray spectrometer with an acquisition time of 123 087.7 s (live time; Section 4.3.3) and an energy calibration of 0.541 87 keV ch^{-1}.

Channel no.	No. of pulses	Channel no.	No. of pulses	Channel no.	No. of pulses
1094	34	1106	70	1118	38
1095	47	1107	102	1119	40
1096	35	1108	120	1120	49
1097	28	1109	108	1121	98
1098	44	1110	59	1122	167
1099	36	1111	41	1123	208
1100	41	1112	40	1124	190
1101	39	1113	35	1125	124
1102	26	1114	38	1126	59
1103	41	1115	40	1127	40
1104	37	1116	37	1128	30
1105	45	1117	33	1129	36

4.8 When acquiring a pulse height distribution from a flask containing ^{131}I using a NaI(Tl) detector a count rate of 1450 cps is obtained. The energy interval of the ROI is 350–380 keV. What is the source strength of this sample given that the absolute efficiency, ε_{abs}, in that particular energy window is 37%?

CHAPTER 5

Sampling and Sample Preparation for Radiometry

CONTENTS

T O OBTAIN A comprehensive and reliable understanding of the radiation situation and to verify the actual presence of radionuclides in various environments samples must be collected and prepared for radionuclide assessment. Modelling and computer simulations can never fully replace the need for environmental sampling due to inherent uncertainties in the analytical dispersion models of radionuclides. Sampling for radiological assessments can be made for various purposes, as will be described in this chapter. Great care must be taken when carrying out sample collections, and several procedures must be considered before radiometry can be carried out. This is especially the case when assessing the presence of alpha- and beta-emitting particles as the short range of the emitted radiation requires laborious procedures to enable the detection, either by radiometry or by mass spectrometry. The following chapter will also briefly describe the theoretical concept of sampling, some common types of environmental sampling performed within the field of environmental radioactivity and emergency preparedness, and the procedures connected with preparing these samples for radionuclide assessment.

5.1 PRINCIPLES OF SAMPLING FOR RADIOMETRY

5.1.1 General Aspects of Radiological Sampling

In the modern era, the creation and dispersion of anthropogenic radioactivity has led to environmental contamination, with radioactivity appearing in the soil and in freshwater and marine systems. Any of these can act as the starting point for exposure pathways affecting humans, and an overview of these kinds of pathways was given in Chapter 3. For accurate prediction of the environmental consequences of any release of radioactivity, compartment modelling is often used (as described in Section 3.1). It is also important to collect samples that provide a representative measure of the contamination levels in a given compartment at a particular time. This can be a challenge, and experience has shown that dispersion of the three major releases that have occurred to date—global fallout from nuclear weapons tests in the 1950s and 1960s, fallout from Chernobyl in 1986, and the release from the Fukushima nuclear power plant in 2011—have all resulted in highly inhomogeneous depositions on the ground. Over bodies of water where a higher degree of mixing takes place, such as the Baltic Sea, dispersion of radioactive substances has resulted in less local inhomogeneity than that found in soil and samples from the biosphere (UNSCEAR 1977; UNSCEAR, 1982). Fallout mainly consists of *fission products* (such as ^{131}I, ^{132}Tl, ^{137}Cs) and *neutron-activated fission products* (such as ^{134}Cs), which can have a wide variety of chemical forms. The level of deposition at a particular site depends on the following:

- Dry or wet deposition. The fraction of airborne radionuclides in a plume that is deposited on the ground increases considerably if there is washout by precipitation, for example, ground deposition rates for ^{137}Cs increase by more than a factor of 3.

- The chemical state of a given radionuclide, and its physical form (aerosol, gas or solid particle). A given radionuclide can exist in a wide variety of states, and their relative prevalence varies with distance from the release point, and on weather conditions during deposition.

- The chemical and physical characteristics of the receiving medium. For example, partition of a particular radionuclide between soil and its surrounding media will be different for different soil types (organic material, clay or sand), leading to considerable inhomogeneity in the activity concentration.

To illustrate the high degree of variation observed in activity concentrations in the environment, Figure 5.1 shows spatial variations in the ground deposition of ^{137}Cs on the national and regional scale (a macroscopic 1000-km scale), and Figure 5.2 shows variations in ^{137}Cs concentration on the mesoscopic (1–1000 m^2) and microscopic length scale (<1 m^2 surface) (Bernhardsson 2011).

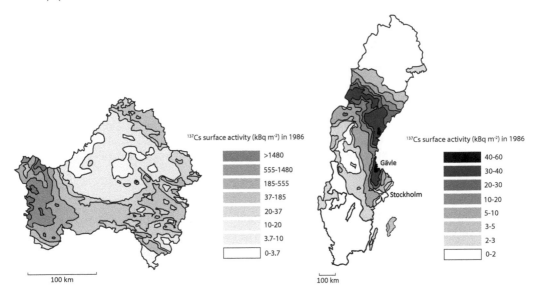

FIGURE 5.1   Left: Ground deposition of ^{137}Cs the Bryansk region, Russia, from the Chernobyl fallout in 1986. Modified from Bernhardsson et al. (2015). Right: The corresponding deposition in Sweden. Modified from Rääf et al. (2006).

In order to investigate the presence of radionuclides, it is necessary to select sampling matrices to represent a specific species (e.g. grass on pasture land, mussels in shallow waters, wild game in a boreal forest, perch in a freshwater lake, and so on), or a particular medium (e.g. fly ash from a coal-fired power plant, surface water from a river, topsoil from a field). Both species and media can be represented as compartments in environmental modelling. Within the fields of environmental radioactivity and emergency preparedness, sampling may also refer to direct detector readings acquired on site (*in situ*) using stationary or mobile equipment. The radionuclide concentration data as well as the detector readings obtained *in situ* are raw data, and in many cases must be refined and corrected in order to obtain the type of data needed. For example, the number of pulses within an energy interval of a portable NaI(Tl) detector must be converted into a value that corresponds to an average surface deposition using predetermined calibration coefficients.

Precautions must be taken to ensure that the samples collected will provide data that are relevant for assessing general levels of contamination among affected human populations and other species, and that the inputs to environmental models are reliable. Relying on biased data can lead to gross misinterpretations of the radiological situation, and can result in errors being made when deciding on remedial action to prevent further radiation exposures. Sometimes, composite samples must be collected to make a particular sampling scheme more representative, especially in circumstances where a large variation in radionuclide concentration can be anticipated. Composite sampling means that multiple subsamples are collected over an area or volume, to provide a representative overall view (see Section 5.1.2 for further details).

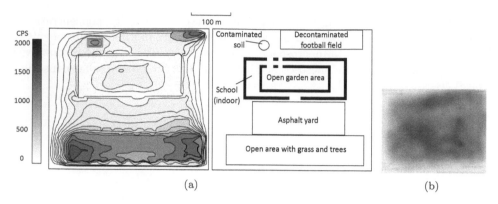

FIGURE 5.2 (a): Typical isodose (ambient dose rate) curves measured in 2008 at a height of 1 m around a building contaminated by the Chernobyl fallout in 1986. (b): Scan of a radiochromatic film (30×24 cm^2) after exposure for one week to a 6-mm-thick layer of soil taken from a Russian village in 2008. Modified from Bernhardsson (2011).

The sample size, and the number of subsamples, if any, should be chosen to ensure a given confidence level. If accurate quantitative determinations of the radionuclide concentration are required, a larger number of subsamples is needed to ensure that the variability of the radionuclide concentration in the medium is accounted for. If certain materials are to be subject to legal exemptions, the samples in the area investigated should be as representative as possible. Repeated sampling of a species or material over time provides information on the time dynamics of the radionuclides. Time series of environmental samples are therefore important for tracking historical releases of radioactivity, and for predicting the future development of contamination from past and existing sources. Data from these time series can also be used to predict the development of the radionuclide concentration in a given medium in case of a new release event.

Environmental sampling can be made for various reasons, and three principal applications are given below. In all these applications, the various aspects of sampling techniques must be considered and adapted to be fit for purpose.

- *As a basis for decision making.* Sampling must be performed such that a known relative uncertainty (e.g. 10%), at a certain confidence level (e.g. a confidence level of 95%, corresponding to about 2 standard deviations for a normally distributed set of data) can be associated with the final value. This is necessary in order to judge how the value obtained compares to a reference value. A typical example of this kind of sampling and comparison to a reference value is the widely employed exemption level measurements for classifying various kinds of radionuclide content in radiological and nuclear waste. In these cases the assessed material can only be exempted from regulation if the value obtained is below the exemption level at a confidence level of, for example, 95%. It is therefore necessary to employ rigorous criteria for the confidence levels of measurements, and to ensure that sampling of large quantities of bulk materials is representative.

- *For monitoring* of a long-term or large-scale trend. In this case an assembly of data is created where each additional data point contributes information about either the long-term trend, or the spatial pattern of radiation over a given geographical region. Examples include airborne surveys of radionuclide deposition over a given area or region, or regular monitoring of the radionuclide concentration in the vicinity of a nuclear facility to survey trends in operational releases with time. For this type of data sampling, the relative uncertainty associated with each data point is not necessarily fixed at a certain value, but the data should be accurate enough to contribute valuable information to the existing database.

- *For research.* The investigation of fundamental processes associated with radiation may involve the collection of environmental samples. Radiation data can also be used in research as an instrument for studying other biological, physical, geological or environmental processes, where radiation or radionuclides can be used as tracers to follow the movements and behaviour of various stable elements. Examples include the use of a radioactive tracer to study how ingested uranium is incorporated into systemic tissues. Another example is the detection of ^{99}Tc in marine species for studying the movements of shallow water in seas. In such applications the quality or confidence level of the data is maximized given the level of resources (typically, time and funding).

5.1.2 Sampling Theory

Radiometric data obtained from the final stage of analysis of a sample can usually only be associated with that particular sample. However, the value obtained is often used as a basis for estimating the radionuclide concentration in other parts of the original set of samples, or in surrounding materials or species living in the area close to where the sample was collected. This will introduce uncertainties, since the estimate (or generalization of the results) may involve qualified guesses or generic estimates based on previous studies with more comprehensive sampling.

Let us first assume that we wish to quantify the concentration of radioactive substances in a particular entity, e.g. the pasture on a farm used for dairy cows. The entity in this case, i.e. all the pasture on the farm, is also referred to as the "lot". Let us then assume that that the value of the concentration of a certain radionuclide obtained from γ-ray spectrometry of a grass sample is X Bq kg^{-1}, and that the relative uncertainty at the 95% confidence level (± 2 SD) *associated with the spectrometer output* is 10% of X. However, if we then make a generalization and infer that the value X is valid for the whole lot (i.e. the entire surface of the pasture), without considering the high variability of radioactive deposition, we may err by a factor of two or more. Hence, more samples must be taken from the pasture to account for the inherent variability in the radionuclide concentration over the surface studied, thereby giving a better representation of the radionuclide concentration in the pasture. The truth (within statistical limits determined by the stochastic nature of radioactive decay) can only be experimentally obtained if all the grass in a given field is collected, homogenized and assessed with a radiation detector appropriate for the radionuclide whose level we wish to determine. This task would be impossibly laborious and time consuming, given the limited value this exact knowledge would contribute to public health.

As a compromise between single sampling (with limited value) and the *truth*, samples should be collected following a so-called "qualified sampling process" in which equal sub-samples, referred to as *increments*, are extracted from each sample. These increments can be merged to form a general sample that is representative of the whole lot, as illustrated in Figure 5.3. Samples collected randomly, or when this process is not followed, should be referred to as *specimens* rather than samples.

The following description of the steps used in a qualified sampling process is taken from Petersen et al. (2005). Assume that the true activity concentration of a particular

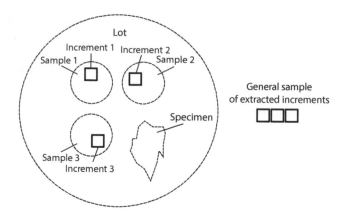

FIGURE 5.3 Schematic view of the steps in the analysis of the concentration of a radionuclide.

radionuclide is a_m [Bq kg^{-1}] in the lot of interest, e.g. the grass in a certain area of a pasture. A full assessment of the sample following the scheme set out in Figure 5.3, results in an observed value, a_s, with an associated statistical uncertainty, s_s. The sampling error is defined as

$$e_s = \frac{(a_s - a_m)}{a_m}, \tag{5.1}$$

where a_m is the true value in the lot. A sampling procedure is accurate to within a predetermined criterion, Δe, which is chosen by those carrying out the investigation; for example, Δe=0.05, which means that the error is 5% of the true value a_m. After a number, i, of repeated samplings, we obtain a mean value for $e = < e >$. In order for a sampling procedure to be considered *accurate* within the preset criterion, the following must be fulfilled:

$$\langle e_s \rangle^2 = \left\langle \frac{a_{s,i} - a_m}{a_m} \right\rangle^2 \langle \Delta e \rangle^2, \tag{5.2}$$

where $< e_s >$ is the relative coefficient of variation in the repeated sampling. The selection of samples, with an observed standard uncertainty, s_s, is considered *reproducible* within a given predetermined criterion s_0. An example of such a criterion is s_0 =10% of the true value a_m. Hence,

$$s_s^2 = \text{Var} \langle e_s \rangle \leq s_0^2 = 0.1^2 \text{ (if criterion of } s_0 \text{ is 10\%).} \tag{5.3}$$

Finally, the parameter r_s^2 is related to how representative the selection of the lot is by means of the following relationship:

$$\left(r_s^2 = \langle e_s \rangle^2 + s_s^2 \right) < \left(r_0^2 = \Delta e^2 + s_0^2 \right), \tag{5.4}$$

that is, the sum of the square of the relative error and the square of the relative coefficient of variation in the repeated samples is less than the corresponding sum of the square of the predetermined relative tolerance/error and the square of the relative coefficient of variation. The example above, with $\Delta e = 0.05 \cdot a_m$ and $s_0 = 0.01 \cdot a_m$, would yield a representability criterion of $r_s = \left(0.05^2 + 0.1^2 \right) = 0.11 \cdot a_m^2$.

A lot can be described as having several dimensions in terms of the representability,

r_s, achieved when taking a set of samples (Fig. 5.4). A zero-dimensional lot obeys two conditions: first, that the whole lot is included in the sample; and second, that the expectation value of the radionuclide concentration, $< a_s >$, in samples taken from the lot is independent of their location within it (see Fig. 5.4). The presence of inhomogeneities in the activity in the lot, represented in the theory of sampling as groups or fragments, will inherently lead to a sampling error called *the fundamental sampling error*. This is the minimal theoretical sampling error that can be achieved, assuming the entire lot is divided into samples. However, another basic sampling error arises from the sampling process. For example, consider a lot that consists of a number of bags of sugar. If only the top layer of the sugar in each bag is sampled for assessment of the radionuclide concentration, the sample will contain a disproportionate fraction of the lighter sugar crystals as the heavier ones will have settled during transportation. If the radionuclide concentration is related in some way to the effective adhesive surface area of the sugar crystals, sampling from the top of the bag will cause a bias in the results, which will translate into a so-called grouping or segregation error in the data.

The term *groups* refers to structures or elements in the lot that are inhomogeneously distributed but whose radionuclide concentrations are correlated to each other. The presence of these groups may contribute to the bias in the analytical estimate obtained from the sample. A correct sampling procedure is one in which all the groups (fragments of the material) have the same, non-zero probability of contributing to the total amount of the merged sample. Furthermore, similar elements that do not belong to the lot material, must also have a non-zero probability of ending up in the sample. A symptom of a biased sampling procedure is that the estimated concentration in samples collected in the lot appears to be correlated to those in samples taken nearby. The amplitude of this bias is also referred to as *autocorrelation*, meaning that samples from one lot resemble others nearby. This phenomenon is also very important in data series such as mobile gamma surveys (Section 7.3.6), where each consecutive data point is partially correlated with its neighbours due to the partly overlapping fields of view of the detector system.

For a more detailed discussion on the theory of sampling, the reader is referred to Petersen et al. (2005).

5.2 SAMPLING FOR RADIONUCLIDE ASSESSMENT

5.2.1 Air Sampling

Sampling air to determine radionuclide concentrations is important for tracing large-scale airborne releases, to predict inhalation doses to unshielded inhabitants in the vicinity of the sampling area, and to detect the detonation of nuclear weapons for the purpose of non-proliferation monitoring. In an air sampling instrument, air is pumped through a filter which collects the airborne substances. A variety of filter types may be needed to collect different types of airborne matter efficiently: for example, glass fibre filters for particulates and charcoal filters for highly volatile products such as noble gases (Fig. 5.5). The filter is removed after a certain sampling time, and then analysed using alpha or gamma-ray spectrometry (as discussed in Section 4.5 and 4.6). Figure 5.6 shows an example of a sample and activity report for air filter sampling.

When low activity concentrations are expected, or when detailed information is required on the radionuclide composition and the concentrations of the individual components in the release, air sampling instruments with capacities on the order of $1–1000$ m^3 h^{-1} are required (see the example in Fig. 5.7). Smaller units, with an air flow capacity of $10–100$ dm^3 min^{-1},

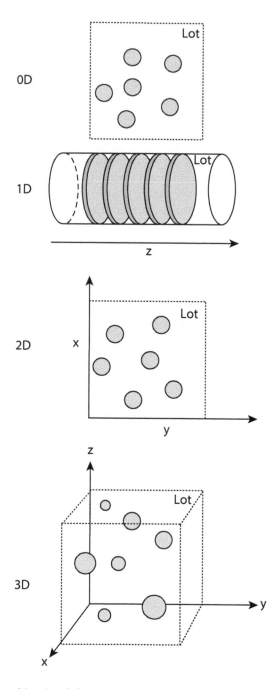

FIGURE 5.4 Illustration of levels of dimensions in sampling. The grey shapes represent volumes that are part of the lot, and whose radionuclide concentrations are to be determined in order to represent the corresponding value of the whole lot. In case 0D each sample is collected independent of any dimension (0-dimensional sampling). In case 1D the samples are collected at specific depths in the lot (1 dimensional). In cases 2D and 3D, specific sampling locations in 2D (x and y) and in 3D (x, y and z) are done to represent the lot. Note that e.g. parameters such as time, temperature, age etc. can be treated as a dimension describing the sampling of a lot.

(a)   (b)

FIGURE 5.5   A typical air filter used in the Swedish emergency preparedness system for air sampling. The air filter is composed of a particle filter on top (a), and a charcoal column behind the particle filter (b).

are used in emergency preparedness to determine the air concentration of radionuclides associated with early releases from nuclear power plants, mainly ^{131}I, ^{132}I, ^{132}Te.

Materials suspended in air can exist either as particulates or as free gas molecules. Particulates in air are referred to as aerosols, and typically have diameters in the range 0.1–10 μm, although dust and mist can consist of droplets up to 100 μm in size (Baron 2006). These are small enough to remain in the air without being deposited on the ground through gravitational settling (see Section 3.2.4). Suspended particles are present close to the ground in most environments, and will contain natural radioactive elements such as ^{7}Be and Rn daughter products such as ^{212}Bi. When particulates are deposited, they form a sediment that can be re-suspended in the air close to the ground. An example of this is visible dust, which may be stirred up and become airborne, but is relatively quickly re-deposited on the ground. Particulates can be solid fragments or liquid droplets, or form colloids.

Naturally occurring aerosols include fog, forest exudates (resins, gums and oils) and dust. Man-made particles arising from human activities are dust, haze (combined with fog), smoke, and other particulates resulting from combustion. Releases of nuclear material may consist of a combination of these particulates, as well as gas molecules, and noble gases. Operational releases into the air from a nuclear power plant consist mainly of noble gases and gas molecules. However, during accidental releases the elevated temperature of the reactor will cause volatile fuel fragments and organic substances to be released as aerosols. The high temperature of the radionuclides in these releases increases the probability of their reacting with organic plastics and materials in the power plant to form volatile organic substances, which can be transported many hundreds of kilometres away from the release site.

Air samplers consist of air pumps connected to a filter whose main purpose is to accumulate airborne particulates. The particulates, and hence the radionuclides carried by them, will be relatively easily collected by glass fibre filters. However, in order to extract gaseous radionuclides, such as noble gases, and iodine in the form of methyl iodide (CH_3I) or elemental iodine, I_2 (g), from the airflow, charcoal filters are connected in series after the particle filter. Charcoal filters consist of porous granulates of graphite in which the pores increase the effective surface area available for adhesion. In contrast to most noble gases

Report form for air sampling

Data from measuring site

Sampling point:			
Sampling data: .... Name:			
Charcoal filter ID.	Air sampling start time	Air sampling stop time	Comments (preciptation, wind direction etc.)

Filled in by the measuring laboratory

Laboratory:				
Operator:				
Mearuring date:				
Radionuclide	Chemical form	Average air concentration [Bqm^3]	Comitted dose* adult [µSv]	Comitted dose*, infant (1y) [µSv]
I-131	Gas			
	Particle			
Te-132	Gas			
	Particle			
Other	Gas			
	Particle			
Other	Gas			
	Particle			

* Comitted dose due to inhalation of radionuclides during 1 hour.

FIGURE 5.6   A sample and activity report for air filter sampling using a combination of a glass fibre particle filter and a charcoal filter.

(apart from radon isotopes such as ^{222}Rn), the inhalation of gaseous iodine has considerable dosimetric impact (ICRP 1995). Polar molecules such as CH_3 (I) preferentially adhere to surfaces. ^{133}Xe is a key nuclide in proliferation monitoring, and its presence in the air may indicate illicit detonation of nuclear weapons. ^{133}Xe is a rare gas, and can be collected more efficiently in charcoal filters cooled to liquid nitrogen temperatures (Saey and De Geer 2005).

The air flow velocity must be controlled and regularly checked, as the filter material becomes clogged as the measurement proceeds. The duration of sampling or the intervals at which the air filter is changed must be chosen so as to avoid clogging. The air filter must be located away from sources of dust. According to the Swedish emergency preparedness plan for accidental atmospheric releases from nuclear power plants, fire brigades within a 50-km radius of the plant are required to set up air samplers at fire stations. It is stipulated that the air filters should be changed every 4 hours, and then sent to a laboratory for analysis with γ-ray spectrometry. Since air filters are not usually changed more than a few times a day, for practical reasons, the sampling time must be carefully recorded in order to allow corrections to be made for the decay of short-lived radionuclides between the start of sampling and analysis, and to correct for air flow perturbations due to filter clogging.

5.2.2 Precipitation and Water Sampling

Devices for the collection of rainwater already exist for meteorological surveying. Rainwater collectors should preferably be designed to prevent the evaporation of water. If the interior surfaces of the container collecting precipitation, are smooth, the loss of radionuclides by surface adhesion to the vessel walls will be minimized. Rainwater collectors should be located away from vegetation, preferably on the roofs of buildings. Meteorological institutes have defined routines for the collection of snow in a temperate climate with snowfall in the wintertime. For the surveying of radionuclides, melted snow should be collected in vessels prepared in such a way as to avoid absorption on the walls.

Various methods can be used to distinguish between wet deposition and dry deposition during a deposition event. During the first few weeks after the release of the radioactive plume from the Chernobyl accident in 1986, Mattsson and Vesanen (1988) used rainwater collected in a standard precipitation gauge, in combination with sheets of paper laid out on a horizontal copper roof, free from perturbing nearby objects, to estimate wet and dry deposition. The sheets of paper were replaced regularly to obtain a time series for the dry deposition as the deposition event proceeded.

Samples of water at various depths can be made by using e.g. a Nishkin bottle (Fig. 5.7, right). Water sampling from rivers and lakes should not be done in stagnant areas or near the shoreline. As for other liquid samples, care should be taken to prevent absorption of radionuclides on the container wall. If the water samples must be stored for more than a few days before radiometric analysis, small amounts of hydrochloric acid can be added to prevent adhesion on the vial walls.

5.2.3 Soil Sampling

Soil is sampled in emergency preparedness to determine the extent to which radioactive fallout deposited on the ground has migrated into the soil. There are several reasons why it is important to establish the penetration of man-made radionuclides, for example: (i) to predict the radioecological transfer of man-made radionuclides to crops and pastures, (ii) to estimate the air kerma (K_{air}) and the effective dose (E) to humans living in contaminated areas (determination of the total activity deposition per unit area requires knowledge of

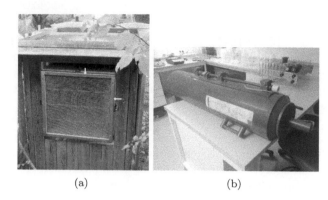

(a) (b)

FIGURE 5.7 (a): Combined air and precipitation sampler, with a 0.5×0.5 m^2 frame in front of a circular air inlet. (b): A water sampler, referred to as a Nishkin bottle, with a collection capacity of up to 40 dm^3.

the ground penetration), and (iii) to correct initially performed field gamma spectrometric measurements of fresh fallout by taking the depth distribution into account. Penetration of fission products into ground vegetation and soil can be expected by wet deposition. This is in contrast to dry deposition, where radioactive particles adhere to vegetation and the top layers of soil, resulting only in very shallow ground penetration. The deeper the penetration, the more the energy distribution of the emitted gamma radiation is shifted towards lower energies, which will affect the dose to residents in the area (see e.g. Finck 1992).

Soil cores are usually obtained by driving a cylindrical metal tube into the ground. The diameter of the cores obtained can vary from 5 to 12 cm, and depth from a few cm to 50 cm, depending on the purpose of the survey. In the immediate aftermath of a fallout event, the top layer of soil is the most important, and simple corers with lengths of only 5–10 cm are sufficient (Fig. 5.8). To facilitate rapid soil sampling, the diameter of the cores should be the same as the kind of plastic containers commonly used for γ-ray spectrometry.

(a) (b) (c)

FIGURE 5.8 (a): Soil corer with support. (b): Soil corer pressed into the ground. (c): Soil column is extracted from the corer.

For most soil types found in temperate zones, core depths greater than 20 cm are rarely warranted when assessing global fallout from weapons tests and reactor releases. The soil core is sliced into sections, typically 2–3 cm thick, which are dried and homogenized, and then either investigated using γ-ray spectrometry, or preconcentrated for alpha spectrometry measurements. As with other environmental samples, materials very different from the main sample matrix should either be completely discarded (for example, stones and organic debris in soil), or incorporated, and dried together with the main sample matrix. It is important to note the weight of discarded matter. If the reason for the measurement is to predict crop availability, the radionuclide concentration in the soil matrix is of the greatest importance. However, if the main purpose is to calculate external dose rates, the radionuclide concentration in the whole sample is of most interest.

Fresh soil cores should be sliced and weighed to determine their raw mass before being processed further. This is to determine the density of each slice, the importance of which will be explained later. Ideally, the soil should then be dried and homogenized so that the main sample can pass through a 2-mm mesh. The density of dried soil usually varies from 1 to 2 g cm^{-3}. For many soil types the density varies considerably with depth. The uppermost layer, consisting of litter, has a density ranging from 0.3 to 1.0 g cm^{-3}, whereas the density of deep, undried soil may be as high as 1–3 g cm^{-3}. In most cases, no information is available on the atomic composition of the soil, and in the absence of other data, a reference model, so-called Beck soil (Beck 1972), can be used for fresh soil. This model soil has a density of 1.6 g cm^{-3} and is composed of 67.5% SiO_2, 13.5% Al_2O_3, 10.0% H_2O, 4.5% Fe_2O_3 and 4.5% CO_2. A fresh deposit of fallout penetrating less than 5 cm into the soil can lead to a concentration of 10–50 Bq kg^{-1} per unit kBq m^{-2} deposition in wet soil.

Soil samples should preferably be collected in flat, open areas with little runoff, to ensure that the samples are representative of the fallout in the area. Several attempts have been made to define spatial sampling patterns for the collection of soil cores in order to guarantee the representativeness of the ground penetration in the area under consideration. In Europe, after the fallout of ^{137}Cs from the Chernobyl accident, a triangular sampling pattern was employed by some researchers in Sweden (Isaksson et al. 2001), whereas a hexagonal pattern was suggested by researchers in Finland and Scotland (Sanderson et al. 1995). Regardless of the sampling pattern, the variation in the fallout concentration within a small flat surface can be substantial. For example, the ^{137}Cs activity concentration per unit area was found to vary by 80% among 18 sampling points taken within an area of 80×80 m^2 on a small airfield (Finck 1992).

Gamma-ray or alpha spectrometry is performed to obtain the radionuclide concentration, $c(z)$ [Bq kg^{-1}] as a function of soil depth, z [cm] (see Section 3.5.2). A typical concentration profile after fresh fallout is often assumed to follow the expression given in Eqn. 3.80, where $1/\alpha$ is the relaxation length, a parameter associated with the physico-chemical properties of the soil [cm]. Due to the variation in soil density with depth it is sometimes better to plot the activity concentration of a given radionuclide against the mass depth, z, defined in Section 3.5.2. The mass depth [g cm^{-2}] of layer i is the product of the average density, ρ_i and layer thickness, z_i. The total mass depth is hence the sum of the mass depths of the individual layers along the depth profile

$$z'(z) = \int_0^z \rho(z')\,dz = \sum_{i=1}^{n} \rho(i) \cdot z_i, \qquad (5.5)$$

where n is the layer whose mid-point is at depth z. For fresh fallout, plotting the activity concentration, $c(\rho \cdot z)$, at a particular layer depth z_i, against the mass depth at the same layer depth, that is, $\sum \rho \cdot z_i$, will yield an estimate of the relaxation length $1/\alpha$.

5.2.4 Pasture Grass

Pasture grass is sampled to predict the potential transfer of radionuclides to grazing animals such as dairy cows. In contrast to live monitoring of radiation or air sampling, samples should not be collected from pastures during an ongoing release event. The activity concentrations in grass should be mapped at various locations in regions affected by large-scale radioactive fallout, especially the fission products ^{131}I, ^{134}Cs, ^{137}Cs and ^{90}Sr. These measurements can then provide the basis for decisions by the relevant authorities concerning remedial action such as stabling of animals, provision of contamination-free fodder, or emergency slaughter. Circular or rectangular areas, typically 1 m^2 in size, are sampled using metal frames or a pole and a rotating arm (Fig. 5.9). The grass should be cut at a height of approximately 5 cm, corresponding to the height at which the cows graze.

When measuring the concentration of man-made radionuclides in a pasture, several sampling areas should be chosen, for example, in a diagonal pattern, and the individual samples from each area should be combined. When combining grass samples, foreign materials such as soil or stones should be discarded, as well as plant species known not to be consumed by the grazing animals (e.g. buttercups, *Ranunculus spp*). The fresh weight of the grass is then determined, after which the sample is dried to obtain the corresponding dry weight. Typically, the density of dried grass is between 0.15 and 0.3 g cm^{-3}. Figure 5.9 shows an example of how grass is sampled by Swedish emergency preparedness organizations.

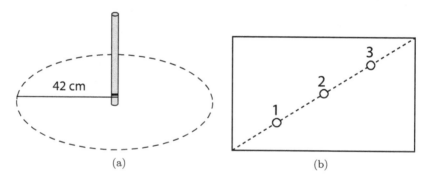

(a)   (b)

FIGURE 5.9 Examples of grass sampling procedures used by Swedish emergency preparedness organizations. Left: a grass sampling area defined using a central pole with a 42-cm-long rotating arm, giving a circular area of 1 m^2. Right: three circular areas are sampled along a diagonal line across a pasture. The sampling areas are placed at a quarter, a half, and three-quarters of the way along the diagonal.

5.2.5 Foodstuffs

Plant species known to be of key importance in the food chain should be collected to monitor the radioecological transfer of a radioactive deposition to various ecosystems. Crops such as rye, barley, wheat and oats are of great importance for both animal fodder and human consumption. Animals accumulate radionuclides by eating contaminated plants or other feed, and by drinking contaminated water. Thus, meat and milk from livestock living in contaminated areas must be sampled for assessment of their radionuclide concentration before being consumed.

Only the edible parts should be used in the assessment of radionuclide concentrations in samples of animal origin, such as beef, pork, lamb, shellfish, freshwater fish, or marine fish, or samples from plants, including salad, vegetables and fruit. After careful selection, the samples should be dried and homogenized. Analysis of fresh meat and fish samples is not recommended, unless an emergency makes this necessary, since migration into layers of the meat matrix that have undergone different degrees of decomposition occurs relatively soon after samples have been collected.

Processed foods such as milk and cheese, or fabricated foods which are canned, dried, or frozen, can be sampled in such a way as to represent the typical consumption pattern for a given country and age group. In many countries, the national authorities responsible for regulating food have data on typical consumption patterns. Because of the high proportion of milk in the diet among children, cow's milk is an exceptional case, and the concentration of the radionuclides ^{131}I and ^{137}Cs must be monitored carefully after any major nuclear power plant accident or nuclear weapons fallout (Fig. 5.10).

The radioecological transfer to various foodstuffs from artificial ecosystems, such as areas of intensive agriculture with a high degree of artificial fertilization and high biomass output per unit area, tends to be less important than food from natural and semi-natural ecosystems. This is partly due to dilution effects: for example, deep ploughing can transfer contaminated topsoil to depths beneath those where crop root uptake occurs, and such foods are often mixed with other items as part of the manufacturing process. Also, artificial fertilizers containing potassium compete with the fission product ^{137}Cs for uptake by the roots.

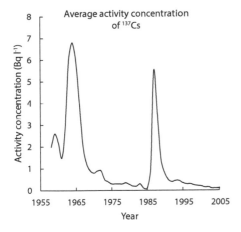 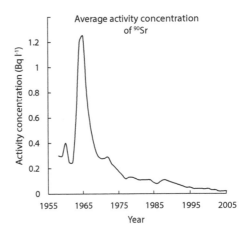

FIGURE 5.10 ^{137}Cs in Swedish cow's milk intended for consumption between 1955 and 2005. Right: Corresponding data for the fission product ^{90}Sr. (Note that the two scales are different.)

Modern agriculture is highly regulated, and must follow national and international restrictions on the handling and distribution of food products. Therefore, foodstuffs containing amounts of radionuclides exceeding the exemption levels (see Table 7.6 in Section 7.2.2), will generally be discarded and will never reach consumers.

For foods originating from semi-natural and natural ecosystems, such as game, freshwater fish, wild berries and mushrooms, there is no competition from fertilizers nor any dilution effects. Hence, these species are thus much more sensitive to radioactive fallout. In some countries, hunters must submit samples of meat to national authorities for determination of radionuclide concentrations, as well as other elements. In Sweden, for example, this is the case for wild boar. For general aquatic monitoring, local health authorities can also carry out regular sampling of freshwater fish, such as perch or pike, which both have a tendency to accumulate high concentrations of the fission product ^{137}Cs. However, authorities can only recommend exemption levels, and cannot forbid domestic consumption by hunters and their families (see also Section 7.2.1).

5.2.6 Biological Human Samples

The main purpose of monitoring the radionuclide concentration in human excreta and biological tissue is to estimate the body burden of radionuclides and the associated absorbed dose. Measurements are used to make retrospective dose assessments of individuals suspected of having internal contamination, as discussed in Section 7.3. Diurnal excreta, such as faeces and urine, will reflect the body burdens of a given radionuclide, if the assay is combined with standard biokinetic data on the radionuclide concerned. The ICRP has compiled exhaustive biokinetic data on radionuclides (ICRP 2012). Urinary sampling is often easier to carry out than faecal sampling, both because it is easier for the subjects, and because sample preparation is easier. Some radionuclides are in fact released mainly through urine; about 80–90% of daily excretion of ^{137}Cs is through urine. However, smaller samples of urine, often called spot urinary samples, must be corrected to standard 24-h excretion before a biokinetic model can be applied for dose assessment (see Section 7.3.4).

For some radionuclides, such as ^{210}Po and ^{60}Co, faecal excretion is the predominant excretion pathway following internal contamination in humans. Here, stool sampling may be necessary for a meaningful quantification of body content. Faeces can be collected in double plastic bags or plastic bottles. It is important that the faeces are kept cool or frozen until they are transported to the laboratory for sample preparation. Since it is possible to carry out gamma spectrometric measurements on radionuclides emitting photons above 100 keV in relatively large bulk samples (100–2000 mL), faeces samples can simply be diluted with distilled water and then homogenized in a mixer. The solution is then poured into plastic vials (<100 mL) and kept frozen until analysis.

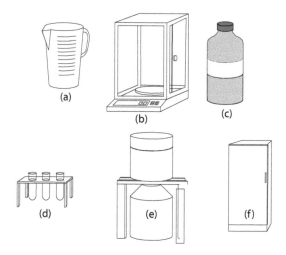

FIGURE 5.11 Collection beaker for urine sampling (a) and a precision scale (b) with an accuracy of at least 0.1 g. Hydrochloric acid (c) is added to the urine samples in order to avoid crystallisation of urea and adhesion of radionuclides on the vial walls. Portions of urine are distributed in test tubes (d) for determination of e.g. stable potassium concentration or creatinine concentration. Radiometry, such as γ-ray spectrometry (e) is performed, and then the samples may be stored in a refrigerator or deep freezer (f) to enable additional measurements.

Special plastic vials are used for the collection of diurnal urine in hospitals and clinics (Fig. 5.11). In cases where it is known beforehand which radionuclide is of interest, and if there is no urgency in assessing the individuals thought to be exposed, the collection vials should be pretreated with a carrier solution to reduce adhesion of radionuclides on the vial wall. When the urine has been collected, it should be either stored cold, at 6–8 °C, or a small amount of HCl should be added. Creatinine, $C_4H_9N_3O_2$, is produced relatively continuously in the body's muscles. For a given individual, its diurnal excretion in urine varies little over a period of days. It can be used to normalize for incomplete urine sampling and is therefore considered a biomarker. After storage at room temperature for 24 hours or more, crystallization of urea makes the sample inhomogeneous, and the creatinine used to normalize to 24-h urine excretion decomposes (Fig. 5.12). If HCl is added at the time of sampling, the sample must be deep frozen below the freezing point for HCL (−36 °C) if the urine is to be freeze dried, as otherwise the HCl will remain in a liquid phase.

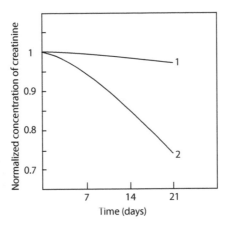

FIGURE 5.12 Creatinine concentration in urine samples as a function of time for: (1) urine treated with 1 mL 37% HCl per 100 mL urine and stored in a refrigerator, and (2) urine with no additives stored at room temperature.

Another biological sample of interest in monitoring internal contamination is hair. Radionuclides have been found to accumulate in hair during protracted internal exposure. For radionuclides emitting only α- or β-particles, such as ^{210}Po or ^{14}C, the concentration found in the hair 1 cm from the scalp can be related the donor's history of internal exposure (Stenstr öm et al. 2010, Rääf et al. 2015). Hair contains the element sulphur, and it can be of interest to carry out an analysis of ^{32}P from neutron activation ($^{32}S + n \rightarrow {}^{32}P + p$). In some cases, this allows an external exposure to neutron radiation to be detected. Typically, around 1 g of hair is needed for alpha spectrometry of ^{210}Po and 10–20 mg for β-particle analysis of ^{14}C.

Other bioindicators such as teeth, bone and nails have been investigated with regard to retrospective dosimetry. Teeth have been used in many retrospective dose applications using EPR or OSL (see Section 7.3.3). Depending on the measurement techniques used, these samples must sometimes be handled very carefully. For example, if samples are to be evaluated using luminescence techniques it is often necessary to protect them from light, to avoid a loss of signal in the detector. Bone and other tissues obtained at autopsies have been used in research for the assessment of internal contamination by radionuclides such as ^{137}Cs.

5.2.7 Bioindicators

Some species are of particular interest in environmental surveying. This may be because of their availability and abundance in a given ecosystem, the ease and accessibility of sampling, or their population stability during changing ambient conditions. These species are referred to as *bioindicators*, and it is their property of accumulating a radionuclide present in their surroundings that makes them interesting. Bioindicators can be repeatedly sampled over time to assess the time evolution of the radionuclide concentration (see Fig. 5.13) and such time series can then be used to demonstrate ongoing and previous aquatic, atmospheric and terrestrial releases of radionuclides. The concentration factor, CR, is defined by the IAEA (IAEA 2010) as follows:

$$CF_i = \frac{\langle c_i \rangle}{c_{med}}, \tag{5.6}$$

where $< c_i >$ is the average activity concentration of nuclide i [Bq kg^{-1}] normalized to the average activity concentration in the surrounding medium, such as shallow marine waters, deep estuarial water, or deep groundwater. When reporting the values of CR, it must be specified whether c_i refers to the wet or dry mass.

(a) (b)

FIGURE 5.13 (a): Bladderwrack, *Fucus vesiculosus*, which grows in shallow saltwater. (b): Annual discharges from the nuclear fuel reprocessing facility at Sellafield, UK (bars), and the activity concentration in *F. vesiculosus* off the west coast of Sweden (solid line). The timescale specifies the year of the measurement of activity in the seaweed, whereas the bars represent the annual release rate from the Sellafield plant 4 years prior to the measurements in the seaweed. Data from Bernhardsson et al. (2008).

A number of terrestrial species have been shown to accumulate important radionuclides such as the fission product ^{137}Cs, the neutron activated fission product ^{134}Cs, and neutron activation products such as ^{54}Mn, ^{58}Co, ^{60}Co and ^{110m}Ag. Moss and lichen growing in temperate and subarctic climates act as accumulators of past atmospheric fallout. In mosses such as white moss (*Sphagnidae*), which has no root system, three layers can be distinguished that should be analysed separately for the determination of the radionuclides present: (i) the upper layer, representing fresh annual growth, (ii) the second layer representing the growth of previous years, in which fallout will be accumulated, and finally, (iii) the bottom layer, which is mixed with soil. It has been found that little exchange occurs between radionuclide contaminants in the soil and in the moss, making moss a suitable indicator for atmospheric fallout.

Examples of bioindicators used for monitoring long-term variations in radionuclide concentrations in aquatic environments, especially in shallow waters, include seaweed, especially bladderwrack (*Fucus vesiculosus*), molluscs, edible mussels (*Mythus edulis* and *Macoma balthica*), as well as seasonal biofilms consisting of various species growing on Fucus, on mosses, or on the walls of water outlets. Fucus generally accumulates heavy metals from the surrounding waters, including the fission product ^{137}Cs and the neutron activation products ^{54}Mn, ^{58}Co, ^{60}Co, and ^{110m}Ag. Bladderwrack has a particularly high uptake of the fission product ^{99}Tc which, in the form of pertechnetate in seawater, has a reported average concentration factor, CF (Eqn. 5.6), of 10^5 in open ocean waters. Two Fucus species *F. vesiculosus* and *F. serratus*, which are common in the coastal waters of the North Sea and the Baltic Sea up to a latitude of around 61 °N, have been used as bioindicators of radionuclide contamination in Northern Europe since the 1960s.

Figure 5.14 illustrates how *F. vesiculosus* is collected in the shallow (<1 m) waters close to the former nuclear power plant at Barsebäck in southern Sweden.

FIGURE 5.14 Sampling of *Fucus vesiculosus* and subsequent preparation for drying in the laboratory before performing beta spectrometry or γ-ray counting.

5.2.8 Waste Water and Sludge

Another material that can be used to assess the environmental transport of released radionuclides is sewage sludge. Sewage sludge is a by-product of the treatment of domestic sewage and water runoff from urban areas, and constitutes the total outflow of radionuclides from the serviced area (Erlandsson et al. 1989). Samples of sludge are usually collected at a municipal sewage plant when the sludge has been digested after the last of the three treatment stages (mechanical, chemical and biological), as shown in Figure 5.15. The mean residence time of sludge in the sewage works depends on the type of plant, and usually varies from 23 days to 34 weeks. An example of a series of measurements of ^{137}Cs concentration in sewage sludge is given in Figure 3.78 in Section 3.5.5.

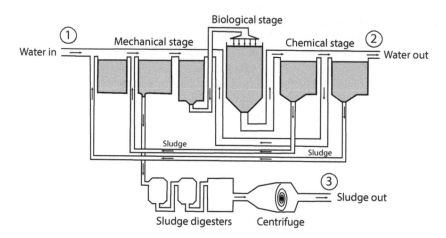

FIGURE 5.15 Samples can be collected from the incoming water (1), outgoing water (2), and accumulated sludge (3) (only sludge from the mechanical stage is shown) and used to estimate the radionuclide outflow from the serviced area, as well the residence time (time delay) before release into the outflow.

5.2.9   Sediment Samples

Sedimentation is a process in which particles are deposited on a surface under the force of gravity. This may take place in the sea or freshwater systems, or on the ground. If runoff and lateral mass transport are negligible, a sedimentation layer is created that forms a basin for deposition in that area. When surveying radioactive deposition in terrestrial environments (such as peat bogs) or aquatic ones (river and estuarine bottoms, the sea floor), sediment samples should be taken from relatively undisturbed areas. As with soil sampling, corers are used, typically with somewhat smaller diameters. However, core depths greater than 20 cm are necessary, especially if layers from fallout from the 1950s and 1960s are to be investigated (Fig. 5.16). The whole core should preferably be frozen before slicing, if the sampling location is close to the radiometric laboratory. Otherwise, the sediment should be sliced on location, and the slices placed in plastic bags. If the concentration of ^{222}Rn is to be determined, the slices must be stored in tightly sealed vials.

Peat bogs are formed when organic material from plants is not fully decomposed due, for example, to anaerobic conditions. Peat bogs are often created from former lakes with poor drainage that have been filled with partially decomposed organic material. Sedimentation takes place as growing moss and other plants create bottom layers where decomposition occurs, adding material to previous layers of partially decomposed growth. Radionuclides and other forms of environmental pollution are continuously deposited from the atmosphere onto the ground. In many cases, it can be assumed that radionuclides such as fission products or actinides will be relatively immobile in the decomposed matrix of peat bogs. Hence, historical fallout may be stored in an immobile layer whose depth increases annually due to the deposition of newly decomposed organic material (Fig. 5.17.)

(a)   (b)

FIGURE 5.16   Core sampling from a raised peat bog in southern Sweden (Vakö mire).

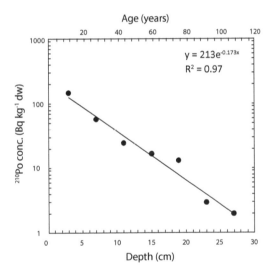

FIGURE 5.17   Concentration profile for ^{210}Po as a function of depth [cm] and age, measured in a peat bog profile from Madagascar. Modified from Holm et al. (2015).

Typically, the thickness of this fixed layer increases by 1–10 mm year^{-1}. The deposition rate can be determined by using the fact that deposition of airborne radon and radon daughters from continental air masses is relatively constant over time. The ^{222}Rn daughter ^{210}Pb is also considered to be immobile in peat layers. However, a daughter nuclide of ^{210}Pb, ^{210}Po, in the same decay chain is more readily determined by radiometry than the β-decay of ^{210}Pb. Assuming equilibrium between ^{210}Pb and ^{210}Po, the ^{210}Pb concentration as a function of mass depth [kg cm^{-2}] can be determined by alpha spectrometry of ^{210}Po in the dried sediment profile. The age of the various segments of the profile and the annual mass deposition rate in the peat bog can be extracted from the concentration profile of ^{210}Pb using the physical half-life of the nuclide, $T_{1/2} = 22$ years.

5.2.10   Preconcentration of Samples

For all the sample matrices described above, the drying process for volume reduction should be adapted to suit the radionuclide is of interest. For halogens, such as iodine (^{131}I), a large fraction will become gaseous and evaporate at temperatures as low as 60 °C. Therefore, drying at temperatures just below 60–70 °C is preferred to make sure volatile elements such as I and Po remain in the matrix. However, if the elements of interest are more refractory, drying temperatures up to 110–140 °C can be used. The samples should be dried until most of the water has evaporated. The samples should be weighed every 2–3 h during drying to monitor the gradual decrease in weight. Drying can be discontinued when the mass of the sample no longer decreases. However, if the sample is weighed immediately after removing it from the drying oven, there will be an initial increase in mass due to the absorption of moisture from the ambient air. The weight should stabilize after a few minutes at room temperature. The wet and dry masses of the sample are then recorded. The concentration of the radionuclide in the original fresh state of the sample can then be determined using alpha spectrometry, total β-particle counting, or gamma-ray spectrometry.

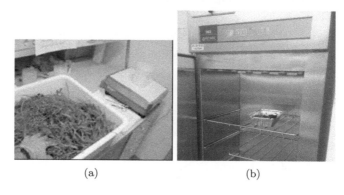

(a)   (b)

FIGURE 5.18   A fraction of the fresh grass sample (left) is dried in an oven at 60 to 90 °C (right).

For large bulk samples (>1 kg) it is important that some form of homogenization is carried out before drying, as only a fraction of the original sample can be contained in standard beakers used for γ-ray spectrometry (typically ranging from 30 to 200 mL in size; see below). For sliced soil and sediment samples, which often have a fresh weight of less than 250 g per slice, homogenization can take place after drying the separate slices. The objective is to achieve a homogeneous sample that can be accommodated in standard beakers, reproducing the instruments calibration geometry as closely as possible. This will ensure agreement with the standard sample used to derive the absolute efficiency, ε_{abs} (Eqn. 4.51), of the gamma spectrometer, and high accuracy in the estimate of the activity concentration.

From the discussion on sampling in Section 5.1, it is clear that it is important that the dried fractions are truly representative of the overall sample (or lot). Milling and grinding dried samples before transferring them to the measurement beakers further increases the homogeneity. In some cases, when the dried sample does not fill the measuring beaker, granulated sugar, which is usually considered to be completely free from naturally occurring radionuclides such as ^{40}K, can be used as a filling matrix by mixing it with the sample. Note that sugar may sometimes contain traces of antropogenic radionuclides such as ^{137}Cs if originating from areas with high atmospheric deposition. It is, however, recommended that a measuring beaker be chosen so that the dried sample completely fills it.

Using γ-ray spectrometry, it is possible to make quantitative assessments of the activity concentration in bulk liquid samples without any preconcentration or homogenization. However, for accurate determinations it is recommended that measures are taken to maintain the homogeneity of the sample. An example of this is the addition of HCl to fresh urine, as described in the previous section. For milk, which contains large numbers of lipid molecules, volume reduction by means of wet or dry ashing (see Section 5.3.1) is recommended instead. Large freshwater samples (>1000 mL) should be passed through a coarse filter to remove large particulates and small biota from the water. The filtered water can then be passed through ion-exchange resins to extract the various radionuclides in ionic form.

It is recommended that a number of standard beakers are selected that are compatible with the standard geometries used for measurements in a particular radiometry laboratory, and to process samples in similar ways. For liquid samples, which are often treated with acids such as HCl, it is important that the beakers containing the sample can withstand the long-term storage of solutions with low pH. Examples of suitable beakers are plastic tubs or containers that can withstand a low pH and storage conditions in cooled or room temperature for the desired time are shown in Figure 5.19. However, many plastic beakers are not suitable for storage in a deep freezer if the sample contains a large amount of water (urine), as these will not withstand the volume expansion of the frozen sample.

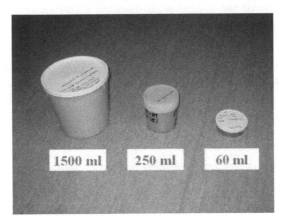

FIGURE 5.19 Three typical beakers and standard sample geometries used in a radiometry laboratory for γ-ray spectrometry.

It is recommended that sample containers of at least the following types and geometries are available at a radiometry laboratory.

- Large beakers, 1000–2000 mL, are recommended for urine or faeces that must be analysed soon after collection. Large beakers are also needed for bulk water samples that have not been preconcentrated. This beaker size is also preferable for fresh grass samples. However, a different calibration standard must be used for fresh grass, as its density is much less than that of liquid and solid samples, and this will lead to significantly different values of the absolute efficiency in gamma-ray spectrometry systems.

- Intermediate beakers between 100 and 1000 mL in volume: A beaker with a radius of 5–10 cm and a length of 6–8 cm is recommended for unsliced soil samples collected in the field. Fractional urine samples can also be analysed in a beaker of 100–250 mL (this is discussed further in Section 7.3), but with a screw lid rather than a snap lid

FIGURE 5.20 Standard beakers and sample holder geometries used for high-resolution γ-ray spectrometry.

to prevent leakage. For liquid samples such as urine it must be possible to fill the remaining volume of the beaker with distilled water to achieve a standard geometry in the gamma-ray spectrometry system.

- Small beakers: 10–100 mL beakers are preferable for dried vegetation samples, or sliced, dried soil or sediment samples. Dried vegetation usually has densities in the range 0.2–0.5 g cm^{-3}, while dried sediment samples can have densities in the range 0.5–2 g cm^{-3}. As with the large sample geometry, it is recommended that different calibration standards are used for the two ranges of density.

5.2.11 General Factors Regarding Environmental Samples

Sample labelling and identification are crucial in situations where large numbers of samples are collected and analysed. A specimen arriving at a laboratory with no information on the collection date, or on the exact sampling location will be of little use in most applications within radiation protection and emergency preparedness. Pre-printed sheets should be made available that can easily be completed in the field. These should contain information on collection date and time, location of sampling (which can now be determined very exactly thanks to commercially available portable GPS systems), and the name of the person carrying out the sampling. Ambient factors, such as precipitation, wind, temperature and humidity that may influence the analysis of the sample should also be recorded. The samples should be labelled with a code that makes it traceable to an individual entry in the datasheet describing each sample collected. Digital images or hand-drawn maps of the sampling location can easily be attached to a particular measurement or sample. Tagging systems using naming conventions to encode sample-specific information can be adopted, similar to the digital tagging systems used for commercial products in the retail trade.

It is also important to prevent cross contamination. This may pose a problem when large, wet samples, such as fresh pasture vegetation or moist soil columns, have to be collected and analysed urgently. For emergency soil sampling, where no further sample processing is necessary before γ-ray spectrometry, the samples collected in containers or beakers can easily be wrapped in plastic bags. The samples should first be unpacked at the facility in a location away from where the charged particle or γ-ray spectrometry equipment is located. The exterior surface of the beakers should be treated as potentially contaminated until verified as uncontaminated. This can be done by hand-held survey meters such as GM counters or total gamma counting scintillators with a β-particle-sensitive entrance window.

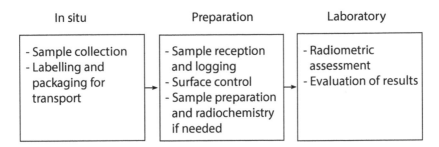

In situ	Preparation	Laboratory
- Sample collection - Labelling and packaging for transport	- Sample reception and logging - Surface control - Sample preparation and radiochemistry if needed	- Radiometric assessment - Evaluation of results

FIGURE 5.21 Example of a sample collection and analysis chain, from sample collection in the field to analysis and potential long-term storage.

A sample collection and processing chain should be decided upon for different types of samples, and should be consistent with the reason for collecting and analysing the sample. It is especially important to determine whether samples should be placed in long-term storage in a sample library, or if they are to be discarded after radiometric assessment.

The sampling chain illustrated in Figure 5.21 is suitable for environmental sampling (soil, water, vegetation, etc.). Transportation may present a problem if samples are collected from a region with highly elevated radioactivity levels. Soils samples containing fresh fallout may have to be transported as radioactive goods if, for example, each package contains a total γ-ray-emitting activity of 10 kBq or more. In other cases, transportation times may be long if the samples are collected in remote areas (e.g. sediments and glacier columns in the Arctic and Antarctic). Both the radioactive decay and the physical storage conditions during transport must be considered in order to ensure that accurate corrections can be made once the samples are analysed in the laboratory.

Caution must always be exercised when handling samples of unknown origin. For example, samples submitted by concerned members of the public must be verified to be free from external contamination, which could otherwise be introduced into the sample treatment laboratory. In addition, a radiometry laboratory may also carry out the analysis of samples from tracer studies, in which case it is desirable to separate the sampling procedures of environmental samples from samples that have been intentionally contaminated in conncetion with e.g. tracer studies. Such studies can e.g. consist of biological samples from subjects that have ingested a known amount of ^{57}Co to trace the turnover of radioactive cobalt in humans. This is especially true if the tracer radionuclides consist of pure α or β emitters, such as ^{209}Po or ^{14}C, as it is more difficult to monitor surface contamination on laboratory surfaces or on instrumentation when these radionuclides are involved, by means of commonly used γ-sensitive contamination survey meters.

Finally, there must be a well-planned procedure for the storage of samples and their associated records and data logs. If the documentation on samples is lost, the value of continued storage can be questioned. An important principle regarding storage is to retain samples belonging to extended time series of measurements for as long as possible. On the other hand, samples associated with emergency preparedness may be stored for a relatively short time, or discarded immediately. Samples submitted by private enterprises or as part of commissioned analyses must be stored according to the bilateral agreement governing their collection and use. This is especially true if the party submitting the samples has its own procedures, which are certified and accredited according to international accreditation guidelines, which demand that the method of storage of samples and access to them must be specified.

5.3   RADIOCHEMISTRY

5.3.1   Sample Preparation for Radiometry

For charged particle radiometry, many chemical procedures are necessary following initial preconcentration. These are referred to collectively as radiochemistry. The radiochemical techniques applied within radiation protection and environmental radioactivity are mainly intended to concentrate the samples to facilitate the detection and quantification of α-emitting and pure β-emitting radionuclides. Radiochemistry is also required for pure electron capture elements, very-low-energy γ-ray emitters, and for samples with low concentrations of γ-ray-emitting nuclides. In contrast to γ-ray spectrometry, where the presence of γ-ray-emitting radionuclides can be quantified by an external spectroscopic device, such as a NaI(Tl) or HPGe detector, the physical range of α-particles, and often also of β-particles, is too short to allow the radiation to escape the sample matrix. The radionuclides in the sample must therefore be extracted from their original matrix, and either dissolved in liquid scintillators, or used to form thin samples on a support. It is essential that these processes maintain traceability to the radionuclide concentration in the original sample matrix. For a detailed description, see Jaakkola in Holm (ed.) (1994).

5.3.2   Preconcentration and Tracer Addition

Environmental samples collected for radiometry consist of bulk samples, such as soil columns, sediment columns, or large samples of water. To detect and quantify α-emitting radionuclides, such as the transuranic elements ^{237}Np, $^{239+240}$Pu, or ^{241}Am, preconcentration, also referred to as volume reduction, is necessary, as is digestion of the original sample matrix. Radiometry of α- and β-emitting radionuclides is necessary when the number of atoms of a radionuclide is too small for accurate mass spectrometry to be implemented. The number of atoms, N_i, is a function of the activity, A [Bq], and the physical half-life, $T_{1/2}$ [s], of the radionuclide, as can be seen in Eqn. 5.7. As mentioned earlier, mass spectrometry is preferred over alpha or beta radiometry for radionuclides with a half-life longer than 400 years (see Section 4.2.6).

$$N_i = \frac{A}{\left(\frac{\ln 2}{T_{1/2}}\right)} \tag{5.7}$$

The initial preconcentration of fresh samples was described in Section 5.2.10, where the volume reduction of vegetation, soil and sediment samples for γ-ray spectrometry was described, i.e., homogenization and drying at 55–110 °C to remove water. For biological and environmental samples intended for α- and β-radiometry, preconcentration is also followed by sample digestion to facilitate the removal of the radioactive element(s) from the original sample matrix into a homogenous liquid form. This preconcentration and sample digestion are often performed together, for example, by means of dry or wet ashing (see Section 5.3.3).

Almost all the radionuclides measured with charged particle spectrometry have very low concentrations, i.e. trace quantities, with molarities of 10^{-10} [1 M = 1 mol L^{-1}] down to extremely low concentrations, sometimes less than 10^{-18} M. In some cases, very low concentrations of γ-ray-emitting radionuclides need to be assessed, for example ^{137}Cs in seawater, and similar radiochemical procedures to those used for charged particle spectrometry must be used. The very small elemental concentrations of these radionuclides mean that, if traceability is to be maintained, the radiochemical procedures applied when digesting, dissolving and separating them must be carefully considered and performed. The main problems associated with handling very low concentrations of elements are deviation from first-order reaction kinetics, and adsorption onto surfaces, such as the walls of the vessel containing the sample.

Assume that we have a radioactive element, such as ^{238}Pu, ($T_{1/2}$ = 87.7 y), in an aqueous solution with a molar concentration of 10^{-17}–10^{-18} M. This particular radioactive element may exist in several oxidation states (III-VII). Under normal ambient conditions, exchange occurs between oxidation state IV and state VI, and an equilibrium is established between their concentrations. Imagine that a sample of seawater is collected and brought in a sample vessel to the laboratory. In this sample, the Pu atoms may exist in any of the allowed oxidation states. The atoms are thermally excited according to Boltzmann statistics, but their kinetic energies mean that a long time must elapse before a new equilibrium can be reached in the sample vessel, as the average distance needed to attain uniform dilution in the sample matrix or the solution in the vessel is relatively large. Until such an equilibrium is established, predictable and uniform reaction behaviour of the element in chemical procedures cannot be expected. A proportion of the Pu ions in oxidation state IV may be more likely to attach to colloids and particles in the solution, whereas Pu ions in oxidation state VI may remain in solution. For U atoms in oxidation states IV and VI, a period of about 400 days is required to reach equilibrium. Reduction–oxidation cycles are used to speed up this process.

Adsorption is a problem when preconcentrating and digesting samples that contain low mass concentrations of the element. Ions of the radionuclides can be attached to the walls of vials and to other phases present in the solution. To reduce this problem, it is necessary to acidify the solution, or to add a stable carrier in order to provide enough mass to decrease the relative losses resulting from adsorption. Adsorption of trace elements of a radionuclide can also occur with liquid phases or if a precipitate is present in the solution.

Adsorption is obviously undesirable in radiochemical separation where economy in handling the sparse number of atoms is crucial, not only to ensure traceability, but also to obtain enough atoms to provide a sufficient number of counts in the charged particle spectrometer to overcome quantum uncertainty in a reasonable measurement time. A summary of the methods used to reduce adsorption is given below.

- Maintain a high acid concentration (low pH).

- Use complex-forming agents.

- Avoid the formation of finely divided precipitates with large surface areas.

- Avoid storing dilute solutions for long periods.

- Add stable carriers.

In some cases, no radioactive isotope is available for use as a yield determinant. A stable isotope can then be used, for example, Ni or Fe for the analysis of ^{63}Ni and ^{55}Fe. Chemical analogues can also be used, for example, Am for Cm and Re for Tc.

The role of the carrier is not only to prevent the adsorption of radionuclides on the vial walls, it is also important as it increases the number of atoms available in the chemical separation steps, for example, forming a precipitate from an aqueous solution. Furthermore, the addition of a carrier can prevent or reduce accidental losses of the radionuclide, through sputtering, or evaporation during a so-called wet ashing process (described in the next section). Another very important role of the carrier is to act as a *yield tracer*. A yield tracer is a radioactive carrier, which is added to the preconcentrated sample matrix at well-known amounts, and is used to trace the measured amount of α- or β-emitting nuclei back to the concentration in the original sample matrix (see Eqns. 4.37–4.42). This is necessary for accurate quantification, since even carefully designed separation processes will inevitably lead to losses of the radionuclide of interest, and these must be accurately accounted for in the final calculation of the original sample concentration. The yield determinant must be added before the sample is subjected to any kind of digestion involving reactive chemicals or elevated temperatures. A yield tracer is often in a well-defined chemical state, whereas the radionuclide of interest can exist in one or more oxidation states. This must be considered when attempting to obtain exchange equilibrium between a radiotracer and the radionuclide of interest.

Hold-back carriers are added to prevent unwanted co-precipitation of other radionuclides in the solution. An example of a hold-back carrier is ferric hydroxide, which is used together with the radiotracer and the concentrated uranium in uranium analysis. A scavenger is a strongly adsorbing precipitate that removes unwanted radionuclides from the solution, and therefore acts as a purifier.

5.3.3   Sample Digestion

After initial volume reduction and tracer addition, organic matter, such as foodstuffs, biological samples, soil, and sediment samples, must be decomposed to ensure that the sample is in a chemical form that enables the separation of one or more radioactive elements. Before sample digestion, the radionuclides can be in one of three states.

Ionic form. An anion is an atom or molecule with one or more excess electrons. It is thus negatively charged, and in a dielectric medium it moves towards the positively charged electrode (the anode). A cation, on the other hand, is an atom or a molecule from which one or more electrons has been removed. It is positively charged and will move towards the negative electrode (the cathode) in a dielectric medium. Radioactive elements often exist as cations, and many radionuclides of interest are actinides with high oxidation states, such as Pu(IV) and Am(III). Table 5.1 gives an overview of some ionic states commonly found in radiochemical samples.

Molecular or compound form. Many of the radioactive elements measured using charged particle spectrometry are metals. Metals have a tendency for one or more electrons per atom to be outside the outer filled shell, making them likely to bind to elements with unfilled outer shells. This is called covalent bonding, and is more stable than van der Waals bonding. The combination of two or more elements is referred to as a compound. Non-metals, on the other hand, often have an electron deficit in their outer shell, and they form ionic compounds which often consist of a combination of a metal and a non-metal, often in a solid lattice form (commonly referred to as salts). Table 5.2 gives some examples of metal compounds commonly found in solutions suitable for charged particle spectrometry.

Colloids or pseudo-colloids. Some solutions of trace quantities of elements behave as colloids rather than true solutions. Such species are termed *radiocolloids* and are aggregates of 10^3–10^7 atoms. The size of these aggregates ranges from 0.1 to 1500 nm. They are formed during hydrolysis, especially by actinides in high oxidation states. Two kinds of colloids can be differentiated: true radiocolloids and pseudo-colloids, where in the latter a radionuclide is adsorbed onto an existing colloid particle, such as humic acid or $Fe(OH)_3$. The formation of real colloids can be prevented by reducing the pH or by the addition of complexing agents. It is difficult to predict the chemical behaviour of these radiocolloids as the elements involved are not at equilibrium.

TABLE 5.1 The ions of radioactive elements commonly encountered.

Element and type of ion	Chemical formula
Simple cations	
Hydrogen	H^+
Potassium	K^+
Silver	Ag^+
Sodium	Na^+
Cobalt	Co^{2+}
Polyatomic cations	
Ammonium	NH_4^+
Hydronium	H_3O^+
Simple anions	
Chloride	Cl^-
Fluoride	F^-
Oxide	O^{2-}
Bromide	Br^-
Oxyanions	
Carbonate	CO_3^{2-}
Nitrate	NO_3^-
Phosphate	PO_4^{3-}
Sulphate	SO_4^{2-}
Hydroxide	OH^-
Organic anions	
Acetate	CH_3COO^-
Cyanide	CN^-
Oxalate	$C_2O_4^{2-}$

There are two principle means of decomposing organic matter: wet and dry ashing (oxidization).

Dry ashing (also called *dry oxidation* or *thermal decomposition*). The organic sample matrix is burned to ash (charred) in an oven called a *muffling furnace* (see Fig. 5.22), at temperatures high enough to cause oxidation (combustion) of organic matter. Combustion leads to a dry ash, and thus further reduction of the sample volume. Common ashing temperatures are 550–600 °C. Dry ashing is often used when the radionuclides to be analysed are actinides, such as ^{241}Am, or alpha emitters such as ^{226}Ra, ^{234}U and ^{238}U. The ashing process can be stimulated by adding magnesium nitrate ($Mg(NO_3)_2$). However, dry ashing cannot be used for volatile elements such as Po, or highly volatile ones such as iodine (which evaporates at 100 °C) due to losses during heating. Wet ashing can be used for

TABLE 5.2 Alpha-particle-emitting radionuclides commonly analysed using charged particle spectrometry.

Element	Compound	Chemical formula
Americium	Americium(IV) oxide	AmO_2
	Americium(II) chloride	$AmCl_2$
	Americium(III) chloride	$AmCl_3$
	Americium(II) iodide	AmI_2
Uranium	Uranium(IV) oxide	UO_2
	Uranium(V, IV) oxide	U_3O_8
	Uranium(VI) fluoride	UF_6
	Uranyl(VI) hydroxide	$UO_2(OH)_2$
	Uranyl(VI) carbonate	$UO_2(CO_3)$
Plutonium	Plutonium(IV) oxide	PuO_2
	Plutonium(III) chloride	$PuCl_3$
	Plutonium(IV) phosphate	$Pu_3(PO_4)_4$
Neptunium	Neptunium(IV) oxide	NpO_2
	Neptunium(III) chloride	$NpCl_2$
	Neptunium(III) fluoride	NpF_3
Thorium	Thorium(IV) oxide	ThO_2
	Thorium(II) hydride	ThH_2
	Thorium(IV) sulphate nonahydrate	$Th(SO_4)_2 \cdot 9H_2O$
Polonium	Polonium(IV) sulphide	PoS_2
	Polonium(II) phosphate	$Po_3(PO_4)_2$
	Polonium(IV) thiocyanate	$Po(SCN)_6$

Po (see below), but samples containing I must be dissolved in an alkali solution of, for example, KOH and KNO_3 at 450 °C (referred to as alkaline fusion) after which dry ashing is attempted. In both these separation methods, losses over 20% of the elements I and Po can be expected. It is generally recommended that unnecessarily high temperatures be avoided to prevent the evaporation of volatile elements such as Po.

Wet ashing (also termed wet oxidation). In wet ashing the organic matter is immersed in an acid bath and heated. Wet oxidation can either be performed in open vials in a heated water bath or in a closed liquid acid system. In the former system, the exhaust fumes from the acid solutions must be removed by a venting system such as a fume hood. The acid solution is heated to about 60–110 oC. The most common solution used in wet ashing is a solution of HNO_3 and $HCl + H_2O_2$. Wet ashing is preferable when quantifying the concentration of volatile radionuclides such as [210]Po. Ideally, when the process is complete all the organic matter should be dissolved in the wet ashing solution. Carbon residues can be removed by filtration, to obtain a particle-free solution of the wet oxidized sample.

The advantages of wet ashing over dry ashing are that the low pH prevents adsorption of the radioactive elements on the vessel walls during digestion, and that the lower temperature leads to less loss of volatile fractions of the radioactive element. Furthermore,

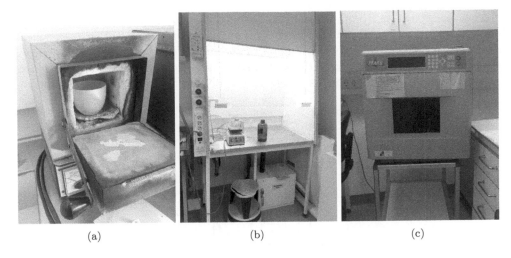

(a)　　　　　　　　　(b)　　　　　　　　　(c)

FIGURE 5.22 (a): A muffling furnace used for dry ashing of preconcentrated organic samples. (b): An open fume hood used for e.g. wet ashing processes. (c): A microwave oven designed for the combustion of organic samples.

more isotropic exchange between the radiotracer and the radionuclide can be achieved in a strongly oxidizing solution. The disadvantages of wet oxidization are that the chemical procedure is more labour intensive, and the throughput of samples is lower. Also, the formation of non-soluble components may require removal by filtration. Finally, the high acidity leads to a high reagent background, which can interfere with the subsequent separation of the radioactive elements.

It is also important to consider the venting of the fumes. Excessive use of heated HNO_3 and H_2O_2 can cause corrosion in ventilation systems, and these should be clad with a non-reactive material such as Teflon, and must be able to withstand a total fume production of 0.5–1 litres HCL or H_2SO_4 per week.

5.3.4 Radiochemical Separation

After volume reduction and sample digestion, the radionuclides from the original sample are distributed in an acid solution, and then undergo radiochemical preparation. Methods of separating various compounds or elements from each other are essential in radiochemistry in order to isolate the element of interest for charged particle spectrometry. The separation methods used in radiochemistry are described below.

5.3.4.1 Precipitation

The most common chemical separation technique is precipitation (Fig. 5.23). This is usually carried out by adding a chemical agent, referred to as the *extractant*, to a dilute aqueous solution of the radionuclide, the *analyte*. This allows the slow growth of large crystals formed by the reaction between the radionuclide and the extractant. The pH of the solution is adjusted so as to minimize colloid formation. The radioactive elements Pu, Am, U and Th, for example, are co-precipitated with Fe in aqueous solutions. Actinides, such as Am and Pu, in the +3 and +4 oxidation states can be precipitated as fluorides and then dissolved using boric acid in concentrated HNO_3.

After precipitation, the precipitate is carefully washed to remove impurities, and then dissolved and re-precipitated to aid further purification. The precipitate is then collected by filtration or centrifugation. As the amount of the radionuclide present may be very small, the addition of macroscopic quantities of carriers and yield tracers is usually required to ensure that the radioactive species will be in kinetic and thermodynamic equilibrium.

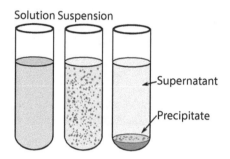

FIGURE 5.23 A schematic showing how solution, suspension, precipitate, and supernatant are defined.

Even when a stable carrier is added to ensure maximal yield, unwanted co-precipitates should be minimized. An example of such co-pecipitation is when a radionuclide is incorporated into, or adsorbed onto, the surface of a precipitate that is not formed from the same element as the radionuclide, or when the radionuclide isomorphously substitutes one of the elements in the precipitate. The absorption of radionuclides by $Fe(OH)_3$ and the co-precipitation of the actinide fluoride LaF_3 are examples of this kind of behaviour. Further purification is generally necessary.

5.3.4.2 Solid Phase Extraction Using Ion-Exchange Resins

Radionuclides can be separated from other compounds in a solution by passing the solution over an ion-exchanger (often called a resin). Depending on the chemical properties of the radionuclide and that of the resins, solutions and resins can be combined so that the desired radionuclide is separated either by becoming attached to the resin, or by being transmitted through the resin, while other unwanted compounds are removed from the solution by filtration.

Ion-exchange resins consist of an insoluble matrix (also termed a support) containing a so-called *extractant*, which is an electronegative or electropositive non-metallic element that can be used to attract correspondingly electropositive or electronegative substances. Reversible chemical reactions take place in which there is an exchange of ions between the resin and the surrounding solution. A solution containing two or more elements can be passed through a column containing the resin, in order to exchange the electronegative or electropositive elements in the solution, depending on the type of resin. A positive ion, such as Pu(IV), can be made to behave like a negative ion by forming a complex in strong HCL or HNO_3. The elements in the solution, also referred to as counter-ions, become bound to the exchange sites on the resin, as illustrated in Figure 5.21. A resin has a finite number of exchange sites, which limits the amount of sample that can be treated per unit mass of the resin.

Solid ion-exchangers can be produced synthetically or derived from naturally occurring material. The most common type of resin consists of synthetic organic matrices, which can be tailor-made for ion-exchange in a number of solutions. A distinction is made between

Cation-exchange resin

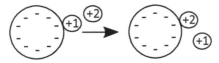

Anion-exchange resin

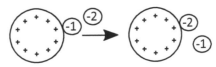

FIGURE 5.24 Illustration of cation and anion exchange. The small circles indicate positive (cation exchange) and negative (anion exchange) counter-ions, and the large circles the resin beads.

weak and strong ion-exchange resins, in that a strong exchanger is able to remain ionized in solvents over a broader pH interval (2–12). A wide variety of resins is commercially available for radiochemistry, as well as for many other chemical applications.

Ion-exchange resins can be used in so-called *ion-exchange chromatography*. A single column of resin can be used to absorb an analyte consisting of multiple elements, or of a radionuclide in several chemical states, by passing the solution or suspension through a resin in which the components move at different rates, as illustrated in Figure 5.25. These components (or analytes) can then be separated by adding so-called elutes in successive steps to eluate the various analytes separately (Fig. 5.25). Extraction resins have the advantage over traditional precipitation methods of being more environmentally friendly, in that they usually generate less liquid waste. They can also be used with weaker acids, and they do not create hazardous waste in the form of organic solvents.

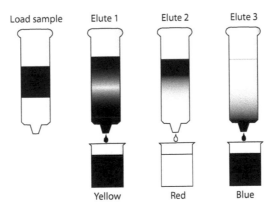

FIGURE 5.25 Principle of extraction of several analytes from one solution, showing how various analytes can be extracted (eluted) from a sample using the same resin column. See also colour images.

5.3.4.3 Solvent Extraction

Another method of separating compounds in radiochemistry is to use their relative solubility in two different immiscible liquids, usually an organic solvent and water. This type of separation is well suited to the separation of metal ions and has therefore been used extensively for the separation of fission products and actinides in samples from environmental monitoring programmes. Because of its selectivity, solvent extraction is also the method used in large-scale nuclear and radiochemical processes, for example, the reprocessing of spent nuclear fuel to retrieve the fissile elements. The technique is based on the formation of uncharged organic metal complexes, which have a higher solubility in organic solvents, enabling extraction of the radioactive elements of interest. Three main types of organic metal compounds can be formed with this kind of extraction: (i) organic chelates, (ii) inorganic metal complexes forming adducts with solvating organic compounds such as hexone, and (iii) ion–pair complexes between large organic cations and negatively charged inorganic complexes. Solvent extraction is similar to anionic and cationic exchange, but in the liquid phase.

The naturally occurring isotope ^{210}Po is commonly quantified in both environmental and biological samples, for example, when analysing the amount of ^{210}Pb (the low-energy beta-emitting mother nuclide of ^{210}Po) for the dating of sediments, or when monitoring the presence of alpha and beta emitters in urine or faeces from industrial workers. When mixed in a solution in ionic form, Po can be extracted into organic solvents as it forms complex halides. The rate of extraction of the element can be increased by solution in hydrochloric acid. Generally, if extracted in a high concentration of mineral acid, back-extraction is performed with a weak concentration, and vice versa.

5.3.4.4 Combined Methods of Extraction

In practice, the methods mentioned above are combined in the laboratory to achieve the highest possible recovery of the analyte from the original solution. Extraction chromatography is an example of a method that combines solvent extraction, with its diversity of selective extractants, and ion-exchange chromatography. Figure 5.26 presents a flow chart describing the successive procedures used for the extraction of Pu from a water sample. Further details on radionuclide extraction can be found, for example, in Sidhu (2004).

5.3.5 Source Preparation for External α-Particle Spectrometry

The radioactive elements separated using radiochemistry must be deposited onto discs or slides for α-particle or low-energy β-particle spectrometry by means of an external detector. (This is in contrast to LSC, where the sample is dissolved in the liquid detector). Charged particle spectrometry also requires that the samples be measured in a vacuum, as air between the sample and the detector causes straggling or absorption of the emitted charged particles (see Fig. 4.41). Ideally, the radioactive elements deposited on the disc should be fixed, but should not penetrate too deeply below the surface. It is important that the yield tracer and the radionuclide of interest penetrate the disc to a similar degree. The prepared sample should also be free from interfering elements not relevant to the analysis of the radionuclide. The more material is deposited on the sample disc, the greater the mass depth, which in turn, increases the energy straggling of emitted α- and β-particles. Two main methods are used for sample deposition.

- *Spontaneous electrochemical deposition.* The radioactive element in an acid solution is

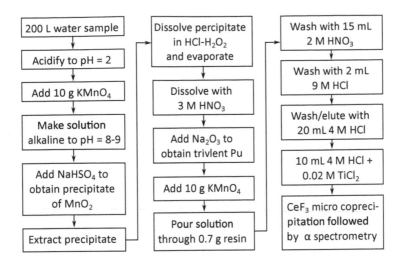

FIGURE 5.26 Flow chart illustrating the extraction procedures used for the separation of plutonium in a 200-L seawater sample. A resin designed for the extraction of actinides is also used for the extraction of Th, U, Am, Cm and Fe.

electrochemically deposited onto a metal surface. It is important that the electrochemical potential between the radioactive element and the metal is sufficient to stimulate deposition, and that there are no other ions in the solution that could compete with the radioactive element and be co-deposited on the metal disc. The type of acid in which the radioactive element is dissolved in the separation process affects the plating rate of the element for a given type of disc. For example, Po is more smoothly deposited on silver discs when dissolved in HCl than in HNO_3, H_2SO_4 or acetic acid solutions. A common problem is interference by ions in the same ionization state as the radionuclide of interest: for example, Fe ions can inhibit Po deposition. In addition to the acid solution and the presence of interfering ions, the efficiency of the plating procedure is also affected by the temperature; optimum temperatures being around 60 °C.

- *Electrolysis deposition.* When an electric field is created between two electrodes placed in an acid solution, it acts as an electrolyte, and the ions of the radioactive element are reduced and deposited on the anode. If the anode is a disc (of platinum (Pt), nickel (Ni), or gold (Ag)), it can be used as the sample disc for charged particle spectrometry.

There are other methods for sample preparation using precipitation techniques. For further reading we suggest, for example, Sidhu (2004).

5.3.6 Sample Preparation for Liquid Scintillation Counting

As mentioned in Section 4.3.4, radiometry using liquid scintillators is a process in which the sample is dissolved in the detector itself, in order to achieve an almost 100% absolute detection efficiency. However, various measures are needed to achieve the optimum detection efficiency. The first is to separate the radioactive elements from their natural matrix or, if that is not possible, to prevent the matrix from interfering with the detection of the radioactive elements. This means that for several sample matrices, radiochemical processing is needed before the radioactive elements in the sample can be dissolved in the liquid scintillator. This section is in part based on R. Edler (Edler 2015).

5.3.6.1 Liquid Scintillator Materials

Liquid scintillators are often referred to as cocktails (see also Section 4.5.3 and Fig. 4.44). There are two groups of liquid scintillator cocktails: *lipophilic* cocktails and *emulsifying* cocktails. Lipophilic cocktails consist of a solvent, often an aromatic organic compound, and a scintillator (or fluorescing agent). Emulsifying cocktails are also a combination of a solvent and a scintillator, but a surfactant or detergent is also added, which allows otherwise insoluble species to be suspended in the solvent. The suspended particles are so small that they are sufficiently uniformly distributed. This suspension is also referred to as a micro-emulsion, and is similar to the suspension of fat globules in milk.

Many environmental and biological samples analysed with radiometry are in aqueous solutions, which are not easily mixed with an aromatic solvent, and thus require an emulsifying cocktail for liquid scintillation radiometry. As described in Section 4.5.3, the fluorescent molecules that afford the solution its radiometric capability must be bound to the main molecular solvent. The main solvent is usually an organic solvent such as toluene (C_7H_8), xylene (C_8H_{10}), or an alkyl benzene ($C_{18}H_{30}$). When the emulsifier, or surfactant (for example, ethoxylated alkylphenol, which has a lipophilic organic tail connected to a C_9H_{19} molecule) is added to the cocktail, it ensures that the radioactive aqueous solution is uniformly distributed throughout the mixture, providing stable measuring conditions. It is necessary to use aromatic molecules as the main solvent since they have a high density of electrons in their π-orbitals, making them efficient at capturing energy from the charged particles emitted by the dissolved radionuclides.

The beaker containing the cocktail and the radioactive solution must be transparent to visible light. These are often 4–20 mL in size, corresponding to the volume of the PMT counting cavities of the liquid scintillation counter. Glass vials not only have the high optical clarity necessary to transmit the visible light generated by the scintillation process to the PMTs (Fig. 4.44), but they are also chemically inert, and provide stable containment of cocktails and radiation samples. Plastic beakers are sometimes preferable as they give less contribution to the detected signal from natural background radiation. They are also combustible, and thus easier to dispose of. Plastic beakers are also shatterproof, improving safety during laboratory handling.

5.3.6.2 Mixing Various Samples with Liquid Scintillator Materials

The benefit of liquid scintillation counting of radioactive substances in solution is that certain liquids are miscible and can therefore be rapidly and completely mixed with LSC cocktails. However, in aqueous solutions, where the radioactive elements may form a compound with inorganic anions such as chlorides, nitrates, sulphates and phosphates, problems may arise due to divalent and trivalent anions which may cause the mixture in the LSC cocktail to become chemically unstable. Mixing is, however, dependent on both the volume and concentration of these anions. Some metallic salts form coloured solutions that cause colour quenching of the LSC process. The remedy is to use liquid scintillation cocktails designed for the anions in question, such as Ultima Gold™ (PerkinElmer) and the HiSafe™ series (Sigma Aldrich). Some cocktails also have a higher resistance to colour quenching. Another way of obtaining a more chemically stable sample and liquid scintillation mixture is to dilute the original radioactive sample with distilled water before mixing with the cocktail.

Care must also be taken when acids such as HCl or organic acids are present with the radioactive elements in the solution, as some strong mineral acids can interact with components in the LSC cocktail causing the development of colour, and thus colour quenching. Furthermore, some acids inherently introduce their own luminescence into the LSC cocktail. Therefore, concentrated acids should not be added to the cocktail, and if it is known that metal acids are present, the samples should be diluted with distilled water before mixing with the LSC cocktail. Some cocktails have been specially developed for use with such acids. If, on the other hand, the liquid sample medium is alkaline, consisting, for example, of sodium or potassium hydroxide, chemical luminescence due to reactions between the cocktail and the sample will be induced in many cocktails, and colour development may also occur, degrading the LSC counting efficiency. These samples must be slightly acidified or diluted with distilled water before mixing with the LSC cocktail. The mixture should be stored in room temperature or slightly above for some days before liquid scintillation counting in order to promote the decay of chemiluminescence.

Substantial chemical quenching effects can be expected in most LSC cocktails when adding radioactive samples dissolved in organic solvents, and the LSC cocktail should be carefully chosen so as not to reduce the counting rate. Radioactive elements in organic sulphide and halide solutions will be more strongly chemically quenched during liquid scintillation counting than when dissolved in alcohol or ether. If possible, the structure of the organic radioactive substance should be changed into one that causes less quenching before being mixed with the LSC cocktail.

Biological samples for liquid scintillation counting can be handled in three ways.

- *Direct counting*. This means the raw sample is added directly to a suitable cocktail. Cocktails are commercially available that can be used with biological matrices such as urine, serum, plasma and water-soluble proteins. When measuring, for example, ^{226}Ra or ^{210}Po in urine using direct addition, the samples should first be diluted to reduce colouring and thus the quenching effects of urine when added to the LSC cocktail. About 24 mL of commercially available cocktails is added to 4–10 mL urine samples. About 1 mL ethanol or isopropyl alcohol can be added per 10 mL of cocktail to suppress precipitation of proteins in the urine.

- *Solubilization*. The sample is digested and dissolved before addition to the LSC cocktail. Suitable biological samples are blood, plasma, serum, faeces, homogenates and bacteria. For alkaline samples, solution in tetravalent ammonium hydroxide can be used (alkaline hydrolysis) to reduce the chemical luminescence in the cocktail. For acidic sample solutions, acidic oxidation is used to convert the sample into a form with less chemical luminescence, for example, by converting the acid into perchloric or nitric acid.

- *Combustion*. The samples are oxidized or burnt to convert organic species into gases, which are then trapped and dissolved in the LSC cocktail. Common examples are the combustion of ^{14}C samples to form $^{14}CO_2$, or tritium samples to form ^3H_2O.

Although no synthetic chemical can be considered environmentally safe when released or disposed of, commercial biodegradable alternatives are available for LSC cocktails, which make waste disposal less problematic.

5.3.7 Concerns Regarding Radiochemistry

Radiochemical processing of samples is necessary in order to accurately quantify low concentrations of many neutron activation and fission products, as well as to assay transuranic elements (actinides) that essentially emit only alpha and beta radiation. It is also necessary for the analysis of Ra and U in environmental samples, such as drinking water, or urine, where disequilibrium due to ^{222}Rn emanation must be accounted for. Before the advent of high-resolution γ-ray spectrometry at the beginning of the 1970s, radiochemical preparation and subsequent α- or β-particle counting of the separated elements were the only techniques available to separate and identify many radionuclides, since some γ-ray-emitting radionuclides have too complex a γ-ray spectrum to be resolved by NaI(Tl) detectors. Radiochemistry is still necessary when analysing samples for radionuclides with very low-energy gamma emission and radionuclides that decay through electron capture. Table 5.3 give an illustration of the various measurement methods used for assessing α- and β-emitting radionuclides in a given application (in this case the decommissioning of a research reactor).

Radiochemistry is associated with a greater investment in laboratory equipment, as well as higher consumption of disposable equipment and materials than γ-ray spectrometry. The equipment needed includes a fume hood with at least two separate vented compartments, one for interim storage of heated acid solutions when performing wet oxidation, and one for the operator to carry out manual tasks. The venting system connected to the fume hood often requires special acid-resistant protective cladding on the inner surfaces. Other equipment required includes ovens with high-precision temperature ramping, automatic blenders and mixers, specialized storage for hazardous chemicals, often strong acids such as perchloric acid ($HCLO_4$) and HCL. Many consumables are also required, such as acid solutions (HCL, HNO_3, $HCLO_4$), and ion-exchange resins. To reduce the risk of cross contamination it is also necessary to use disposable glass vials, disposable tools for liquid transfer, disposable gloves and protective clothing, and disposable shrouds to protect surfaces and other equipment. Regular wipe tests of the laboratory surfaces are necessary for α and β counting since cross contamination cannot always be detected quickly, unlike γ-ray emitters, where a simple hand-held γ-ray detector is sufficient to detect contamination.

TABLE 5.3 Examples of radionuclide assays and the mode of assessment for α and β emitters of interest during the decommissioning of a research nuclear reactor.

Sample type	Nuclides	Measurement method	Sample mass (g)	MDA (Bq g^{-1})
Water				
	^{3}H	LSC	5–100	<0.05
	^{14}C	LSC	5–100	<0.05
	^{55}Fe	LSC	5–100	<0.05
	^{63}Ni	LSC	5–100	<0.05
	^{36}Cl	LSC	5–100	<0.05
	^{129}I	LSC	5–100	<0.05
	^{90}Sr	β-counting	100	0.002
	^{99}Tc	β-counting	100	0.002
	^{94}Nb	γ-ray spectrometry	100	0.1
	^{239}Pu	Mass spectrometry	100	<0.07
	^{240}Pu	Mass spectrometry	100	<0.07
	^{242}Pu	Mass spectrometry	100	<0.07
	^{238}Pu	α spectrometry	100	<0.07
	^{241}Am	α spectrometry	100	<0.07
	^{243}Cm	α spectrometry	100	<0.07
Soil, sediment, vegetation, animal tissue				
	^{3}H	LSC	2	0.05–0.1
	^{14}C	LSC	2	0.05–0.1
	^{55}Fe	LSC	20	0.05–0.15
	^{63}Ni	LSC	20	0.05–0.15
	^{36}Cl	LSC	20	0.05–0.15
	^{41}Ca	LSC	20	0.05–0.15
	^{90}Sr	β-counting	20	0.05–0.15
	^{99}Tc	β-counting	20	0.05–0.15
	^{94}Nb	γ-ray spectrometry	20	0.5
	^{238}Pu	α spectrometry	20	0.01
	^{239}Pu	α spectrometry	20	0.01
	^{240}Pu	α spectrometry	20	0.01
	^{241}Am	α spectrometry	20	0.01
	^{244}Cm	α spectrometry	20	0.01
	^{99}Tc	Mass spectrometry	20	0.01–0.3
	^{239}Pu	Mass spectrometry	20	0.01–0.3
	^{240}Pu	Mass spectrometry	20	0.01–0.3
	^{241}Pu	Mass spectrometry	20	0.01–0.3

Source: Data from Hou et al. (2015).

5.4 REFERENCES

- Baron, P. 2006. *Generation and Behavior of Air-Borne Particles (Aerosols).* National Institute of Occupational Health and Safety, http://www.cdc.gov/niosh/topics/aerosols/pdfs/Aerosol101.pdf. Retrieved on August 2015.

- Beck, H. L., De Campo, J. and Gogolak, C. V. 1972. *In Situ Ge(Li) and Na(Tl) Gamma-Ray Spectrometry.* New York, US Department of Energy, Environmental Measurements Laboratory, HASL-258.

- Bernhardsson, C., Erlandsson, B., Rääf, C.L. and Mattsson, S. 2008. Variations in the ^{137}Cs concentration in surface coastal water at the Swedish west coast during 40 years as indicated by Fucus. In: Strand, P., Brown, J. and Jolle, T. (eds.) *Poster proceedings Part 2: International Conference on Radioecology and Environmental Radioactivity,* 15–20 June 2008, Bergen, Norge, pp. 91–94.

- Bernhardsson, C. 2011. *Radiation Exposure of Human Populations in Villages in Russia and Belarus Affected by Fallout from the Chernobyl Reactor: Measurements Using Optically Stimulated Luminescence in NaCl, Tl Dosemeters and Portable Survey Instruments.* ISBN 978-91-86871-11-6, Lund University, Faculty of Medicine Doctoral Dissertation Series 2011:62, 10 June 2011.

- Bernhardsson, C., Rääf, C.L. and Mattsson, S. 2015. Spatial variability of the dose rate from ^{137}Cs fallout in settlements in Russia and Belarus more than two decades after the Chernobyl accident. *J. Environmental Radioactivity,* 149, 144–149.

- Edler, R. 2015. PerkinElmer Inc. Rodgay-Jgesheim, Germany. Web page: http://www.perkinelmer.com/liquidscintillation/images/APP_Cocktails-for-Liquid_tcm151-171743.pdf. Accessed 20 March 2016.

- Erlandsson, B., Bjurman, B. and Mattsson, S. 1989. Calculation of radionuclide ground deposition by means of measurements on sewage sludge. *Water, Air, and Soil Pollution,* 45(3–4), 329-344.

- Finck, R.F. 1992. *High Resolution Field Gamma Spectrometry and Its Application to Problems in Environmental Radiology.* Depts. of Radiation Physics, Malmö and Lund, Lund University, Dissertation.

- Holm, E. H., Rääf, C. L., Rabesiranana, N., Garcia-Tenorio, R. and Chamizo, E. 2015. Fallout of Pu-238 over Madagascar following the Snap 9A satellite failure. In *Environmental Radiochemical Analysis V,* ed. P. Warwick, 44–49. Royal Society of Chemistry. Cambridge. ISBN: 978-1-78262-155-3.

- Hou, X., Olsson, M., Vaaramaa, K., Englund, S., Gottfridsson, O., Forsström, M. and Togneri, L. 2015. *Progress on Standardization of Radioanalytical Methods for Determination of Important Radionuclides for Environmental Assessment and Waste Management in Nordic Nuclear Industry.* Report NKS-327. ISBN 978-87-7893-408-6.

- IAEA, International Atomic Energy Agency. 2010. *Handbook of Parameter Values for the Prediction of Radionuclide Transfer in Terrestrial and Freshwater Environments.* Technical Reports Series No. 472. International Atomic Energy Agency. Vienna.

- ICRP 1995. *Age-Dependent Doses to Members of the Public from Intake of Radionuclides—Part 4 Inhalation Dose Coefficients.* ICRP Publication 71. Ann. ICRP 25 (3-4).

- ICRP 2012. *Compendium of Dose Coefficients Based on ICRP Publication 60.* ICRP Publication 119. Ann. ICRP 41(Suppl.).

- Isaksson, M., Erlandsson, B. and Mattsson, S. 2001. A 10-year study of the ^{137}Cs distribution in soil and a comparison of Cs soil inventory with precipitation-determined deposition. *Journal of Environmental Radioactivity*, 55(1), 47–59.

- Jaakkola, T. 1994. Radiochemical separations. In *Radioecology. Lectures in Environmental Radioactivity*, ed. E. Holm, 233–253. Singapore: World Scientific.

- Mattsson, S. and Vesanen, R. 1988. Patterns of Chernobyl fallout in relation to local weather conditions. *Environment International.* Vol. 14, 177–180.

- Petersen, L., Minkkinen, P., Kim, H. and Esbensen, K. H. 2005. Representative sampling for reliable data analysis: Theory of sampling. *Chemometrics and Intelligent Laboratory Systems*, 77, 261–277.

- Rääf, C. L., Hubbard, L., Falk, R., Ågren, G. and Vesanen, R. 2006. Transfer of ^{137}Cs from Chernobyl debris and from nuclear weapons fallout to different Swedish population groups. *Science of Tot. Env*, 367, 324–340.

- Rääf, C. L., Holstein, H., Holm, E. and Roos, P. 2015. Hair as an indicator of the body content of polonium in humans: Preliminary results from study of five male volunteers. *Journal of Environmental Radioactivity*, 141, 71–75.

- Saey, P. R. J. and De Geer, L. E. 2005. Notes on radioxenon measurements for CTBT verification purposes. *Applied Radiation and Isotopes*, 63 (56), 765–773.

- Sanderson, D. C. W., Allyson, J. D., Toivonen, H. and Honkamaa, T. 1995. *Gamma Ray Spectrometry Results from Core Samples Collected for RESUME 95.* Scottish Universities Research & Reactor Centre, September 1995.

- Sidhu, R. 2004. *Extraction Chromatographic Separation of Sr, Pu and Am in Environmental Samples.* Dissertation. Faculty of Mathematics and Natural Sciences, University of Oslo.

- Stenström, K., Unkel, I., Nilsson, C. M., Rääf, C. and Mattsson, S. 2010. The use of hair as an indicator of occupational C-14 contamination. *Radiat and Environmental Biophysics.* 49(1), 97–107.

- UNSCEAR, United Nations Scientific Committee on the Effects of Atomic Radiation. 1977. *Sources and Effects of Ionizing Radiation, Annex C.* New York: United Nations.

- UNSCEAR, United Nations Scientific Committee on the Effects of Atomic Radiation. 1982. *Ionizing Radiation: Sources and Biological Effects, Annex E.* New York: United Nations.

5.5 EXERCISES

5.1 A reference material, considered as a "lot" as defined in Section 5.1.2, has a well-determined activity concentration of 22.3 Bq kg^{-1}. A laboratory that purchased 2.4 kg of the material extracted 7 samples from the lot, and analysis with γ-ray spectrometry yielded activity concentrations of 23.2, 22.1, 24.0, 21.9, 25.0, 22.8 and 21.5 Bq kg^{-1}. Are the seven samples taken from this lot representative?

5.2   A glass fibre filter, 0.5×0.5 m^2, is squeezed into a 350-mL plastic bottle and analysed using γ-ray spectrometry. After appropriate corrections for internal absorption, the measurement gave an estimated activity of 35.3 Bq ^{131}I, with a reference date set to the onset of air sampling. What is the estimated air concentration [Bq m^{-3}] of ^{131}I if the average air flow rate was 0.3 m s^{-1}, and the sampling time was 24 h?

5.3   100 mL of an aqueous solution containing 100 Bq of ^{89}Sr is contained in a glass vial (of diameter 5 cm) that has an affinity for Sr ions (0.2 nm in diameter) and binds them. Assume an inter-atomic distance of 1 nm between the Sr atoms when bound to the glass. How large a fraction of the original content of ^{89}Sr is lost to wall adsorption? Using the law of mass action, and assuming that the probability of a given chemical reaction is proportional to the concentration of the element, how much stable Sr must be added to reduce the loss to less than 1%?

5.4   A grass sample collected at midday from a 3-m^2 area at a dairy farm is received at a laboratory a few hours later. The gross weight of the grass sample is determined to be 1105.7 g. The sample is homogenized by chopping the grass and mixing it thoroughly in a large plastic bin. From this bin, 187.5 g of fresh homogenized grass is extracted. It is placed on a sheet of aluminium foil and dried at 65 °C for 48 h. After drying, this fraction of the sample weighs 41.2 g. Three days after collection of the original sample at the farm, the dried grass is placed in a 200-mL beaker and analysed using γ-ray spectrometry for 2 h. The net count rate in the ^{131}I window at 364.5 keV is 0.185 cps, and the corresponding net count rate in the ^{137}Cs window at 661.6 keV is 0.987 cps. The absolute efficiency of the gamma spectrometer (Eqn. 4.51) is 0.00185 [cps s^{-1}] at 364.5 keV and 0.00107 [cps s^{-1}] at and 661.6 keV. What are the concentrations of the two radionuclides in the grass per unit wet weight and per unit area of pasture?

5.6   FURTHER READING

Holm, E. (Ed.). 1994. *Radioecology. Lectures in Environmental Radioactivity.* Singapore: World Scientific.

Lehto, J. and Hou, X. 2010. *Chemistry and Analysis of Radionuclides: Laboratory Techniques and Methodology.* Wiley, Weinheim, Germany. ISBN: 978-3-527-32658-7.

Nuclear and Radiological Safety

CONTENTS

THE concept of risk and strategies for risk communication are important topics in radiation protection. Based on the knowledge of the risks associated with exposure to ionizing radiation, the ICRP has issued recommendations, which are also implemented in the safety standards issued by the IAEA. The basic principles for radiation protection will be discussed, as will the implementation of these principles in practical radiation protection. The radiological impact of the nuclear fuel cycle will be discussed, regarding occupational exposure, as well as exposure of the public due to these practices.

6.1   RISK CONCEPTS AND RISK COMMUNICATION

6.1.1   Risk Concepts

A risk is generally perceived by the public as the possible occurrence of an undesired event. For example, it can be argued that the risk of serious injury is high when diving into unknown water. However, in order to quantify a risk, it is necessary to establish the probability of the undesired event, as well as its consequences. The probability of the event may be composed of several probabilities; for example, the probability of a fault in a technical system and the probability of the automatic shutdown function failing. The consequence of the event can be specified as the number of people injured, or the years of life lost, if the injury is fatal. If there are several consequences, C_i, each of which has a given probability, P_i, the risk can be quantified by the expectation value

$$\sum_i P_i \cdot C_i. \tag{6.1}$$

The risk associated with an accident can be defined as the frequency of the accident [events/unit time] multiplied by the consequence per event. The risk will then be expressed as the consequence per unit time. For example, if we assume that there are 70 000 road traffic accidents in a certain country in a particular year, and that the consequence of one in 250 was a human fatality, then the risk of death from road traffic accidents is given by $7 \cdot 10^4$ accidents/year × 1 death/250 accidents, which is 280 deaths/year. This figure represents the societal risk associated with road traffic accidents, and the average individual risk can be found by dividing by the population of that country. Assuming that the population is 8 000 000, the average individual risk is given by 280 deaths/year divided by $8 \cdot 10^6$, which is $3.5 \cdot 10^{-5}$ deaths per person-year.

The individual risk is thus expressed as the probability of death per person per year, and this quantity can be used as a mathematical tool for risk analysis. Although the fractional form is often used in public risk comparisons, it has no physical meaning since a whole person either dies or survives. However, this can be used to express the risk of dying from a road traffic accident during the course of a year in terms of 3.5 per 100 000 inhabitants.

Risks cannot be completely avoided, but one can often choose which risks can be taken in planned practices. The benefit of the practice should then preferably outweigh the risk, i.e. the practice should be justified. A systematic evaluation should include the characterization of the nature and the probability of the undesired events (*risk assessment*), as well as characterization of the benefits. Following the risk assessment, a decision must be made as to whether the risk is worth taking, given the anticipated benefit(s) of the practice (Hansson 2011).

Characterization of the risks and benefits should ideally be based on scientific consensus, while *risk management* involves weighting of the risks and benefits, and is value based. However, some argue that risk assessment is also value based to some extent, due to the large uncertainties associated with characterizing the risk, and the fact that risk-related decisions are often made on a rather short timescale (Wikman-Svahn 2012). Even in science, value judgements are often used, for example, in hypothesis testing, when making scientific decisions. These are often denoted *epistemic values* and include simplicity and predictive power.

Example 6.1 *Risk assessment*

A new road is being planned and the authorities are considering two possible routes. One is close to the sea, where hurricanes may cause flooding of the highway, and the other is in a mountainous area. It is estimated that the number of deaths from car accidents on the road close to the sea will be 20 per year if no flooding occurs, but will increase to 500 if the highway is flooded. On the narrower mountain road, it is estimated that 40 deaths per year will occur. The probability of flooding is estimated to be 0.001. The expected number of deaths per year for the two alternatives can be calculated as the expectation value for each alternative, i.e.:

Sea highway: $500 \cdot 0.001 + 20 \cdot 0.999 = 20.5$ *deaths per year*
Mountain highway: $40 \cdot 0.001 + 40 \cdot 0.999 = 40$ *deaths per year*

This example shows that although the number of deaths will be high if flooding occurs, the low probability of flooding makes the risk smaller than in the case of the mountain road. The authorities must then decide whether to base their decision on the consequences (the high number of deaths if flooding occurs) or on the expectation values. This may be difficult, especially if an event with an extremely low probability results in catastrophic consequences. In these situations, the concept of expectation values may lack meaning, as the value will consist of the product of zero and infinity.

Hypothesis testing is used in science to minimize to probability of false claims. Consider, for example, the measurement of two radioactive sources to determine whether the activities are equal. The numerical results will differ due to measurement uncertainties, and hypothesis testing can be used to determine whether the difference is within the statistical limit given by the level of significance. If the difference between the two measurements is greater than the level of significance, it can be concluded that the activities are not equal. If the level of significance used is 5%, the probability of rejecting a true null hypothesis is 5%. This means that even if the activities are equal, they may be considered different, due to statistical fluctuations causing the measured difference to be greater than the level of statistical uncertainty. This can be very important in risk assessment, since the difference being determined is often an increase in risk, and precaution is therefore generally more important in risk assessment. There is, however, no objective way of determining the appropriate level of precaution.

Risk management and emergency preparedness are closely related since the nature of the risk determines the planning of preparedness measures. If both the probability and the consequences of an undesirable event are known from observations and previous events, the risk can be determined, and preparedness planned accordingly. In the case of events that have not occurred previously, the probabilities and consequences can be estimated, but these estimates will be associated with a degree of uncertainty. The risk associated with them will thus also be uncertain, and plans for preparedness will have to be based on plausible scenarios. There may also be events for which the probabilities and consequences are completely unknown. The risk will then be unknown, and preparedness plans would have to deal with the unexpected, for example, by maintaining high levels of competence and skills.

6.1.2 Risk Communication

The communication of risk is a crucial part of emergency management since the public response to an accident or an undesirable event has considerable impact on the way in which the situation is handled. Effective communication requires an understanding of how people perceive various kinds of risks. Communication concerning risks, for instance radiation risks, differs to some extent from other kinds of communication since strong emotions are often involved. The risk perceived by the general public will then often differ from the scientific assessment of the risk (Perko 2014), and this must be taken into consideration by emergency managers. The general public's perception of risk is not necessarily due to a lack of understanding of quantities such as probability, consequence or expectation value. In many cases it will be a matter of accepting the practice causing the risk, rather than the risk itself.

Key psychological factors that affect the public's perception of risks have been listed by the IAEA (IAEA 2012) and include personal, as well as societal factors.

- Public *trust* in communicators and organizations will reduce the perceived risk. Several factors are important in gaining the trust of the public. A key factor is honesty; refraining from admitting mistakes or trying to suppress the truth will damage trust. Consideration of people's feelings, respecting their fears and not telling them how they should feel, are also important in ensuring trust. Furthermore, real risks should not be diminished in order to try to calm the public: although the level of fear may increase, the increased trust in officials may prevent panic.

- Little *media attention* will lessen the perception of risk, while public fear will increase with increased media attention. Communicators must establish close relationships with the media in order to ensure that accurate and relevant information is communicated.

- *Clear communication*, without unnecessary scientific concepts, will increase the *understanding* of the situation, and hence decrease the perception of risk.

- *Familiarity* with the hazard will reduce the public's perception of the risk. Radiological emergencies can arise in areas where the inhabitants are normally unfamiliar with radiological or nuclear practices, and the perception of risk will then be greater.

- *Scientific certainty* and consensus decrease the perception of risk, while a lack of scientific consensus will increase the perception of risk due to intuitive judgements.

- Accidents that have already occurred will increase the perception of risk. For example, the perceived risk from a nuclear facility will be increased by the *history/stigma* associated with previous accidents at nuclear facilities, e.g. Chernobyl.

- The perception of risk will be greater if the *onset of effects* occurs with little warning, or if the hazards are considerable and have immediate effect.

- Hazards that are not *reversible* will increase the perception of risk.

- The perception of risk will be reduced if sufficient and authoritative *information is available*.

- The perception of risk will be less if the public has a choice in how to behave in the situation, e.g. *voluntary* participation.

- The perception of risk will also be less if members of the public feel they have *control* over the situation. This can be achieved, for example, by giving instructions on what they can do during an emergency.

- The *fairness of risk distribution* will affect the perception of risk. If the distribution of costs and benefits is perceived as fair, the perception of the risk will be reduced.

- Hazards resulting from natural events will generally be perceived as representing less risk than hazards arising from human activities (*origin of risks*).

- An accident with a considerable *catastrophic potential*, i.e. a large number of people may be injured or killed, will be perceived as a greater risk.

- *Personification* of the risk leads to an increase in the perceived risk. The risk is perceived as greater if specific victims can be identified, compared with a hazard that could potentially affect an unspecified population.

- The risk will be perceived as lower by a person who is not *personally involved* in the event, e.g. when the event is happening far away.

- *Awareness* of risks tends to decrease the perception of risk.

- *Fear* of considerable pain and suffering will increase the perception of risk.

- The perception of risk tends to be greater if the hazard can be expected to affect *children and future generations*.

A communicator should thus be honest and open, and acknowledge people's concerns. Their concerns are real, regardless of whether any real hazard exists, and are based on the risk perception factors given above. It is therefore important that the communicator is aware of these, and the ways in which they may affect people's behaviour. Honesty and openness will help maintain trust between the emergency response organization and the public. Statements such as "No comment" may seriously erode trust, since it implies that the expert doesn't know, or is trying to hide information. Trust is also generally created if the communicator admits some degree of uncertainty, rather than claiming that the situation is under control, when this is clearly not the case. Further communication will be difficult if it later turns out that the communicator was not telling the truth, and this may also lead to people questioning the competence of the authorities.

Members of the general public often tend to distinguish between two types of radiation: "good" radiation and "bad" radiation. Radiation used in medical diagnostics and treatment is often regarded as "good", since it is intended to be beneficial to the individual. On the other hand, the radiation from nuclear facilities, nuclear weapons tests and nuclear accidents is "bad". Accidents are obviously not beneficial, but for nuclear facilities it is often a question of whether the practice is perceived as being beneficial or not. This distinction can be explained by several of the key factors listed above, e.g. understanding and voluntariness.

Increasing the public's understanding of radiation and its effects by providing information is a complex task. The terms used are very seldom encountered in daily life, and must be used with great caution. The IAEA (2012) gives useful guidelines for communicating the basics of radiation. According to these guidelines, the concept of radiation can be explained by referring to the two main sources of radiation: naturally occurring and man-made sources. Everyone is exposed to naturally occurring radiation, and most of us are also exposed to artificial sources during medical examinations or treatment. The main types of radiation can be described without discussing the subatomic features of α-, β- or γ-radiation. These types of radiation can instead be described by their range in various materials, and their effects on human tissue. It is also important to distinguish between external and internal radiation sources since the hazard, as well as the protection measures, will depend on the location of the source.

It is a common misconception that someone contaminated by radiation can "infect" other people. However, radioactive contamination cannot be compared with a contagious disease since the total number of radioactive atoms will remain the same, even when redistributed among several subjects. Radioactive contamination and the transfer of radionuclides to others can instead be compared to dirt on clothes or hands that may be transferred by direct contact. It is also important to stress that removing the contamination will also remove the radiation source. Furthermore, a person exposed to external radiation will not become radioactive.

Quantities and units should not, if possible, be used when communicating with the public. It may, however, be necessary to mention the units for (effective) dose and dose rate, mSv and mSv h^{-1}. These units are then given with the prefix "milli", since it will make them directly comparable to the background radiation. If larger doses have to be communicated, these should preferably be given in the same units, e.g. 1500 mSv and not 1.5 Sv in order to avoid confusion. Relating dose and dose rate quantities to potential health effects requires that the source of radiation and the exposure scenario are known, and that we also know who was exposed (IAEA 2012).

The long half-life of a radionuclide is often thought to be responsible for the potential detrimental effects of radioactive releases to the environment. However, the half-life itself can only be related to the outcome if the exposure pathway is known. Consider, for example, the ingestion of a radionuclide with a very short half-life. Unless the clearance is very fast, practically all of the ingested radioactive atoms will decay while in the body, and the radiation dose will be determined by the ingested activity. If, on the other hand, the half-life is very long, a considerable fraction of the ingested activity will be cleared from the body before the atoms decay. The radiation dose will then be lower (assuming that the type and energy of the emitted radiation are the same for the two radionuclides considered). For radionuclides released into the environment, the situation may be the reverse: a radionuclide with a long half-life will be available for uptake in plants and animals for a longer time.

The comparison of risks using numbers should be avoided, since each member of the public makes their own evaluation of the risk, based on the psychological factors affecting his or her perception of the risk. Risks should therefore be compared using their perceived similarity. The IAEA (2012) exemplifies this by suggesting that the risk of an individual exposed during a radiation incident developing cancer should be compared to the same risk for a radiation worker, instead of comparing it to the risk of developing cancer from smoking. The latter is not related to radiation, and also involves a degree of voluntariness. If unavoidable, or deemed necessary, communication using numbers should be as simple as possible. Risk perception is based on both feelings and facts, and numbers only reflect the factual dimension. The IAEA (2013) has listed common errors made when assessing the radiological health hazards, together with possible consequences of the errors. This report also gives instructions on how to place the health hazards in perspective.

Informing the public of countermeasures to reduce the risk from exposure to ionizing radiation may create undesired effects if the rationale behind the measures is not understood by the public. An example of this is the recommendation to discard parsley and chives, which was issued by the Swedish authorities after the Chernobyl accident. It was found that parsley and chives were quite heavily contaminated, and it was considered a simple measure to throw them away. However, some people had already eaten parsley and chives, and were genuinely concerned about their health. When questioned, the authorities stated that there was absolutely no risk of harmful effects from consuming parsley and chives. Why then, did the authorities recommend discarding them when there was no risk?

The reason for giving this recommendation was to decrease the collective dose (see Section 2.4), since the cancer incidence is assumed to be proportional to the collective dose (this will be discussed in more detail in Section 6.2.3). Let us assume that the probability of an individual who has eaten parsley developing cancer is one in a million, or 10^{-6}. If eight million people have eaten parsley, the expectation value of the number of cancer cases will be 8. Thus, the recommendation of a simple countermeasure that would, statistically, prevent 8 cases of cancer is probably reasonable in this situation, although the risk to an individual is not large. This is why the authorities could say that there was no risk to *individuals* who had eaten parsley. However, since the public did not understand the difference between a statistical expectation value applicable to the whole population, and a probability applicable to an individual, the recommendation was misunderstood.

6.2 RADIATION PROTECTION

6.2.1 The ICRP Recommendations

The ICRP was established in 1928 and has published both recommendations and numerous supporting documents on radiation protection. The first recommendations concerned avoiding effects on tissue and organs among medical professionals working with radiation sources by restricting their working hours. This restriction led to an annual individual dose roughly equivalent to 1 Sv. Limits based on dose measurements were gradually introduced in recommendations published in 1934 and 1951. These recommendations were still based on thresholds for tissue effects. However, in the 1950s, evidence of excess malignant disease among radiologists and excess leukaemia among the survivors of the atomic bombings of Japan began to emerge from epidemiological studies. This led to recommendations of lower weekly and accumulated doses, for both radiation workers and the public, in order to reduce stochastic effects. Since it was not possible to determine whether a threshold exists or not, it was recommended that exposure be reduced to the "lowest possible level" (ICRP 1955). This was later rephrased as the ALARA principle: As Low As Reasonably Achievably, which takes economic and social considerations into account (ICRP 1973).

The risks of stochastic effects were first quantified in Publication 26 (ICRP 1977), where a system for dose limitation was proposed. The three principles: *justification*, *optimization* and *individual dose limitation*, which will be discussed later in this section, were then introduced. In the 1990 recommendations (ICRP 1991), ICRP also distinguished between different exposure situations. The two situations "practices" and "interventions" were defined, to differentiate planned from unplanned exposures. The 1990 recommendations remain essentially unchanged in the 2007 recommendations, given in ICRP Publication 103 (ICRP 2007), but some changes were made regarding weighting factors (see Section 2.4) and risk estimates. The system of radiation protection is still based on the three principles mentioned above, and the ICRP recommendations are intended to apply to all sources and all individuals exposed to radiation. For this reason, three exposure situations were defined in the 2007 recommendations (ICRP 2007).

1. *Planned exposure situations* involve the deliberate use of radiation, for example in medical examinations and treatment, and in nuclear power plants, where protection can be planned before exposure, and includes all categories of exposure: occupational, public and medical. Planned situations are intended to lead to exposures that can be anticipated beforehand, but may also lead to potential exposures, in the case of deviations from the planned procedure, causing unintentional exposure. An example of a potential exposure is the possible loss of a radiation source, normally used in a planned operation.

2. *Emergency exposure situations* are unexpected situations requiring urgent action. These situations may be the result of a planned operation or of a malicious act. The aim of the protective actions taken during the emergency is to avoid or reduce the consequences of the exposures caused by the emergency, and these situations should, as far as possible, be considered at the planning stage of emergency preparedness.

3. *Existing exposure situations* include naturally occurring exposures, for example, exposure to radon at workplaces and in dwellings, exposures due to past events/accidents (e.g. prolonged exposures due to land contamination), and exposures that have previously been disregarded. The radiation source is thus already present when decisions concerning protective measures are taken.

Practices continuing over a period of several years may lead to the build-up of annual doses to the individual, from one year to the next. To ensure that the annual dose resulting from past and present releases does not exceed a specified threshold, the *dose commitment* can be calculated. If the same amount of radioactivity is released each year, it can be shown that the dose commitment from one year's release is equal to the maximum future annual dose.

Let us consider the following example (Lindell 2003). Assume that an atmospheric release of radionuclides from a nuclear power facility gives rise to fallout that will cause the annual dose to be 100 units in the first year. The dose will decrease with time due to weathering and decay, and we can assume that the doses during the three subsequent years will be 70, 40 and 10 units, respectively. This is depicted by the black bars in Figure 6.1. Let us now assume that the practice releases the same amount of radionuclides during the second year. The dose during this year will then be 170 units: 100 units due to the release during the second year and 70 units due to the release during the first year. Continuing this line of reasoning, we find that the dose during the third year will be 210 units, and the dose during the fourth year will be 220 units. However, the dose during the fifth year will still be 220 units, since the release from the first year no longer contributes to the dose.

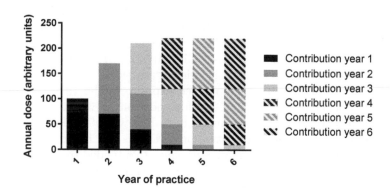

FIGURE 6.1 The dose commitment from one year's release, shown by bars of the same appearance, is equal to the maximum future annual dose, provided the releases continue unchanged from one year to the next. In this example, the committed dose is 220 units. The highest future annual dose can thus be controlled by restricting the annual dose commitment, instead of the annual dose.

Radiological protection measures can be directed towards the source of radiation, or an exposed individual, or group of individuals. *Source-related* approaches originate from the assumption that, provided the exposure is below the threshold for tissue effects, the contribution to the radiation dose from one source is independent of those from other sources. A source-related restriction in planned situations can be formulated as a *dose constraint*, defining the highest contribution to the individual radiation dose from a particular source. For potential exposures, which are not planned (but can be anticipated), it will be difficult to define a dose constraint, and a *risk constraint* is used instead.

For emergency and existing exposure situations, the dose constraints are replaced by *reference levels*. These reference levels are used to guide the optimization process and can be defined as intervals, within which measures to reduce the exposure can be taken. If no action is taken, the anticipated radiation dose is called the *projected dose*, while the exposure when protective measures are taken is called the *residual dose*. The residual dose should not be confused with the *averted dose*, which is defined as the difference in dose with and without the implementation of a specific protective measure. For example, the radiation dose to inhabitants in a village affected by fallout is estimated to be 50 mSv during a period of two weeks. If the population is evacuated to a new location for two weeks, where the radiation dose is estimated to be 1 mSv, the averted dose resulting from this measure, will be 49 mSv. The averted dose is still considered a useful concept by the ICRP, although the Commission recommends that the management of an emergency focus on the overall strategy, instead of individual measures (ICRP 2007).

The ICRP makes a distinction between *occupational exposure* and *public exposure*. Since we are all exposed to ionizing radiation of natural origin, occupational exposure is restricted to exposures that can be considered the responsibility of the employer. Public exposures include those that are neither occupational nor medical. The embryo and foetus are also considered members of the public and should be regarded as such when considering the radiation dose to pregnant workers. For the protection of the public, the ICRP recommends the use of a "representative person", which has replaced the previously used "critical group". The representative person is used to characterize a more highly exposed individual, for example, an imaginary person living close to a nuclear power plant and eating locally produced food. Guidance on how to characterize the representative person is given by the ICRP in Publication 101 (ICRP 2006).

The principle of justification is formulated as follows by the ICRP (ICRP 2007):

> any decision that alters the radiation exposure situation should do more good than harm.

The radiation exposure could be increased by the introduction of new radiation sources in a practice. It could also be decreased by measures that reduce the exposure from existing exposure situations, or that reduce the risk from potential exposure situations. Any detriment caused by such actions should be balanced by the benefits of the action. This principle can thus be regarded as the minimal requirements for any action that involves radiation: i.e., whether a practice should be permitted or not. However, the principle of justification raises some questions, for example, the definitions and use of the terms *benefit* and *harm*, and how these can be compared. Other issues concern how situations should be dealt with when harm and benefit do not concern the same individual, and who should be responsible for making the decision (Wikman-Svahn 2012).

The principle of optimization of protection is formulated as follows (ICRP 2007):

> the likelihood of incurring exposures, the number of people exposed, and the magnitude of their individual doses should all be kept as low as reasonably achievable, taking into account economic and societal factors.

The ICRP explains this principle by saying that the margin of benefit over harm should be maximized, but that there should also be restrictions on doses to individuals in order to avoid severely inequitable distributions of the doses among individuals. The optimization principle thus consists of two parts: the ALARA principle, and individual restrictions, expressed as dose or risk constraints and reference levels (Wikman-Svahn 2012). An optimization option should only be implemented if the second part is fulfilled, i.e. optimization is applicable to doses below the restrictions. These two principles apply in all exposure situations and are source related, in contrast to the principle of the application of dose limits, which is individual related, and is applicable only in planned exposure situations.

The principle of the application of dose limits is expressed as (ICRP 2007):

> the total dose to any individual from regulated sources in planned exposure situations other than medical exposure of patients should not exceed the appropriate limits recommended by the Commission.

The (individual-related) dose limits should be decided by the authorities, and they apply to workers as well as members of the public. However, dose limits can only be controlled and maintained for occupational exposures, and dose limits for the public can be regarded as a theoretical tool for authorities when deciding dose constraints on, for example, a certain practice. The use of dose constraints enables the authority to take several sources into account. Medical examinations and treatments are exempt from the application of dose limits and, in such cases, the professional responsible for the medical treatment should optimize the use of radiation by, for example, considering the risk arising from an x-ray examination and the risk associated with the lack of diagnostic information if the x-ray examination is not performed.

6.2.2 Basic Safety Standards

The scientific basis for the ICRP recommendations can often be found in reports from the United Nations Scientific Committee on the Effects of Atomic Radiation (UNSCEAR). However, neither the UNSCEAR reports nor the ICRP recommendations are legally binding, and the IAEA has therefore developed safety standards, for example, the Radiation Protection and Safety of Radiation Sources: International Basic Safety Standards, BSS, (IAEA 2014), which apply to occupational, public and medical exposure. The aim was to base the IAEA basic safety standards on the ICRP recommendations, as much as possible.

Safety standards issued by the IAEA are sometimes developed in cooperation with other UN bodies or specialized agencies, for example, the European Commission (EC), the Food and Agriculture Organization of the United Nations (FAO), the United Nations Environment Programme (UNEP), the International Labour Organization (ILO), the OECD Nuclear Energy Agency (NEA), the Pan American Health Organization (PAHO) and the World Health Organization (WHO). These organizations have also developed standards concerning radiation safety. The ILO established a convention and recommendations concerning the protection of workers as early as 1960 (ILO 1960a, ILO 1960b). Guidelines for levels of radionuclides in foodstuffs have been published by the Joint FAO/WHO Codex Alimentarius Commission (http://www.codexalimentarius.org/), while guidelines for drinking water are issued by the WHO (WHO 2011). Euratom has developed Basic Safety Standards, laid down in Council Directives, which are binding in European Union member states.

The development of standards by more than one body means that they reflect an international consensus regarding the protection of humans, as well as the environment, against the harmful effects of ionizing radiation. The IAEA's safety standards are not legally binding in IAEA member states, but may be adopted in national regulations. However, if a state enters into an assistance agreement with the IAEA, that state is required to follow those parts of the standards that are related to the activities in the agreement. The binding regulations regarding radiation protection are issued by the government in each state or, as is the case in the EU, by the Council Directive (Euratom).

The IAEA Basic Safety Standards include a number of requirements that must be fulfilled. Application of the principles of radiation protection is to be ensured in all exposure situations by the parties responsible for protection and safety, primarily the person or organization responsible for the practice that gives rise to radiation risks. The government shall ensure that a legal and regulatory framework is established by an independent regulatory body. This regulatory body shall be given the legal authority, competence and resources required to fulfil its assignments, which are to establish regulations and a system for their implementation. Finally, protection and safety shall be integrated into the overall management system of the organization.

Dose limits for planned exposure situations are given by the IAEA (IAEA 2014) for occupational, as well as public, exposure. These dose limits coincide with the ICRP recommendations (ICRP 2007). For occupational exposure of adult workers, the limit on the annual effective dose is 20 mSv, when averaged over five consecutive years. However, the effective dose in a single year must be below 50 mSv. For apprentices or students aged 16 to 18 years, who can be exposed in job training or studies, the limit is 6 mSv y^{-1}.

Organ doses are specified as equivalent doses for the lens of the eye, the extremities (hands and feet) and skin. For an adult radiation worker, the annual equivalent dose to the lens of the eye shall not exceed 20 mSv, averaged over five consecutive years, or 50 mSv in a single year. The corresponding dose limit for apprentices or students aged 16 to 18 years is 20 mSv y^{-1}. The maximum annual equivalent doses to the extremities and the skin are 500 mSv and 150 mSv, respectively, for both adult workers and apprentices or students. The skin dose is the average dose to 1 cm^2 of the most exposed area of the skin. The working conditions for pregnant or breast-feeding female workers shall be adapted to ensure that the dose to the foetus or breastfed infant is below the dose limits that apply to members of the public.

The limit on the effective dose to members of the public is 1 mSv y^{-1}. However, this limit may be exceeded in a single year if the average effective dose during five consecutive years does not exceed 1 mSv y^{-1}. The limit on the equivalent dose to the lens of the eye is 15 mSv y^{-1}, and the limit on the equivalent dose to the skin is 50 mSv y^{-1}.

The limits on the effective dose are the sum of the effective doses from external and internal exposure. The contribution from external exposure is the effective dose from external exposure in the specified time period, e.g. one year, while the contribution from internal exposure is the committed effective dose from the intake of radionuclides during the same time period. Although the dose from the intake of radionuclides is protracted, the committed effective dose is included in the dose limit for the period during which the intake occurred. The committed effective dose is calculated for a 50-year period for adults, and up to 70 years of age for children.

6.2.3 Radiation Risk

Nominal risk coefficients, expressed, for example, as the number of cases per 10 000 persons per Sv, can be found from epidemiological data, and in radiation risk assessment, the risk estimates are largely based on the findings of the Life Span Study, LSS (Preston et al. 2007), including survivors of the atomic bombings in Japan, although other important sources of information are patients with ankylosing spondylitis and patients treated with partial-body irradiation, uranium miners and radium dial painters. These data are averaged by sex and age at exposure for representative populations, and the lifetime risk is calculated for various tissues and organs. The sum of the nominal risks for each of these tissues or organs then gives the total nominal risk for a specific population. To account for differences between populations, the total nominal risks are calculated as the means of the population-specific average risks, generally across the Asian and Euro-American populations (ICRP 2007).

The radiation risk within an exposed population can be described by either the EAR or ERR model (see Section 2.3.2.1) if variation with, for example, sex and age at exposure is taken into account. However, when applied across populations, the results may differ substantially due to variations in baseline rates between the populations. The ICRP has defined population risks based on weighted averages of risk estimates from both types of models. For most tissues, a weight of 0.5 was given to both the EAR and the ERR estimates, while an EAR model was used for breast and bone marrow, and an ERR model was used for thyroid and skin. For the lungs, the ERR model was assigned a weight of 0.3. These models are used as the basis when deciding the tissue weighting factors used to calculate the effective dose.

The estimate of lifetime risk is also adjusted by the ICRP by applying a dose and dose-rate effectiveness factor (DDREF), which reduces the risk by a factor of two. This factor takes into account the fact that the epidemiological findings were largely based on people who were exposed to doses of about 200 mSv or more, most of them acute exposure. Since protracted or fractionated exposure is associated with a smaller risk, it is reasonable to assume that the risk coefficients are less in the case of the exposure situations usually encountered in radiation protection.

The radiation detriment for a tissue, T, is defined in terms of the nominal risk of fatal disease, $R_{F,T}$, and the risk of non-fatal disease, $R_{NF,T}$, as (ICRP 2007)

$$D_T = (R_{F,T} + q_T R_{NF,T}) l_T, \qquad (6.2)$$

where q_T is a weighting factor that accounts for the reduced quality of life. q_T depends on the lethality of the disease and also takes into account pain, suffering and the adverse effects of treatment. The risk of cancer is thus not only based on lethality, but also on the burden associated with the illness. The factor varies between 0 and 1, and details on how it is determined can be found in ICRP Publication 103 (ICRP 2007). The average life lost (average number of years of life lost), l_T, is expressed relative to normal life expectancy, and is calculated from the number of deaths in various age groups of a population. This equation is an example of the multiplication of a probability by a consequence, discussed previously in Section 6.1.

In the risk assessment presented in ICRP Publication 103, data on cancer *incidence* are used instead of mortality data, which were used in previous risk assessments (ICRP 1991). Eqn. 6.2 then becomes slightly more complicated since the nominal risk is given by $R_{I,T} = R_{F,T} + R_{NF,T}$. The risk of fatal disease is then given by: $k_T R_{I,T}$, where k_T is the lethality fraction.

The detriment-adjusted nominal risk coefficients proposed by the ICRP, are given in Table 6.1. Since the risk coefficients are determined from averages over sexes, ages and populations, they are only intended to apply to populations, and must not be used to estimate the risk to an individual. For example, a mean dose of 100 mSv to an adult population will result in a probability increment in fatal cancer of: $4.1 \cdot 10^{-3}$, or 0.41%, which should be *added* to the baseline probability of fatal cancer in that population.

The practical system of radiological protection recommended by the ICRP is based on the assumption that there is a linear relationship between the radiation dose and the probability of incurring cancer or heritable effects, i.e. a linear dose–response relationship, at doses below about 100 mSv. This dose–response model is denoted the *linear-non-threshold* or the LNT model; see Figure 6.2. The LNT model thus describes the relation between a small *dose increment* and the corresponding *probability increment* of incurring cancer or heritable effects attributable to radiation. This is given by the slope of the dose–response curve at a point corresponding to the total accumulated radiation dose (Lindell 2011). It is

TABLE 6.1 Detriment-adjusted nominal risk coefficients.

Population	Cancer (10^{-2} Sv^{-1})	Heritable effects (10^{-2} Sv^{-1})	Total (10^{-2} Sv^{-1})
Whole population	5.5	0.2	5.7
Adult population	4.1	0.1	4.2

Source: Data from ICRP (2007).

thus not relevant, in this context, whether the assumption of linearity holds down to zero dose. No one is ever exposed to such small doses, and most people accumulate doses on the order of 100 mSv or more during their lifetime.

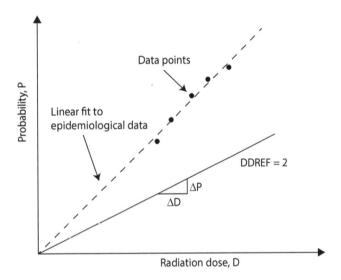

FIGURE 6.2 According to the linear-non-threshold model, each dose increment will give rise to a proportional increase in probability.

The excess risk from low dose increments will cause too few cases to be discernible in epidemiological studies, and the number of radiation-induced cases of cancer may even be less than the yearly variation in the baseline number of cases in a population. However, the ICRP considers the LNT model combined with a DDREF to be a good tool for the management of the risks arising from exposure to low doses, although the Commission emphasizes that it is unlikely that unambiguous verification of the LNT model will be possible in the near future. Due to the uncertainties, the ICRP recommends that the LNT model is not used to calculate a hypothetical number of cases (cancer and heritable effects) in large populations exposed to low doses over very long periods of time.

These hypothetical values could, in theory, be found by multiplying the collective effective dose by the nominal risk coefficient. However, the ICRP states that it is not appropriate to use collective effective dose in risk projections, since this quantity is not intended as a tool for epidemiology. It is not deemed reasonable to calculate cancer deaths from small exposures to individuals, although when received by a large population, these can give rise to a considerable collective effective dose. Due to the large biological and statistical uncertainties associated with the calculation of collective effective dose (e.g. in the tissue weighting factors), the number of cases will be very uncertain. Collective dose is instead intended for optimization, e.g. for comparing strategies for radiation protection.

However, other models have been proposed to describe human response to ionizing radiation, some of which are illustrated in Figure 6.3. Biological mechanisms, such as cellular adaptive response, induced genomic instability and bystander signalling, as well as immunological phenomena, may affect the shape of the dose–response curve, and hence the cancer risk, at low doses. Genomic instability and bystander signalling may both lead to an increased risk of carcinogenesis, while adaptive response (i.e. increased ability of DNA repair) and immunological responses leading to apoptosis may decrease the risk. However, the uncertainties in these mechanisms, and the consequences regarding cancer development, are still significant. Furthermore, the ICRP notes that the estimates of the nominal risk coefficients are based on human epidemiological data, which should include the contributions from these mechanisms.

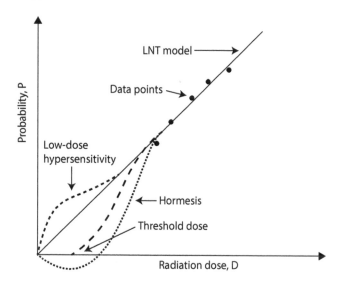

FIGURE 6.3 Alternative models describing the relation between radiation dose and the probability of harm at low radiation doses. The term hormesis refers to a favourable biological response to low exposures of a stressor.

6.3 THE NUCLEAR FUEL CYCLE

The nuclear fuel cycle describes the chain of processes from uranium mining to spent nuclear fuel. It includes the extraction of uranium from uranium ore, conversion to a form suitable for fuel production, fabrication of nuclear fuel elements, use in a nuclear reactor, storage of spent nuclear fuel, treatment of waste, and transport of material between installations. These stages are shown in Figure 6.4. Depending on the type of reactor, enrichment of ^{235}U may also be included in the nuclear fuel cycle, as may recycling of spent nuclear fuel by reprocessing.

All steps in the nuclear fuel cycle can potentially lead to the exposure of workers and members of the public. The doses to exposed individuals vary widely, depending on the installation (UNSCEAR 2008). The impact from each stage can be evaluated by normalizing the collective effective dose to the electrical energy generated, expressed in the units GWa

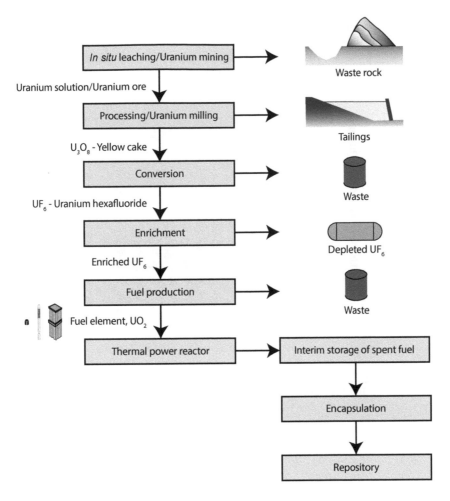

FIGURE 6.4 Steps included in the nuclear fuel cycle, excluding reprocessing of the spent nuclear fuel. The various steps are explained in the text.

(i.e. the product of the electrical power in gigawatts multiplied by the duration of operation in years). The normalized collective dose is then given in the units man Sv $(GWa)^{-1}$.

6.3.1 Mining and Milling

"Reasonably assured resources" are defined as resources where the tonnage, density, shape, physical characteristics, grade and mineral content of the resource can be estimated with a reasonable level of confidence, and this depends on the current price of uranium on the world market. It was estimated in 2012 that the reasonably assured resources of uranium amounted to 4 587 200 tonnes at a price of up to USD 260 $(kg\ U)^{-1}$, while only 507 400 tonnes are reasonably assured if the price is below USD 40 $(kg\ U)^{-1}$ (OECD 2014). When tonnage, grade and mineral content can only be estimated with a low level of confidence, the occurrence of uranium is considered an "inferred mineral resource". These resources range from 3 048 000 tonnes (at USD 260 $(kg\ U)^{-1}$) to 175 500 tones (at USD 40 $(kg\ U)^{-1}$).

Prior to 2012, more than two million tonnes of uranium had been produced, mainly in Canada and the USA. In 2012, about 59 000 tonnes were produced, with Kazakhstan, Canada and Australia being the main producers (OECD 2014). The distribution of historical and recent (2012) production is shown in Figures 6.5 and 6.6. Estimates of future production and demands have been made by the OECD (OECD 2015).

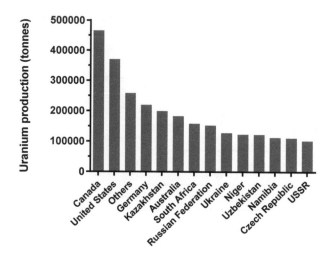

FIGURE 6.5 Total historical production of uranium prior to 2012, according to country. Countries with a historical production exceeding 100 000 tonnes are shown individually. The total production of uranium amounts to 2 709 434 tonnes prior to 2012. Data from OECD (2014).

Uranium can be extracted from the bedrock by several methods. In conventional mining, either as open-pit mining or mining below ground, the uranium ore is crushed and milled. The fine grains are then mixed with water to form a slurry, and the uranium is leached out, for example, by sulphuric acid, producing the sulphate complex $UO_2(SO_4)_2^{2-}$. This complex can be removed from the water solution by an organic solvent or using anion-exchange resins. Since uranium ores vary in composition, production methods differ according to the elements that have to be removed from the ore during processing. The final product is denoted *yellow cake*, commonly consisting of ammonium diuranate, containing 65–70% U (Choppin et al. 2002). Further purification leads to almost 100% pure U_3O_8. This product is practically free from the radioactive daughters of uranium and can be safely handled in steel drums without radiological concern.

An alternative to conventional mining is *in situ leaching*, ISL, where a solvent (sulphuric acid, sodium bicarbonate or CO_2), is pumped into boreholes drilled into the uranium ore deposit. The extracted uranium solution is then fed through an ion-exchange resin, which becomes loaded with uranium, and is used to produce yellow cake. In 2012, 45% of the world's uranium was produced by ISL, 26% by underground mining and 20% by open-pit mining (OECD 2014). Uranium is also produced, to a lesser extent, as a by-product, for example, in copper mining.

Mining and milling generate waste in the form of waste rock and *tailings*, Fig. 6.7, (a slurry from which water is extracted to an extent depending on the method used). It is estimated that one tonne of extracted ore generates about one tonne of tailings, and that about 5–10% of the uranium is retained in the tailings. The piles of waste rock may also cause exposure of the public. Apart from liquid residues, the use of ISL may also lead

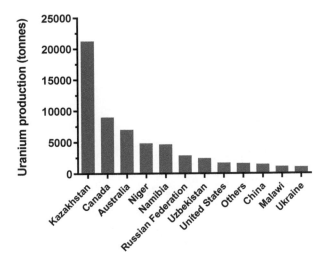

FIGURE 6.6   Total production of uranium in 2012, according to country. Countries with a production exceeding 1000 tonnes are shown individually. The total production of uranium in 2012 amounted to 58 816 tonnes. Data from OECD (2014).

to dispersion of the solvent outside the ore deposit and subsequent contamination of the groundwater.

The tailings are of concern from a radiological point of view due to the high abundance of radioactive elements. The tailings contain about 85% of the activity in the extracted ore, and the activity concentration is 40–100 kBq kg^{-1} of ^{238}U and 1–20 kBq kg^{-1} of ^{226}Ra (UNSCEAR 2008). The presence of ^{230}Th in the tailings will lead to the continuous production of ^{226}Ra, which will lead to the exhalation of ^{222}Rn. Since tailings are often located in dry areas, ^{226}Ra (and other elements) may also be eroded from the surface and dispersed in the environment from deposits above ground. Furthermore, dam failure may cause the dispersion of radioactive effluents, as well as other hazardous substances, in the environment. Older ponds for the storage of tailings were constructed without regard to the possible environmental impact, and many have no effective form of containment or only low embankments (IAEA 2004).

6.3.2   Enrichment and Fuel Fabrication

The uranium used as fuel in light-water reactors has to be enriched in the fissile uranium isotope ^{235}U, as discussed previously in Section 1.3.1.3. After enrichment to 3–5% ^{235}U by mass, the uranium then consists of 95–97% ^{238}U and 0.03–0.04% ^{234}U, where the activity is dominated by ^{234}U due to its shorter half-life. However, both commercial processes used for isotope separation, *gaseous diffusion* and *gas centrifugation*, require that the uranium is in the gaseous form. This is achieved during conversion, where the yellow cake is combined with fluorine to form uranium hexafluoride (UF_6) gas. According to the UNSCEAR (2008) conversion to UF_6 was performed in eight countries, as shown in Figure 6.8. Estimates of future conversion capacities have been made by the OECD (OECD 2015).

The gaseous diffusion process utilizes the fact that the average velocity of a molecule depends on its mass; lighter molecules having a higher average velocity. Isotope separation

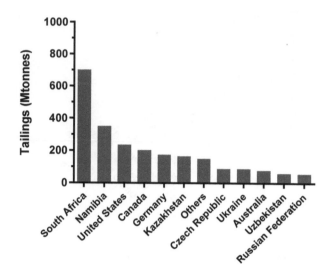

FIGURE 6.7 Total amount of tailings from uranium mining and milling, according to country, reported by the UNSCEAR in 2008. Countries with tailings exceeding 50 Mtonnes are shown individually. The total amount of tailings was about 2350 Mtonnes. Data from UNSCEAR (2008).

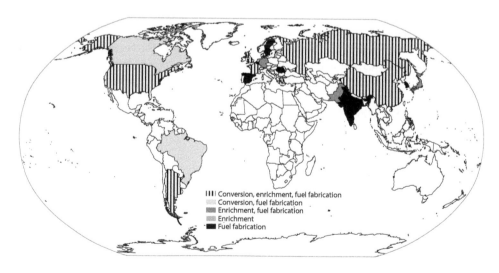

FIGURE 6.8 Countries where conversion, enrichment or fuel production is performed. Data from UNSCEAR (2008).

is achieved by allowing the gas to diffuse through a membrane. The maximum theoretical separation factor for $^{235}UF_6$ and $^{238}UF_6$ using this technique is 1.0043.

Higher separation factors can be achieved using the gas centrifugation process, as depicted in Figure 6.9. Centrifugation utilizes the fact that light molecules are enriched at the centre of the centrifuge, whereas heavier molecules are enriched at the periphery. The gas is fed through a series of vacuum tubes, each containing a rotor, and the enriched gas from one tube is then fed into the next tube for further enrichment.

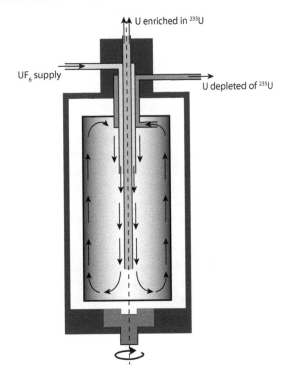

FIGURE 6.9 Rotation forces the heavier ^{238}U towards the periphery of the rotating cylinder (rotor), while the lighter ^{238}U remains at the centre. The scoops extract the circulating material by the pressure created by rotation. Thermally driven circulation can be achieved by heating the lower end of the cylinder, while keeping the upper end cool. In this case, the lighter material will be collected at the top of the cylinder.

The fabrication of nuclear fuel involves three steps: the production of uranium dioxide (UO_2) from UF_6, fabrication of fuel pellets, and the assembly of fuel elements. Uranium dioxide may also be produced from uranium trioxide (UO_3) if the uranium is not enriched. UF_6 can be converted by heating and vaporizing UF_6, followed by mixing with steam and hydrogen to form UO_2. A more elaborate process, in which it is possible to control the microstructure of the fuel pellets, is to mix the UF_6 with water, followed by reaction with ammonia or ammonium carbonate, filtering, drying, and finally, heating in a reducing environment.

The UO_2 powder can then be homogenized and additives such as Gd (a burnable poison, see Section 1.3.1.3) are added before pressing into fuel pellets. These pellets are then sintered, i.e. heated in a reducing atmosphere at about 1750 °C, to decrease their volume and porosity. Before being loaded into the fuel elements, the pellets are given their exact shape and checked for quality.

MOX fuel, i.e. mixed uranium oxide and plutonium oxide, can be made from the depleted uranium produced by the enrichment process, and the plutonium oxide produced during the reprocessing of spent nuclear fuel (see below). The plutonium used is *reactor-grade* plutonium, which contains more than 18% ^{240}Pu (and thus a lower amount of the fissile ^{239}Pu than *weapons-grade* plutonium).

An important safety aspect related to fuel fabrication is that criticality (Section 1.3.1.2) should be controlled by reducing the probability of neutron interaction and fission. The critical parameters are the mass, volume and geometry of the uranium that is handled in processing.

6.3.3 Operation of Nuclear Power Plants

In order to minimize the risks associated with the operation of nuclear power plants, they must first be properly identified, then assessed and managed. The identification of risks relies on previous experiences, research and scenarios. The use of *probabilistic safety assessment* (see below) is a valuable tool in identifying combinations of events that may have a negative impact on the operation of the plant. The risks identified must then be assessed to ensure their optimal management; see Section 6.1.1. Risk management consists of documents such as written instructions for normal operation and emergencies, activities such as education and training, maintenance, and a good "safety culture".

If, for example, core meltdown is not identified as a possible event, the risk of this event can be neither assessed nor managed. This was the case in the accident in 1979 at Three Mile Island (Harrisburg), which, however, led to only minor releases to the environment due to appropriate handling of the situation. At Chernobyl in 1986, the consequences of rapidly changing reactivity were identified and assessed, resulting in administrative and technical safety measures. However, these instructions and security systems were overridden, which rapidly led to total destruction of the reactor and reactor building. Finally, in the Fukushima accident in 2011, the risk of Tsunamis was identified, but the frequency and possible amplitude of such waves was not properly assessed.

6.3.3.1 Defence in Depth

In order to deal appropriately with the risks associated with the operation of a nuclear power plant, the concept of *defence in depth* has been developed in which the first priority is to prevent accidents (IAEA 1996). If prevention fails, the strategy should limit the potential consequences, and prevent the accident from becoming more serious. The effective implementation of a defence-in-depth strategy requires that the basic prerequisites: appropriate conservatism, quality assurance, and safety culture, are incorporated into the policy governing the design and operation of the plant.

The strategy of defence in depth is structured in five levels, where the next level is activated if the previous level has failed.

1. *Prevention of abnormal operation and failures* through a conservative design and by maintaining a high quality regarding construction, as well as operation. This includes a clear definition of operating conditions that makes it possible to distinguish between normal and abnormal conditions. It also includes robustness and resistance to accident conditions, and adequate operating instructions to trained personnel. Level 1 also includes measures for protection against external and internal hazards, for example, earthquakes and fire. Should this level fail, abnormal operation and failures are dealt with by the second level.

2. *Control of abnormal operation and detection of failures* by, for example, technical surveillance systems and administrative measures. Systems to control abnormal operation and to take corrective actions are included in this level, with the aim of restoring the plant to normal operation. If this level fails, and an accident situation occurs, other safety systems and features are activated in the third level.

3. *Control of accidents within the design basis*, which can be achieved by effective safety features. The aim is to prevent the development of the accident situation into a severe accident, especially core damage. High reliability of the safety systems is ensured, for example, by redundancy, automation and physical or spatial separation of the various safety systems. Failure to maintain the structural integrity of the core (i.e. core damage has occurred) due, for example, to flooding or complete failure of the safety system, will cause level four to come into play, in an attempt to limit the progression of the accident and mitigate accident conditions.

4. *Control of severe plant conditions, including prevention of accident progression and mitigation of the consequences of severe accidents* by accident management. The objectives at this level include monitoring plant status, maintaining cooling of the reactor and protecting the confinement. The main aim is to reduce the likelihood of severe conditions and restrict the magnitude of releases to the environment, if releases should occur. Should the situation be out of control and significant releases occur, the fifth level will be activated.

5. *Mitigation of radiological consequences of significant releases of radioactive materials*, for example, by planned measures to inform and protect the population in the areas surrounding the plant. This level is thus concerned with the off-site emergency response, which will be discussed in more detail in Section 7.2.1.

The likelihood of releases to the environment in the case of severe accidents can be decreased by physical barriers confining the radionuclides produced in the plant. The barriers in a light-water reactor used for power production are the fuel matrix, the fuel cladding, the boundary of the reactor coolant system (reactor vessel), the containment system and the filtered containment venting system. These are schematically depicted in Figure 6.10.

6.3.3.2 Safety Assessment

The safety functions at a nuclear power plant can be analysed by deterministic analyses or probabilistic assessments. A deterministic analysis is based on predetermined disturbances, which are assumed to be managed by the safety functions. Examples of such disturbances are loss of external power or breakage of a tube carrying coolant. Each barrier is analysed separately to determine whether it can withstand all the possible disturbances. Safety should be maintained regardless of the failing component. The components are assumed to fail one after another, and multiple faults are thus not considered. The advantage of deterministic analysis is that it is robust and easy to understand.

A *probabilistic safety assessment*, PSA, offers the possibility to analyse the effect of failure of all the barriers at the same time and to determine how they work when subjected to various disturbances. PSA enables the description and analysis of the safety of the whole plant, providing a more realistic model of an accident scenario. PSA thus considers multiple faults. The method is resource-demanding and complex, since the reliability of thousands of components is modelled. The first modern PSA was published in the "Rasmussen Report", WASH-1400, in 1975 (NUREG 1975). Following the Chernobyl accident, PSA became mandatory for all nuclear power plants.

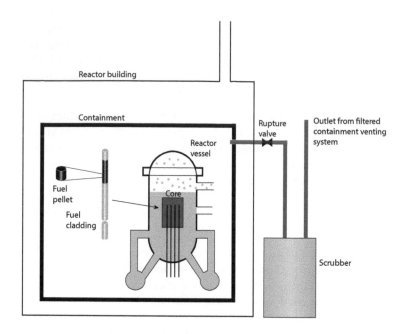

FIGURE 6.10   Physical barriers for the confinement of radioactive material in a nuclear power plant. In the case of a severe accident, leading to increased pressure in the containment, the pressure can be released through a scrubber, which may reduce the activity of iodine and caesium in the release to less than 1%. However, 50–100% of the rare gasses may pass through the scrubber and be released.

The outcome of a *Level 1 PSA* is an estimate of the probability that the plant will remain in a safe state after an initiating event, or if some kind of core damage should occur, in terms of the frequency of core damage $[y^{-1}]$. The availability and dependencies of the safety systems that are assumed to respond to the initiating event, for example, a loss of coolant, are then modelled by an initiating event tree, schematically depicted in Figure 6.11.

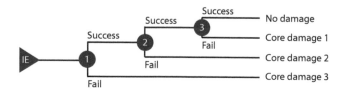

FIGURE 6.11   The frequency with which the initiating event, IE, occurs and the probability that a certain safety system will be unavailable can be estimated from previous experience of operating the plant. If all three safety systems (1–3) are available, the initiating event will not cause any core damage. However, if some, or all, of the safety systems are unavailable, some kind of core damage will occur, exemplified in the figure by three different types of core damage.

The frequency of an initiating event can be estimated from the number of times this event has occurred in the past, divided by the total operation time of all reactors. The latter is the product of the number of reactors and the duration of operation of each reactor. Let us assume that a particular initiating event occurs with a frequency of $2 \cdot 10^{-6}$ per reactor-year. If the probability that safety system 1 fails is $3 \cdot 10^{-4}$, the frequency of core damage type 3 can be estimated to be $2 \cdot 10^{-6} \times 3 \cdot 10^{-4}$, which is $6 \cdot 10^{-10}$ per reactor-year. The estimated frequencies of the other types of core damage, and the probability that the initiating event will not generate any core damage, can be calculated similarly, with knowledge of the probability of failure of the other two safety systems.

The probability of a particular safety system being unavailable due to an initiating event can be modelled from a fault tree, as shown in Figure 6.12, which is used to deduce the likelihood of a system failure based on logical combinations of simpler events. If the probabilities of the unavailability of the various parts of the system are properly estimated, the probability of failure of the safety system will be given by the fault tree. This probability is then used as input to the initiating event tree, in an approach denoted *fault tree linking* or *event tree linking*, depending on the details in the combination of the fault trees and the event tree (see e.g. Nusbaumer and Rauzy 2013). The application of Level 1 PSA is further discussed in IAEA Specific Safety Guide No. SSG-3 (IAEA 2010a).

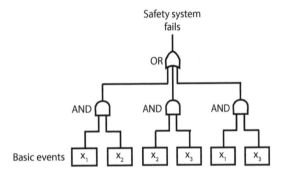

FIGURE 6.12   A sufficient flow of water to a cooling system will be maintained as long as at least two of the pumps (x_1, x_2, x_3) start on demand. The basic events indicate failure of a certain pump. Example from Nusbaumer (2015).

The probability that the safety system depicted in Figure 6.12 will fail, i.e. that fewer than two pumps will start, can be calculated approximately using *minimal cutsets*. These are defined as the set of minimal combinations that results in failure of the system, in this example $\{x_1, x_2\}$, $\{x_2, x_3\}$ and $\{x_1, x_3\}$, i.e. all combinations of basic events (pump failure) that will generate a failure of the whole system. If the probability of failure of each pump is denoted $p(x_i)$, the probability of system failure, $p(system)$ is given by

$$p(system) = p(x_1) \cdot p(x_2) + p(x_2) \cdot p(x_3) + p(x_1) \cdot p(x_3). \tag{6.3}$$

The exact solution is given by the truth table, Table 6.2, which also includes the success terms. These terms are normally omitted, and the use of minimal cutsets thus provides an upper limit for the estimate of the probability of system failure. Assuming that $p(x_i) = q$ for each pump, the probability given by the minimal cutsets, Eqn. 6.3, equals $3q^2$, while the truth table gives $3q^2 - 2q^3$.

The consequence of each type of core damage is then modelled in a *Level 2 PSA*. The modelling then focuses on the unavailability of the safety systems, e.g. the containment and its associated safety systems, which are intended to prevent the release of radioactive material. The result of modelling using a containment event tree is a number of accident scenarios, depending on the status of the containment. Examples of these are intact containment with very low releases (in spite of severe core damage), filtered releases and the

TABLE 6.2 Truth table for a "two-out-of-three" system, depicted in Figure 6.12.

x_1 fails	x_2 fails	x_3 fails	System failure	Probability
No	No	No	No	
No	No	Yes	No	
No	Yes	No	No	
No	Yes	Yes	Yes	$(1-q) \cdot q^2$
Yes	No	No	No	
Yes	No	Yes	Yes	$q \cdot (1-q) \cdot q$
Yes	Yes	No	Yes	$q^2 \cdot (1-q)$
Yes	Yes	Yes	Yes	q^3

Source: Data from Nusbaumer (2015).

release of radioactive material into the containment. The outcome is given in terms of the frequency and amount of radioactive material released [Bq y^{-1}].

The consequences for the environment are modelled in a *Level 3 PSA*. The estimated released activity and frequency for each type of release are then used as source terms in environmental models, which give an output in terms of the frequency and severity of environmental and health effects. Various kinds of environmental models have been discussed in Chapter 3. Level 3 PSA thus provides insights into accident prevention and mitigation measures, as well as into various aspects of accident management.

6.3.4 Reprocessing

Nuclear power plants generate long-lived radionuclides as the result of various reactions. The fission of ^{235}U will generate fission products, while neutron capture in ^{238}U produces ^{239}Pu (following β-decay of ^{239}U and ^{239}Np). Furthermore, neutron capture in ^{239}Pu generates ^{240}Pu, which captures neutrons to produce ^{241}Am (following β-decay of ^{241}Pu). The actinides other than uranium and plutonium are commonly referred to as *minor actinides*.

After 10 years of cooling, spent nuclear fuel (UO$_2$) from a PWR with a burn-up of 33 GWd (gigawatt–days) per tonne, consists of 95.5% uranium and 3.2% stable fission products. Other, minor constituents are Pu (0.8%), short-lived Cs and Sr (0.2%), minor actinides (0.1%), long-lived I and Tc (0.1%), and other long-lived fission products (0.1%). The minor actinides in 1 tonne of spent fuel consist of ^{239}Np (0.5 kg), and isotopes of Am (0.6 kg) and Cm (0.02 kg). The long-lived fission products are ^{129}I (0.2 kg), ^{99}Tc (0.8 kg), ^{93}Zr (0.7 kg) and ^{135}Cs (0.3 kg), and the short-lived fission products consist of ^{137}Cs (1 kg) and ^{90}Sr (0.7 kg) (OECD 2006).

The radiotoxicity, determined by the activity and the dose coefficients for ingestion and inhalation, of spent nuclear fuel from light-water reactors is dominated by the fission products during the first 100 years, while the transuranic elements, mainly isotopes of Pu and Am, dominate the long-term radiotoxicity. Figure 6.13 shows the contribution to the radiotoxicity from the various constituents of spent fuel, compared with the radiotoxicity of the uranium used to produce 1 tonne of enriched uranium. Due to the ingrowth of uranium daughters, the activity of natural uranium, and hence the radiotoxicity, increases with time until it reaches the radiotoxicity of natural uranium in equilibrium with its daughters.

Spent nuclear fuel can be managed using three different strategies. After interim storage for cooling, the fuel can be encapsulated and deposited deep in a bedrock repository, as is planned, for example, in Finland, Sweden and Switzerland. This is referred to as the

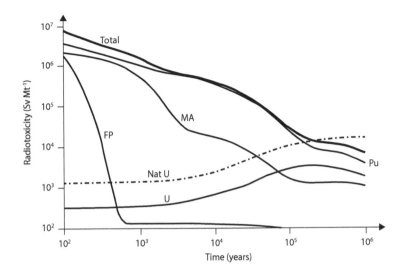

FIGURE 6.13 Contributions to the total radiotoxicity from Pu, minor actinides (MA) and U, including their decay products, and from fission products (FP). The radiotoxicity of uranium with a natural isotopic composition (Nat U) is shown for comparison. This level is reached by spent nuclear fuel after more than 100 000 years. Data from OECD (2006).

uranium once-through fuel cycle (UOT). Another strategy is to reuse the spent fuel after reprocessing, as is done, for example, in France, India and the UK. Reprocessing is currently performed in France, Japan, the Russian Federation and the UK (UNSCEAR 2008). The third strategy is to shorten the storage time by transmuting the long-lived radionuclides in the spent fuel to radionuclides with short half-lives. This method is currently only at the research stage.

The steps involved in fuel reprocessing are shown schematically in Figure 6.14. The uranium and the plutonium are separated at the reprocessing facility. The uranium is converted, enriched and used as uranium oxide fuel, while the plutonium is recovered and mixed with uranium to produce MOX fuel. Thus, reprocessing will remove a large part of the actinide activity from the spent fuel.

The recycling of plutonium will change the isotopic composition due to neutron interactions with the fuel. The fraction of ^{238}Pu increases after a few recycles, which will increase the specific activity due to the short half-life of ^{238}Pu (88 y), while the amount of the fissile fraction (^{239}Pu and ^{241}Pu) decreases. The spontaneous fission of ^{238}Pu and ^{242}Pu, makes recycled Pu less suitable for use in nuclear weapons (Choppin et al. 2002).

The separation of U and Pu from the spent fuel is performed by liquid–liquid extraction (solvent extraction), as depicted schematically in Figure 6.15, where a dissolved substance (the solute) is transferred from one liquid phase to another in a two-phase liquid–liquid system. This liquid–liquid system consists of two immiscible, dissolving liquids (solvents), usually an aqueous and an organic phase. The organic phase is often a mixture of an extractant (a substance capable of extracting an element)) and a diluent (a diluting substance).

The PUREX (Plutonium Uranium Redox EXtraction) process starts with dissolution of the spent fuel in nitric acid. The cladding and the gasses released (I, Kr and Xe) are collected and stored. Uranium and plutonium are then separated from the fission products by solvent extraction and purification in three cycles, where more than 99% of the fission products are separated in the first cycle. These fission products are retrieved and vitrified.

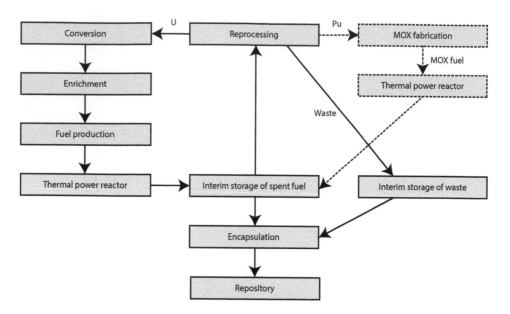

FIGURE 6.14 Reprocessing of spent nuclear fuel generates new uranium oxide (UOX) fuel, as well as MOX fuel, for use in light-water reactors.

The first cycle also separates almost 100% of the uranium and plutonium from the nitric acid into the kerosene-TBP (tributyl phosphate) phase. The uranium is then separated from the plutonium in the partition stage, also utilizing solvent extraction. After reduction from valency 4+ to 3+, the plutonium will then be transferred to the aqueous phase, and the uranium will remain in the organic phase, from where it is extracted by nitric acid. The uranium and plutonium are then processed separately (in the second and third cycle) to increase the purity of the elements. Finally, the Pu nitrate and U nitrate solutions remaining after purification are converted into oxides, UO_2 and PuO_2.

Some problems associated with the PUREX process are, for example, that Tc and Np will be present in the uranium fraction, and that Am, Cm and Np will leave the process together with the fission products, which will increase the radiotoxicity, and hence the storage time, of this fraction. However, the improved PUREX process can decrease the amount of Tc extraction and also allow the removal of Np.

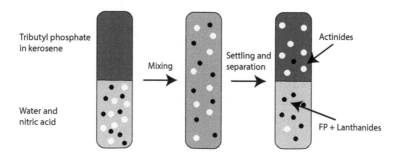

FIGURE 6.15 Separation of actinides from lanthanides and fission products (FP) by an extractant, mixed in one of the two immiscible liquids.

Reprocessing has historically resulted in large releases to the environment (see, for example, Fig. 3.50), however, these have been reduced substantially in recent years. The total discharges of α emitters and β emitters, excluding tritium, from the Sellafield reprocessing plant are shown in Figure 6.16 and Figure 6.17, respectively. Since tritium is of rather low radiological significance, it is often excluded from these types of summaries. The annual discharge of tritium from Sellafield was below 2000 TBq until the middle of the 1980s, with the exception of peaks of about 5500–6000 TBq in 1967, 1969 and 1971, and a maximum of about 12 000 TBq in 1970. Between 1986 and 2000, the annual discharge varied between about 1000 and 3000 TBq.

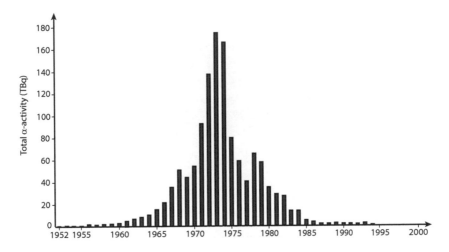

FIGURE 6.16   Releases of α emitters from Sellafield 1952–2000. Data from European Commission (2003).

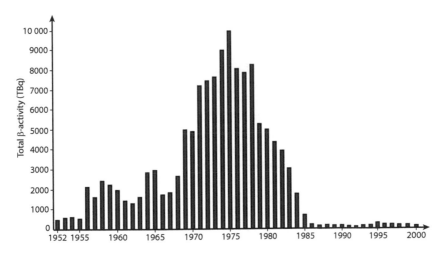

FIGURE 6.17   Releases of β emitters, excluding tritium, from Sellafield 1952–2000. Data from European Commission (2003).

Emission-free spent fuel recycling is possible but comes at a considerable economical cost, and is therefore not fully utilized. The main contaminants, ^{85}Kr, ^{99}Tc and ^{129}I, can be collected in cooling traps or on zeolites, or removed by solvent extraction or ion-exchange filters.

6.3.5 Waste Management and Storage

Guidance on the classification of radioactive waste has been published in the IAEA Safety Guide (IAEA 2009), which covers the whole range of waste, from that exempted from regulations (and thus no longer regarded as radioactive waste), to spent nuclear fuel. The classification scheme is based on six classes of radioactive waste, as illustrated in Figure 6.18.

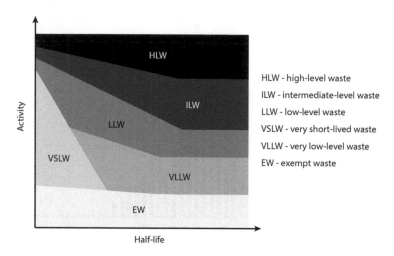

FIGURE 6.18 Classification of radioactive waste into six classes, depending on the content of long-lived radionuclides and the activity of the waste. The activity may refer to total activity, specific activity or the activity concentration of the radionuclides in the waste. The characteristics of each class are described in the text. Modified from IAEA (2009).

Waste that is no longer regarded as radioactive is denoted *exempt waste* (EW); exemption levels for various radionuclides can be found in the IAEA Basic Safety Standards (IAEA 2014). Waste that can be stored for decay over a few years (primarily waste from research or medical applications, containing short-lived radionuclides) and then cleared from regulatory control is termed *very short-lived waste* (VSLW). *Very low-level waste* (VLLW) consists of radioactive material that does not meet the criteria for EW, but is of sufficiently low activity to allow disposal in shallow landfills. The activity of VLLW is often expected to decrease significantly with time due to generally very low concentrations of radionuclides with long half-lives. The fourth class, *low-level waste* (LLW), also contains limited amounts of long-lived radionuclides, but is of such activities that isolation and containment are required for up to a few hundred years. This waste may be stored in specially designed near-surface facilities, and does not require deep burial. *Intermediate-level waste* (ILW) is defined as waste that contains long-lived radionuclides (i.e. α emitters), in quantities that require a greater degree of containment and isolation than is provided by near-surface disposal. ILW should therefore be buried at depths from a few tens to a few hundred metres. *High-level waste* (HLW), which contains high levels of both short- and long-lived radionuclides, requires special storage facilities. Plans for these facilities include disposal in deep geological formations at depths of several hundred metres. Examples of HLW are spent nuclear fuel and waste from reprocessing plants. Typical activity concentrations in HLW are 10^4–10^6 TBq m^{-3} (IAEA 2009).

Some of the most important factors that must be considered when classifying radioactive waste include the origin of the waste, the risk of reaching critical mass, and the radiological, physical, chemical and biological properties. Examples of radiological properties are half-lives, the heat generated, type of radiation emitted and dose factors. Physical and chemical parameters include the physical state, volatility, solubility and corrosiveness, while biological properties include bio-accumulation (IAEA 2009).

Examples of radiation sources and equipment in the waste classes described above are (IAEA 2009):

- **VSLW**: Brachytherapy sources with half-lives shorter than 100 d and activities below 100 MBq (^{90}Y, ^{198}Au), or below 5 TBq (^{192}Ir).

- **LLW**: Various sources of ^{3}H (tritium targets), ^{60}Co or ^{85}Kr with half-lives shorter than 15 y and activities below 10 MBq, or ^{137}Cs brachytherapy sources and moisture density gauges with activities below 1 MBq.

- **ILW**: Irradiators (^{60}Co, below 100 TBq or ^{137}Cs, below 1 PBq), thickness gauges and RTGs (^{90}Sr, below 1 PBq), static eliminators (Pu, Am or Ra, below 40 MBq) and gauges (^{241}Am or ^{226}Ra, below 10 GBq).

Radioactive waste should be stored so that human health and the environment are protected. This requirement applies both now and in the future. Radioactive waste should also be stored such that no undue burdens are imposed on future generations (IAEA 2006).

6.3.6 Public and Occupational Exposure

The radiological impact of the nuclear fuel cycle can be expressed as the normalized collective effective dose, given as man Sv GW^{-1} y^{-1}, where normalization refers to the total amount of electrical energy generated by nuclear power. In the period 1998–2002, the normalized collective effective dose to members of the public was 0.72 man Sv GW^{-1} y^{-1} (UNSCEAR 2008). The relative contributions from the various practices in the nuclear fuel cycle are shown in Figure 6.19.

The dose commitment from the operation of reactors and reprocessing plants is dominated by globally dispersed ^{14}C, which accounts for 70% of the dose commitment to the world's population. Approximately 25% of the collective dose due to ^{14}C will have been delivered by the year 2200. Until then, ^{14}C will contribute 19%, the remainder being due to all other radionuclides, which will then have delivered almost all of their doses (UNSCEAR 2008).

The normalized collective effective dose for radiation workers has decreased significantly since the 1970s, as can be seen in Figure 6.20. The total normalized collective effective dose decreased from about 20 man Sv GW^{-1} y^{-1} in the period 1975–79 to about 1 man Sv GW^{-1} y^{-1} in the period 2000–2002. The contribution from each practice in the nuclear fuel cycle is shown in Figure 6.21. Mining and reactor operation are the major contributors to the collective dose to workers. The collective doses from these practices decreased significantly between the late 1970s and the early 1990s, while the collective dose from reprocessing increased during this period.

Compliance with regulatory requirements, for example, discharge limits, protection of the public and the environment, and normal operation of nuclear installations, is verified by monitoring programmes. These programmes should provide data for dose estimates and

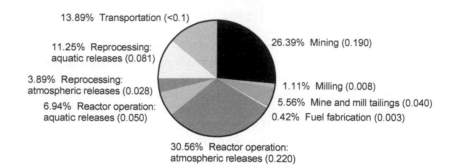

13.89% Transportation (<0.1)

11.25% Reprocessing: aquatic releases (0.081)

3.89% Reprocessing: atmospheric releases (0.028)

6.94% Reactor operation: aquatic releases (0.050)

26.39% Mining (0.190)

1.11% Milling (0.008)

5.56% Mine and mill tailings (0.040)

0.42% Fuel fabrication (0.003)

30.56% Reactor operation: atmospheric releases (0.220)

FIGURE 6.19 Fractions of normalized collective effective dose (man Sv GW^{-1} y^{-1}) to members of the public, 1998–2002, from all practices in the nuclear fuel cycle. Values of the normalized collective effective dose are given in brackets. UNSCEAR estimates the contribution from transportation to be less than 0.1 man Sv GW^{-1} y^{-1}, but the value 0.1 was assigned to this practice in the figure. Data from UNSCEAR (2008).

should also reveal deviations from normal practice. This monitoring can be divided into *source monitoring*, *environmental monitoring* and *individual monitoring*. In very special, but rare cases, individual monitoring of members of the public can be performed (IAEA 2005).

Source monitoring involves monitoring of a particular source or the discharges from a nuclear installation or other practice in which radioactive materials are used. The quantities measured are usually dose rates at predetermined locations at a site and/or discharge rates. Source monitoring is required for planned exposures, such as medical examinations, and emergency exposure situations. The way in which monitoring is performed depends on the exposure situation in question.

Environmental monitoring should be performed at installations from which radiologically significant releases of radionuclides can be anticipated. This can be performed both on site and in the surroundings, and these monitoring programmes are designed to collect data on activity concentrations in air and in products intended for human consumption, such as drinking water and foodstuffs, as depicted in Figure 6.22. Environmental monitoring can be subdivided into source-related and person-related environmental monitoring, where the latter is utilized if several sources can be expected to contribute to public or environmental exposure. Environmental monitoring is required in all three exposure situations (planned, emergency and existing) except when using weak sources in planned exposures (IAEA 2010b).

The procedures for individual monitoring of workers depend on the radiation conditions in the working environment and the type of work being performed. According to the European Commission (2013), workers liable to receive an effective dose above 6 mSv y^{-1}, or an equivalent dose to the lens of the eye above 15 mSv y^{-1}, or an equivalent dose to skin and extremities above 150 mSv y^{-1} (category A workers) shall be equipped with personal dosimeters calibrated for the operational quantities relevant to the work in question. Other radiation workers do not need to wear a personal dosimeter, if it can be assured by other types of measurements that their work is such that they do not belong to category A.

The dose to the worker is estimated by retrospective assessment of the doses resulting from both external and internal exposure, and is given approximately by the sum of the personal dose equivalent ($H_p(10)$) and the committed effective dose ($E_{(50)}$) (ICRP 2015).

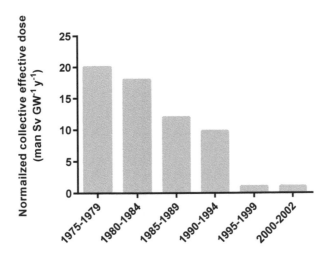

FIGURE 6.20 Normalized collective effective dose (man Sv GW^{-1} y^{-1}) to workers, from 1975 to 2002, resulting from all practices in the nuclear fuel cycle. Data from UNSCEAR (2008).

The committed effective dose is assessed using the dose coefficients (committed effective dose per unit intake) for ingestion and inhalation, and an estimate of the intake, based on whole-body measurements, measurements on excretion samples, or measurements in the working environment, for example, air sampling and surface smear samples (IAEA 1999). Revised dose coefficients for the occupational intake of radionuclides will be published by the ICRP (ICRP 2015)[1]. These coefficients are based on the Human Alimentary Tract Model and the Human Respiratory Tract Model, as well as on revised biokinetic models (see Section 2.8). Technical recommendations for monitoring individuals for occupational intakes of radionuclides are currently being prepared, and it is planned to issue them in 2016 (Etherington et al. 2015).

The monitoring shall be carried out by an approved dosimetry service, and personal dosimetry should be used when possible. If personal dosimetry is not possible, the individual dose should be assessed, for example, by area monitoring or personal dosimeters carried by other workers performing similar work tasks. The monitoring period is usually about one month, but special tasks may require shorter monitoring periods. Direct-reading dosimeters, which are often equipped with an alarm function, may be desirable for workers performing tasks in highly varying radiation fields. When a non-uniform exposure of the body can be anticipated, additional dosimeters should be used to estimate the dose to the fingers and the lens of the eye.

The use of biokinetic modelling and monitoring data to estimate the occupational intake of uranium has been discussed by Leggett et al. (2012). The default absorption types for various forms of airborne uranium include all three types, F, M and S (see Section 2.8.1.1). Type F (fast) is applicable, for example, to uranyl nitrate (at fuel fabrication and reprocessing plants) and uranium hexafluoride (at conversion plants), type M (moderate) to yellow cake and uranium ore dust, and type S (slow) to uranium dioxide. The committed effective dose and the peak kidney concentration vary slightly (within about 35% compared to the values at the default AMAD of 5 μm) with aerosol size between 0.1 and 20 μm AMAD for

---

[1]This ICRP Publication, 130, is the first in a series replacing Publications 30, 54, 68 and 78.

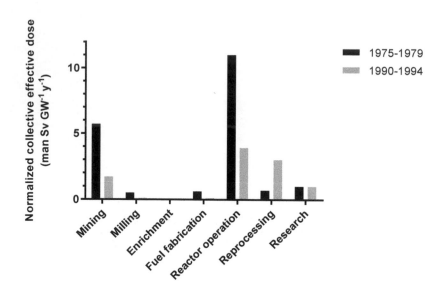

FIGURE 6.21 Normalized collective effective dose (man Sv GW^{-1} y^{-1}) to workers resulting from each practice in the nuclear fuel cycle, for the two periods 1975–1979 and 1990–1994. Data from UNSCEAR (2008).

absorption type F. The peak kidney concentration resulting from an acute intake of uranium of absorption type M also varies slightly (about 35%) with particle size in this interval, but for absorption type S, the variation is greater, about 65%. For absorption types M and S, both the committed effective dose and the peak kidney concentration resulting from chronic intake are more sensitive to variations in aerosol size, and generally increase with decreasing particle size, particularly between 0.5 μm and 0.1 μm. At 0.1 μm, the values are about 3 to 4.5 times those found for the default particle size (5 μm AMAD).

Particle sizes of uranium aerosols in working environments vary depending on the type of practice, but are typically in the range 1–20 μm. However, some practices may generate particles smaller than 1 μm, and special measures should be considered in order to ensure that neither chemical nor radiological guidance levels are exceeded (Leggett et al. 2012). The intake of airborne uranium in the workplace can be easily monitored by the analysis of urine samples. However, in order to estimate the intake it is important to have reliable knowledge on the type of exposure (acute or chronic), the route of intake (inhalation, ingestion, wound), enrichment, particle size and absorption type. Figure 6.23 shows a simulation of the large variation of the uranium concentration in urine with time.

6.4 PROTECTION OF THE ENVIRONMENT

In order to prevent and reduce the frequency of radiation effects on the biosphere, a system for the protection of the environment has been developed and is outlined in several ICRP publications (ICRP 2003, ICRP 2008, ICRP 2009, ICRP 2014). According to ICRP Publication 103 (ICRP 2007), the aim is that the level of the effects should have negligible impact on e.g. ecosystems and conservation of species. In some exposure situations, the decisions may be based primarily on protection of the environment, while protection of people may

	Nuclear power	Fuel fabrication or enrichment	Reprocessing	Waste management and repository
External radiation	Dose rate - γ	Dose rate - γ Dose rate - n	Dose rate - γ Dose rate - n	Dose rate - γ (Dose rate - n)
Air sampling	Aerosols Gasses	Aerosols Gasses	Aerosols Gasses	Aerosols (Gasses)
Food and feedstuffs	Leafy vegetables Other vegetables Fruits Drinking water Milk Grain Meat Game	Leafy vegetables (Other vegetables) (Fruits) (Drinking water) (Milk) (Grain) (Meat) (Game)	Leafy vegetables Other vegetables Fruits Drinking water Milk Grain Meat Game	Leafy vegetables Other vegetables Fruits Drinking water Milk Grain Meat Game
Terrestrial media and bioindicators	Grass Soil Lichen, mosses	Grass	Grass Soil Lichen, mosses	Grass Soil Lichen, mosses
Aquatic media and bioindicators	Water Sediment Fish Shellfish Seaweed Benthic animals	(Water) (Sediment) (Fish) (Shellfish)	Water Sediment Fish Shellfish Seaweed Benthic animals	Water Sediment Fish Shellfish Seaweed Benthic animals

FIGURE 6.22 The measurements and samples shown are recommended as part of environmental monitoring at various nuclear facilities. Items in brackets should be considered for inclusion in an environmental monitoring programme. Items not listed are regarded as not recommended, not necessary, or not practicable. Modified from IAEA (2010b).

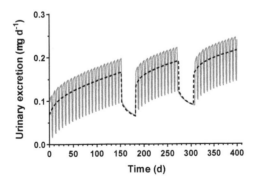

FIGURE 6.23 Simulated daily urinary excretion of uranium for shift workers (working weekdays) exposed to a chronic intake (dashed curve). The calculations are based on a yearly intake of 30 mg U by inhalation (ICRP default type S, AMAD 5 μm) and equal intakes each working day. Two vacation periods of one month with no uranium exposure is assumed for both intake patterns. Westinghouse Electric Sweden AB is acknowledged for simulation of the excretion pattern.

be of primary concern in other situations. There may also be situations where an integrated approach can be used (Larsson 2016).

Similar to reference levels used in optimization of protection of people, *Derived Consideration Reference Levels*, DCRLs, have been developed for protection of the environment in planned and existing exposure situations. Each DCRL is defined for a certain *Reference Animal and Plant*, RAP, for which radiation induced effects and dose conversion factors are given in ICRP Publication 108 (ICRP 2008). A DCRL should be interpreted as a range of absorbed dose rates where an assessor or a regulator should consider if the situation may cause environmental effects and if counteractions are needed. They are thus not intended to be regarded as dose limits. In planned exposure situations, ICRP recommends that the lower end of the DCRL for a relevant RAP is used as a reference dose rate for the sum of the contributions from all sources. In existing exposure situations, it is recommended that exposures should be brought to within the DCRL.

A RAP is defined as a hypothetical animal or plant that can be assumed to have the basic characteristics of its real counterpart. The generalization is made on the taxonomic level of family, for example, the RAP "deer" represents the species within the family *Cervidae*. The representative organism in a given exposure situation may be found among the defined RAPs, which permits the corresponding DCRL to be used directly. However, if this is not the case, the differences have to be assessed. At present, the following RAPs have been defined (wildlife group represented given in brackets): Deer (*Cervidae*; large terrestrial mammals), Rat (*Muridae*; small terrestrial mammals), Duck (*Anatidae*; aquatic birds), Frog (*Ranidae*; amphibians), Trout (*Salmonidae*; pelagic fish), Flatfish (*Pleuronectidae*; benthic fish), Bee (*Apidae*; terrestrial insects), Crab (*Cancridae*; crustaceans), Earthworm (*Lumbricidae*; terrestrial annelids), Pine tree (*Pinacea*; large terrestrial plants), Grass (*Poaceae*; small terrestrial plants), and Brown seaweed (*Cyclosporea*; seaweeds).

As an example, the DCRL for deer, rat and duck is currently 0.1–1 mGy d^{-1}. For the first two RAPs, the probability for effects is very low if the dose rate is within this interval, while no information on effects is available for ducks. However, in the next band of dose rates, 1–10 mGy d^{-1}, there is a potential for reduced reproductive success (for all these three RAPs).

The dose rate to animals and plants can be modelled from measured activity concentrations of radionuclides in the environment (e.g. soil or water). The activity concentration in the organism is then estimated using concentration ratios and distribution coefficients (see Section 3.4.1.2 and 3.3.2, respectively), and the corresponding dose rate from internal exposure is calculated by dose coefficients, which give the dose rate per unit activity concentration in the organism. The dose rate from external exposure can be found by dose coefficients relating the dose rate to the activity concentration in the surrounding environment. Concentration ratios for reference animals and plants are given in ICRP Publication 114 (ICRP 2009), while dose conversion factors are presented in Publication 108 (ICRP 2008). These recommended transfer parameters are based on data in the *wildlife transfer database*, WTD, that was established to collate parameter values for estimating transfer of radioactivity to wildlife—http:www.wildlifetransferdatabase.org/ (Copplestone et al. 2013).

In order to facilitate assessment and management of environmental risks from exposure to ionizing radiation, various initiatives have been taken. For example, the ERICA Integrated Approach was developed between 2004 and 2007 and comprises assessment, risk characterisation and management (Larsson 2008); see also Further reading below. A computer-based tool, the ERICA tool, has been developed to facilitate the calculations and has been made freely available from http://www.erica-tool.com and http://www.erica-tool.eu/ (Brown et al. 2008, Brown et al. 2016).

6.5 REFERENCES

- Brown, J. E., Alfonso, B., Avila, R., Beresford, N. A., Copplestone, D., Pröhl, G., and Ulanovsky, A. 2008. The ERICA tool. *J. Environ. Radioact.* 99, 1371–1383.

- Brown, J. E., Alfonso, B., Avila, R., Beresford, N. A., Copplestone, D., and Hosseini, A. 2016. A new version of the ERICA tool to facilitate impact assessments of radioactivity on wild plants and animals. *J. Environ. Radioact.* 153, 141–148.

- Choppin, G., Liljenzin, J-O. and Rydberg, J. 2002. *Radiochemistry and Nuclear Chemistry.* Woburn: Butterworth-Heinemann.

- Copplestone, D., Beresford, N. A., Brown, J. E., and Yankovich, T. 2013. An international database of radionuclide concentration ratios for wildlife: development and uses. *J. Environ. Radioact.* 126, 288–298.

- Etherington, G., Bérard, P., Blanchardon, E., Breustedt, B., Castellani, C. M., Challeton-de Vathaire, C., Giussani, A., Franck, D., Lopez, M. A., Marsh, J. W. and Nosske, D. 2015. Technical recommendations for monitoring individuals for occupational intakes of radionuclides. *Radiation Protection Dosimetry*, doi:10.1093/rpd/ncv395.

- European Commission 2013. Council Directive 2013/59/EURATOM.

- European Commission 2003. *MARINA II. Update of the MARINA Project on the Radiological Exposure of the European Community from Radioactivity in North European Marine Waters.* Radiation Protection 132. Annex A: Civil Nuclear Discharges into North European Waters.

- Hansson, S. O. 2011. Radiation protection—sorting out the arguments. *Philos. Technol.*, 24, 363–368.

- IAEA, International Atomic Energy Agency. 1996. *Defence in Depth in Nuclear Safety.* INSAG-10. International Atomic Energy Agency. Vienna.

- IAEA, International Atomic Energy Agency. 1999. *Assessment of Occupational Exposure Due to Intakes of Radionuclides.* Safety Guide No. RS-G-1.2. International Atomic Energy Agency. Vienna.

- IAEA, International Atomic Energy Agency. 2004. *The Long Term Stabilization of Uranium Mill Tailings.* IAEA-TECDOC-1403. International Atomic Energy Agency. Vienna.

- IAEA, International Atomic Energy Agency. 2005. *Environmental and Source Monitoring for Purposes of Radiation Protection.* Safety Guide No. RS-G-1.8. International Atomic Energy Agency. Vienna.

- IAEA, International Atomic Energy Agency. 2006. *Storage of Radioactive Waste.* IAEA Safety Standards Series No. WS-G-6.1. International Atomic Energy Agency. Vienna.

- IAEA, International Atomic Energy Agency. 2009. *Classification of Radioactive Waste.* IAEA Safety Standards Series No. GSG-1. International Atomic Energy Agency. Vienna.

- IAEA, International Atomic Energy Agency. 2010a. *Development and Application of Level 1 Probabilistic Safety Assessment for Nuclear Power Plants.* IAEA Safety Standards Series No. SSG-3. International Atomic Energy Agency. Vienna.

- IAEA, International Atomic Energy Agency. 2010b. *Programmes and Systems for Source and Environmental Radiation Monitoring.* Safety Reports Series No. 64. International Atomic Energy Agency. Vienna.

- IAEA, International Atomic Energy Agency. 2012. *Communication with the Public in a Nuclear or Radiological Emergency.* Emergency Preparedness and Response. EPR-Public Communications 2012. International Atomic Energy Agency. Vienna.

- IAEA, International Atomic Energy Agency. 2013. *Actions to Protect the Public in an Emergency Due to Severe Conditions at a Light Water Reactor.* Emergency Preparedness and Response. EPR-NPP Public Protective Actions 2013. International Atomic Energy Agency. Vienna.

- IAEA, International Atomic Energy Agency. 2014. *Radiation Protection and Safety of Radiation Sources: International Basic Safety Standards.* General Safety Requirements Part 3, No. GSR Part 3. International Atomic Energy Agency. Vienna.

- ICRP 1955. Recommendations of the International Commission on Radiological Protection. *Br. J. Radiol.,* (Suppl. 6).

- ICRP 1973. *Implications of Commission Recommendations that Doses Be Kept as Low as Readily Achievable.* ICRP Publication 22. Pergamon Press, Oxford, UK.

- ICRP 1977. *Recommendations of the International Commission on Radiological Protection.* ICRP Publication 26, Ann. ICRP 1 (3).

- ICRP 1991. *1990 Recommendations of the International Commission on Radiological Protection.* ICRP Publication 60, Ann. ICRP 21 (13).

- ICRP, 2003. *A Framework for Assessing the Impact of Ionising Radiation on Non-human Species.* ICRP Publication 91. Ann. ICRP 33 (3).

- ICRP 2006. *Assessing Dose of the Representative Person for the Purpose of the Radiation Protection of the Public.* ICRP Publication 101a. Ann. ICRP 36 (3).

- ICRP 2007. *The 2007 Recommendations of the International Commission on Radiological Protection.* ICRP Publication 103. Ann. ICRP 37 (2–4).

- ICRP, 2008. *Environmental Protection—the Concept and Use of Reference Animals and Plants.* ICRP Publication 108. Ann. ICRP 38 (4-6).

- ICRP, 2009. *Environmental Protection: Transfer Parameters for Reference Animals and Plants.* ICRP Publication 114, Ann. ICRP 39(6).

- ICRP, 2014. *Protection of the Environment under Different Exposure Situations.* ICRP Publication 124. Ann. ICRP 43(1).

- ICRP 2015. *Occupational Intakes of Radionuclides: Part 1.* ICRP Publication 130. Ann. ICRP 44(2).

- ILO, 1960a. *R114—Radiation Protection Recommendation, 1960 (No. 114).* International Labour Organization.

- ILO, 1960b. *C115—Radiation Protection Convention, 1960 (No. 115), Convention Concerning the Protection of Workers against Ionising Radiations.* International Labour Organization.

- Larsson, C. M., 2008. An overview of the ERICA Integrated Approach to the assessment and management of environmental risks from ionising contaminants. *J. Environ. Radioact.* 99, 1364–1370.

- Larsson, C. M. 2016. Overview of ICRP Committee 5: protection of the environment. In: *ICRP, 2016. Proceedings of the Third International Symposium on the System of Radiological Protection. Ann. ICRP 45(1S).*

- Leggett, R. W., Eckerman, K. F., McGinn, C. W. and Meck, R. A. 2012. *Controlling Intake of Uranium in the Workplace: Applications of Biokinetic Modelling and Occupational Monitoring Data.* Oak Ridge National Laboratory. ORNL/TM-2012/14.

- Lindell, B. 2003. *Herkules storverk* [In Swedish]. Atlantis, Stockholm, Sweden.

- Lindell, B. 2011. *Sisyfos möda* [In Swedish]. Atlantis, Stockholm, Sweden.

- NUREG, 1975. *Reactor Safety Study. An Assessment of Accident Risks in U.S. Commercial Nuclear Power Plants.* WASH-1400 (NUREG-75/014). United States Nuclear Regulatory Commission, Washington D.C. USA.

- Nusbaumer, O. and Rauzy, A. 2013. Fault tree linking versus event tree linking approaches: A reasoned comparison. *Proc IMechE Part O: J Risk and Reliability.* 227(3), 315–326.

- Nusbaumer, O. 2015. *Introduction to Probabilistic Safety Assessments (PSA).* http://nusbaumer.tripod.com/resources/publications/nusbaumer_introduction_to_probabilistic_safety_assessments.pdf (2015-12-25)

- OECD 2006. *Physics and Safety of Transmutation Systems. A Status Report.* Nuclear Energy Agency. Organisation for Economic Co-operation and Development. NEA No. 6090.

- OECD 2014. *Uranium 2014: Resources, Production and Demand.* Nuclear Energy Agency. Organisation for Economic Co-operation and Development. NEA No. 7209.

- OECD 2015. *Nuclear Energy Data.* Nuclear Energy Agency. Organisation for Economic Co-operation and Development. NEA No. 7246.

- Perko, T. 2014. Radiation risk perception: A discrepancy between the experts and the general public. *Journal of Environmental Radioactivity*, 133, 8691.

- Preston, D. L., Ron, E., Tokuoka, S., Funamoto, S., Nishi, N., Soda, M., Mabuchi, K. and Kodama, K. 2007. Solid cancer incidence in atomic bomb survivors: 1958–1998. *Radiation Research*, 168, 1–64.

- UNSCEAR, United Nations Scientific Committee on the Effects of Atomic Radiation. 2000. *Sources and Effects of Ionizing Radiation, Vol.I: Sources, Annex C.* New York: United Nations.

- UNSCEAR, United Nations Scientific Committee on the Effects of Atomic Radiation. 2008. *Sources and Effects of Ionizing Radiation, Vol.I: Sources of Ionizing Radiation, Annex B.* New York: United Nations.

- Wikman-Svahn, P. 2012. Radiation protection issues related to the use of nuclear power. *WIREs Energy Environ.* doi: 10.1002/wene.22.

- WHO, World Health Organization. 2011. *Guidelines for Drinking-Water Quality, 4th* Ed. World Health Organization, Geneva.

6.6 EXERCISES

6.1 Evaluate the consequences of opening a uranium mine in your neighbourhood. Make a simple risk assessment concerning the opening and operation of the mine.

6.2 A company wants to build and operate a small reactor for power production. Make a simple risk assessment concerning the operation of the power plant.

6.3 Which types of monitoring are adequate for the power plant in the previous exercise?

6.4 Compare light-water reactors and heavy-water reactors regarding their operational safety, handling of spent fuel and nuclear non-proliferation issues.

6.7 FURTHER READING

Castellani, C. M., Marsh, J. W., Hurtgen, C., Blanchardon, E., Berard, P., Giussani, A. and Lopez M. A. 2013. *IDEAS Guidelines (Version 2) for the Estimation of Committed Doses from Incorporation Monitoring Data.* EURADOS Report 2013-01.

Howard, B., and Larsson, C. M. (Eds.). 2008. *The ERICA Project, Environmental Risk from Ionising Contaminants: Assessment and Management.* Journal of Environmental Radioactivity, Volume 99, Issue 9, Pages 1361–1518. Elsevier.

IAEA 2002. *Management of Radioactive Waste from the Mining and Milling of Ores.* Safety Guide. Safety Standards Series No. WS-G-1.2. International Atomic Energy Agency. Vienna.

IAEA 2004. *Methods for Assessing Occupational Radiation Doses Due to Intakes of Radionuclides.* Safety Reports Series No. 37. International Atomic Energy Agency. Vienna.

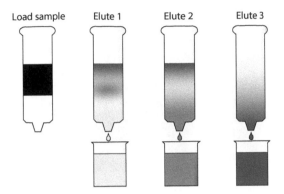

FIGURE 5.25: Principle of extraction of several analytes from one solution, showing how various analytes can be extracted (eluted) from a sample using the same resin column.

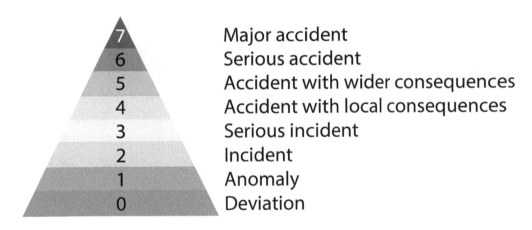

FIGURE 7.1: The IAEA International Radiological and Nuclear Event Scale (INES).

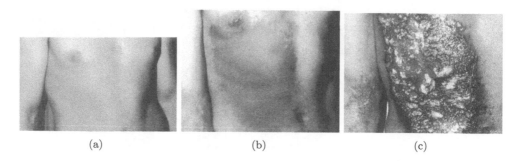

(a)　　　　　　　(b)　　　　　　　(c)

FIGURE 7.5: Illustration of tissue damage caused by a 2-hour exposure to a Category 2 source (0.18 TBq of ^{192}Ir) used for radiography. Reproduced with permission by the IAEA. International Atomic Energy Agency, The Radiological Accident in Gilan, IAEA, Vienna (2002c).

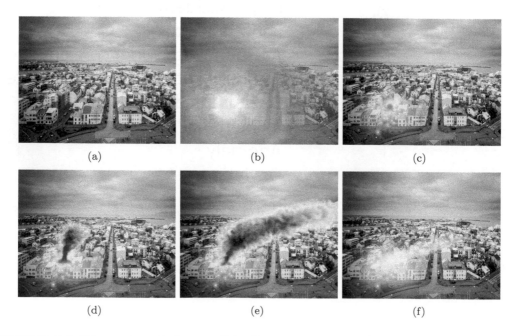

(a)　　　　　　　(b)　　　　　　　(c)

(d)　　　　　　　(e)　　　　　　　(f)

FIGURE 7.6: Illustration of the possible dispersion pattern of a radiological dispersion device (RDD) deployed on the roof of a multi-storey car park at a shopping mall.

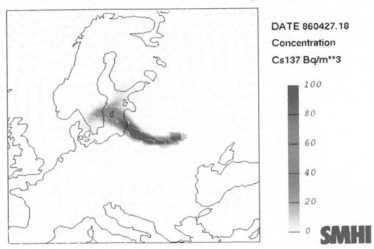

MATCH simulation of the Chernobyl accident

DATE 860427.18

Concentration

Cs137 Bq/m**3

100

80

60

40

20

0 **SMHI**

FIGURE 7.20: A mathematical simulation of the Chernobyl release entering Sweden in 1986. The simulation was carried out by the Swedish Meteorological Institute (SMHI) using quality-checked data from a large number of weather observation stations.

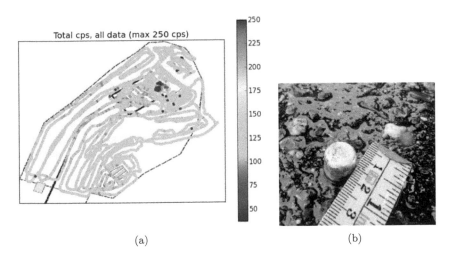

Total cps, all data (max 250 cps)

(a) (b)

FIGURE 7.43: (a): Map of the γ dose rate intensity recorded by a HPGe spectrometer (123% relative efficiency) carried in a backpack. (b): Example of a radiation source (^{110m}Ag) found by mapping an abandoned radiological waste repository on foot.

FIGURE 7.46: Dose rate contributions from surface contamination of ^{137}Cs in an urban environment in the city of Gävle, Sweden.

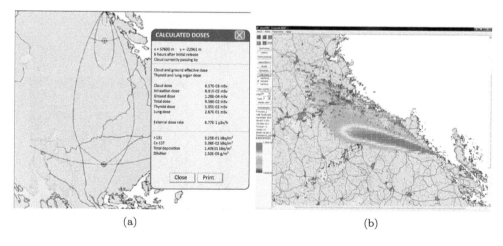

(a)   (b)

FIGURE 7.50: Isodose maps of NPP release plumes using a simple rectilinear transport model (a), and a particle trajectory based model incorporating more authentic metrological (b). Figures are obtained from test runs conducted by the Swedish Radiation Safety Authority.

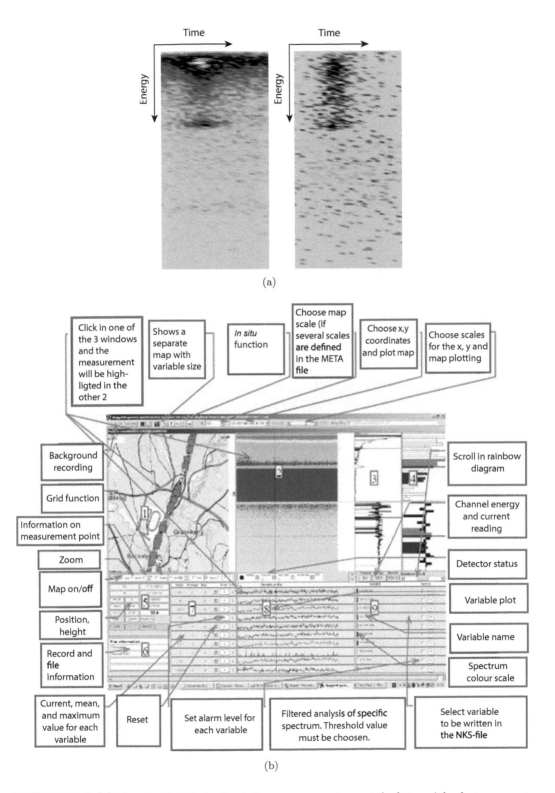

Time

Energy

Time

Energy

(a)

Click in one of the 3 windows and the measurement will be highligted in the other 2

Shows a separate map with variable size

In situ function

Choose map scale (if several scales **are defined** in the META **file**

Choose x,y coordinates and plot map

Choose scales for the x, y and map plotting

Background recording

Scroll in rainbow diagram

Grid function

Channel energy and current reading

Information on measurement point

Zoom

Detector status

Map on/off

Variable plot

Position, height

Variable name

Record and **file** information

Spectrum colour scale

Current, mean, and maximum value for each variable

Reset

Set alarm level for each variable

Filtered analysis **of specific** spectrum. Threshold value must be choosen.

Select variable to be written in **the NKS-file**

(b)

FIGURE 7.70: (a): A waterfall plot of mobile gamma spectrometric data, with photon energy on the downward y-axis, and time on the x-axis. (b): Various graphical representations of time series and spectrometric data acquired by a mobile gamma spectrometric system. This window is the working tool for the operators carrying out the mobile survey.

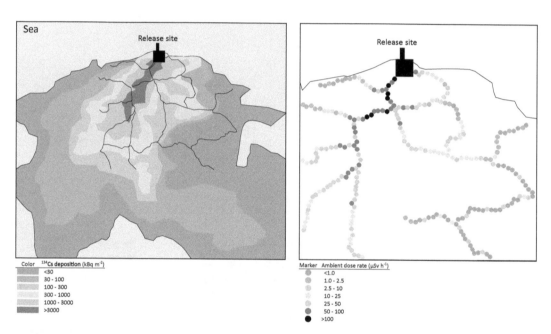

FIGURE 7.71: Left: Schematic activity deposition data map from airborne surveys of region close to an affected NPP releasing fission products and neutron activated fission products such as [134]Cs. Black lines symbolize road network close to the site. Right: Schematic dose-rate map from car-borne measurements close to the affected NPP site, indicating the large gradients and highly inhomogeneous nature of the ground deposition.

Emergency Preparedness

CONTENTS

RADIATION-generated radioactive sources (R) or from fissile materials (N) constitutes a hazard to the health of anyone exposed. The exposed persons can be members of the public, patients undergoing therapy or diagnostic procedures, or workers occupied within practices involving these materials. However, if used as planned, these sources will pose a relatively moderate hazard to exposed individuals compared with situations when the radiation source is not under control or if intentionally used for malevolent acts. Therefore emergency preparedness against radiological and nuclear incidents is of key concern for the general crisis management. A first step in building up an emergency preparedness against RN events is to identify and assess the various threats and scenarios leading to uncontrolled radiation exposures. The next step is to design remedial actions to obtain control of the exposure situations, to mitigate continued exposure and to remediate the exposure and harm already endured. The remedial actions, however, come with a price; not only by means of economic costs to the society but also in terms of the harm inferred to the affected individuals. Therefore actions taken must be subject to similar radiation protection principles as for practices done under controlled situations. Although experts and decision makers from a wide range of disciplines and specialties are involved in the crisis management of RN events, the radiation protection expert has the central role in that they are basically the only ones who can quantitatively assess the radiological situation by means of radiometry. Various methods of detecting the presence of enhanced radiation must be developed. All parties involved must then be regularly trained on these methods including the radiation protection experts themselves. The most important feature in the radiological assessment by the expert is the interpretation of the detector readings into tangible risks to the affected individuals and the other actors involved in the emergency management. A large variety of situations in which such interpretations must be done can be expected in an RN event, from the local damage site up to the expert panel directly advising the director generals of the affected authorities.

7.1   EXPOSURE SCENARIOS

The radiological and nuclear threats that have arisen since the first self-sustaining man-made nuclear fission reaction was created by Szilard and Fermi in 1942, can essentially be divided into three eras. The first era, from 1945–79, was dominated by the Cold War, and the threat of the use of nuclear weapons. In this period, tension between the superpowers led to extensive programmes of atmospheric nuclear weapons testing, resulting in global exposure of humans and ecosystems to man-made radionuclides that had not previously been encountered. The second period, between 1979 and 2001, was initiated by the Three Mile Island accident in Harrisburg, Pennsylvania, US in 1979, the largest civil nuclear energy accident up to that date. This event increased awareness of the threats associated with non-optimal use of commercial nuclear power, with the result that many countries using nuclear power, and their neighbours, began to develop emergency preparedness plans for use in the event of accidents at nuclear power plants (NPPs). This period also coincided with the break-up of the Soviet Union, which led to large-scale dispersion and illicit trafficking of orphan radioactive sources. In 2001, the 9/11 terrorist attacks in the USA raised concern about radiological and nuclear (RN) incidents including the malevolent use of radioactive sources, as well as chemical (C), biological (B) and explosive (E) hazards. The event was followed by extensive mobilization of homeland security in many countries throughout the world. Table 7.1 presents some milestones and events relevant to the use of nuclear power and radiological threats that have occurred since 1945.

At the beginning of the 1990s, the IAEA (IAEA 2015a) suggested a grading system for the severity of RN incidents: the International Nuclear and Radiological Event Scale (INES). It is intended to be applied to civil, non-military events taking place worldwide. The scale ranges from 0 to 7, with each step representing a tenfold increase in the severity of the incident. A value is assigned to an event based on: (i) health effects on humans and biota; (ii) the impact on the barrier and control systems of the nuclear and radiological sources involved; and (iii) the effect on the fundamental security system of the site or organization involved. Events belonging to classes 03 are termed "incidents", while levels 4–7, which have consequences for the surrounding environment, are called "accidents" (Fig. 7.1).

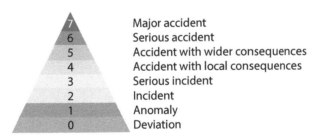

FIGURE 7.1   The IAEA International Radiological and Nuclear Event Scale (INES). See also colour image.

7.1.1   Radiological Threats

7.1.1.1   General Aspects of Radiological Accidents

An analysis of the number of reported events (both incidents and accidents) in terms of the type of source and activity involved shows that the non-nuclear industry dominates, especially regarding incidents and accidents leading to tissue damage and acute radiation

TABLE 7.1   Important milestones and events in the civil and military use of radiological and nuclear materials.

Year	Event
1938	Discovery of induced disintegration of an atomic nucleus (fission) by Hahn and Meitner.
1942	First self-contained fission chain reaction, achieved by Szilard and Fermi, in Chicago.
1945	First nuclear detonations: first test at Los Alamos, USA, followed by detonation of two bombs over Japan, at Hiroshima and Nagasaki.
1952	Reactor core damage at the First Chalk River research reactor in Canada (INES rating 5). Although there were no fatalities, the lessons learned initiated many features of modern-day nuclear safety.
1957	Kysthym accident in the USSR involving the explosion of a storage tank containing liquid nuclear waste. This released a total of 740 PBq of fission products, of which about 10% (mainly ^{137}Cs and ^{90}Sr/^{90}Y) was dispersed over an area of up to 20 000 km^2 (INES rating 6). The contaminated land resulted in the protracted release of ^{137}Cs and ^{90}Sr/^{90}Y into the Techa River between 1957 and 1962.
1957	Fire at the military reactor at Windscale, UK, releasing, e.g., ^{131}I (INES 5).
1961	The peak of atmospheric nuclear weapons testing with a 50-Mton thermonuclear bomb being detonated above Novaya Zemlya in the Arctic Ocean.
1960s to 1980s	Peacetime tests of nuclear explosives by the USA and USSR. Tests in the USSR led to delayed dispersion of long-lived fission products among people living near the Kraton 3 test site in subarctic Russia.
1961	Discovery of elevated ^{137}Cs concentrations in lichen, reindeer and reindeer herders in Scandinavia, prompting the realization that Arctic and subarctic ecosystems are highly sensitive to dispersion of radioactive elements.
1961	A prompt criticality and steam explosion at the small SL-1 reactor in the US kills 3 operators, the first victims of a nuclear accident.
1960s and 1970s	Initiation of civil nuclear power programmes in many industrialized countries, such as the USA, Canada, Japan, Germany, France, the USSR and Sweden.
1966	Mid-air collision of a tanker plane and a B52 bomber during airborne refuelling outside Palomares, Spain, leading to a non-nuclear explosion of the B52's hydrogen bombs, causing the dispersion of fissile plutonium over a 2-km^2 area of land (not INES rated).
1978	The nuclear-reactor-powered Soviet satellite Cosmos 954 fell out of orbit and crashed in a remote area of Canada (not INES rated).
1979	Partial core meltdown at the Three Mile Island NPP in the USA, leading to atmospheric releases of, e.g., radioactive noble gases and ^{131}I. This precipitated the construction of an extra barrier at many NPPs worldwide, involving an external filter device connected to the pressure-relief system of the inner reactor containment system (INES rating 5).
1980	Introduction of modern emergency preparedness in the event of NPP accidents in Sweden and many other countries with nuclear power. The last atmospheric nuclear detonation to date was conducted by China.

TABLE 7.1 (Continued.)

Year	Event
1983	A ^{60}Co therapy source was mistakenly salvaged by a local worker in Cuidad, Mexico, which contaminated his truck. When later scrapped and recycled, the contaminated vehicle contaminated more than 5000 m^3 of steel. Unknown to the manufacturer, the contaminated steel was used in furniture that was purchased by many US households.
1986	Core meltdown and explosion of the graphite-moderated Reactor Number 4 at Chernobyl, leading to the largest accidental release of fission products and transuranic elements from a NPP. Up to 28 fatalities resulted from combined acute radiation exposure and fire. UNSCEAR predicts 50 000 excess cancer cases over a 50-year period following the accident (INES rating 7).
1987	Theft of an abandoned ^{137}Cs therapy source in Goiania, Brazil, leading to 4 fatalities with many more suffering significant internal and external exposure (INES rating 4).
1987–1990s	Further development of NPP emergency preparedness. Some countries decommissioned their graphite-moderated NPPs (e.g. two reactors in Lithuania), and others implemented modifications of existing NPPs.
1994	Theft of a ^{137}Cs source, mistaken for scrap metal, from poorly guarded storage in Tammiku, Estonia, resulting in two fatalities and severe exposure of a young child. This case, which received international attention, was the first of several related to the gradual loss of control over radiation sources during the fall of the former USSR.
1995/1996	Although possibly a publicity stunt, Chechen rebels planted a primitive radiological dispersion device at a park in Moscow. This was the first attempt to use a "dirty-bomb" against civilians.
1999	Criticality accident at the Tokaimura nuclear fuel factory in Japan, leading to two fatalities and significant exposure of other staff at the site (INES rating 4).
2001	Non-nuclear 9/11 terrorist attacks in the USA, resulted in awareness of the need for better preparedness against the malevolent use of hazardous substances.
2006	Assassination of former USSR agent Alexander Litvinenko by poisoning with ^{210}Po, raising concern about the illicit circulation of ^{210}Po from various sources.
2011	The Great East Japan earthquake and tsunami, leading to the flooding of the Fukushima Daiichi NPP and subsequent nuclear accident, with meltdowns in three reactors and atmospheric releases of radioactive material (INES rating 7).

syndrome (Fig. 7.2). However, after taking into account the global impact of late effects and anticipated cases of induced cancer, the number of people affected by the two major nuclear reactor accidents—at Chernobyl in 1986 and at Fukushima in 2011—is greater than the total number of people affected by all other RN events.

UNSCEAR (UNSCAER, 2000) has divided radiation and nuclear accidents into three categories: (i) criticality accidents involving exposure caused by loss of control over nuclear reactions or fissile materials (critically >1; see Section 1.3.1); (ii) accidental exposures from radiological devices such as sealed sources, x-ray units, or accelerators; and (iii) accidents caused by exposure from unsealed radioisotopes, such as radiopharmaceuticals, or uranium and actinides used in civil and military nuclear fuel processing. The first category accounted for the dominant share of events during the pioneering era of nuclear power in the 1940s and the beginning of the 1950s. These events were often related to research facilities, and involved operators with little experience of radiation safety. However, since then, accidents related to radiological devices have accounted for the majority of reported incidents and accidents due to the extensive introduction of radiological devices in healthcare and industry.

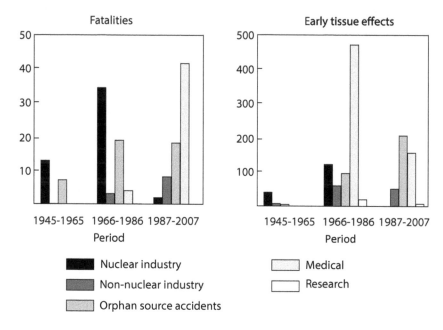

FIGURE 7.2   Fatalities (left) and early tissue effects (right) resulting from radiation accidents. Data from UNSCEAR (2008).

Apart from the major nuclear power accidents at Chernobyl and Fukushima (both rated 7 on the INES), examples of accidents (INES rating >3) that have occurred since 1945 are the release of the fission products ^{137}Cs and ^{90}Sr/^{90}Y from the Kysthym accident in 1957, the accident at the Three Mile Island NPP in 1979 (both INES rating 6), and the criticality accident at the nuclear fuel factory in Tokaimura in 1999 (INES rating 4). Table 7.1 gives important examples of milestones in the history of nuclear research, together with RN emergencies since 1945.

In addition to accidents, there are also a number of examples of the malevolent use of radioactive materials. In the mid-1990s a student at a research department in Taiwan using ^{32}P (a pure β emitter, $E_{Max} = 1.70$ MeV) was poisoned with the radionuclide by an adversary. In Russia in 1993, the director of a packaging company succumbed to acute radiation-induced illness after being unknowingly exposed for weeks by, among other things, ^{137}Cs sources inserted into his office chair. The assassination of the former USSR agent Alexander Litvinenko attracted considerable attention in the autumn of 2006, since administration of the ^{210}Po was discovered to have been accompanied by severe contamination of bystanders, such as passengers on an international flight carrying the suspected assassins. The specific activity of ^{210}Po is 166 TBq g^{-1}, which means that it is highly radiotoxic when ingested. An activity of 100 MBq, or less than 1 mg, can be fatal. This almost pure alpha emitter (there is a low probability of photon emission: $E_\gamma = 803$ keV with $n_\gamma = 0.000012$) is very difficult to detect using external detectors, which require long-range radiation particles for detection. Another political figure suspected to have been poisoned with ^{210}Po was Yasser Arafat, although this has been refuted by some experts (Froidveaux et al. 2013).

Two factors govern the use of radiation sources or other nuclear material: (i) security, which refers to how well the source is protected from access by unauthorized persons, and (ii) safety, which is related to the risk of harm to people handling the source. In most countries, a license is required guaranteeing that the user or operator conforms to the safety and security regulations for radiation sources and nuclear materials before an organization or company is allowed to handle and use radioactive sources. International organizations such as the IAEA issue guidelines (e.g., IAEA 2009) to national regulatory authorities which govern the procurement, handling, storage and dismantling of radiation sources. Often, international guidelines emphasize the need for operators to establish a safety culture in connection with their practices. A safety culture means that employees at all levels recognize the potential threats and risks associated with radiation. This means that, for example, responsibility for security is shared throughout the whole organization, that continuous training of new and existing staff members is carried out, and that there is a well-functioning system for reporting incidents and feedback of experience.

In spite of the rigorous regulations imposed on operators in connection with the use of radiological or nuclear devices, there are many cases where control over radiation sources has been lost (see Fig. 7.2), or they have been damaged or physically lost. The most common reason for the inadvertent exposure of workers and the public is unintentional loss of control by the operator, leading to misuse of the source (as in the Tammiku accident in Estonia in 1994). Another reason for the accidental exposure of humans to radioactive or nuclear sources is that they have been abandoned by the operator (as in the Goiania accident in Brazil in 1987). Such sources are termed *orphan sources*. The impact of accidental exposures, and the necessary countermeasures, will be similar if the exposure was the result of the malevolent use of radiological or nuclear devices. Experience from radiation accidents can therefore be vital when planning emergency preparedness in the event of intentional malicious exposures to radiation.

7.1.1.2 *Hazardous Radioactive Sources*

The IAEA (IAEA 2005a) has defined five categories of radioactive sources in terms of their hazard rating. The categories are: 1. Extremely dangerous; 2. Very dangerous; 3. Dangerous; 4. Unlikely to be dangerous; and, finally, 5. Most unlikely to be dangerous (Fig. 7.3).

The category *extremely dangerous* includes radioisotope thermoelectric generators, RTGs, one type of which uses ^{90}Sr/^{90}Y sources with activities between 330 and 2500 TBq. Although a pure beta emitter, the high-energy beta particles emitted from ^{90}Y generate substantial bremsstrahlung that can penetrate the beta-particle shielding and cause very high absorbed dose rates in the immediate vicinity of the source. There are cases of strong ^{90}Sr/^{90}Y sources being mistaken for gamma sources when measured by simple non-specific nuclear radiometry instruments. RTGs are used for the generation of electric power independently of the distribution grid, and are especially common in remote areas such as Siberian lighthouses and in spacecraft. An RTG consists of two metals with different electrothermal potentials heated by a radioactive source (Fig. 7.4). Such a device can deliver between 10 and 1000 W of power. It is cheaper to manufacture RTGs using ^{90}Sr as a source instead of ^{238}Pu, but the former require more shielding due to bremsstrahlung. It is estimated that more than 1000 discarded RTGs are being stored in the former USSR, requiring maintenance and security to prevent theft.

Another example of Category 1 sources is ^{60}Co sources used in sterilization units. These sources are designed to deliver up to 1000 Gy min-1 and are used for the preparation of food and spices, as well as for the sterilization of medical and surgical instruments. When fully exposed, such a source can deliver a lethal radiation dose at a distance of 1 m in a time of 0.1 s to 10 min.

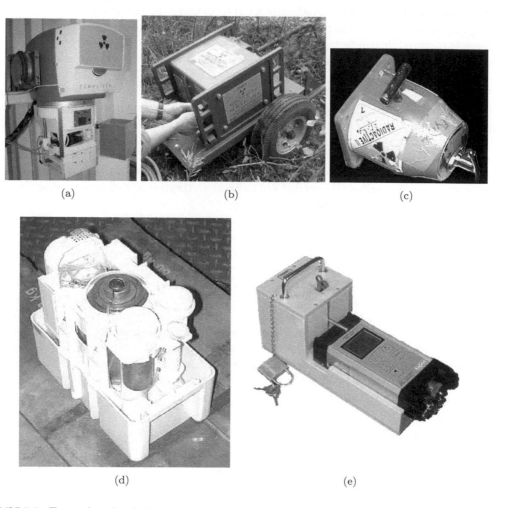

(a)   (b)   (c)

(d)   (e)

FIGURE 7.3 Examples of radiation sources in the five categories defined by the IAEA (IAEA 2009). (a) ^{60}Co source used in therapy. (b) ^{60}Co source used for radiography. (c) ^{60}Co source used as a level gauge. (d) ^{99}Mo/^{99m}Tc generator with the upper casing removed. (e) ^{137}Cs source used for calibration of dose rate meters.

Category 1 also includes sources used for radiation therapy in cancer treatment, often using ^{137}Cs or ^{60}Co. Typical source strengths are 10–1000 TBq for ^{60}Co, and 10–100 TBq for ^{137}Cs. Although they are being phased out as gamma therapy sources in standard radiation therapy, these sources are still required at many medical centres for dose calibration and reference measurements. Persistent tissue damage results within minutes of exposure at a distance of 1 m from the source, and lethal damage occurs between 10 minutes and an hour.

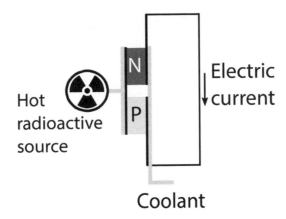

FIGURE 7.4 Schematic diagram of a radioisotope thermoelectric generator (RTG) consisting of a radioactive heat source such as ^{90}Sr, ^{238}Pu or ^{210}Po. Ideal radionuclides are pure alpha emitters such as ^{238}Pu, which can generate 0.56 W g^{-1}. However, the cheapest and most easily manufactured source, ^{90}Sr, can emit bremsstrahlung, requiring substantial radiation shielding.

To illustrate the enormous activity of Category 1 sources, a 500 000-TBq ^{60}Co source for sterilization would, in theory, when dispersed homogeneously, be capable of covering a surface area of 500 000 km^2 with 1 MBq m^{-2} of ^{60}Co activity. (In practice, the dispersion of such a source would be very inhomogeneous, and deposition would be governed by local weather conditions). Such a surface deposition would require temporary relocation of residents in the area. If dispersed uniformly, the activity from an RTG containing a ^{90}Sr source would be sufficient to cover a surface of 2500 km^2 with 25 MBq m^{-2} of ^{90}Sr, also requiring temporary relocation of the residents.

The second category, *very dangerous* radioactive sources, includes sources used for industrial radiography, a technique used to visualize structures in construction materials. Industrial radiography can be regarded as the analogy of medical radiography of humans and living species. However, the source strength required for radiography and the associated potential exposure rates are much higher. Usually, such radiography sources consist of gamma-ray-emitting radionuclides such as ^{60}Co, ^{192}Ir or ^{75}Se, depending on the type and thickness of the structures to be investigated. The source strength ranges from 0.1–10 TBq. A bare source of 10 TBq of ^{60}Co can lead to lethal dose levels at a distance of 1 m within a few minutes. The corresponding time for the lower activity range is approximately an hour. For ^{192}Ir the corresponding times range from 10 h to a few weeks, as ^{192}Ir has a lower average photon energy.

An example of an accident involving a Category 2 source is illustrated in Figure 7.5. It shows the tissue effects (acute radiation damage) caused by a person carrying a 0.18 TBq ^{192}Ir source in his shirt pocket for 2 h.

The third category, *dangerous sources*, includes industrial fixed gauges used in applications such as monitoring the flow, level and volume of materials present in various systems. They can also be used for assessing moisture content, or monitoring mineral processing.

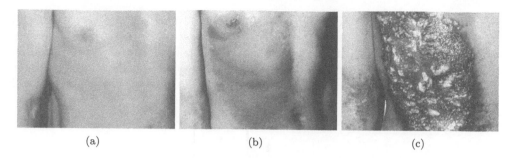

(a)   (b)   (c)

FIGURE 7.5 Illustration of tissue damage caused by a 2-hour exposure to a Category 2 source (0.18 TBq of ^{192}Ir) used for radiography. Reproduced with permission by the IAEA. International Atomic Energy Agency, The Radiological Accident in Gilan, IAEA, Vienna (2002c). See also colour images.

Most of these gauges consist of gamma emitters such as ^{60}Co, ^{137}Cs and ^{133}Ba. Typical source strengths range from 0.1–1 GBq. A bare source of 0.1 TBq of ^{60}Co would cause a lethal absorbed dose at 1 m after a few weeks. However, if brought into close physical contact with the human body, such a source can cause permanent tissue damage (such as necrosis of the epidermis), within hours of exposure. Level gauges have been accidentally recycled with other scrap metals and then mixed with molten metal in the furnace, resulting in steel contaminated with ^{60}Co, which was then exported to many different countries worldwide. In other cases, disused Category 4 sources have been stolen or lost, and have ended up on the scrap metal market. Thus, contamination by ^{60}Co and ^{137}Cs has been found in metal parts manufactured from recycled metal. Nowadays, many smelters have radiation monitors to detect contaminated metals before they are introduced into the furnace.

Category 4 sources include the open sources used for medical radiotherapy, such as ^{131}I and ^{99m}Tc. Another example is the short-lived sources of positron emitters used for PET-CT imaging, such as ^{18}F, which are often transported from the cyclotron where they are produced to various medical centres. Typical source strengths range from 1–10 GBq for ^{131}I and 10–1000 GBq for ^{99}Mo/^{99m}Tc sources. Moisture gauges, which use neutron emitters such as combined ^{241}Am-Be sources (with an approximate activity range of 1–10 GBq), or ^{252}Cf (0.1–0.3 GBq) are also classified as Category 4. For external exposures it is not likely that these sources will give rise to permanent tissue damage. Nevertheless, dispersion of such open sources would be of great concern, especially if internal contamination by inhalation occurred at an accident site, for example, in a transportation accident involving severe damage to packages containing medical radionuclides. If such sources were to be mixed with conventional explosives to form what is often termed a "dirty bomb", or *radiological dispersion device, RDD*, this would have significant psychological impact, although radiation injuries would be minor.

Hypothetically, a homogeneous dispersion of a 0.37 TBq ^{99}Mo/^{99m}Tc source would cause a deposition density of 1 MB m^{-2} over several square kilometres of land. For open sources, where the integrity of the damaged package is in question, dispersion to nearby soil and into the air must be considered by emergency responders. In a worst-case scenario, malevolent use of such an open source would be sufficient to contaminate a medium-sized shopping centre. Extended close contact with grains of ^{99}Mo can give rise to minor skin injuries, as emitted beta particles can penetrate the first few millimetres of the epidermis.

The criterion for a source to belong to Category 5 is that it is unlikely to cause any harm if dispersed by fire or an explosion. Radiation sources in Category 5 include sealed

^{57}Co sources used for x-ray fluorescence imaging, lightning preventers (^{241}Am or ^{226}Ra), and unsealed medical sources such as ^{32}P. Typical source strengths range from 0.01–1 GBq for gamma emitters.

In addition to the above categories, a number of very weak radiation sources are used in society today. Sources above about 10 kBq may be subject to transport regulations (Moseley 2010), depending on their physical form and the packaging of the source. Other industrial products for commercial use, such as smoke detectors and density gauges, may be exempt from certain transport restrictions. Many sealed γ-ray-emitting sources used for the calibration of radiological and nuclear detectors have a source strength below the 10-kBq level in order to allow for easier administration when the source and instrument are transported to different locations. It is not uncommon for many of these very weak sources to end up in scrap yards when the instruments become obsolete and are discarded. However, the total activity from these sources is on a par with many low-level NORM-associated materials (see Section 1.2), and their radiological health hazard is therefore almost negligible. Another notable example of such weak sources is the ^{241}Am used in domestic smoke detectors, with a typical source activity of 37 kBq.

As mentioned in Section 1.3.3, medical radionuclides are transported routinely. Even in a sparsely populated country such as Sweden, more than 10 000 packages of Category 4 sources are transported by road every year, and at least one incident involving a damaged package is reported in Sweden each year. Compared with transport accidents involving hazardous chemicals, the rate of transport accidents involving RN sources is very low, and local rescue teams therefore have more experience in the risks associated with chemical exposure at an accident site than radioactive materials. This is important to bear in mind when planning training and exercises in emergency preparedness for local rescue teams. It is vital that transportation regulations include a system of signage and proper declaration of goods for packages containing radioactive material, as well as other hazardous substances. Clear labelling on the packaging can save lives in the event of a severe traffic accident or a fire at a site where radioactive materials are handled and stored.

7.1.1.3 Malevolent Use of Radioactive Substances

Radiological terrorism involves procuring radiation sources or fissile nuclear materials for the purpose of intentionally exposing people to radiation. A non-state terror organization can obtain these materials illicitly by theft from industries and other RN operators, or by purchasing materials on the black market. The extent of such a market is not known for certain. However, cases have occurred, and up to 124 kg of low-enriched uranium (2% ^{235}U) is known to have been stolen, of which only 100 kg has been recovered. More than 5 g of weapons-grade plutonium (^{239}Pu) originating from Russia was retrieved in Munich in 1994. At the end of the 1990s, at least one case of illicit trafficking of radiological or nuclear substances was discovered in Russia each year.

A nuclear or radiological terror attack can be carried out by the dispersion of radioactive material by an explosion or by fire (see Fig. 7.6). Mechanical devices can be used to distribute the source or fragments of the source in public places. Radiation sources can be placed at locations such as train stations or airports, where the density of people is high. Unsealed sources can be distributed by various means, or dispersed in freshwater systems, or as a poison in food at a stage in the distribution chain. The result will be radioactive contamination of surfaces, similar to that discussed for accidental releases of radiation. Economic damage to real estate, and to nearby food producers will be severe. However, the most serious effect will be the considerable apprehension among those exposed and affected, although the radiation doses incurred by individuals may still be relatively low

(<100 mGy). In some extreme scenarios, individual doses far above 100 mGy will occur, causing radiation-induced tissue effects.

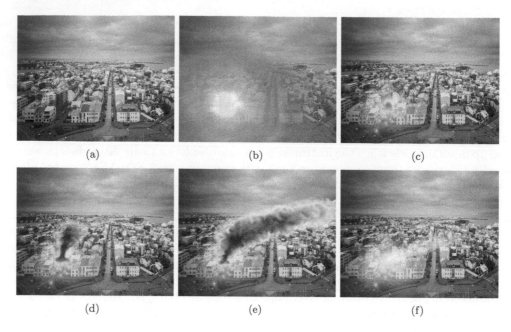

FIGURE 7.6 Illustration of the possible dispersion pattern of a radiological dispersion device (RDD) deployed on the roof of a multi-storey car park at a shopping mall. See also colour images.

In conclusion, the fear associated with radiation and exposure to radioactivity is often considerably greater than the actual health impacts of public exposure. It appears that the perception of risks and the fear associated with radiation by the public and decision makers is disproportionate compared to those associated with dangerous chemicals such as ammonia or chlorine gas, and other hazardous substances. Sensitive instrumentation such as nuclide-specific detectors and highly sensitive dose rate instruments make it possible for a radiation protection expert to detect elevated levels of radiation much lower than the levels that will result in significant health effects, and considerably lower than those that will cause acute tissue damage. This creates a pedagogic and psychological dilemma in communicating the risks associated with elevated levels of radiation to the public and to various stakeholders. The adverse economic effects on farming and tourism are real factors that should be taken into account when developing strategies for remedial action (see Section 7.2). The malevolent use of relatively harmless sources, such as those in Categories 5 and 4, is sufficient to create massive public concern.

7.1.2   Military and Antagonistic Nuclear Threats

7.1.2.1   Nuclear Terrorism

To create a nuclear detonation or a major criticality incident, an antagonistic perpetrator can either steal a compete nuclear weapon, or steal enough fissile material to sustain a nuclear reaction. It is also possible that a perpetrator may attempt to manufacture a primitive nuclear bomb. The illicit manufacture of a nuclear bomb is feasible if the perpetrator has access to highly enriched (>90% ^{235}U) uranium. Suspected terror attempts on military nuclear plants in Pakistan, and other incidents in the Middle East, have raised concerns about nuclear terrorism. Due to the extensive logistics and technical skills required to make and detonate a nuclear weapon, the probability of this scenario is expected to be very low. However, the mere threat of hijacking nuclear weapons can cause widespread alarm. It is also questionable whether the threat of nuclear terrorism is greater than the fundamental threat arising from the proliferation of nuclear weapons in countries in political conflict with other states.

Although the probability of nuclear terrorism is small, the consequences of such a scenario would be disastrous. The detonation of a small nuclear device in a densely populated area would cause thousands of fatalities, and it would be impossible to implement life-saving measures close to the epicentre of the detonation (so-called ground zero). (This term was also applied to the site of the non-nuclear Al-Qaeda terror attacks on the World Trade Center in New York in 2001).

Nuclear terrorism may also include attacks on NPPs used for the commercial production of electricity. Terrorists may attack a nuclear reactor building using aircraft, or try to kidnap a hostage to gain entry to a nuclear plant, or they may attempt to steal fresh or spent nuclear fuel during transportation. Attacks on fresh nuclear fuel could lead to criticality, or cause the release of fission products and neutron activation products from the NPPs inner water cooling systems leading, in turn, to release into the environment. Suicide attacks using aircraft would not easily penetrate the vessel containing the reactor core, but significant damage could be caused to other parts of the reactor.

7.1.2.2 Nuclear Weapons Testing

Nuclear weapons testing can be performed below ground or above ground (atmospheric tests). Atmospheric tests are responsible for most of the global fallout of long-lived fission products (such as ^{137}Cs, ^{90}S/^{90}Y, ^{238,239}Pu and ^{14}C). In the Northern Hemisphere alone, the total release of ^{137}Cs from nuclear weapons testing is estimated to have been at least ten times higher than that from the Chernobyl accident in 1986 (Aoyama et al. 2006). Test detonations have been carried out with the nuclear device in high towers, in balloons, on small islands, or being dropped from an aircraft. In some cases, atmospheric tests were performed by ejecting the nuclear device from a missile at high altitude. Most nuclear weapons tests in the 1950s and 1960s were atmospheric, because of the relatively low cost, and because the outcome of the test could be more easily assessed than for terrestrial or underground tests. It was anticipated before the tests that local fallout would be high, but the degree to which radioactive elements were ejected into the higher layers of the stratosphere and dispersed globally was not predicted. In particular, the process of rain-out (see Section 3.2.4.2) of suspended fission products during the spring resulted in elevated activity deposition levels, even at locations that were geographically very far from the test site. The deposition densities of ^{137}Cs, ^{90}S/^{90}Y and ^{239,240}Pu are correlated with latitude and with the annual precipitation of a given area.

Well-controlled underground tests lead to much less dispersion of transuranic and fission products. However, leakage through cracks in the ground is possible, and could lead to substantial atmospheric releases. Since the partial test ban treaty in 1963, the majority of nuclear weapons tests have been carried out underground. Nuclear weapons have also been tested under water. The nuclear device is then mounted on a vessel before detonation. The purpose is to evaluate the effects on ships or naval bases caused by the detonation of nuclear devices beneath sea level, for example, submarine-borne nuclear weapons. This method generates enormous amounts of contaminated rainfall and mist which can contaminate ships and other vessels near the test site. Such tests also have serious consequences for the marine biota at the test site. An example of a submarine nuclear detonation is Operation Crossroads, carried out at Bikini Atoll in 1946.

Testing nuclear devices in space, or, rather, in the exo-atmosphere (>12 000) m altitude), would lessen the impact of radiation. However, the electromagnetic pulse (EMP) generated by the detonation would be detrimental to many satellites and vital satellite-borne communication systems, as well as to terrestrial electromagnetic communication systems over a large land area. An atmospheric nuclear detonation will generate a cloud of contaminated dust and fragments of the nuclear device. The extension in altitude of this cloud depends on the strength or yield of the nuclear detonation (see Fig. 7.7). The yield of a nuclear explosion is measured in a unit that corresponds to the amount of energy released by 10^6 kg (a kiloton, or kton) of conventional explosive (TNT trinitrotoluene). The Hiroshima atomic bomb corresponded roughly to 16 kton.

A distinction is made between surface and altitude detonations (see Section 1.3.2.2). The height of a surface detonation is typically 150 m, while an altitude detonation takes place at about twice that height. The choice of detonation height depends on the target: i.e., whether the intent is to destroy aircraft and satellite communications, or to destroy infrastructure on the ground (for examples, see Glasstone and Dolan 1977). One characteristic of a surface explosion is that delayed radiation is much more important. This is because the low altitude of the fireball draws in dust particles from the ground in its back flow, and these dust particles are then irradiated by neutrons and contaminated by fission products and actinides. The heavier fraction of these contaminated particles (in the mm range) is then deposited locally giving rise to highly elevated dose rates close to the ground. Some nuclear devices and detonation modes are designed to achieve maximum biological damage with a minimum blast effect.

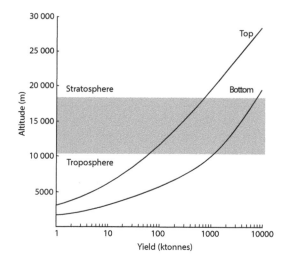

FIGURE 7.7   Vertical extension of a radioactive cloud (top to bottom) as a function of the explosive yield of a surface detonation.

At the end of the 1950s and beginning of the 1960s, atmospheric testing of high-yield (>1 Mt) nuclear weapons was carried out in the stratosphere. These detonations resulted in radioactive plumes that reached into the upper layers of the stratosphere and were then carried over the entire global hemisphere. After a delay of up to a year, the Northern Hemisphere spring saw fission products in isolated higher layers becoming mixed with deeper layers. On re-entering the lower troposphere, fission products become subject to rain-out and are deposited on the ground, with a peak deposition rate at a latitude of 45°.

7.1.2.3 Nuclear Detonation in Warfare

Although described in more detail in Section 1.3.2.2, we will here recapitulate the three principal effects of a nuclear weapons detonation.

i The blast effect: A large part of the energy released in a nuclear explosion is directed into a blast that causes the destruction of physical structures on the ground. A pressure wave follows at the speed of sound, which not only causes further physical damage, but can also destroy the inner organs of unprotected humans, even outside the blast area. The blast radius caused by a nuclear weapon depends on the altitude of detonation (Glasstone and Dolan 1977).

ii The heat wave: The second largest amount of energy following a nuclear detonation is the dissipation of heat. An initial thermal radiation wave increases the temperature to such an extent that combustible materials catch fire, and can create firestorms, as occurred in Hiroshima.

iii Initial radiation effects: The remaining energy dissipated in a nuclear explosion is divided between prompt and delayed radiation. Neutron and prompt gamma radiation caused by the explosive yield of fission neutrons and their associated prompt gamma photons, create an instantaneous high absorbed dose near the detonation point.

The secondary effects of a nuclear detonation are residual nuclear irradiation and the EMP.

i Residual nuclear irradiation: This refers to the radiation present at the detonation site more than 1 min after the detonation. The yield of fission products and actinides, similar to those found in an NPP accident, will be deposited along a plume, whose direction is governed by the prevailing wind at the time of the explosion. As above, the fraction of locally deposited radionuclides increases with reduced detonation altitude. Hence, surface detonations are considered to be radiologically the most serious in terms of health and environmental effects (see Fig. 7.8).

ii Electromagnetic pulse: The irradiation of the surrounding air causes an EMP that will destroy most electromagnetic and radio-wave communication systems in the area.

A map illustrating the effect of a small surface nuclear detonation (1 kton) is presented in Figure 7.9. It is assumed that the wind is directly from west to east. The blast will create a crater with a radius of 20 m and a depth of about 9 m. The blast will also cause a shockwave that will kill 50% of unprotected or unsheltered individuals within a radius of 200 m. The shockwave is still powerful enough to cause permanent lung damage in humans at a radius of 270 m from ground zero. The thermal radiation will be sufficient to cause third-degree burns at a radius of 700 m. This is approximately the same radius at which the initial irradiation is sufficiently high (>3 Sv) to kill unshielded individuals. Up to 1 km away from ground zero, the initial irradiation will cause acute radiation effects such as tissue damage. Delayed radiation from the deposition of suspended fission products will cause a lethal external radiation dose to unshielded persons up to 200 m downwind of ground zero.

In order to put the immense biological effects of a nuclear detonation into perspective, we can consider a surface detonation of a 100-kt nuclear device (around 6 times the yield

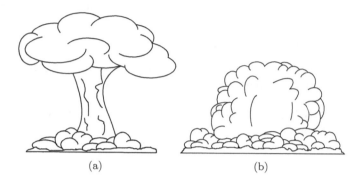

FIGURE 7.8 (a): Detonation at altitude with little or no local delayed radiation in terms of deposition of fission products. (b): Surface detonation with extensive local ground deposition of fission products.

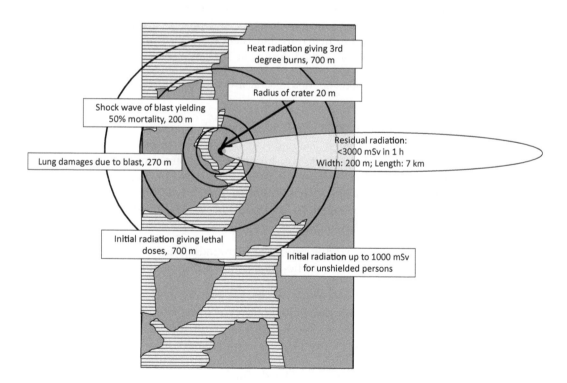

FIGURE 7.9 Map illustrating the damage radii from ground zero at a fictional geographical location. Inner circles are the radii of the blast and pressure waves. Outer circles are the radii for thermal irradiation, lethal direct ionizing radiation, and radiation-induced tissue damage. A plume of suspended fission products downwind of the nuclear detonation is also shown. The isodose shown here is for *unprotected* individuals.

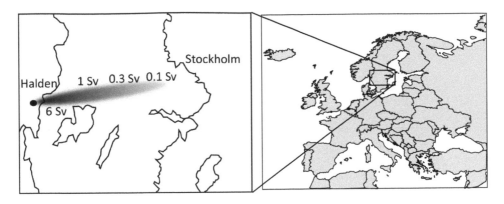

FIGURE 7.10 Isodose shading for a deposition plume from a nuclear detonation at Halden in Norway close to the Swedish border. It is assumed that the wind is blowing towards Stockholm, the capital of Sweden, at a speed of 11 m s^{-1}.

of the Hiroshima bomb) at the border between Norway and Sweden. It can be predicted that this will result in absorbed doses above 100 mSv over 300 km into Swedish territory 2 days after detonation. Lethal doses will be absorbed by unshielded individuals as far away as 150 km from ground zero (Fig. 7.10). Such levels call for countermeasures such as rapid evacuation of residents from the area affected. Evacuation may, however, not be feasible if it is not possible to predict the path of the plume. If evacuation is not carried out quickly, the radiation exposure will be higher than if people remained in the area in shelter. If it is possible to survive in a shelter, it is thus better to remain in the area for the first 23 weeks, rather than attempting to evacuate the inhabitants. This example shows that even neutral countries not involved in armed conflicts are very likely be seriously affected by a nuclear detonation in the region close to them (<200 km).

The dose rate from ground deposition of fission products from nuclear detonations consists of contributions from various radionuclides whose half-lives vary greatly; see Fig. 7.11. The yield of neutron-activated fission products is very low compared with releases from an accident at a NPP. For example, following the ground detonation of a nuclear weapon the ^{134}Cs component is likely to be negligible compared to ^{137}Cs, whereas for NPP fallout the ratio between ^{134}Cs and ^{137}Cs in the initial fallout ranged from 0.56 following the Chernobyl accident to around unity after the Fukushima accident. As a rule of thumb, for times longer than an hour after detonation, the external dose rate follows a power law decrease over time, with an exponent of -1.2. In other words, a factor of seven increase in the elapsed time, t_e, leads to a tenfold decrease in the dose rate. Eqn. 7.1 can be used to calculate the absorbed doses to rescue workers in an affected area, and may be compared with the accumulated dose to the public in the same area. Note, however, that internal exposure due to inhalation and ingestion of the fission products produced (delayed radiation) were not considered in these examples. Bearing in mind the severe disturbances to society that will result from a nuclear detonation, it is not feasible to make even rough models of how these radionuclides will be transferred to humans.

$$\dot{D}_{ext}(t) \sim t^{-1.2} \rightarrow D_{ext} \sim \int_{t_0}^{t_0+t_e} t^{-1.2} dt \rightarrow D_{ext}(t_e) \sim \left(1 - t_e^{-0.2}\right) \qquad (7.1)$$

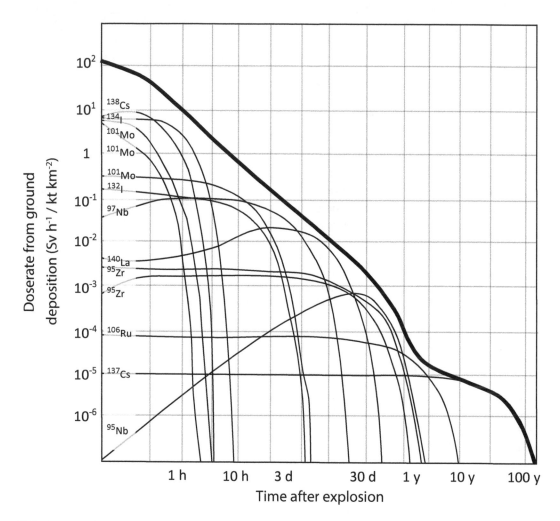

FIGURE 7.11 External dose rate from ground deposition following a nuclear detonation as a function of time, and the principal radionuclides contributing to the external dose rate as a function of the time elapsed since deposition. Note that for ^{137}Cs, gradual migration into the soil and the associated decrease in the dose rate contribution was not taken into account in this figure.

7.1.3 Civil Nuclear Threats

Nuclear reactors for the commercial production of electricity to be fed to national power grids are typically operated at an electric power output of 500–1000 MW. Research reactors are usually operated at 10–100 times lower power. The reactor inventories of radionuclides during normal operation of a nuclear power plant were described in Chapter 1. It should be noted that a nuclear power reactor contains larger amounts of long-lived fission products such as ^{137}Cs, than are produced by the detonation of a nuclear weapon. Therefore, a number of physical barriers are constructed to protect the surrounding environment from exposure (as discussed in Chapter 6). An accident causing rupture of the cooling system or the reactor vessel will release an immense amount (> 1PBq $= 10^{15}$ Bq) of fission products and neutron activation products, as well as neutron-activated fuel fragments, if the reactor fuel is damaged. The total inventory of a 1000 MW nuclear reactor with an average fuel renewal rate of five years is approximately 100–1000 PBq of ^{137}Cs. This corresponds to a sterilization source, or the combined activity of more than 100 000 Category 1 therapy units (see Section 7.1.1.2 above).

Releases of radionuclides and elevated radiation resulting from failures at NPPs can take several forms: (i) the escape of fission products and evaporated fuel fragments from a damaged reactor core (as in the Chernobyl accident in 1986), (ii) the escape of volatile fission products vented from a reactor core which is intact but at too high a pressure (Three Mile Island, 1979), (iii) releases from a transport accident involving used nuclear fuel or other high-level nuclear waste, (iv) the escape of fission products and criticality in the reactor plant's interim storage for nuclear fuel (Fukushima, 2011), and (v) releases resulting from accidents at research reactors (Halden in 2001). The IAEA (IAEA 2007a) describes five categories of threat levels for nuclear and radiological facilities, in analogy with the five threat categories for radiation sources. Category I consists of nuclear power plants. Although the probability of an accident is very low, the consequences could be very severe, as is evident from the major events that have occurred to date. Category II sites involve smaller nuclear reactors (<100 MW) often used for research. The lower categories involve non-nuclear industrial sites using strong sources, as discussed in Section 7.1.1.2. It is noteworthy that criticality accidents that had only a limited impact on the surrounding environment, such as that at the Tokaimura fuel fabrication plant in 1999, may result in more acute radiation-induced fatalities than a major INES rating 7 accident such as that at the Fukushima Daiichi plant in 2011 (see Table 7.1).

The consequences of a release from a Category I nuclear site would be fatal, and radiation-induced tissue damage to workers at the accident site would be extensive (>100 mGy). External exposure of the rescue workers, site personnel and members of the public living close to the site would be high. Releases into aquatic and terrestrial ecosystems would lead to elevated levels of fission products in fish, game, wild berries, mushrooms, milk, crops, seasonal vegetables, and other foodstuffs (see Fig. 7.12). Internal exposure would result from the ingestion of contaminated food or the inhalation of fission products suspended in the radioactive plume. The sum of the internal and external exposures to staff and the affected local population will result in a net increase in the risk of radiation-induced cancer. Radiation is, in fact, a small risk factor for the occurrence of cancer compared with many other well-established carcinogens such as tobacco smoke. In large accidents such as the Chernobyl accident, statistically significant increases in the incidence of cancer have only been detected in regions with high ground exposures (such as the Bryansk region of Russia and in parts of Belarus), and for cancers with a short latency time, such as leukaemia and thyroid cancer (UNSCEAR 2008). The social and economic consequences of discontinued farming, forestry, fishing and other practices that rely on soil, air and water may be just as serious as the incidence of radiation-induced cancer. If the affected areas lack adequate health care and welfare systems, the loss of employment and income has the potential to cause psychosomatic diseases whose morbidity is on a par with radiation-induced cancer.

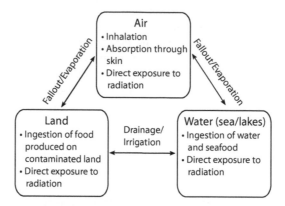

FIGURE 7.12 Schematic representation of various radiation exposure pathways to humans, and the contribution from airborne radioactive releases.

In 2014, approximately 440 nuclear power reactors were in operation worldwide, and about 70 were under construction (Fig. 7.13). An increasing proportion of nuclear reactors is sited in large Asian countries such as China and India, which are currently experiencing a rapid increase in energy consumption. However, in Europe, the nuclear power industry has stagnated since the Chernobyl accident in 1986, and it had already stalled in the USA after the accident at Three Mile Island in 1979. This has led to a progressively ageing worldwide inventory of nuclear power plants as operations are extended at existing plants, as is illustrated in Figure 7.14.

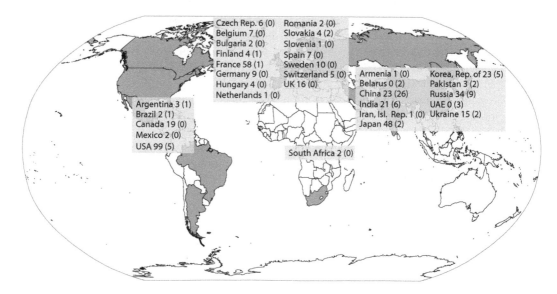

FIGURE 7.13 Nuclear power reactors in operation and under construction (in brackets) 31 Dec 2014. Data from IAEA (2015b).

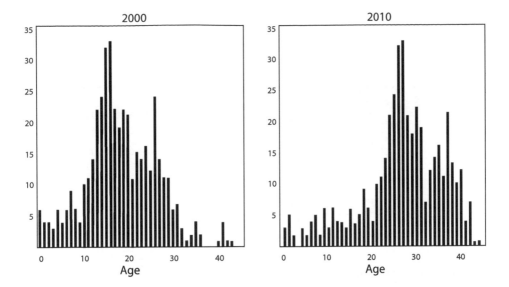

FIGURE 7.14   Age histograms of the nuclear power plants in operation worldwide in the years 2000 and 2010.

A signature in terms of the time evolution and the extent of releases from a nuclear power accident can be visualized in a so-called time-release plot (Fig. 7.15). The fraction of the nuclear inventory of a given radionuclide that is released is plotted on the y-axis, and the time after the start of the incident precipitating the release is plotted on the x-axis. The definition of this starting time is essentially the time at which the operators of the control system acknowledged that an anomaly had occurred at the reactor. It is also the point at which the operators of the plant are forced to notify the relevant authorities that an elevated level of alert is required. The time elapsed from the initial alert and the onset of the release to the environment outside the nuclear plant represents the time window for the initial emergency response workers to prepare their resources and initiate action. Time-release plots for a number of hypothetical accident scenarios can be estimated theoretically. Such plots are a valuable tool for the authorities and operators concerned in the planning of emergency preparedness, as well as for decision makers who must allocate resources for remedial actions.

To illustrate the amount of activity in the reactor inventory of a typical light-water reactor, we can consider the complete release of the inventory of ^{137}Cs, see Figure 7.16. This would be enough to cover half the surface area of Sweden $(450\,000\ \text{km}^2)$ with a ground deposition of 1 MBq m^{-2}. At this level the external dose contribution to humans would be about 25 mSv y^{-1} in the initial phase and, in some cases, will warrant permanent relocation of the residents. However, as was illustrated in Figure 5.1, the fallout following a real NPP release will be extremely inhomogeneous, which means that some areas will suffer a deposition of activity that rules out human habitation for generations, as was the case in the 30-km radius inner zone near the Chernobyl plant.

The release pattern of the Three Mile Island accident was a release of about 10% of the noble gas inventory (mainly ^{133}Xe) about 3 h after the initial alarm that an anomaly was taking place (Fig. 7.17). A small fraction of the reactor inventory of ^{131}I ($>$0.00002%) also escaped into the atmosphere. The individual who received the highest exposure received

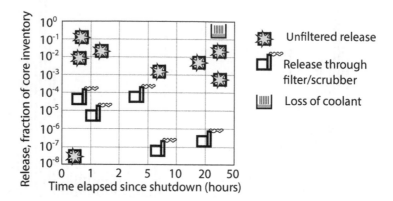

FIGURE 7.15 Event chart for hypothetical accident scenarios such as pipe rupture of the cooling water system, loss of external auxiliary power, etc. The time elapsed from the initial emergency shutdown of the reactor is plotted on the x-axis.

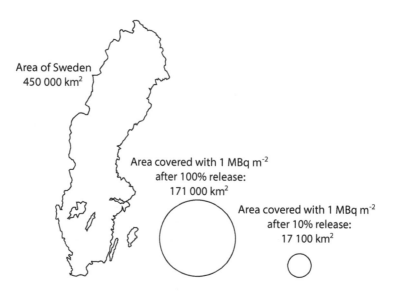

FIGURE 7.16 Illustration of the activity deposition following the release of 100% and 10% of the typical ^{137}Cs inventory of a light-water reactor ($1.7 \cdot 10^{17}$ Bq) over Sweden.

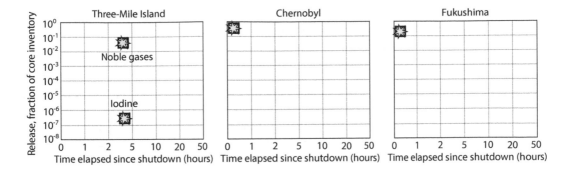

FIGURE 7.17 Event charts for the three main nuclear power accidents to date: Left: the Three Mile Island accident in 1979 (noble gases 9%, iodine 0.00002%). Middle: Chernobyl accident in 1986 (noble gases 100%, iodine 55% and caesium 33%). Right: sum of atmospheric releases from reactors 1–4 in the Fukushima accident in 2011, normalized to the inventory of one reactor (10% ^{137}Cs over about 5 days). Modified from Katata et al. (2015).

an effective dose of around 0.7 mSv, and about 260 people received absorbed doses in the range 0.2–0.7 mSv. This can be compared with the estimated natural background radiation dose to local residents of 1.2 mSv y^{-1}. Of the 760 individuals living within a 5-km radius of the plant who were examined by whole-body counting (see Section 7.3.4.3), none was reported to have sustained an elevated dose. Neither was any elevated level of ^{131}I found in dairy milk. The main remedial action carried out was the evacuation of 140 000 children and pregnant women from the area.

The Chernobyl accident is the largest nuclear power disaster that has occurred to date. Acute radiation syndrome was found in 134 liquidators (clean-up workers) and rescue workers, 28 of whom died within weeks of the accident (partly also because of burns). There was an increased incidence of thyroid cancer, in total about 5000 cases of childhood thyroid cancer, in the period 1986–2006. The incidence of leukaemia was observed to be double the expected natural rate. Other effects, such as malformed foetuses or miscarriages were, however, not seen to increase. The number of cancer deaths that can be attributed to the fallout from the accident has been estimated by UNSCEAR to be around 50 000. A number of organizations and sources dispute this estimate, however, none of these has demonstrated to the scientific community how their estimates were derived from currently accepted linear no-threshold models. It should be noted that most of the childhood thyroid cancers were curable, and that survival is expected to be normal. The contribution of the total absorbed dose from Chernobyl fallout was initially dominated by internal exposure from radioecologically transferred fission products such as ^{131}I in milk, or ^{134}Cs and ^{137}Cs in crops, wild game, milk and forest products. Internal doses were much more important in the initial years following the accident. Over the long term, external exposure from the ground deposition will dominate dose contributions to the affected population.

The Fukushima accident in 2011 led to a total release to the atmosphere of ^{137}Cs which was about 11% of that from Chernobyl. However, a major part of the airborne releases were transported into the Pacific Ocean and did not affect terrestrial ecosystems. An additional 27 PBq is estimated to have leaked into the Pacific Ocean from the contaminated reactor water. In all, 300 000 individuals were evacuated. The highest absorbed dose received by an individual (a worker at the plant) was around 700 mGy. No acute radiation syndrome was observed among those carrying out initial rescue operations. It should be pointed out that the natural disaster that precipitated the nuclear power failure, a Tsunami caused by an earthquake with a magnitude of 9.0 on the Richter scale, killed more than 18 500 people. Table 7.2 gives a brief overview of the timelines of the above three NPP accidents.

The exposure pathways of an atmospheric release from a nuclear power plant can be divided into the following components: (i) external exposure from the passage of the radioactive plume (so-called cloud shine), (ii) external exposure from gamma and beta radiation from fission products and transuranic elements deposited onto the ground from the plume (ground shine), (iii) internal exposure due to inhalation of radionuclides by people present in the area during the passage of the plume, (iv) internal exposure due to ingestion of contaminated foodstuffs (either directly contaminated, or contaminated through root uptake, or by transfer through nutrient chains in the ecosystem), (v) thyroid dose: in much of the software used to assess the consequences of a radioactive release, the thyroid dose due to inhalation is the key parameter determining whether iodine prophylaxis should be recommended. Internal exposure due to inhalation of re-suspended radioactive deposition, even after the passage of the plume, can sometimes be important and should be considered for personnel who carry out remedial decontamination work on heavily contaminated surfaces.

The consequences of a major release where all safety barriers and mitigating systems have failed can be exemplified by assuming a scenario in which 10% of the reactor inventory of the principal fission products (such as ^{137}Cs and ^{131}I) is released over a time interval of 50 h (Fig. 7.18). The wind direction is assumed to be 45° towards the southeast. In the first scenario we assume dry weather conditions, meaning that the fallout is dry deposited from the radioactive plume. The estimated radiation doses to an unshielded individual at various distances away from the failed reactor during the first 24 h are given in Table 7.3. The radiation dose consists of external doses from cloud shine and ground shine. In addition to the more long-lived ^{131}I, ^{134}Cs, ^{137}Cs and ^{90}Sr/^{90}Y, a large number of short-lived fission products dominate the contribution to the dose, such as ^{140}La, ^{106m}Ru, ^{132}Te (which decays into ^{132}I and accumulates in the thyroid if inhaled). The reference situation is thus an unshielded adult who remains outdoors. At a distance of 25 km from the plant the 24-h external dose from ^{131}I and ^{137}Cs is 50 mSv for an unshielded adult. However, if we assume that it rains, at a rate of 1 mm h^{-1} during the first 24 h after release, the 24-h dose will increase substantially due to enhanced washout of the radioactive plume by the rainfall. At shorter distances (<3 km) the external dose will be more than tripled compared with dry weather. However, for those living farther away along the trajectory of the radioactive plume, a lower dose will result because of depletion of radionuclides in the cloud by washout near the release point. In the next section, a discussion is presented concerning the source terms assumed here, and their distribution between the various types of radioactive elements in a nuclear power plant accident.

Nuclear reactors in Sweden are connected to an external filter system by pressure-relief valves in the inner cooling system (see Fig. 6.10). These filter systems are designed to filter out more than 99.9% of the long-lived fission products (corresponding to a fractional release of 0.001 of the reactor inventory), which could otherwise cause long-term radioactive ground contamination over large areas. Noble gases will not have nearly as severe a radiological impact as other radionuclides. In a similar accident scenario to the one illustrated in Figure 7.18, but with a filtered pressure-relief system that functions as intended, the release to the atmosphere will take place through the filter device (Fig. 7.19). In theory, this would lead to very moderate or minor exposures, with a factor of up to 1000 times less external exposure during the first 24 h near the release point.

The quality of the prognosis made in real time following an accident is very dependent on the quality of the weather data provided by the meteorological institute, both regionally and globally. This will be discussed further in Section 7.3.2. The Chernobyl accident, where a

TABLE 7.2   Approximate timelines of the three major NPP accidents to date (2015).

Date and time	History
	Three Mile Island, Harrisburg, 1979
28 Mar 79, 04.00	The cooling system of Reactor Unit 2 is affected by a minor malfunction in the non-nuclear part of the reactor, resulting in a relief valve being opened (time-zero). The valve, however, does not completely close again, leading to leakage of coolant water. The operators misinterpret the situation and are under the impression that coolant *is* flowing back into the system. For the next two hours, reactor coolant is instead leaking out of the system.
28 Mar 79, 06.48	The loss of coolant results in a partial meltdown of one of the exposed fuel rods.
28 Mar 79, 07.00	Local authorities and rescue operators are alerted. However, the operators still consider that the risk of widespread release is small, and that elevated radiation will be limited to inside the reactor building.
28 Mar 79, 13.30 ca	A hydrogen gas explosion occurs but is not initially recognized. The operators then attempt to obtain a prognosis of the potential releases, whereas the authorities, not experienced in RN scenarios, begin to initiate conflicting actions. Real public concern does not begin until two days after time-zero when it becomes apparent that general confusion exists among the authorities concerned regarding both the nature of the situation and how to proceed with remedial action.
	Chernobyl, Pripjat, 1986
24 Apr 86, midday	Precondition: in connection with a temporary shutdown of Reactor 4 for maintenance on 25 Apr 86, it is decided to take the opportunity to make an (ill-fated) experimental test of the potential to use the slowdown turbine as a backup power supply to complement the conventional diesel emergency backup.
25 Apr 86, 23.00	The experiment is delayed when the power dispatch system refuses to shut down in order to be able to feed electric power to the external grid. The operators insist on proceeding with the experiment, and somehow bypass the automatic inhibit functions in the control system.
26 Apr 86, 00.30	There is a dramatic fall in power (to 30 MW). The operators are, however, unaware of the risk of the so-called negative void effect, which occurs in reactors in connection with very rapid changes in power. Thus the experiment progresses.
26 Apr 86, 01.03	The experiment commences with an (unbeknownst to the operators) unstable reactor.
26 Apr 86, 01.24	The negative void effect causes a sudden uncontrolled increase in the neutron flux and thus in the power dissipated in the reactor, which in turn causes rupture of the fuel, followed by two explosions which blow the lid off the reactor vessel some tens of centimetres into the air. An open passage is established between the burning nuclear fuel and the atmosphere.

TABLE 7.2 (Continued.)

Date and time	History
About 26 Apr 86 to 10 May 86	Initial efforts are made to extinguish the chemical fire. After two days, attempts are made to quench what are assumed to be remaining nuclear reactions in the damaged reactor by dumping neutron-absorbing materials into the leaking reactor vessel. Evacuation of local residents does not begin until about 27 Apr 86, 15.00.

Fukushima Daiichi, 2011

Date and time	History
11 Mar 11	Start of crisis: Tsunami at 15.27. Twenty minutes later, the main electric power supply is lost due to flooding of the backup diesel generator at Reactor 1. Loss of integrity of the storage pool for spent nuclear fuel at Reactor 2.
11 Mar 11, 21.00	Initiation of evacuation from a 3-km radius zone nearby.
12 Mar 11, 05.50	Venting of over-pressurized Reactor 1 is carried out, there is a trade-off between the risk of reactor fuel damage and hydrogen explosion in the vessel. Start of the atmospheric release.
12 Mar 11, 10.58	Venting needed in Reactor 2, resulting in further release. Yet more releases in the two remaining reactors follow suit because of the lack of cooling to the remaining nuclear fuel, including explosions in various parts of the four affected reactors.

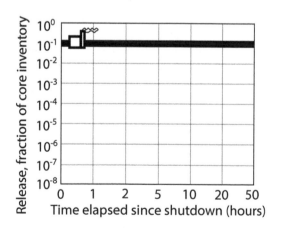

FIGURE 7.18 Event chart of an assumed accident scenario in which 10% of the volatile and refractory elements (see Tab. 7.3) of the reactor inventory of a light-water reactor is released over 50 h.

TABLE 7.3 The radiation dose 3 km from the release site in the wind direction, assuming a 10% release of iodine and caesium.

Condition	Dose	Unit
Effective dose during 24 hours		
Rain, dose outdoors	3000	mSv
Rain, dose indoors	1500	mSv
Evacuation	400	mSv
Averted dose 1100 mSv		
No rain, dose outdoors	900	mSv
No rain, dose indoors	300	mSv
Evacuation	250	mSv
Averted dose 50 mSv		
Thyroid dose, 1-year-old child		
Indoor stay, no KI administered	6000	mGy
Indoor stay, KI administered	300	mGy
Averted dose 5700 mGy		

nuclear reactor exploded, and hot fuel fragments were ejected into the atmosphere, provided the general RN emergency preparedness community with new experience and knowledge. It became clear that NPP releases could also be dispersed over very large distances, similar to those observed from the high-nuclear-yield weapons tests at the beginning of the 1960s. In particular, the local deposition maxima associated with heavy rainfall in areas as far as 1000 km from the release point had not been considered in emergency preparedness before the Chernobyl event. In some local areas as far away as in Sweden, ^{137}Cs concentrations as high as 0.1 MBq m^{-2} were deposited on the ground (Fig. 7.20).

The main lesson learned from the Chernobyl accident was that even a foreign NPP accident, located more than 1000 km away, could lead to radioactive fallout that has serious effects on practices such as farming and forestry. Animal husbandry such as reindeer herding and freshwater fisheries can be affected especially severely due to loss of competitiveness of their products on the international market. Although an initial emergency response such as evacuation may not be warranted, more long-term remedial actions must be planned to prepare for an accidental release from a foreign or distant NPP.

In addition to nuclear power reactors for commercial energy production, nuclear reactors of somewhat lower power output (10–100 MW electric power) are used in civil and military marine vessels. It should be noted that military submarines and naval ships may contain large inventories of fission products, without necessarily being equipped with nuclear weapons. The advantage of using nuclear fuel at sea is that it is an independent and reliable fuel supply, compared with conventional fuel, which has a higher dependence on international logistics and must be replenished more frequently. This is especially true for icebreakers and ships undertaking polar operations. Nuclear fuel for naval use typically has a higher degree of enrichment of ^{235}U than that used in NPPs producing electrical power. Fuel more highly enriched in ^{235}U is more expensive, but has an advantage that the reactor fuel rods have to be replaced less often. A number of incidents connected with the naval use of nuclear power have occurred: eleven reactor accidents, such as coolant leakage (K_- 192 Echo in 1989) and four fires. Up until 2003, five nuclear-reactor-powered submarines have sunk. Several wrecks from nuclear powered submarines and ships can be found in the Barents Sea off the Kola peninsula (see Fig. 7.21).

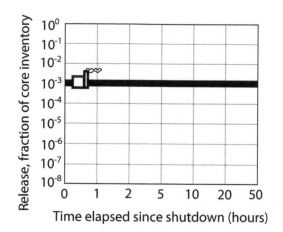

FIGURE 7.19 Event chart of a release passing through a pressure-relief filter where less than 0.001 of the reactor inventory of ^{131}I and ^{137}Cs escapes into the atmosphere.

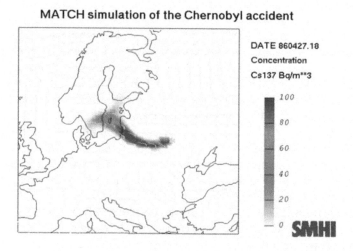

FIGURE 7.20 A mathematical simulation of the Chernobyl release entering Sweden in 1986. The simulation was carried out by the Swedish Meteorological Institute (SMHI) using quality-checked data from a large number of weather observation stations. See also colour images.

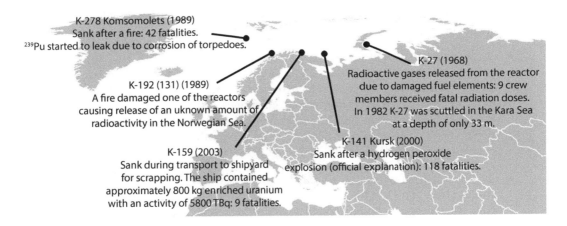

FIGURE 7.21 Nuclear-fuelled submarines and ships that have sunk in the Arctic.

Nuclear reactors are also used for deep space missions such as the unmanned probes Galileo and Cassini. These nuclear reactors are even smaller than those used for naval applications, often in the range 0.5–2 MW_{th}. However, as moderation and heat exchange must be designed differently in a spacecraft, the electric power is only on the order of 5% of the thermal power. Today, more than thirty satellites equipped with inactive nuclear reactors launched by Russia orbit the earth. In 1964 a United States navigational satellite, Transit 5BN-3, failed to achieve orbit and its Snap-9A thermoelectric generator (RTG), containing ^{238}Pu, re-entered the atmosphere and burnt up over the western Indian Ocean near Madagascar. The nuclear fuel was designed to vaporize during re-entry and was dispersed worldwide. Traces of ^{238}Pu can still be measured today in the marshlands of Madagascar, although the potential dose contribution and the associated health impact are negligible.

7.2 REMEDIAL ACTIONS IN RADIOLOGICAL AND NUCLEAR EMERGENCIES

7.2.1 Emergency Preparedness

7.2.1.1 Strategies for Remedial Action

Every independent country has a responsibility to appoint authorities charged with the general regulation of radiation protection issues and crisis management during national radiological and nuclear emergencies. A number of international organizations issue guidelines to aid states in drafting regulations and laws governing the provision, use and disposal of radioactive or nuclear material. These guidelines can be seen as a framework, within which each country can adjust and adapt regulations to make them fit for purpose. In Europe, the following organizations have contributed to regulations and guidelines for radiation protection and the handling of radiological or nuclear crises.

- The International Atomic Energy Agency (IAEA) was established on the initiative of the United Nations in 1957 with the aim of spreading the peaceful use of atomic and nuclear energy, and of preventing the proliferation of nuclear weapons and their use for illicit or malevolent purposes. The IAEA issues technical reports and provides guidelines and recommendations for issues such as the secure and safe operation of nuclear power production, the management of nuclear and radiological waste from industrial, medical and research applications (IAEA 2007b), and nuclear safety in general (IAEA 2008). In addition, the IAEA has also issued manuals and guidelines for first responders in an RN crisis (IAEA 2006).

- The International Commission on Radiological Protection (ICRP) is an international, independent, non-governmental organization that provides guidelines and recommendations for radiation protection to experts, workers and decision makers. Since 2000, the ICRP has issued three important guidelines. Publication 103 is the most recent version of the important ICRP recommendations for general radiological protection (ICRP 2007). Two more detailed guidelines regarding emergency preparedness are provided as Publications 109 and 111 (ICRP 2009a, 2009b).

- The Nuclear Energy Agency (NEA) is a subdivision of the Organization for Economic Co-operation and Development (OECD), which issues guidelines for the peaceful use of nuclear energy, including quality assurance, research and development, and waste management. It also provides instructions on nuclear safety and nuclear energy production. Together, the NEA member states represent 85% of the global nuclear energy production.

- The European Atomic Energy Community (EAEC, now called Euratom) is a European organization whose original aim was to create a European platform for commercializing the distribution of nuclear energy. Today, it is closely related to the EU, and provides directives to member states that affect national legislation on issues such as radioactive waste management and radiation protection for workers. For example, a recent directive set out basic safety standards for the protection of radiation workers and suggested the introduction of worker categorization based on the anticipated annual dose contribution attributable to their workplace (Euratom 2013).

In many countries, such as Sweden, national authorities are responsible for the central coordination of RN emergency preparedness. These authorities adopt international recommendations and directions, such as the Euratom directive of 2013 (Euratom 2013). It is recommended that emergency preparedness and remedial actions should follow three basic principles formulated by the ICRP. These principles were discussed in Section 6.2.1 and can be summarized as follows.

- Justification: Introducing a proposed activity which will cause the exposure of humans or the environment to ionizing radiation, or altering an existing radiation situation, must do more good than harm. This means that, by introducing a new radiation source, by reducing existing exposure, or by reducing the risk of potential exposure, one should achieve sufficient individual or societal benefit to offset the detriment it causes. The risks in terms of adverse health effects are assumed to be proportional to the dose contribution from the proposed activity. The basic assumption is that the linear non-threshold model applies to exposed individuals.

- Optimization: Once a planned acitivty or interevention has been found to be justified, the likelihood of incurring exposures, the number of people exposed, and the magnitude of their individual doses should then be kept as low as reasonably achievable (ALARA), taking into account economic and societal factors. This means that the level of protection should be optimimal under the prevailing circumstances, maximising the margin of benefit over harm. In order to avoid severely unequal outcomes of this optimisation procedure in terms of incurred individual radiation exposures, there should be restrictions on the doses or risks to individuals from a particular source (dose or risk constraints and reference levels). The ALARA principle is the basic tool for any operation requiring the exposure of a number of individuals. The continuous reduction of collective exposure must be achieved at a reasonable cost (for example, 1 million euro per averted collective dose of 1 Sv). The linear non-threshold model is also applied in these ALARA procedures.

• Dose limits: The total dose to any individual from regulated sources in planned exposure situations other than medical exposure of patients should not exceed the appropriate limits. These are set to protect individuals from excessive detriment. Dose limits, such as the five-year limit of 100 mSv effective dose to any single worker, are imposed by national regulatory agencies.

As mentioned in Section 6.2.1, the concept of "dose limits" is not used in the management of emergency and existing radiation exposure situations. Instead, the term "reference level" is used when justifying and optimizing measures and actions taken to decrease *collective* radiation exposures. The priority is to reduce the risk of acute effects or tissue damage by preventing any member of the *public* from receiving an effective dose higher than 100 mSv, by taking appropriate measures. This principle must be modified for rescue workers performing life-saving actions (see discussion below). The measures taken to reduce individual exposures to below 100 mSv must be justified: that is, they must bring more net benefit than the harm incurred by the measure. Remedial actions that are found to be justified must be optimized according to the ALARA principle, to yield the maximum benefit per unit of harm caused and money invested in carrying out the measures.

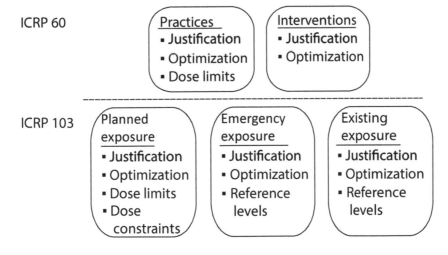

FIGURE 7.22 Principles of radiation protection applied in planned and unplanned irradiation situations. The difference between the new ICRP recommendations (2007) given in ICRP 103 and those given in ICRP 60 (ICRP 1991) is illustrated.

The ICRP thus recommends a system for limiting the dose received by people in unplanned exposures (these were formerly called interventions, a concept still used by many regulators and operators within the radiation protection community: Fig. 7.22). This system has two main features. ICRP publication 109 (ICRP, 2009a) describes recommendations for handling emergency situations, where the aim is to establish strategies for remedial actions to be carried out by national authorities to gain control of the exposure situation. In ICRP publication 111 (ICRP, 2009b), the ICRP describes corresponding strategies for long-term remedial actions, which follow the principles applied to existing exposure situations, such as reducing exposure from naturally occurring radon, or striving to decrease the collective dose originating from medical exposures.

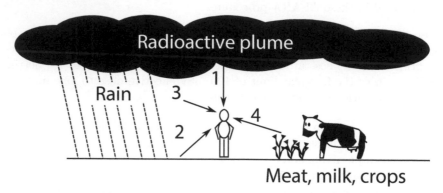

FIGURE 7.23   Main exposure pathways following the atmospheric release of radionuclides.

Figure 7.23 illustrates the main exposure pathways to residents in the area affected by an atmospheric release from a nuclear power plant. They are (1) external gamma radiation from the plume (cloud shine), (2) external gamma and beta radiation from radionuclides deposited on the ground over areas affected by the radioactive cloud (ground shine), (3) internal exposure from radionuclides inhaled during the passage of the plume, or from re-suspended radionuclides from ground deposition, (4) internal exposure to radionuclides from the ingestion of directly or indirectly contaminated foodstuffs, and (5) (Not shown in the figure) external exposure from external contamination on the skin (IAEA 2007a).

The primary aims of emergency preparedness and planned countermeasures is to avoid deterministic radiation effects on members of the public, and preferably also to rescue workers carrying out countermeasures and to limit the stochastic effects on the affected population caused by exposure to radiation. UNSCEAR clearly states (e.g., in UNSCEAR 2000) that these effects are predominantly manifested as an increased risk of induced cancer over a considerable time period (up to seventy years).

Various remedial actions must be taken during the various phases following the onset of exposure, resulting, for example, from extensive radioactive fallout over a country or region. In the first phase (*the rescue phase*: Fig. 7.24), life-saving measures have the highest priority. However, other measures must also be carried out, such as immediate evacuation of those threatened with exposure from a passing radioactive plume; issuing prophylaxis such as stable iodine tablets (especially for children), and imposition of immediate restrictions on grazing cattle to prevent the initial transfer of direct contamination to dairy and beef herds. This phase ends when the radiation levels in the affected areas cease to increase, and when control has been gained over the radioactive release (the source). The latter condition means that the operators concerned must have sufficient control over the source that any additional radioactive releases from the plant can be ruled out.

In the second phase (*the recovery phase*), when the exposure is under control and the risk of further exposure is mitigated, the priority is to reduce the risk of late effects in individuals exposed to radiation. Preparations must be made for large numbers of evacuees to be permanently relocated for a long period, or, in some cases, permanently. This phase may last several days or years following the onset of exposure. In the third phase (*the restoration phase*), when the exposure is under control and the risk of further exposure is mitigated, the aim is to restore normal functions in the affected area. People living in the affected area must be adapted to the remaining radiation. The guiding principle is to justify and optimize protective measures according to the prevailing radiation situation, in the same way as measures implemented in medical procedures or when dealing with natural background radiation.

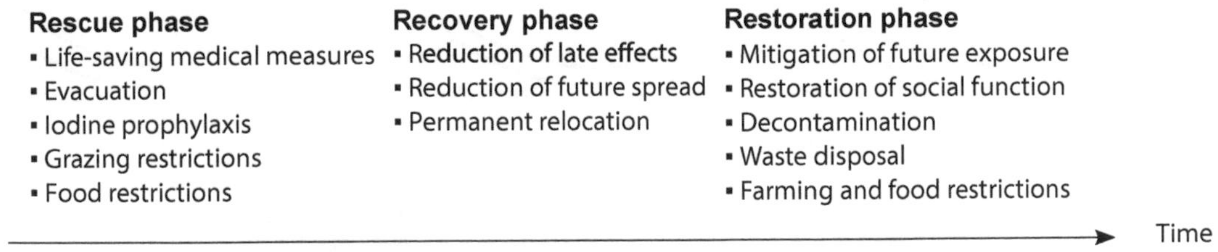

Rescue phase
- Life-saving medical measures
- Evacuation
- Iodine prophylaxis
- Grazing restrictions
- Food restrictions

Recovery phase
- Reduction of late effects
- Reduction of future spread
- Permanent relocation

Restoration phase
- Mitigation of future exposure
- Restoration of social function
- Decontamination
- Waste disposal
- Farming and food restrictions

→ Time

FIGURE 7.24 The three phases after the onset of exposure resulting from a radiological or nuclear emergency situation. In the case of release from a NPP, the imminent threat and subsequent release of a plume constitute the rescue phase. When the risk of further release is negligible, the recovery phase begins, in which measures are taken to avoid exposure from ground contamination. This phase may last several months or years after the incident. During the restoration phase, measures are taken to make the affected areas habitable, such as decontamination of the ground, buildings and public areas.

In contrast to planned and existing exposures, where the aim is to minimize the collective dose, it is more suitable to base strategies for remedial action in connection with RN emergencies on estimated *individual* radiation doses. In the former ICRP 60 concept of strategies for interventions in terms of countermeasures, the term *averted dose* was used. The averted dose is the reduction in the average individual dose to a population achieved by a particular countermeasure. It was calculated by taking the difference between the time-integrated average individual dose in the absence of any intervention, i.e., the *projected dose*, and a similar estimated dose if a particular measure, such as evacuation or iodine prophylaxis, were to be taken. Dose estimates were to be provided by experts drawn from the radiation protection community, and decision makers were expected to relate this quantity to the benefits and costs of taking this countermeasure. This approach embodies the principle that a measure should be justified. In the next stage, the measure would be optimized by adjusting and modifying the measure with regard to cost and available resources, infrastructure, social factors, etc. Figure 7.25 illustrates how the projected dose and averted dose can be represented graphically. The remaining dose was termed the *residual dose*.

Furthermore, the IAEA has suggested so-called operational intervention levels (OILs) for a number of important remedial actions (see Section 7.2.2). OILs are based on actual measurable quantities such as ambient dose rates recorded by stationary monitors located in the area surrounding a nuclear power plant. Before 2007, OILs were defined as a range between a lower and upper level. An upper and a lower operational intervention level were defined for each countermeasure in terms of the averted dose, D_{OIL} (Fig. 7.26). The upper OIL represented the averted dose above which the considered countermeasure would always be justified. The lower interventional level represented the dose below which the considered countermeasure would rarely or never be justified. The interval between them ($D_{lower,OIL}$ to $D_{upper,OIL}$) thus represented the range of averted doses in which it would be possible to apply the considered countermeasure, provided it was optimized according to the ALARA principle. This operational interval, intended to give decision makers and experts flexibility

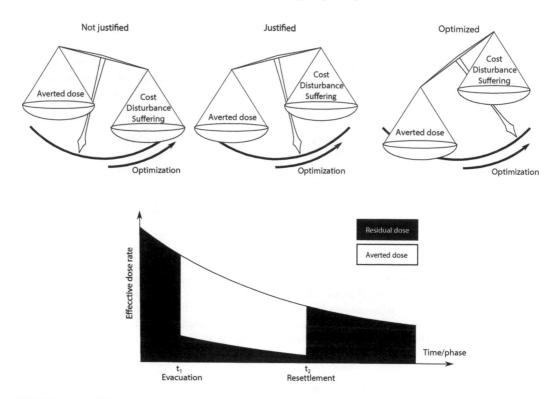

FIGURE 7.25 Schematic representation of how optimization changes the relationship between averted dose and cost, disturbance and suffering. Also shown is the estimated averted dose and residual dose resulting from evacuating an exposed population from an area at time t_1, and then allowing residents to return to their homes at time t_2.

in choosing how countermeasures should be applied, apparently led to considerable confusion. It was found that this concept was not very useful for decision makers in the early phase of an RN emergency, as detailed data on the actual exposure situation tended to be very scarce until full control of the source had been achieved.

The concept of OIL was thus substantially modified in 2007, when the ICRP instead suggested the use of residual dose as a key parameter for decision makers to justify a countermeasure in an exposure situation. Today, the OIL is given as the value of a particular dose that will trigger a countermeasure. Any countermeasure that reduces the residual dose below the reference level (levels range from 20100 mSv y-1) is essentially always justified, provided it is optimized according to the ALARA principle. A strategy consisting of several countermeasures can be applied, which may result in annual residual doses below the 20–100 mSv range. Figure 7.27 shows the principle governing use of the reference dose and residual dose.

It is often necessary to take immediate measures based on what is known about the situation at the source, without having the opportunity to make measurements in the environment. This implies that the action taken is based on predetermined rules appropriate to the threat situation. Average individual exposures and averted doses are calculated in advance for possible scenarios, and are then included in emergency management plans. Radiation doses above 100 mSv will almost always warrant drastic dose reduction measures. However, in the dose range of 0–100 mSv, dose reduction measures that are simple and cost

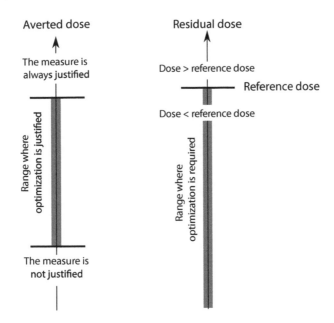

FIGURE 7.26 Left: the application of upper and lower operational intervention levels for a given type of countermeasure as formerly suggested in ICRP 60 (ICRP 1991). Right: the application of residual dose and reference dose now suggested by the ICRP (ICRP 2007).

effective, and which involve least suffering for those affected are primarily selected. Measures affecting everyone in the potentially affected area are thus implemented before people are subjected to radiation, or as soon as possible after the onset of exposure. The area over which measures are taken must be selected with a sufficient safety margin in terms of the meteorological conditions, such that the area actually affected is smaller.

The choice of reference dose levels for the selected countermeasures is based on the projected scenarios, in which it is assumed that justification and optimization have been applied as far as is possible. In the next stage, when radiometry data from the affected area are available, remedial action can be taken based on the measured values. However, it is often necessary to base individual doses on models, and the aim when selecting countermeasures must be to reduce the dose to those people receiving the highest individual doses, as this is almost always the strategy that is most cost-effective per unit averted dose.

The cost of a countermeasure almost always increases proportionally with the number of individuals affected. Therefore, individuals with approximately equal projected doses will be combined into a group. The cost per unit individual averted dose of carrying out a given countermeasure for this group can then be calculated. The countermeasures that result in the lowest cost per unit averted dose for a given group of individuals are the most justified, and should be implemented first. When fully optimized, it is also possible to theoretically estimate the averted collective dose. The cost per unit averted collective dose, expressed in years of human life saved, making use of the risk coefficients linking effective dose and the risk of fatal cancer (Section 2.4), should not be significantly greater than the amount society is prepared to pay for other measures that save lives and reduce injuries (for example, when building motorways).

In emergency exposure situations the ICRP and the IAEA acknowledge that it is justifiable to accept dose exposures to rescue workers that are higher than the normal dose

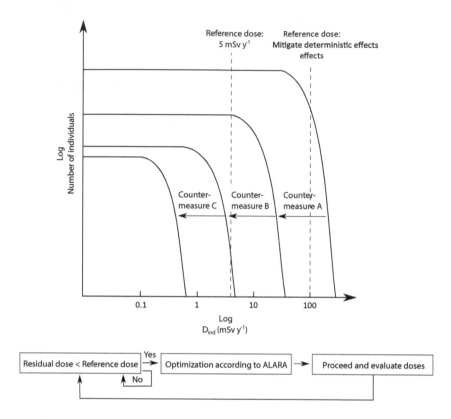

FIGURE 7.27   Above: the principle of using the residual dose to justify a countermeasure. Examples of countermeasures are A) evacuation of residents from the area affected by the airborne plume, B) relocation, and C) decontamination of the most-affected residents. Below: flow chart illustrating the selection of justifiable countermeasures after the early phase of an emergency or an existing exposure situation.

limits for radiation workers. A reference level for rescue workers above 100 mSv (but no higher than 500 mSv) is mentioned by the IAEA (IAEA 2013), and refers to life-saving rescue actions where the number of life-years saved heaviliy outweighs the risk to the single individual.

The Swedish rescue authorities have chosen to interpret the guidelines for an RN emergency as illustrated in Figure 7.28. For a single rescue action resulting in an absorbed dose of up to 50 mSv, which is the legal dose limit for a single year, provided that a total of 100 mSv is not exceeded during a five-year period, the action must be justified and protection must be optimized. If it is believed that this restriction is satisfied, the on-scene commander is permitted to demand or order personnel to carry out the rescue action. If the planned intervention or rescue action will result in an individual dose in the range 50–100 mSv, all staff involved in the action must be informed of the risks associated with the anticipated radiation exposure, and their participation must be voluntary. Women who cannot rule out being pregnant are not permitted to participate in such a rescue action. Radiologically, a foetus shall be given the same level of protection as a member of the public.

A second reference level of 100 mSv is used, above which the Swedish rescue authorities regard only life-saving actions as justified. Furthermore, the participating rescue workers must be volunteers, and must also be fully informed of the associated radiological risks.

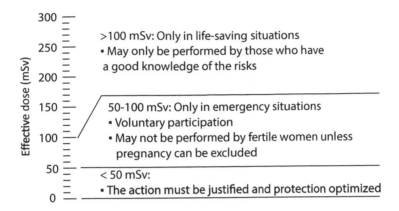

FIGURE 7.28 Interpretation by the Swedish rescue authorities of international recommendations on reference levels of absorbed dose for workers in connection with emergency exposures.

7.2.1.2 Emergency Planning and Organization

The United Nations, together with other organizations such as the OECD, have formulated legislative requirements for countries in which the responsibilities for preparedness for, and response to, an RN emergency are clearly allocated (IAEA 2002b). These requirements stress that each country should (i) have a governmental body that coordinates threat assessments, (ii) identify and assign necessary functions in the case of an RN event, and (iii) allocate responsibilities for establishing response organizations and maintaining necessary functions. A national coordinating authority is advocated that has the ability to play a major part in responding to an RN emergency. It is further suggested that this authority should have the following responsibilities.

- Ensure that threat assessments are conducted regularly to identify new practices (Fig. 7.22) or events that could call for an RN emergency response. This assessment must also include an exchange of information with neighbouring countries.

- Determine or outline the tasks that each organization involved in an RN emergency response should perform. This outline should also include an assessment of the existing resources and the resources necessary for each organization to perform their respective tasks.

- Ensure that the organizations involved in RN emergency response have agreed to their assigned responsibilities. The IAEA publication *Method for Developing Arrangements for Response to a Nuclear or Radiological Emergency* (IAEA 2003), presents an overview of the tasks that are necessary for a successful RN emergency response. It is also stressed that the organizations involved should themselves be committed to improving and developing their RN emergency response capability.

- Coordinate the development of an "all-hazards" plan (an example of which is described briefly below).

- Foster the implementation of measures required to fulfil relevant international or bilateral obligations with other countries.

It is further suggested that within this national coordinating authority, a single national RN emergency planning coordinator should be appointed with the task of leading the preparedness process. This coordinator should preferably have in-depth technical and operational knowledge of all issues associated with RN emergency responses. Multi-year funding is recommended to finance such a coordinator, in order to provide resources for long-term development and maintenance of response capability.

The IAEA (IAEA 2007a) mentions three levels when assigning responsibility for emergency preparedness and response: (i) operator level, (ii) off-site level, and (iii) international level. An operator should mainly be responsible for identifying an emergency and for taking immediate action to mitigate its consequences. This is done by protecting individuals at the affected site and other areas under the operator's control. Communication channels should be established with off-site officials, who should be promptly notified of the extent of the emergency. The operator is also expected to keep off-site officials and the general public informed about the evolution of the accident at the site, and provide technical and radiation monitoring advice.

The off-site level of RN emergency responsibility refers to those organizations involved in carrying out countermeasures in the areas outside the plant and beyond the control of the operator. The IAEA refers to local officials who are assigned by the government to provide immediate support and protection of the public in the emergency zone. The term local officials includes organizations such as the emergency services responding at the scene, i.e. the police, fire brigade, medical personnel, and civil emergency services. Several of these will also be classed as first responders (see discussion below).

For emergency response on the international level, the IAEA refers to the *Convention on Early Notification of a Nuclear Accident* (Early Notification Convention) and the *Convention on Assistance in the Case of a Nuclear Accident or Radiological Emergency* (Assistance Convention). The parties who have signed these conventions have committed themselves to informing states that may be affected by any trans-boundary release. Notification can either take place directly between the states involved or via the IAEA. Furthermore, signatories are committed to providing technical assistance in emergency response to the affected countries. Such technical assistance may include large-area mapping of the ground deposition, and medical assistance to affected individuals. Other international UN-associated organizations involved in the international level of RN emergency include the World Health Organization (WHO) and the Food and Agriculture Organization (FAO), which can both provide technical and advisory assistance in an RN emergency.

Thus, a number of organizations must be involved in an RN emergency response to comply with the IAEA guidelines. During the initial rescue phase it is primarily the so-called *first responders* who will attempt to establish control over the exposure situation. The first responders may be (i) fire brigade officers, whose task it is to restrict the dispersion of hazardous substances by, for example, extinguishing fires, and who carry out life-saving actions on victims at the site; (ii) pre-hospital medical staff (such as paramedics and ambulance nurses), who establish a local facility for casualties at the site; and (iii) police or military personnel who help establish control over the flow of both materials and personnel, including civilians and victims, out of the accident site. If the accident scenario involves a nuclear power plant or a large site containing radiological or nuclear materials, a local rescue organization that includes local officials (emergency management) will supervise workers at the accident site.

Sweden can serve as an example of national coordination of emergency authorities; see Figure 7.29. In Sweden, no single authority assumes sole overall responsibility. Local authorities are relatively autonomous in Sweden, and are independent of the national government. This means that operative decisions are primarily executed by local authorities, rather than the government. In larger and more populous countries such as Russia, several authorities

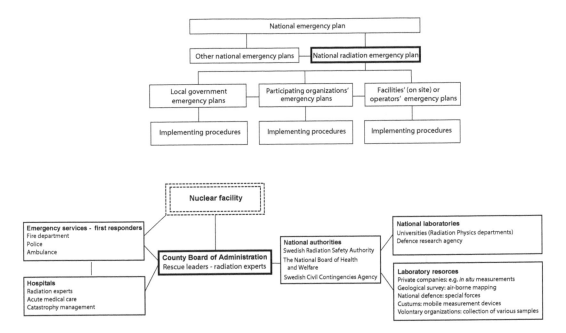

FIGURE 7.29 Above: schematic view of the integrated planning concept for response to hazards suggested by the IAEA (IAEA 2007a). Below: examples of various actors who would be involved in rescue operations following a nuclear power plant accident in Sweden.

may have direct responsibility for RN emergency management. Sweden's all-hazard crisis management system is built on four principles.

1. *Similarity principle regarding responsibility.* All authorities and organizations responsible for functions in society under normal conditions, should also have responsibility for these functions during a crisis. This involves collaboration with other authorities in order to find ways to restore normalcy, using the resources available in the most efficient way possible.

2. *Sector responsibility.* This principle refers to the specific responsibilities associated with a certain service, such as radiation protection and protection against harmful substances. This principle means that all the authorities within this area have a special responsibility to collaborate and find common means of restoring the situation in the case of a crisis.

3. *Maintaining normal functions.* The similarity principle, which identifies the authorities involved in both normal and crisis situations, also implies that these authorities are charged with maintaining and restoring normal functions during a crisis. For example, health care centres and hospitals must be able to deal not only with the extra flow of patients resulting from an RN incident, but also provide and maintain the same patient care for the public as under normal circumstances.

4. *Proximity principle.* In a relatively large but sparsely populated country such as Sweden, emergency response and crisis management must be divided into different geographical levels. An accident should primarily be resolved by local resources at the municipality level, and the local or regional emergency management group should be

located as close to the accident site as possible. If more resources are needed, crisis response should be coordinated at the regional level. A large-scale crisis may demand all the available domestic resources, and sometimes also international aid, and in such cases, crisis management must be coordinated on the national authority level.

It follows from the last of these principles that there is a geographical division of responsibility when managing RN emergencies. In Sweden, the municipalities, which on average cover an area of about 1000 km^2 with a population of approximately 30 000, are the core geographical units defining the remit of local emergency management and rescue teams. This means that a traffic accident involving a cargo of radioactive material and the associated risk of dispersal of volatile radioactive elements, must primarily be coordinated by a local rescue team (usually the fire brigade). It is probable that most local accidents involving radioactive substances will not be interpreted as RN events by the first responders. In the first instance, a traffic accident on a motorway will be reported to the national emergency services operator (called SOS Alarm in Sweden) as an ordinary traffic accident. It may not be until the rescue crew inspects the damaged cargo that they discover that the vehicle contains harmful or radioactive substances, and that assistance and other expertise is required. Local rescue services in Sweden have regular exercises in the detection and notification of harmful substances to the appropriate authorities. This ensures that the officer in command can rapidly obtain assistance from experts with specialized instruments (e.g., radiation detectors).

Since 1981, it has been the responsibility of the relevant regional authority to draw up crisis management and emergency plans for nuclear reactor accidents (see Section 7.1.3). A county in Sweden has a similar administrative function to departments in France, oblasts in Russia or prefectures in Japan, but their size and population can differ considerably between countries. Swedish counties typically have an area of 20 000 km^2 and a population of less than 0.5 million. The regional authorities in Sweden are called the County Board of Administration.

The IAEA guidelines indicate that crisis management is more efficient and emergencies are handled more effectively if there is organized cooperation between all the parties and stakeholders involved, including national authorities, regional authorities, local authorities, health care centres/hospitals, representatives from the industries involved, military personnel, and non-governmental volunteers. Again, using Sweden as an illustration, so-called *cooperational divisions* have been set up to address particular kinds of crises (Fig. 7.30). These are used as planning forums for the authorities involved to develop common emergency plans and monitor national resources in the event of a crisis. Sweden has six such divisions. The Swedish Radiation Safety Authority, which under normal conditions primarily works proactively on matters related to nuclear safety, radiation protection, the handling of radioactive waste and non-proliferation, is mainly involved in the cooperational division for *Hazardous Substances*, which is concerned with the ability to manage CBRNE (chemical, biological, radiological, nuclear, and high-yield explosives) events.

National emergency preparedness for RN incidents can often be enhanced by consolidating radiometry and expert resources within a country, or within an international cooperation framework. Laboratories where radiometric equipment is in regular use and radiometry data are assessed, for example, at universities, governmental research institutes, hospitals or in industry, can be contracted to the authorities responsible for national emergency preparedness to contribute their measurement resources and expertise in the case of an RN incident. Such a network should also conduct regular training and exercises using authentic radiation sources to maintain competence among their experts, as well as to check that the specialized instrumentation required in an RN emergency is in good working order.

Hazardous substances

Swedish Armed Forces	Swedish Board of Agriculture
Swedish Coast Guard	National Forensic Centre
National Food Agency, Sweden	National Veterinary Institute
Swedish Civil Contingencies Agency	Swedish Radiation Safety Authority
Swedish Police	Swedish Defence Research Agency
Public Health Agency of Sweden	Customs
National Board of Health and Welfare	Local authorities

FIGURE 7.30 Swedish authorities involved the cooperational division for hazardous substances in the case of RN incidents.

7.2.1.3 Small-Scale Events

International guidelines have been drawn up for the management of a local radiological or nuclear incident, and for establishing emergency management. An Emergency Preparedness and Response report issued by the IAEA (IAEA 2006) gives guidelines for first responders to a radiological emergency. However, international directives may be adopted in different ways in different countries, depending on how the organizations involved in the initial crisis response are constituted, and on the roles of the various national authorities in a particular country. The IAEA report primarily gives guidelines to rescue teams using their own procedures and equipment to detect ionizing radiation. In many countries, a section of the fire brigade is trained in detecting harmful substances. In Sweden, a basic level of capability, for example, monitoring the gamma dose rate at an accident site, is provided by all municipal fire brigades.

The management of a local incident can be organized according to the scheme in Figure 7.31. On-scene control is established by the so-called response initiator. The initiator alerts the relevant authorities at the local or regional levels, depending on the gravity of the situation. Warning signs on vehicles transporting hazardous substances are therefore vital, to facilitate the appropriate response in the case of an accident. The on-scene commander (sometimes referred to as the *incident commander*) coordinates the practical rescue activities at the site, as well as the expertise and external support from authorities on a broader regional or national level. The on-scene commander may have remote assistance from local and regional officials who coordinate rescue actions, for example, via the regional authority. Ambulance personnel alert not only the nearest hospitals, but also hospitals in the entire region so as to consolidate resources such as hospital beds and emergency medical care.

The incident site is organized into various cordoned areas, or sectors (Fig. 7.32). The purpose of these sectors is to delineate the level of hazard, and indicate which rescue actions are prioritized in each sector. The actual damage site may be very small, only a few square metres, but an inner cordoned area (enclosing the damage site) is defined by the level of hazard or health risk to rescue personnel. Initially, in the absence of data on the gamma dose rates at the site, the IAEA recommends a safety perimeter be set at a radius of 30 m

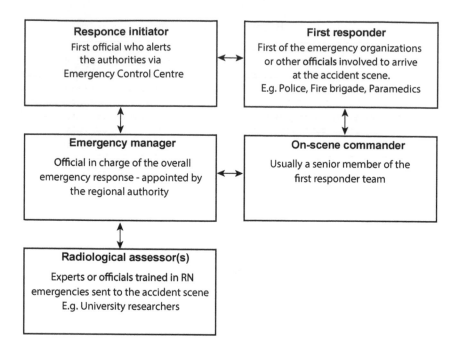

FIGURE 7.31 Schematic diagram of the on-scene chain of command and the relevant actors at the incident site, as well as regional and national backup and decision support.

around an unshielded or potentially damaged source (IAEA 2007a). If there is a spill area, the 30-m perimeter must extend from the outer boundary of the spill. If there is a major spill, the perimeter radius is set to 100 m. Moreover, if there are signs of fire, an explosion or fumes, the safety perimeter must be further extended to 300 m, and to 400 m, if it is suspected that there may be a bomb at the site.

Only life-saving actions are allowed within this cordoned-off inner safety area, and special protective clothing and respiratory protective equipment are to be worn. This area is also referred to as the *hot zone*. Of the first responders, only rescue personnel usually enter this area. Personnel, victims and materials leaving the hot zone must pass through a decontamination unit.

The *warm zone* is the area between the hot zone and the middle cordon, separating the hot zone from the *cold zone*. Medical staff and ambulance personnel can operate and set up triage (see Section 7.4) in this area so that life-saving care can be given. All personnel in this zone must wear standard protective clothing. The middle cordon is defined by an external ambient dose rate of 100 μSv h^{-1}. Passage from the warm zone into the outer cordoned area takes place through a second decontamination unit where more extensive decontamination of patients takes place. In some countries, the police have the task of registering and storing personal belongings that cannot follow the patient to the medical unit, but which must be decontaminated and exempted before being returned to their owners.

In many countries, the outer area, or cold zone, is not marked by an external cordon. In Sweden, however, a cordon is set up, and is guarded by the police. The aim is to prevent members of the public, unauthorized persons and materials from entering the area, or leaving it without passing a security checkpoint. If this is not done, contaminated persons and equipment, that for some reasons have not been decontaminated, will disperse. Some of

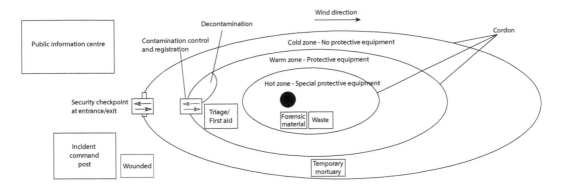

FIGURE 7.32 An example of how the emergency site may be organized based on IAEA guidelines (IAEA 2007a).

these contaminated persons may e.g. visit hospital wards and waiting rooms, thus exposing more people to ionizing radiation. This phenomenon is also referred to as *wild evacuation.* The boundary for this outer cordoned area can be set, for example, at an external ambient dose rate of 1 μSv h^{-1}, although in some countries no specific value has been defined. This zone is intended to act as a working zone for all the on-site rescue teams and personnel involved, and the guarded outer cordon also serves to prevent personnel from being disturbed by journalists or concerned members of the public.

When an operator such as a radiation protection specialist or a crime scene investigator enters the outer cordon of the damage site, a collection point is often set up, supervised by the on-scene commander. The on-scene commander briefs new arrivals on the situation, and instructs external experts on practical matters and protection guidelines. The main goals of a radiation protection expert at the site of an accident involving radiological or nuclear substances are (i) to protect and rescue workers and other personnel at the site from ionizing radiation, and (ii) to protect the public from being exposed to radiation. However, the radiation protection expert and other on-site personnel must be aware of other risks involved in the rescue operation, such as explosions, asphyxiation by gas emissions, fire, dispersion of harmful chemical and biological substances and falling objects. The on-scene commander must apply an all-hazards and overall security and safety approach to the rescue work.

A radiation protection expert should be summoned to the site as soon as possible, and should assume the role of *radiological assessor.* This person recommends protective actions at the site. Examples of recommendations are minimizing external exposure, and ensuring that actions comply with the radiological principle of justification. In Fig. 7.33 is given an illustration of how the roles of the radiological assessor and the on-scene controller are assumed in the inital repsonse to a radiologcal emergency.

The rescue team must be instructed to limit their activities in the inner zone to life-saving actions, and not to collect or pick-up any objects from the site, as these may be fragments of the dispersed radioactive material. Internal exposure and contamination can be mitigated by strict use of respiratory protection equipment. In worst cases, airborne contamination may be so severe that skin contamination, high enough to cause tissue damage, may occur up to 100 m from the suspected radiation source. It is unlikely that the groundwater table will be contaminated to such an extent that it exceeds the activity concentration limit for drinking water, but the fear factor among the public can be substantial. No person who

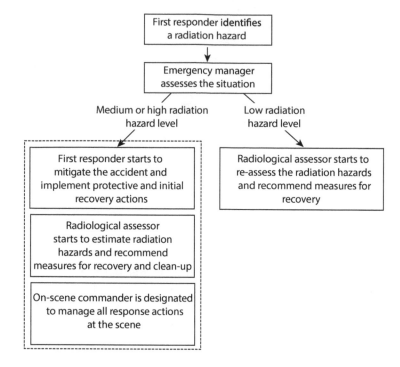

FIGURE 7.33 An example of an initial response to a radiological emergency given by the IAEA (IAEA 2006). Note that the response initiator in many accidents will be the first responder, often the local fire rescue service.

is still alive can be so severely internally contaminated that life-saving operations are not justified. Removing the casualties closest to the radiation source, and then removing their clothing will probably decrease external contamination by more than 90%. As mentioned in Section 7.1, the most hazardous radiation sources may result in lethal exposure to an unshielded victim within one minute. For lower categories of radiation hazard, this time may be up to an hour, allowing time for rescue actions justified by the ICRP radiation protection principle.

Several lessons have been learned from previous major accidents involving radiation and other harmful substances. As in many other kinds of crisis management, it is beneficial if all the participating actors have a well-defined role. These roles must not only be defined in an emergency plan, but there must also be awareness among the various categories of first responders of their function in any CBRNE crisis. Training of first responders, especially in joint exercises, can increase their understanding of the various practical functions at an accident site. A unit specialized in public communication should be established to coordinate information to the public and the media. Specialists and experts should therefore not communicate directly with the media until this unit has been established, in order to prevent ambiguous or conflicting statements that cause further concern and hinder rescue work.

In terms of the medical management of CBRNE events, experience shows that unexposed members of the public suffering from acute anxiety, the so-called *worried well*, will completely dominate the patient flow to medical centres. Considerable resources will therefore be devoted to alleviating public concern by performing clearance measurements, and reassuring individuals that they have not been exposed to harmful radiation. In contrast, there may be a risk that some medical staff will refuse, or will be reluctant, to give medical care to patients suspected of being externally or internally contaminated by radioactive substances. Furthermore, some medical staff may confuse external exposure to ionizing radiation with being externally contaminated, or "still being radioactive". In many countries there is also a perception that radioactive contamination is contagious, similar to biological infections. The radiological assessor must inform medical personnel that although radioactive contamination can be passed from one individual to another, the radioactive material can never proliferate in the way bacteria or viruses do.

The security aspect of the rescue work must also be considered. If an act of terrorism is involved, the antagonist(s) may still be present at the site or be among the victims. If malevolent actions are suspected, the work of the police and crime scene investigators may affect how the casualties are cared for or transported to medical centres. For rescue workers, this means that they should not tamper with, or alter objects at the damage site that may be of forensic interest.

The logistics needed to establish remote communication must be considered. Both stationary and mobile radiocommunications systems can be destroyed by an accident, and auxiliary or backup communications systems must be available to coordinate the rescue work. For a radiation protection expert to be able to function efficiently at the site, it is vital that they are able to rapidly communicate measurement data and status reports on the radiation environment to backup support or to other off-site experts.

7.2.1.4 Large-Scale Events

A distinction is often drawn in emergency preparedness between local incidents and large-scale events involving casualties over a wide geographical area, requiring resources far beyond those that can be provided by local or regional emergency organizations. The IAEA defines various zones for emergency management around large nuclear facilities of Categories I and II. The on-site area refers to the area over which the operator of the facility and the first responders have control over access. The on-site area may be, for example, the fenced area around the facility. Some Category I and II facilities, such as NPPs in Sweden, have an in-house rescue service. The off-site areas refer to those areas beyond the control of the operator, the facility or the first responders. For efficient emergency management, the IAEA suggests that off-site areas be divided into at least two zones: (i) a precautionary action zone (PAZ) where some remedial action may already be taken at a lower alert level, before any atmospheric release occurs, and (ii) an urgent protective action planning zone (UPZ), for which the IAEA recommends that arrangements be made to prepare for an atmospheric release from Category I or II nuclear sites. The outlines and boundaries of these zones should be roughly circular. The recommended radius of the PAZ for a standard nuclear power plant (operating at 1000 MW_{th}) ranges between 3 and 5 km from the site, while the radius of the UPZ may range from 5 to 30 km (Fig. 7.34). No PAZ is defined for Category II nuclear facilities, whereas the UPZ is recommended to be 0.5–5 km from the site (IAEA 2007a).

Thus, if an incident occurs at a large NPP or nuclear facility (Categories I and II), local authorities will implement their plans for the establishment of the required emergency organization. This organization must coordinate several important functions and actors such as regional hospitals, local and national radiation protection experts, local or regional military resources trained to assist in a civil crisis, local and national radiometry resources, assistance from other NPPs, domestically or internationally, assistance from meteorological institutes, and assistance from traffic coordination bodies and the coast guard. Coordination

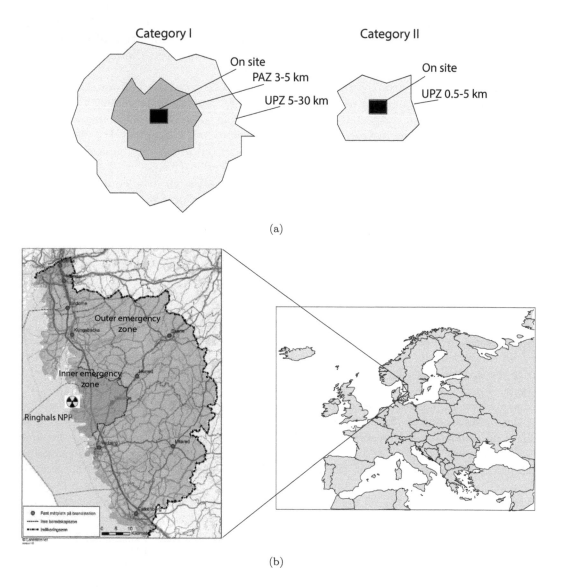

FIGURE 7.34   (a): Emergency zones for response to a large-scale RN emergency as suggested by the IAEA (IAEA 2007a): (b): The Swedish emergency zones defined for the Ringhals NPP, south of Gothenburg on the west coast of Sweden. Map source: Lantmäteriet, Sweden.

of expertise in other authorities, such as the National Board of Health and Welfare, which provides medical expertise in RN-related injuries, is also required. Regional and national organizations responsible for agriculture must be kept informed so that they can provide instructions to farmers and food producers regarding necessary action. The County Boards of Administration, which are the coordinating governmental actors for major crises at the regional level in Sweden, including RN emergencies, developed emergency plans in counties where NPPs are located at the beginning of the 1980s. Figure 7.35 shows the various organizations involved in the rescue and recovery phase after a NPP accident in Sweden.

In the case of an accident at a large NPP, there is a time window between the initial alert of an anomaly in the plant's operation and the onset of radioactive release to the environment. The event must be classified in terms of a predefined emergency action level relating to the abnormal conditions at the plant, or practice concerned (this is discussed in more detail in IAEA 2007a). Should an incident occur at any of the reactors within the country, the regional authorities should be alerted, or, in the case of Sweden (Fig. 7.35), the Swedish Radiation Safety Authority should be alerted directly. This will initiate the deployment of an emergency organization by the regional authority in the affected county. The Swedish Radiation Safety Authority will also alert all other regional authorities in the country so that they can be prepared to offer assistance. The government will also be alerted. The Radiation Safety Authority also notifies a network of laboratories throughout the country. The laboratories participating in this network are referred to as emergency preparedness laboratories, and have the measuring capabilities needed in both the rescue and recovery phases of an NPP accident.

In the case of NPP accidents in other countries, international warning systems coordinated by the IAEA or the EU will alert the Swedish Meteorological Institute and the Swedish Radiation Safety Authority via the SOS Alarm organization. As in the case of a domestic NPP accident, all the County Boards of Administration and the network of radiometry laboratories will also be alerted.

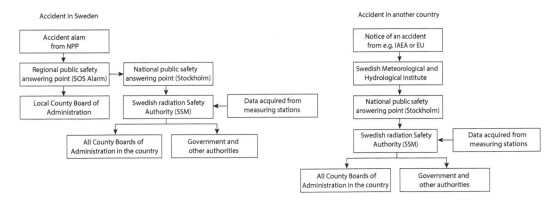

FIGURE 7.35 Example of alert routes in connection with NPP incidents and accidents in Sweden and in other countries.

In an RN emergency, it is the responsibility of the Swedish Radiation Safety Authority to provide advice to the County Board of Administration on the dispersion of radiation, and on radiometry, radiation protection and decontamination after a release of radioactive materials. Expert advice should also be given regarding the technical status of the facility where the abnormal conditions occurred. The authority must also maintain and take command of a national expert organization providing decision support to other authorities involved during an RN emergency. The authority is also required to coordinate and abide by international and bilateral agreements regarding assistance with various resources in an RN emergency.

The emergency alert system on the national level includes (i) an officer in readiness, drawn from staff appointed by the authority, (ii) a supplementary officer in readiness for reactor safety, and (iii) emergency groups for media communication and operative communication, and an expert panel. The authority's emergency preparedness plan includes regulations and procedures for the crisis organization deployed in an RN emergency, including manuals and action lists, as well as a list of personnel suitable for the various positions in the crisis organization. Key contact lists are compiled in a connection directory. Figure 7.36 provides a schematic view of the crisis organization for an RN emergency at the Swedish Radiation Safety Authority.

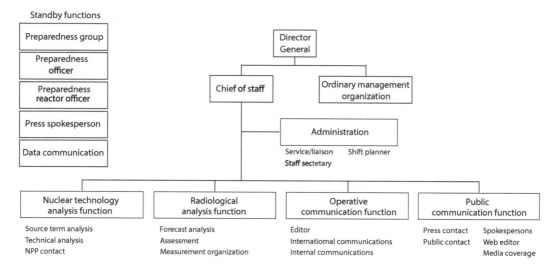

FIGURE 7.36 Crisis organization set up by the Swedish Radiation Safety Authority in the event of an RN emergency (for both domestic and international events).

The expert support network established within the organization plays a vital role in the Swedish crisis organization. Its purpose is to provide advice and information to the radiological assessment group in charge of analytical calculations and prognoses related to the radiological situation, as well as to interpret the data reported by the radiometry laboratories. The aim is to provide sustainable crisis management for at least seven days following the incident. There is a special focus on resources for the large urban areas in the country.

Two levels of alert have been defined in Sweden, corresponding to the emergency classification drawn up by the IAEA: (i) an *enhanced alert*, when there is an anomaly during normal operation, and (ii) a *release alert*, when it is judged that a substantial release from the reactor site is probable within the next twelve hours, or if a release is taking place.

Following an enhanced alert, all the staff and experts listed in the alarm lists of the County Board of Administration will be summoned. At this alert level the following measures are taken:

- All members of the public living less than approximately 15 km from the affected NPP (inner emergency area) are instructed to remain indoors, and close all doors and windows.

- Instructions are given to turn off air conditioning and ventilation systems (this cannot be done by residents themselves in multi-storey buildings).

- Residents in the inner emergency area are instructed to listen to the radio or use other media for official information from the local authorities.

Following a release alert, the crisis management organization of the County Board of Administration will, after communication with the expert group of the national authority, proceed by taking some immediate countermeasures:

- Instructions to house livestock within the inner emergency zone indoors (corresponding to the PAZ defined by the IAEA).

- Instructions to residents in the entire emergency zone (corresponding to the UPZ) to take iodine tablets that have been distributed in advance to those in areas that may be affected by NPP incidents.

- Evacuation of residents up to 3 km away from the affected NPP.

In most cases, it will not be possible to immediately assess the threat associated with the release of radioactive material resulting, for example, from major failure of a vital non-nuclear part of the reactor. In the absence of a complete picture of the potential threat, the radiation protection experts may not be able to give a specific prognosis following an enhanced alert. In such cases, a general release is assumed in the Swedish emergency plan, corresponding to about 1% of the inventory of refractory radionuclides in the affected reactor; Table 7.4. The release is assumed to take place over 24 h. The following measures should then be prepared to take effect by the County Board of Administration:

- Evacuation of parts or whole sectors of the inner emergency zone (or PAZ).

- Instruction to people within a radius of 50 km of the plant (i.e., within the UPZ) to stay indoors.

- Additional distribution of iodine tablets to residents in the entire emergency zone (up to 50 km from the NPP).

- If precipitation is ongoing or anticipated, preparations should be made for a full-scale evacuation of the entire emergency zone.

During the rescue phase, steps must be taken by the County Board of Administration to set up groups that can develop strategies for long-term measures, and predict the resources that will be required in the region during the restoration phase. This is especially important for agriculture as it is likely that a large proportion of the region's livestock will have to be slaughtered. Restrictions on livestock and harvesting of crops may extend over wide areas, and measures must be taken to alleviate the economic burden on the affected farmers.

Another very costly remedial measure is the decontamination of houses and other public spaces, such as school playgrounds, especially in urban areas. In Sweden, a national expert group on decontamination procedures in urban areas has made plans for appropriate strategies to deal with extensive contamination. Such a procedure may only be justified in some parts of the contaminated areas, and if this is the case, measures must be taken to optimize efforts according to the ALARA principle.

TABLE 7.4   Generic release model used in the Swedish emergency preparedness plan to make a prognosis following a potential atmospheric release from a light-water reactor. Radionuclides expected to dominate in airborne releases from NPP accidents are given in boldface.

Category	Radionuclide	Assumed release fraction (%)
Noble gases	^{85}Kr, ^{85m}Kr, ^{88}Kr, ^{131m}Xe, ^{133}**Xe** ^{135}Xe	1–100
Volatile and refractory elements	^{131}**I**, ^{132}I, ^{134}I, ^{135}I,^{86}Rb ^{88}Rb, ^{134}**Cs**, ^{136}Cs, ^{137}**Cs**, ^{127}Sb ^{129}Sb, ^{129}Te, ^{131m}Te, ^{132}**Te**	1
Less refractory elements	^{58}Co, ^{60}Co, ^{99}Mo, ^{99m}Tc, ^{103}Ru ^{105}Ru, ^{89}Sr, ^{90}**Sr**, ^{91}Sr, ^{137m}Ba ^{140}Ba	0.1
Inert elements	^{95}**Zr**, ^{97}Zr, ^{95}Nb, ^{97}Nb, ^{140}La ^{141}Ce, ^{143}Ce, ^{144}Ce, ^{143}Pr, ^{144}Pr ^{147}Pm, ^{147}Nd, ^{238}**Pu**, ^{239}Pu, ^{240}Pu ^{242}Am, ^{242}Cm, ^{244}Cm	0.01

7.2.2   Remedial Actions

7.2.2.1   Countermeasures and Intervention Levels

A number of remedial actions in response to radiological or nuclear dispersion scenarios were mentioned in the previous section. In the Nordic countries, guidelines for protective measures in the early and intermediate phases of a nuclear emergency are summarized in a booklet (*Nordic Flagbook*) issued jointly by the Nordic radiation protection authorities (SSM 2014). The main goal of remedial actions after a radiological or nuclear emergency is to keep the average residual doses (illustrated in Fig. 7.24) below 20–100 mSv during the first year after the onset of the crisis, using the concept of reference doses in emergency situations, as defined in ICRP 103 (ICRP 2007) and 109 (ICRP 2009a), and complying with international safety standards given, for example, by the IAEA (IAEA 2002a) and the EU (Euratom 2013). Strategies must thus be designed by decision makers to achieve this goal, often by combining one or more remedial actions. It should be assessed whether the countermeasures are justified and, in addition to the residual dose, consideration should be given to the cost of the action, its efficiency in reducing dose, the material and human resources required, their long-term impact on the economy, the waste generated, and any possible socioeconomic harm and psychological factors. For example, in many countries the public believe that authorities treat everyone equally, and have little understanding for authorities making exceptions due to cost. However, there is often public acceptance of devoting special resources and efforts to children and the elderly. (For further details, and an in-depth discussion of the choice of countermeasures, see SSM 2014, IAEA 2013 and ICRP 2009a).

A brief summary of the remedial actions available to reduce the exposure of inhabitants in areas affected by NPP fallout is given in Table 7.5.

Considerable efforts must be made by decision makers to communicate the motivation behind protective measures. Radiation protection experts and radiological assessors must not confuse their role with decision making. Their purpose is to present advice based on the

TABLE 7.5 Overview of some key remedial actions that can be taken following a large-scale RN emergency. The operational intervention levels (OILs) are described briefly in Section 7.2.1.1.

Remedial action	Intervention criteria (OIL)	Harm or adverse effects associated with the intervention
Emergency medical procedures	Defined through triage. Patients receiving >5 Sv should receive palliative care. Early symptoms determine further treatment.	Medical staff may be mildly exposed, <10–100 mSv, during the care procedure, which is less than family members receive from thyroid carcinoma patients. Contamination of sophisticated and sensitive medical equipment and tools.
Sheltering	1–10 mSv per 6 h avertable dose per individual, during an airborne release. Prevention of exposure due to inhalation and cloud shine.	Lack of oxygen in crowded environments if ventilation is turned off.
Evacuation (temporary)	3–30 mSv per day	The elderly and severely ill may not be able to cope with the hardship of evacuation, and mortality in these groups may be higher if they are evacuated.
Relocation	5–50 mSv in the first month: repopulation once the avertable dose is <3–30 mSv per month.	Severe psychosocial consequences of moving and establishing a new home in a new region.
Food restrictions	Approximately 300 Bq kg^{-1} γ-ray-emitting radionuclides for commercial food. 1500 Bq kg^{-1} recommended for game for private consumption.	Economic losses for individual farmers.
Iodine prophylaxis	10–100 mSv avertable thyroid dose for children. 10 times higher for adults <40 years old.	Some risk of, e.g., allergic reaction.
Decontamination	1–10 μSv h^{-1} gamma dose rate on surfaces of buildings and vehicles. 10 μSv h^{-1} indicates need for urgent decontamination.	Extremely costly, and generates large amounts of radiological waste.
Access control	1 μSv h^{-1} gamma dose rate on built-up recreational surfaces (such as playgrounds).	

guidelines available, and to ensure that decision makers do not disregard the upper OIL. However, with the exception of the rescue and recovery phases after the Chernobyl event, where authorities deliberately withheld information from many affected groups, such as dairy farmers in Belarus who allowed their cattle to graze in May 1986 unaware of the very high radioactive fallout (>1 MBq m^{-2} ^{137}Cs), the tendency has been for decision makers to want to perform remedial actions at levels far below the lower interventional levels. An example of this is the distribution of iodine tablets to all Swedish citizens in Japan during the initial phase of the Fukushima accident, many of whom resided far beyond the distance recommended by the IAEA as the UPZ. Generic fear of radiation among the public may sometimes cause harm that is not proportional to the radiological risk. It may cause effects such as stigmatization of residents having to live in areas affected by NPP fallout, adversely affecting their social and economic status (for example, people remaining in the rural zones of eastern Belarus and the Bryansk region of Russia). The radiation protection community must also be prepared to confront misinformation and even disinformation from unauthorized persons who claim to be experts in the field, but who lack formal competence in radiation protection. Such disinformation may seriously interfere with the communication of the true risks to decision makers. Radiation protection experts must therefore take all these factors into consideration and be prepared to accept that decision makers may carry out countermeasures that, in terms of radiological harm alone, seem unjustified.

7.2.2.2 First Responders at a Local Accident Site

As mentioned in Section 7.2.1.3, guidelines for first responders and radiation protection experts regarding managing local accidents are provided in the EPR report issued by the IAEA in 2006. The guidelines include instructions on how to obtain a rapid overview of the radiation environment at the site by means of initial measurements, how to organize personal dosimetry, contamination monitoring of victims and rescue workers, contamination monitoring of equipment, personal effects and vehicles at the site, triage of patients, and so on. It is recommended that *the first step* for any category of first responders is to make a number of initial observations of the site in order to rule out malevolent radiological or nuclear acts. This will help determine how to proceed with protective measures, especially for the rescue workers involved, and what kinds of expertise will be needed in the continued work, either at the site or as off-site assistance.

- Has an explosion occurred? This may indicate a bomb, meaning that the accident was caused by malevolent action.

- Have threats been made recently? Has this incident occurred in connection with an important political event.

- Do victims exhibit medical symptoms of acute radiation effects? In many transport accident scenarios it is unlikely that the radiation sources will expose individuals nearby to lethal doses, or even doses causing local tissue effects, unless the integrity of the radiation source and its shield have been severely damaged or deliberately tampered with.

- Does the gamma dose rate exceed 100 mSv h^{-1} over an extended area? This may indicate the use of a dispersion device or deliberate fragmentation of a radiation source.

- Have there been recent reports of stolen or lost radiation sources in the area?

- Are radiation sources missing in connection with a fire, leakage or explosion?

- Further check points (refer to IAEA 2006).

Once the response initiator has alerted the first responders, *the second step* is to transmit the observations made at the site to the radiation protection experts via a national emergency coordination service (such as the Swedish SOS Alarm). An emergency management group must then be established by the first responders together with local or regional authorities as described in Section 7.2.1.3. The latter appoint an emergency manager for the accident, who will manage and coordinate most of the operations at the site. The first responders to reach the site usually will be the local fire rescue service. They will become the so-called response initiator, and will issue alerts if harmful substances (CBRNE) are involved. It is also common for the emergency manager to be in close contact with decision makers. The more serious the accident, the more likely it is that the emergency manager will be stationed with the decision makers, for example, at the County Board of Administration.

In *the third step*, the cordoned-off areas of the accident site described in the previous section are established. However, the first responders often lack specialized instruments for radiation detection, and preliminary areas are cordoned off to enable a functional working organization during the initial rescue phase. The radius of the inner cordoned area is 30 m away from the suspected radiation source. This radius is modified if there is a persistent wind, such that the radius is more than 100 m from the source along the direction of the wind. The initial perimeter is set at 100 m if there is evidence of leakage or a spill from a radioactive source. The radius is further extended, 300 m, if there are fumes, fire or smoke coming from the radiation source.

If a radiation source is located inside a building, and if it is suspected that the containment is damaged, or there is evidence that leakage of the source has occurred, the first responders are instructed to evacuate the floors above and below and adjacent buildings. If there is a fire or another kind of incident that could lead to the risk of wider dispersion, then the whole building, as well as a safety buffer some distance around the building should be evacuated.

As mentioned in Section 7.2.1.3, once the dispersion of radioactive material is confirmed, the staff at the site must use respirators, or in their absence, a handkerchief, to avoid inhalation of soot and contaminated particles. Hands should be kept away from the mouth, and eating, drinking and smoking are not permitted in the hot or warm zones. A log should be kept of all the actions taken, especially the time that individual rescue workers and other staff have spent in the contaminated zone, to enable dose reconstruction if personal dosimeters are lacking. Hair and skin should be protected, eyes should be protected from β-particle exposure using suitable glasses or plastic hoods, and shoe covers should be used. Responders should be aware that some respirators are not intended for outdoor use and can become misted up, reducing the vision of the wearer.

If the gamma dose rate exceeds 100 mSv h^{-1}, no staff member should be present inside the hot zone for more than thirty minutes. Only life-saving actions are justified in the hot zone. If rescue workers must enter a zone where the gamma dose rate exceeds 1 Gy h^{-1}, they should have remote guidance by radiation protection experts.

The radiation protection specialist (radiological assessor) at the site must inform the rescue crew and premedical staff about fundamental principles of γ-ray and ionizing particle exposure, and explain how to economize (or optimize) their actions by taking the inverse square law, the time factor and the shelter factor into account, so as to avoid unnecessary radiation exposure. They should also instruct personnel on the limitations of personal dosimeters, and explain that these cannot detect internal contamination or contamination on the skin.

Rudimentary hand-held instruments are necessary for rapid determination of the iso-dose lines, for contamination measurements, and direct personal dose monitoring. Rescue workers performing tasks in the warm and hot zones must wear personal dosimeters if available. Direct-reading dose monitoring instruments with alarms that can indicate whether a given gamma dose limit has been exceeded should be used. Examples of suitable gamma monitoring instruments are given by Collison (2005) and will also be discussed in Section 7.2.2.3.

Once the incident command post (Fig. 7.32) has been set up, and radiation protection experts have arrived, there will be a number of groups present who have different roles and tasks. It is therefore recommended that all professionals follow their action lists, and coordinate with the on-scene commander if conflict regarding who should carry out which tasks should arise. It is also important that all individuals present in the warm zone wear clearly distinguishable clothing or vests over their protective clothing to improve cooperation and communication. All tasks carried out should be optimized with regard to time so as to minimize exposure to radiation. All personnel leaving the hot or warm zones must pass through the decontamination unit.

If the first responders have access to rudimentary radiometry equipment, such as a dose rate meter and a contamination probe, sensitive to α- and β-particle radiation, the inner cordon is set according to the criteria mentioned in Section 7.2.1.3: (i) a gamma dose rate exceeding 100 μSv h^{-1}, (ii) a total β/γ activity per unit area exceeding 1000 Bq cm^{-2}, or (iii) a total α-particle activity exceeding 100 Bq cm^{-2}.

7.2.2.3   Radiometry by On-Site Radiation Protection Experts

Some accident scenarios will require radiometry by radiation protection experts in a number of environments, including NPP facilities and their surroundings, hospitals (for example, at the accident and emergency (A&E) unit), local incident sites at non-nuclear facilities (where the staff is rarely trained for radiometry under accident conditions), and at road accidents involving radioactive or nuclear cargo. The radiation protection expert may have to carry out any of the following tasks depending on the kind of accident: (i) a survey of gamma dose rates to determine safety perimeters and define cordoned-off areas, (ii) location of the radiation source or areas with elevated radiation levels (hot spots) with sufficient resolution to enable planning the retrieval of the source, (iii) identification of the γ-ray, β or α emitters present at the incident site (nuclide-specific measurements), (iv) quantification of the source activity of the identified radionuclides, (v) a precise contamination survey of all staff, victims and materials being transferred between the cordoned-off areas at the incident site or the hospital unit, and (vi) survey and documentation of (log) the readings of personal dosimeters being carried by staff at the incident site or at the hospital.

When surveying sites where the radioactive sources may expose people to hazardous levels of irradiation, speed is prioritized over precision. The survey meters and survey methods required for the task are seldom optimized for nuclide-specific measurements or for precise determination of the ambient dose rate at the site. For the precise localization of a source, good time resolution in the survey meters is not only necessary, but can be life-saving. A survey meter programmed to display a new reading every ten seconds may be too slow to alert a rescue worker passing through a narrowly collimated beam of gamma radiation. The requirements of the survey instrument can be summarized as follows (see also Sections 4.1 and 4.2).

- High time resolution: In some extreme conditions, for example, when there is a highly variable level of radiation (e.g., locating orphaned Category 4 and 5 sources), it can be extremely dangerous or fatal to rely on lower-resolution survey meters based on gaseous detectors such as GM counters or ionization chambers.

- Optimum counting efficiency: Too low a counting efficiency, as with thin-walled gas-filled detectors, may mean too slow a response to rapidly changing gamma radiation gradients at the site. Solid-state detector elements, such as large NaI(Tl) scintillators, will be subject to dead time effects at elevated gamma radiation levels, which could either paralyze the instrument or lead to underestimation of the true intensity of the radiation.

- Energy resolution: A requirement for nuclide identification is that a detector can distinguish the different energies of the radiation surveyed, whether using β-particle- or γ-ray-sensitive detectors. In emergency situations, some short-cuts can be used to obtain a rough idea of the predominant γ-ray or β-particle energies, for example, by using the body of the operator to test the transmission of a detected radiation beam.

A portable detector with more than one detector material is a powerful instrument that can be used for the simultaneous detection of more than one type of radiation quality, for example, γ-rays and neutrons in a mixed irradiation field. It can also enable rapid triangulation of an irradiation beam, and at the same time provide a means of surveying surfaces in the area for α- and β-particle contamination.

Quantification of γ-ray- and β-emitting sources is possible with relatively simple hand-held instruments, but the uncertainties are large. In some cases, it may suffice to establish the source strength within an order of magnitude before proceeding with the emergency response. Further desirable features of hand-held radiometry instruments suitable for RN emergencies are described below.

- γ-ray dose rate meters: Constant detection response between 50 and 3000 keV, meaning that any event in that energy range will give a response proportional to the absorbed dose to air caused by the incident γ-ray photons. This inevitably requires low-Z detector materials, as high-Z materials have a distinct peak in photon detection probability related to the high cross section of the photoelectric effect.

- β-particle dose rate meters: Thin entrance windows to the sensitive volume of the detector. The beta dose rate must often be determined in an environment with an equally high gamma dose rate. Generally, a high gamma dose rate implies a high beta dose rate at a distance less than 1 m from a contaminated surface, or from a source. This is the reason why eye protection should be worn by staff operating in the inner cordoned areas. Note also that pure beta emitters, such as ^{90}Sr/^{90}Y, may generate bremsstrahlung in the presence of high-Z shielding material, which can be confused with gamma radiation from a γ-ray-emitting source. Swipe tests may be needed to verify the presence of surface-deposited beta emitters, since a high gamma dose rate also implies higher leakage detection of gamma photons in sandwich detectors intended for beta detection.

- Surfaces contaminated by beta or gamma-ray emitters: As with pure β-particle detection this may require detectors with a thin entrance window. Swipe tests may also be needed to detect important radionuclides such as ^{241}Am and ^{125}I.

- α contamination: Detectors with extremely thin entrance windows are needed. Many portable probes used to detect α and β surface contamination are clad with a thin protective foil which transmits alpha particles with energies above 3 MeV. Thin layers of liquid or air may be sufficient to completely quench the α-particle signal. The difficulty in detecting α-emitting radionuclides when contamination is internal is illustrated by the fact that the poisoning of Alexander Litvinenko was identified by detection of 803 keV γ-rays from ^{210}Po contamination, which only has an emission probability of 0.00001 per decay.

- Beta contamination: Detectors with thin entrance windows are needed, as in the detection of beta dose rates. For soft beta radiation (E<300 keV) swipe tests are required. This is of concern for important radionuclides such as ^{3}H and ^{14}C.

- Personal dosimetry: In order to comply with regulations, operators should wear personal dosimeters calibrated to indicate the personal dose equivalent ($H_p(10)$) received.

Gamma detectors with high counting efficiency, such as NaI(Tl) and Cs(Tl) scintillators, are preferred for rapid setting of the cordon perimeters at an incident site, and for surveying the general gamma dose rate in the area. They can rapidly detect a sharp gradient in the gamma dose rate caused by the operator entering a collimated irradiation beam. It is also beneficial, and in some cases necessary, for the detector reading to be connected to an audible signal, so that the operator can concentrate visually on the search for radiation sources, rather than having to repeatedly look at the detector display (Fig. 7.37).

FIGURE 7.37 Exemption measurements of scrap metal in a container found to contain elevated levels of gamma radiation. The operators have chosen to investigate a piece of scrap metal and survey it with hand-held gamma monitors and contamination probes for precise pinpointing of the contaminated areas.

Direct-reading dosimeters must be worn by the operator, preferably with an audible alert indicating when the operator should consider withdrawing from the survey area. Operators must also be equipped with contamination survey probes for passage clearance of personnel and material from the incident site. Finally, a nuclide-specific instrument should be available to identify γ-ray-emitting radionuclides such as ^{241}Am, ^{57}Co, ^{99m}Tc, ^{131}I, ^{133}Ba, ^{16}O/^{18}F (β^+ emitters), ^{134}Cs, ^{137}Cs, ^{226}Ra, and ^{60}Co. In some cases, the energy resolution of NaI(Tl) is insufficient, but LaBr$_3$(Ce) can be useful, despite the perturbation in the ^{40}K photo peak (see Section 4.6.5). Portable, electrically cooled HPGe systems with relative efficiencies up to 50% are now available commercially, however, they are still very much more expensive than the standard 7.54 cm (diameter) by 7.54 cm NaI(Tl) crystals. A summary of hand-held detectors suitable for use by first responders and radiation protection experts at an RN incident site is given below.

GM counters, mainly intended for γ-ray intensity surveys, are designed as thin-windowed detectors (to detect β-particle energies down to 140–200 keV) or as sandwiched GM counters (combined with other detector materials). Energy-compensated detectors provide rough estimates of gamma dose rate over the energy range 50 keV to 3 MeV (Fig. 7.38).

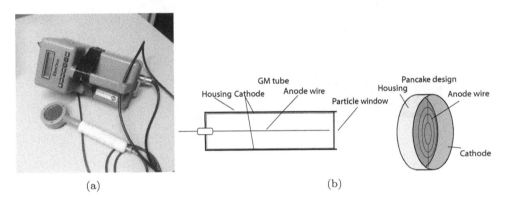

FIGURE 7.38 Photograph of an instrument equipped with a GM counter (a) and a sketch of the basic function of the GM tube (b).

Advantages:

- GM counters require no amplification, which makes them rugged and reliable in the field.

- Central anode does not need to be as thin as for ionization chambers and proportional counters.

- Dynamic range in gamma intensity. Can be further improved by combining two different-sized GM counters to acquire a dynamic range from 1 μSv h^{-1} to 10 Sv h^{-1}.

- Cheap.

Disadvantages:

- Poorest time resolution of all gas-filled detectors. Quenching gas used determines the rate at which a new electric field can be applied over the detector volume.

- Poor gamma counting efficiency, meaning poor sensitivity to changes in the gamma intensity in time and space. This is especially the case for GM counters with a thin entrance window.

- Energy compensation in gamma detection efficiency may not be ideal.

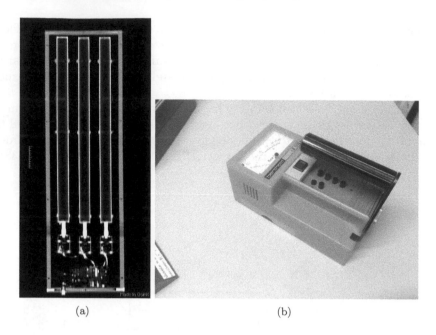

(a)   (b)

FIGURE 7.39 Proportional counter tube (a) and photograph of an instrument designed as proportional counter (b).

Proportional counters: Principal applications are contamination surveys and as neutron survey meters (Fig. 7.39).

Advantages:

- High time resolution, if appropriate quenching gas is used (down to 0.5–1 μs), makes them faster than GM counters.

- Spectrometric, and can be suitable for discrimination between radiation with different linear energy transfer (LET), e.g. γ-rays and neutrons in a mixed field.

- Can be designed for very large surfaces, suitable for contamination surveys of large areas.

Disadvantages:

- To function properly as a spectrometer or stable count rate meter, sophisticated amplification and a high-voltage supply are required. Thus, the electronics associated with a proportional counter are much more sensitive and delicate than those of a GM counter.

- Expensive.

- Quenching and filling gases can be poisonous or flammable. The exception is ^{3}He used in proportional counters for neutron rate surveys, but the global supply of this gas is limited (see Section 4.7.3).

- Gases required severely limit its portability. Proportional counters are not particularly suitable for transportable or vehicle-borne systems.

Ionization chambers, mainly used for detection of ambient dose rate (Fig. 7.40): Typical gas volumes of 100 cm^3, consisting of gas at high pressure (up to 10 atmospheres) to enhance the gamma detection efficiency.

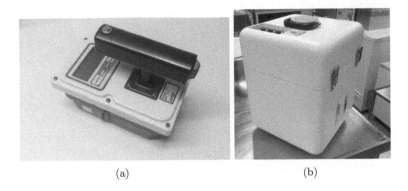

(a)   (b)

FIGURE 7.40   Radiation survey instrument equipped with an ionization chamber (a) and pressurized ionization chamber with Ar as a filling gas (b).

Advantages:

• Low gamma energy dependence for energies > 100 keV makes them ideal for accurate determination of air kerma and ambient dose rate in a high-radiation environment.

• Good time resolution compared with other gas-filled detectors, enabling accurate dose rate detection at very high intensities.

Disadvantages:

• Poor counting efficiency; cannot be used for rapid source localization.

• Relatively sensitive to ambient measuring conditions such moisture, humidity, temperature, etc.

Scintillators: Principal applications are γ dose rate and contamination surveys. γ-ray intensity survey meters may consist of e.g. NaI(Tl), LaBr$_3$(Ce) or CsI(Tl). Contamination probes can often be in a so-called sandwich configurations consisting of a thin α-particle-sensitive ZnS scintillator combined with a plastic scintillator sensitive to soft γ-rays and β-particles (Fig. 7.41).

Advantages:

• Nuclide-specific determinations possible with inorganic scintillators which are robust enough to be used in many challenging measuring conditions.

• High detection sensitivity (both inorganic and solid organic scintillators) makes them ideal for rapid source localization.

• Solid organic scintillators can be configured in various geometrical shapes and sizes, and are very useful for rapid exemption/clearance surveys of, for example, vehicles and containers.

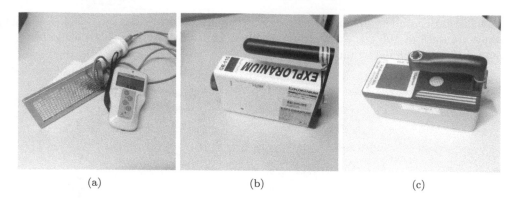

(a)   (b)   (c)

FIGURE 7.41 (a) Survey probe for α and β detection based on a sandwich configuration of a plastic scintillator and a thin ZnS crystal behind a thin protective foil. (b) Typical instrument for γ intensity survey using a 55-cm^3 NaI(Tl) crystal. The instrument has an audible mode for facilitating search without the operator relying on the visual display. (c) A combination instrument based on a NaI(Tl) crystal with both search modality and nuclide identification.

Disadvantages:

- Relatively high variability in the energy calibration due to temperature dependence of photomultiplier tubes.

- Paralyzable in high gamma environments; some common survey meters become saturated at a gamma dose rate of 20 mSv h^{-1}.

Semiconductors: Used in high-resolution γ-ray spectrometry, personal dosimetry and neutron detection (Fig. 7.42). Si diodes have also been used for many years as dosimeters for personal dosimetry. Portable, electrically cooled high-resolution HPGe detectors have been developed and are now commercially available for RN monitoring.

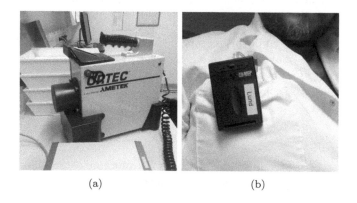

(a)   (b)

FIGURE 7.42 (a) Portable HPGe-spectrometer. (b) Si-based personal dosimeter.

Advantages:

- High-resolution nuclide-specific determinations are possible using new electrically cooled HPGe detectors. These are especially useful when monitoring and searching for shielded nuclear materials, such as weapons-grade plutonium (e.g. the 375 and 414 keV lines of ^{239}Pu).

- Compared with traditional LiF personal dosimeters, Si diodes can easily be re-set between various operations to obtain operation-specific radiation exposures.

Disadvantages:

- Higher demands are placed on battery chargers for portable HPGe devices due to the high power required for cooling.

- Electrically cooled systems are expensive (a 50% detective quantum efficiency model is five times more expensive than a corresponding portable NaI(Tl) spectrometer).

- Non-tissue equivalence of Si diodes in the lower photon energy range (<100 keV).

- Ageing effects if used in a neutron flux.

More sophisticated radiometry equipment can be provided by specialized radiometry laboratories when deemed necessary. Today, nuclide-specific instruments together with high-resolution satellite navigation (GPS) can be used to make very detailed maps of the ground deposition. With foot borne backpack systems or with collimated nuclide specific detectors on trolleys physically very small radiation sources can be located (Fig. 7.43).

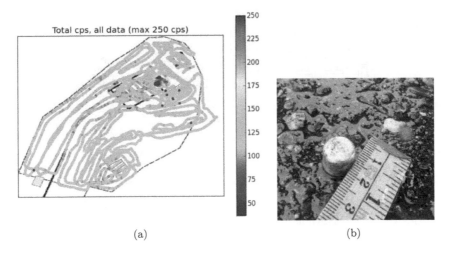

(a)   (b)

FIGURE 7.43   (a): Map of the γ dose rate intensity recorded by a HPGe spectrometer (123% relative efficiency) carried in a backpack. (b): Example of a radiation source (^{110m}Ag) found by mapping an abandoned radiological waste repository on foot. See also colour images.

7.2.2.4   Regional or National Incident Sites

Large-scale RN emergencies inherently involve many more members of the public than small-scale incidents. If a release from a NPP is imminent, sheltering may be necessary for inhabitants within an area up to 50 km away from the plant (corresponding to the UPZ). The anticipated inhalation dose from a passing radioactive plume triggers the recommendation to take shelter. In areas surrounding large nuclear sites, alarms may be sent automatically by telephone to residents in the PAZ, such as the notification system used in Sweden, or given on the radio alerting the public that they and their family must take shelter. The purpose is to avert radiation doses from inhalation of the passing plume, and also from external exposure to the gamma and beta emitters in the plume (cloud shine). The sheltering factor, S, is defined as the γ-ray exposure (ambient dose rate) of a person inside a given building, divided by the radiation exposure that person would receive if unshielded on the ground (Eqn. 7.2).

$$S = \frac{D_{sheltered}}{D_{unsheltered}} \tag{7.2}$$

The sheltering factor can vary substantially depending on the type of building (see Fig. 7.44). The recommended intervention levels in Sweden up until 2014 were 1–10 mSv averted dose over a 6-hour period. *The Nordic Flagbook* now recommends an intervention level of 100 μS h^{-1}. These OILs are also applied in the case of a plume originating from some other RN event occurring outside the predefined emergency areas around NPPs, or from a foreign NPP release. If sheltering of humans is justified, so is sheltering of grazing livestock. Inhalation of, for example, ^{131}I, ^{132}I and ^{132}Te in a passing plume by grazing dairy cows may lead to the transfer of radioiodine to milk. This transfer pathway was the key factor in the contamination of children in Belarus after the Chernobyl accident, where no such protective measures were enforced.

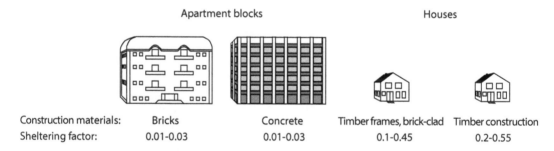

FIGURE 7.44 Sheltering factors in typical Scandinavian dwellings. Left: multi-storey apartment blocks, right: houses with brick cladding and timber cladding.

The recommendation to members of the public to shut off the ventilation in the building is straightforward for private house owners. However, modern ventilation and air conditioning systems are almost all equipped with particle filters which thus prevent a large part of the airborne contamination from leaking in, even when the ventilation is in operation. These systems are also many times automated and remotely controlled, and it has been found to be impractical to shut down these systems, as they can only be shut down by housekeepers or janitors, which would have to be done during the passing of a radioactive plume (e.g. Nordquist 2013).

The aim of sheltering livestock is mainly to avert the inhalation of airborne radionuclides that can easily be transferred to milk, such as ^{131}I and ^{137}Cs. Hence protective measures are especially critical for dairy cattle. In addition, a ban on grazing in the geographical area threatened should be implemented by the veterinary and agricultural authorities to prevent uptake from contaminated feed and water by livestock. Physical protection of local sources of water on farms is also performed in the rescue and recovery phases of the emergency. In Sweden, sheltering of cattle is coordinated by the regional veterinary council of the County Board of Administration. Exemption can be granted from sheltering and limited grazing if local radiometry measurements can verify that the deposition and airborne release are below values that would result in ^{131}I and ^{137}Cs concentration higher than 100 Bq kg^{-1} in dairy milk. Many other measures related to agriculture and food production must be planned in the initial phase of an RN accident. In spite of access to fodder, the sheltering of cattle in barns for more than a few days may not be practicable, as survival of the animals in large-scale barns depends on continuous ventilation. As earlier noted, farming is the sector that, in terms of economic losses, will probably be more affected than any other in the event of a major RN release.

Iodine prophylaxis refers to the oral administration of potassium iodide. Usually, about 130 mg (of which 100 mg is iodine) is usually administered at a time to adults, and half that to children. The trigger for iodine prophylaxis is an anticipated airborne release containing the fission products ^{131}I, ^{132}I and ^{132}Te (which decays to ^{132}I). Inhalation of radioactive iodine by an unshielded subject in the affected area will lead to the accumulation of iodine isotopes in the thyroid, leading to substantial radiation doses. The biological health effects of radioiodine exposure, especially among children and adolescents, were the most evident consequences of the Chernobyl accident (e.g. UNSCEAR 2000). Preventing uptake in the thyroid gland can be achieved by administering stable iodine that saturates the gland and prevents further uptake from the blood. The time window for the administration of iodine to ensure a protective effect is from about 48 hours prior to the onset of internal exposure to about six hours after. The protective effect decreases with time, but remains significant up to 48 hours after intake. For people older than 40, the protective effects of administrating stable iodine are less significant, since their remaining life expectancy is too short to allow development of thyroid carcinoma.

Formerly, OILs for recommending the intake of stable iodine were 10–100 mSv averted equivalent dose to the thyroid among children, and a factor of ten times higher for adults under 40 years of age. The current criterion in the Nordic countries is that potassium iodide should be administered if the anticipated equivalent dose exceeds 50 mSv for adults, and 10 mSv for children. Radiation protection experts must make it clear to decision makers, rescue workers and the general public, that iodine prophylaxis only offers protection against internal exposure by radioactive iodine, and offers no protective effect against external exposure to radioactive iodine. In many countries, iodine tablets have been distributed *en masse* to populations living within a certain radius of NPP sites in preparedness for a potential release.

Evacuation, or temporary relocation, is recommended when an airborne RN release is expected to lead to ground deposition, causing external exposure to local residents from γ- and β-radiation (ground shine). Former operational interventional levels for evacuation were set at an avertable effective dose ranging from 3–30 mSv per person per day. In the Nordic countries the trigger for this intervention is now an effective dose >20 mSv in a week to an unshielded individual. This value is applied if it is anticipated that indoor sheltering must continue for more than two days, and if no other action is taken. As mentioned previously, preparations must be made for the evacuation of people in the PAZ as early as possible after the alert, as a release may be imminent. Local rescue workers in the vicinity of the affected facility will have the main task of coordinating the evacuation of residents. It is recommended that evacuation should be achieved within four hours, but can take days in densely populated areas. As mentioned in Table 7.5, evacuation may not be justified in some individual cases, such as the elderly and seriously ill patients, who may suffer more harm from the stress of physical relocation than the averted risk of late effects of irradiation.

Relocation of residents for long periods (up to fifty years) is the most drastic of the protective measures. It is also the protective action that involves the greatest potential harm in terms of its social and economic consequences. Before 2007, relocation was considered justified when the averted dose to an average individual in the affected area exceeded 5–50 mSv in the first month. In the Nordic countries, relocation for a limited period (up to two years) is now considered justified if the dose rate in the area exceeds 10 μSv h^{-1}. Other criteria for long-term relocation are a ground deposition of a beta- or gamma-ray-emitting nuclide such as ^{137}Cs or ^{90}Sr/ ^{90}Y exceeding 1 MBq m^{-2}. The dose criterion for allowing the return of the population to the affected area, is when the monthly dose after re-habitation is lower than 10 mSv *and* is expected to decrease rapidly because of either intrinsic radioactive decay or decontamination measures, or a combination of both. The Soviet regime suggested a criterion for re-habitation in a fallout-affected area of a lifetime projected effective dose below 350 mSv. Based on IAEA recommendations, the Japanese authorities decided that a deposition level of ^{131}I exceeding 7 MBq m^{-2} was a suitable criterion for evacuation and relocation.

TABLE 7.6   Maximum permitted activity levels of various radionuclides [Bq kg^{-1}] in foodstuffs and drinking water. Data from EU (2016).

Radionuclide	Example	Baby food	Dairy produce	Liquid foodstuffs and drinking water	Other foodstuffs (spices, etc.)
α-emitting radionuclides	^{239}Pu ^{241}Am	1	20	20	80
Strong β emitters	^{90}Sr	75	125	125	750
Iodine isotopes (total)	^{131}I ^{132}I	150	500	500	2000
Miscellaneous nuclides with $T_{1/2} > 10$ d$^{1)}$	^{134}Cs ^{137}Cs	400	1000	1000	1250

$^{1)}$ Except ^{40}K, ^{14}C and ^{3}H.

Restrictions are placed on the movement and consumption of food with the aim of averting collective doses from internal contamination due to the ingestion of contaminated foodstuffs. In the European Union, which strictly regulates the trading of food between its countries, so-called maximum permitted levels are defined for a number of types of radionuclides and consumer, or age cohorts (see Table 7.6). If fallout is deposited during the growing season, direct contamination of crops may exceed the maximum permitted activity levels, requiring the entire harvest to be destroyed. However, the delayed effect of root uptake on crops during following years or seasons, may still lead to significant uptake of radionuclides, leading to further restrictions on their use for food production and fodder. In addition to contaminated soil or directly contaminated crops, freshwater and drinking water reservoirs may be contaminated, and can further contaminate crops if used for irrigation. Very stringent limitations must be placed on the maximum allowable levels of radioactive contamination in all types of foodstuff, since the transfer pathways to humans involve several foodstuff components, and will therefore be additive, and will also depend upon an individual's diet. The overall aim of food regulations is that the internal individual dose does not exceed 1 mSv during the first year after the onset of the emergency. If severe shortages of food arise, authorities may accept a corresponding dose of 10 mSv. Another important aim of this protective action is to comply with international trade regulations and to affirm that all the food products from a country that are sold on the international market have been subject to the same quality control measures as in other countries. It is probable that lower permissible limits will be introduced in many markets in the event of an RN emergency to accommodate both the concerns of the public, and to comply with the needs of the commercial food industry.

Access control is another protective measure authorities can implement for workers and members of the public residing or working in geographical areas affected by high ground depositions from a radioactive release. *The Nordic Flagbook* recommends a similar OIL for the γ dose rate, 100 μSv h^{-1}, to that recommended for cordoning the warm zone at the site of a local incident. The same protective action is recommend if the air concentration of gamma emitters is >10,000 Bq m^{-3}, or that of strong beta emitters is >1000 Bq m^{-3}, or that of alpha-only emitters, such as 239Pu, is >1 Bq m^{-3}.

During the post-rescue phases, measures should be carried out to reduce radiation doses to workers involved in restoration of the affected areas. The Nordic countries recommend that these measures are justified if γ dose rates exceed 10 μSv h^{-1}. The level for lifting protective measures for these workers is suggested to be a γ dose rate below 1 μSv h^{-1}. Remedial measures are considered to constitute a planned exposure scenario, and the same radiation protection principles should be applied as in hospitals or the nuclear industry.

People working in affected areas who are not directly engaged in alleviating the radiation emergency should be treated as members of the public. Hence, measures to reduce their exposure should be undertaken in accordance with the protection measures to limit exposures in other planned exposure situations. Such measures include decontamination of the indoor working environment or coarse decontamination of the outdoor areas in the vicinity of the workplace. In outdoor environments, wetting of the ground can be used to reduce re-suspension of contaminated dust and aerosols and to prevent internal doses due to inhalation.

7.2.2.5 Decontamination

European strategies for decontamination of large surfaces can be found in EURANOS, 2009. The deposition of fission products varies greatly with the type of surface. Dry deposition on building walls is often an order of magnitude lower than that on roofs and mown lawns. Shrubbery and trees have the highest interception of deposition, roughly a factor of three times higher than for roofs. For wet deposition, using mown lawn surfaces as a reference, the relative deposition on the roofs of buildings is about 40%, on trees and shrubbery only 10%, and on the walls of buildings only 1% (Roed et al. 1990). Contaminated areas can be divided into three main types: (i) *urban and developed* areas, (ii) *agricultural/cultivated* areas, and (iii) *natural* areas. Urban and developed areas include residential areas, industrial areas, roads, harbours and airports, and playgrounds and recreational areas. Cultivated areas are surfaces used in connection with food production, including fields and pastures. Natural areas are also referred to as virgin soil, and often have surfaces that are relatively undisturbed by human activities, such as uncultivated woodlands.

There are, in principle, three options for reducing the external dose rates in areas affected by atmospheric fallout of radionuclides: (i) remove the contamination by means of various physical methods, (ii) wait until the gamma dose rate has been reduced by radioactive decay and ecological processes that transport ground deposition deeper into the soil, and (iii) rely on so-called weathering processes in urban areas, which result in removal of the deposition: these include wind, rain, snow, and removal caused by vehicles on road surfaces.

The purpose of an active decontamination measure is to reduce the dose rate to the residents in that area to as low a level as is reasonably achievable, so it can become habitable again. The Nordic countries suggest a reference dose rate of 10 mSv y^{-1}, for the first year after contamination. This value is the sum of the external doses from indoor and outdoor exposures. The choice of decontamination method therefore depends on the type of area involved, the type of indoor environment, and their respective levels of contamination. Decontamination can take only months in some areas, but many decades in others. The decontamination strategies suggested by the authorities must therefore involve many of the affected stakeholders. Furthermore, it is important that the members of the public affected have a realistic notion of the goals achievable with decontamination (Fig. 7.45).

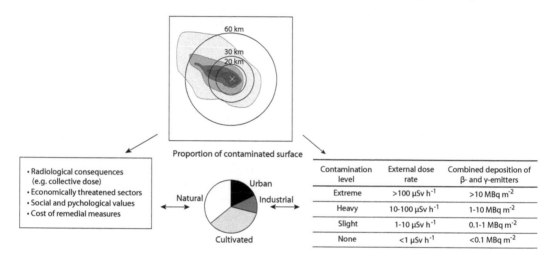

FIGURE 7.45 Illustration of how decontamination measures are selected depending on the type of surface affected, the level of contamination, and the social and economic values affected.

The first step in developing a strategy for decontamination is to define the contamination levels in the affected zones, based on airborne or other mobile deposition mapping. More detailed mapping, with high spatial resolution, is required for densely built urban areas on the boundary between non-contaminated and slightly contaminated zones, where reasonable decontamination measures may be sufficient to reduce the external dose rate below the reference level 1 mSv y^{-1}.

The second step is to thoroughly map the spatial distribution of β-particle and γ-ray emitters in the surface deposition, determine which radionuclides are involved (and their various physical half-lives, $T_{1/2}$), and measure their activity concentrations. Having obtained this data, it is possible to make long-term projections of the exposure levels if no remedial action is taken, and to define the present and future contamination levels. Figure 7.46 provides an illustration of a map of ^{137}Cs contamination in an urban area of Sweden.

In many countries, the government is responsible for decontamination measures based on: (i) the radiological consequences of the release (measured in terms of, for example, the collective dose to affected populations if no remedial action is taken), (ii) the economic and societal value of the threatened area (real estate, industry, cultivated areas, buildings), (iii) the cost of carrying out decontamination, and (iv) other circumstances involving the affected stakeholders in the region. The cost of decontamination is very high. It is estimated that Japan will devote a sum roughly corresponding to about 800 billion euros to decontamination of areas affected by the Fukushima accident. If a similar nuclear accident occurred in the European Union, the effects would involve several countries, and decontamination measures would require unified action at the EU level. These enormous costs, including those for the disposal of waste, together with past experience of the effectiveness of decontamination of urban and cultivated areas (Fogh et al. 1999), mean that justification for this kind of action must be investigated in more detail for areas where the external dose level exceeds the reference dose by a factor or two or more.

FIGURE 7.46 Dose rate contributions from surface contamination of ^{137}Cs in an urban environment in the city of Gävle, Sweden. See also colour image.

The third step in the process of deciding on an appropriate a decontamination strategy is to prioritize the geographical areas and surfaces to be decontaminated. This necessarily involves many social, economic and demographic factors, but it is also important to consider the efficiency of various decontamination methods, which in turn affect the potential reduction in the beta and gamma dose rates and the associated annual exposure of individuals returning to the area. The efficiency of decontamination measures varies greatly, depending on the method chosen and the timing after initial deposition (Ramzaev et al. 2013).

A number of decontamination procedures can be carried out, depending on the type of surface to be remediated. Sweeping dust off streets, car parks and other hard surfaces, or flushing the streets with water, or a combination of both, can remove most of the dry deposited radionuclides. Wet deposition and weathered dry deposition will accumulate in road gutters in urban areas, and gutter rinsing and cleaning can substantially reduce the local dose rate on a street or road. Decontamination of buildings can be performed by wetting the surface (from bottom to top) and then washing with high-pressure water (Fig. 7.47). Sweeping up dust and flushing surfaces in close proximity to the walls of buildings can also reduce the external dose rate to residents in the building. In rural residential areas, gardens can be decontaminated by deep ploughing or by swapping the upper and lower soil layers, which significantly reduces the amount of waste.

Cultivated areas affected by moderate or low levels of contamination can be remediated

(a)   (b)

FIGURE 7.47 Decontamination of a building using wet padding of external contamination on the roof (a), and subsequent high-pressure washing to remove the suspended radionuclides (b). Note that the radionuclides washed off from the roof must be collected on the ground and treated as radioactive waste.

by removal of the topmost 2–3 cm of soil, or by deep ploughing. These tasks can be performed by many of the vehicles used in farming. Scraping also reduces the root uptake of the crops, and consequently averts collective internal doses. Other methods include the use of inversion tilling of the soil to reduce the external dose rate; the suspension of surface soil using water and subsequent removal of the wetted soil; and the addition of a solidifying agent that hardens the topsoil, which is then removed. Many of these methods generate enormous quantities of waste that must be handled as radioactive waste. Scraping a 100-ha field to remove the upper 2 cm of soil will produce 20 000 m³ of waste. In addition to cultivated areas, remedial measures may also be applied to forests used for timber production, and watercourses used for recreational or professional fishing.

7.2.3   Remedial Strategies in RN Emergencies

International guidelines on strategies for remedial measures following RN accidents were developed in the aftermath of the Chernobyl accident in 1986. However, following the Fukushima accident in 2011, the IAEA acknowledged that some vital areas still needed to be improved. Some of the points addressed are given below.

- Clear allocation of the functions and responsibilities of the various stakeholders involved; in other words, the role of the various organizations must be better defined, on both regional and national levels.

- Development and maintenance of national emergency plans, especially for parties who contribute at a distance from the immediate area of the incident.

- Ensuring first responders are better prepared and trained for handling an RN emergency.

- Increasing awareness among those implementing national border controls (customs) and among scrap metal dealers of the radiation issues inherent in a supply of metal whose origin is unknown.

- Improving the capability to monitor the contamination of first responders and victims: this applies especially to the monitoring of internal contamination.

- Improving decision making by establishing generic criteria for when an intervention should be carried out.

- The establishment and development of national strategies to communicate the radiological situation, and the associated protective measures required.

- Increased consideration of the non-radiological consequences of an RN emergency.

The lessons learned after each major RN event contribute to an increased understanding of how an RN emergency should be managed. However, if the core capability and capacity within a society to meet RN emergencies and related threats is to be maintained, it is necessary that training and practical exercises involve not only experts or potential radiological assessors, but also first responders and decision makers. Further information on the lessons learned from RN emergencies can be found at the IAEA web portal (IAEA 2012).

7.3 MEASUREMENT METHODS FOR EMERGENCY PREPAREDNESS

7.3.1 Radiometry Methods for National Emergency Preparedness

Methods for detecting elevated radiation in connection with an RN incident must be designed and evaluated by means of training and intercomparisons before an incident occurs. These methods must, many times, be employed under normal operative conditions to survey the background radiation levels and to establish the effects of earlier incidents in order to detect any signs of a new incident.

Establishing preparedness for radiological and nuclear emergencies on a national level requires a number of resources in terms of radiation protection expertise, radiometric equipment, modelling and measurement methods, as well as meteorological and environmental data. The meteorological institute must be able to provide real-time simulations of the dispersion of atmospheric releases from nuclear power plants, both domestically and in neighbouring countries. Laboratories must be capable of retrospective physical dosimetry using techniques such as thermoluminescence (TL), optical stimulation of luminescent detectors sensitive to ionizing radiation (OSL), and electron paramagnetic resonance (EPR) in biological tissue (see Sections 4.2 and 7.3.2). Hospitals and medical centres must have staff with experience in radiation, who are trained to manage patients suspected to have been exposed to radiation (see Section 7.4).

Laboratories capable of biological dosimetry, such as chromosome aberration counting, are especially important in cases were individuals are thought to have been exposed to radiation at such levels that tissue effects and acute radiation syndrome are anticipated. Stationary units for *in vivo* determinations of the body burden of gamma-ray-emitting radionuclides (whole-body counters) will be needed for mass triage. Mobile radiometry units are needed for (i) in situ gamma-ray spectrometry, (ii) mapping of elevated radiation levels in the environment with backup support, and (iii) airborne surveying of large geographical areas in connection with extensive radioactive fallout (often provided by national geological institutes). Equally important are the availability of radiometry laboratories for alpha and gamma-ray spectrometry of actinides, fission products and neutron activation products, as well as total-beta-counting facilities for beta-only emitters such as ^{3}H, ^{14}C, ^{32}P and ^{90}Sr/^{90}Y.

Computational tools are also of key importance for modelling and prognosis of extensive radioactive plumes based on both domestic and international radiometry and meteorological data. Facilities for stockpiling radioactive waste originating, for example, from localized orphan sources or waste from extensive decontamination of urban or rural surfaces, are often neglected in the initial phase. Forensic resources are also needed to detect traces of crime at an incident site, which could have an impact on the assessment at the incident site. As an example, some specimens may not be eligible for radiometry assay by the radiation protection specialist until analysed by the forensic team. Radiometry results made by the latter team may also be security classified and hence not easily accessed by the radiation protection community. A number of the resources mentioned above are described in more detail in the following sections.

7.3.2 Calculation Tools

It is essential that radiation protection experts have access to generic data and standard dose conversion factors in order to provide guidance and support for decision makers, especially in situations where real data are relatively scarce. It is also useful if guidelines can be presented in a national handbook for radiological and nuclear emergencies, as many different reference levels and OILs have been established by the national authorities of a particular country. These books should contain links to maps and relevant demographic data that are specific to individual regions and counties, especially regarding nuclear and radiological facilities in Categories 4 and 5 (see Section 7.2); note, this should not be confused with the categorization of the hazardousness of radiation sources defined in Section 7.1) to ensure more coherent recommendations from all the experts in a country.

An example of this is a dose handbook, in the form of a compilation of several fact sheets, that has been published on-line by the Swedish Radiation Safety Authority (Fig. 7.48). Text-based guidance and look-up tables accrued over many years were collected into a handbook of approximately 200 pages if printed out in advance. The operator can choose to relying on digital format stored in a field computer or a mobile phone, or print out these pages as a hard copy. Links are also given to international guides relevant for RN emergencies. The dose handbook contains dose conversion factors to be used by national authorities when estimating individual and collective doses from internal and external radiation exposures (McColl 2002; ICRP 1996a; ICRP 1996b; Eckerman and Leggett 1996).

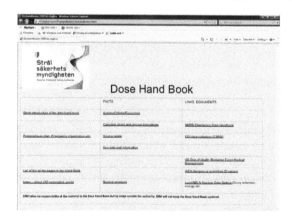

FIGURE 7.48 The first page of the web-based Swedish dose handbook with links to information sheets and look-up tables.

In addition to a compilation of fact sheets, it is beneficial for national radiation protection experts to have access to a range of dispersion models for standard radionuclide releases from the nuclear sites in the country. These dispersion models have various levels of complexity. Some models exhibit higher accuracy for certain weather conditions or for a particular distance from the release point than others. Simple dispersion models are more reliable and are suitable for use by local and regional radiation protection experts. A very simple model used to predict the effects of various releases can be very helpful in providing decision support for emergency management. Advanced dispersion models require more specific knowledge, also by radiation protection specialists, as these models require more details in the meteorological data input. As a last resort, manual calculations of a primitive release plume from a NPP can be made using a simple slice-of-cake model (Fig. 7.49).

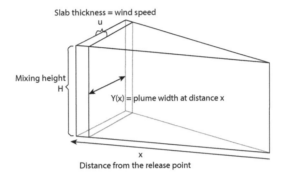

FIGURE 7.49 Cake model of the dispersion of a radioactive airborne release with mixing height H [m], and wind speed u [m s^{-1}].

Let us assume a release at ground level, an air mixing layer of height, H [m], corresponding to the plume height, a release rate of A [Bq s^{-1}], and a wind speed of u [m s^{-1}]. In one second, the radioactivity plume will move a distance $u \times 1$ s= u [m] in the direction of the wind. After a time T [s], the activity of the radionuclide at a distance $x = u \times T$ from the release point, will be distributed in an air volume of $u \times H \times Y$ [m^3], where Y [m] is the width of the lateral mixing layer at distance x from the source (the slab indicated in Fig. 7.49). The activity concentration within the slab at this distance will then be $C = A/(u \times H \times Y)$ [Bq m^3]. The parameter Y is associated with the diffusion of the radionuclide in air, and is heavily dependent on the weather. The ratio Y/x is 0.5 for typically unstable weather conditions and decreases to 0.125 in more stable weather conditions (see Pasquill categories in Table 7.8).

A simple plume model used in a calculation tool can provide estimates of (i) the air concentration at a given position, (ii) the estimated ground deposition using default assumptions on dry and wet deposition, (iii) cloud shine, (iv) the inhalation dose to an unshielded adult using default inhalation rates (which provides the basis for recommending sheltering), (v) ground shine to an unshielded adult (which provides a similar basis for decisions on evacuation or relocation), and finally, (vi) the thyroid equivalent dose, and thus the recommendation of the intake of stable iodine. A simple plume should be illustrated graphically as a superposition on a map. Demographic data on the various villages, towns and cities in the affected area are also useful. A module that includes the effect of precipitation, in terms of the aforementioned dose components, should also be available.

The advantages of a simple model are its robustness, speed and simplicity. Radiation protection experts do not need regular training in the use of such a tool. It is easy to make rapid comparisons between different release scenarios and between the averted dose associated with various remedial actions. Under stable weather conditions, a basic dispersion model can give relatively accurate plume concentrations over a distance of 10 km from the release site. The limitations of a simple dispersion model are that it does not take into account the complexity of the wind and air turbulence, and their effects on the dispersion of the plume. If the atmospheric conditions are unstable, the wind direction may change during the time under consideration, and a simple model will not take this into account. Hence, large errors can be expected in the predicted air concentrations of the plume as little as a few kilometres away from the site.

An example of a simple dispersion model is a straight line trajectory, with a Gaussian distribution in the lateral and vertical directions. The diffusion term, D [s m^{-3}], as a function of spatial location (x, y, z) can be described as in Eqn. 7.3:

$$D\left(x, y, z\right) = \left(2\sigma \cdot \sigma_z \sigma_y \cdot u\right)^{-1} \cdot e^{\left(-0.5 \cdot \left(\frac{y}{\sigma_y}\right)^2\right)} \cdot F, \qquad (7.3)$$

where F is the vertical diffusion parameter given by

$$F = e^{\left(-0.5((h - z/s \cdot s_z)^2\right)} \qquad (7.4)$$

and u [m s^{-1}] is the constant wind speed, σ_z is the standard deviation of the distribution in the vertical direction and σ_y is the corresponding parameter for lateral dispersion. These last two parameters depend on the distance from the release point and the type of weather conditions (see below).

The dispersion parameters in the diffusion coefficient are constant in time. The time-integrated air concentration, C [Bq s m^{-3}] at location $R = (x, y, z)$, with a release point at the origin, is given by the product

$$C\left(x, y, z\right) = Q\left(x, y, z\right) \cdot D\left(x, y, z\right), \qquad (7.5)$$

where Q is the total activity released for a particular radionuclide [Bq]. The plume is thus assumed to move with a constant linear motion. Turbulence and changes in wind direction cannot be accounted for in this simple model.

A more realistic algorithm can be obtained by replacing the dispersion algorithms with a module that calculates the trajectories of single particles. Such models require detailed meteorological data in order to predict the air concentration at a given location. This also places higher demands on the skill and experience of the radiation protection expert using the model. Figure 7.50 shows a comparison between a simple rectilinear plume model and a trajectory model.

Highly complex release prognosis tools also include dose calculations and cost–benefit analysis of remedial actions. These decision-support systems must be managed on a national level, so-called off-site management, involving several stakeholders and experts in fields other than radiation protection. An example of such a decision-support system is the Real-time On-line Decision Support software RODOS (KIT 2016), developed by the Karlsruhe Institute of Technology in cooperation with several European Union countries.

A database and software for visualizing reported radiometry data are essential for both radiation protection experts performing measurements on-site and for off-site emergency management, where an overview is needed of the radiological situation. The database should be able to handle various types of radiological data, such as ambient dose, dose rates, nuclide-specific deposition, and nuclide-specific activity concentrations in various foodstuffs. The database system should include a geographical information system (GIS), containing various mapping layers, down to a local scale of a few hundred metres, so that the radiological

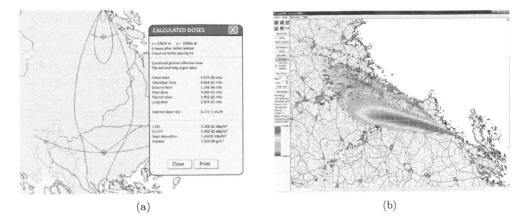

(a)

(b)

FIGURE 7.50 Isodose maps of NPP release plumes using a simple rectilinear transport model (a), and a particle trajectory–based model incorporating more authentic metrological (b). Figures are obtained from test runs conducted by the Swedish Radiation Safety Authority. See also colour images.

environment can be analysed with high spatial resolution. Data gathering should be possible using various web-based interfaces (an example of such is given in Fig. 7.51), allowing simultaneous inputs by a number of actors in the affected country. The various actors must also be able to access the data and maps simultaneously, and each item of data must be tagged with an identity so that it can be traced back to the actors that added it to the database.

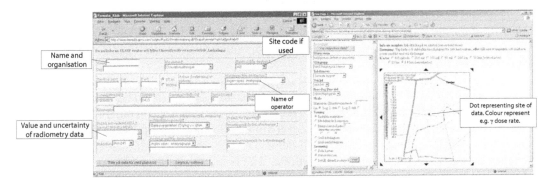

FIGURE 7.51 Left: Web-based template for the insertion of radiological data, including comments and tags identifying the person or team inserting the data and the name(s) of those who made the experimental observation (if different). Right: Visualization of the predicted ambient dose rate at a specific geographical location given by the database and GIS system. See also colour images.

At a very early stage of an RN emergency when there is little or no information on the imminent release of radionuclides, it is useful to make coarse predictions of the possible consequences using standard NPP releases. Table 7.4 in Section 7.2.1.4 listed the various types of radionuclides in a typical light-water reactor inventory, together with their volatilities in connection with an atmospheric release. There are four categories of volatility defined in Table 7.4, ranging from the extremely volatile noble gases to the inert elements in the fuel which are unlikely to be dispersed over long distances in case of a release. The standard releases mentioned in Section 7.2 can vary by several orders of magnitude in terms of the fraction of reactor inventory released.

Table 7.7 gives examples of 7 different release levels from a nuclear reactor that can be used as the source term in a plume prognosis tool. In Sweden, for example, the releases through an external pressure-relief filter from a 3000 MW$_{thermal}$ light-water reactor of 100% of the inventory of the noble gases, 0.1% of the volatile iodine, 0.1% of the refractory radionuclides ^{134}Cs and ^{137}Cs, and 0.01% of the more inert radionuclides such as ^{238}Pu, are used as construction specifications governing permission to operate the plant.

TABLE 7.7   Various release levels, in terms of the fraction of the reactor inventory of various classes of radionuclides, that can be used as source terms in plume release calculation tools.

Type of radionuclide	Release level						
	1	1/10	$1/10^2$	$1/10^3$	$1/10^4$	$1/10^5$	$1/10^6$
Noble gases	1	1	1	1	1/10	$1/10^2$	$1/10^3$
Volatile (iodine)	1	1/10	$1/10^2$	$1/10^3$	$1/10^4$	$1/10^5$	$1/10^6$
Volatile (caesium)	1	1/10	$1/10^2$	$1/10^3$	$1/10^4$	$1/10^5$	$1/10^6$
Inert	1/10	$1/10^2$	$1/10^3$	$1/10^4$	$1/10^5$	$1/10^6$	$1/10^7$
Highly inert	$1/10^2$	$1/10^3$	$1/10^4$	$1/10^5$	$1/10^6$	$1/10^7$	$1/10^8$

Standard dry deposition rates from the plume to the ground can be set at 0.01 [m s^{-1}] for particulate iodine, 0.001 [m s^{-1}] for other radionuclides, 0.0001 [m s^{-1}] for gaseous iodine (such as methyl iodide CH$_3$I), and 0 for noble gases. Similar standard values for wet deposition rates, assuming a precipitation rate of 3 [mm h^{-1}] can be set to $2.4 \cdot 10^{-4}$ [s^{-1}] for particulate iodine and $2.4 \cdot 10^{-6}$ for gaseous iodine.

The sheltering factor (Eqn. 7.2 in Section 7.2.2.4) for residents in the area affected by the plume is set to 0.4–1, as recommended, for example, by UNSCEAR. However, as illustrated in Figure 7.44, the sheltering factor may vary substantially between different types of buildings.

Standard settings have also been developed for weather conditions. The *Pasquill class* (see Section 3.2.2) of a given atmospheric flow pattern describes the stability of the air. A plume released from a stack during Pasquill class A conditions (turbulent air movements) tends to become vertically shaped, whereas a plume formed during Pasquill class F (stable weather conditions) tends to have a well-defined ceiling, above which little particle exchange occurs between the plume and the air mass. The most common weather type in Northern Europe is Pasquill class D, with a typical wind speed of 6 m s^{-1}, an inversion layer at 800 m, and no precipitation. These values can be used if no other meteorological input data are available. The Pasquill classes can be combined with varying ranges of wind speed, inversion layer heights and precipitation rates (Table 7.8).

Plume prognosis tools must be able to estimate the various dose components, for example, from cloud shine and ground shine, using these default values for plume progression, precipitation and sheltering factors. However, as more detailed meteorological data become available, and as the measuring teams deployed, for example, by the national radiation protection authority start to provide real-time radiometry data on air concentrations, nuclide-specific ground deposition, and ambient dose rates at various locations, the prognosis must be updated (Fig. 7.52). Regular updates of the estimated radiological consequences must then be communicated by the expert groups working for the national radiation protection authority to the decision makers on the national, regional and local levels.

TABLE 7.8 Typical ranges of wind speed, inversion layer height and precipitation rate for the Pasquill atmospheric stability classes.

Stability	Wind speed (m s^{-1})	Inversion layer height (m)	Precipitation rate (mm h^{-1})
A	2	500–1600	0
B	2–4	300–1400	0–0.1
C	2–6	150–1400	0–1
D	3–15	150–1200	0–5
E	2–6	150-200	0–0.5
F	2–4	100-150	0–0.2

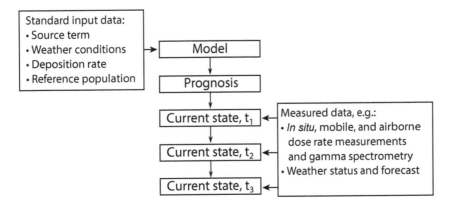

FIGURE 7.52 Schematic flow chart of the development of an increasingly comprehensive understanding of the radiation situation during an RN incident. A prediction is made based on default values, which is then refined and corrected as actual data become available.

7.3.3 Retrospective Dosimetry

Prospective dosimetry refers to determining the absorbed dose, D [Gy] (Eqn. 2.23), or quantities that can be related to the effective dose, E [mSv] (Eqn. 2.51), by issuing dosimeters to individuals who are expected to be exposed to radiation in advance of planned exposures. These individuals are often either industrial or medical radiation workers, or patients undergoing therapy or diagnostics using ionizing radiation. Retrospective dosimetry, on the other hand, is used in emergency situations and at sites with known radiation hazards to determine or verify the exposure received by people who were not originally intended to be subjected to irradiation. Retrospective dosimetry is therefore a central factor in general radiation protection. This is especially true in emergency preparedness, because retrospective dosimetry can be used in low-dose (<1000 mGy) applications, examples of which are given below.

- Reconstruction of the effective dose for assessment of the risk of exposed individuals developing late radiation-induced effects.

- Epidemiological studies of populations subject to radiation exposure in scenarios such as the single exposure caused by the Hiroshima atomic bomb (IAEA 2005b), or the protracted exposure caused by the Mayak accident (Degteva et al. 1994).

- Individual histories of radiation exposure among members of the public, such as iodine contamination of the thyroid in US citizens caused by nuclear weapons testing in Nevada in the 1950s (DHH 2016).

- Previous occupational exposure, such as those of workers in uranium mines (see, for example, the NIOHS studies in the 1990s (Roscoe 1995)).

Dose reconstruction in single patients subjected to high-dose exposures involving tissue effects (also termed *acute*, or *deterministic, radiation effects*) is based on the diagnosis of symptoms and classification of the observed tissue effects, rather than on reconstruction of the absorbed dose to tissue and materials close to the affected person. The prognosis for a person exposed to a high dose can be largely based on the time until development of observable effects (such as lymphocyte counts in peripheral blood, *TMT Handbook*, 2012).

The precision in the reconstructed absorbed or effective dose must be appropriate to the purpose of the investigation. For example, if the aim is to obtain values that are to be used in a legal process, for example, a claim on behalf of radiation exposure victims, high individual precision is necessary. In epidemiological studies of the late effects of radiation exposures, the low probability of a single individual being affected by radiation-induced cancer inevitably means that large cohorts must be studied. This in turn means that it is more important to obtain accurate average dose estimates for the cohort, and then monitor the observed cancer incidence in the cohort. For acute effects that are life-threatening, which hitherto have mostly involved only a few cases per scenario (Table 7.1), it is just as important to thoroughly document the pattern of observed symptoms as to obtain estimates of the actual absorbed dose in various tissues of the affected individuals.

The tool used in retrospective dosimetry is a physically observable marker that can be related to the absorbed dose at a given point in space or in one of the tissues of an exposed person (Eqn. 7.6):

$$D_{med} = f(R, P) \tag{7.6}$$

where D_{med} [mGy] is the absorbed dose in a detector medium, and R [au] is the signal from a dosimeter or another system that is related to the absorbed dose. The quantity P [au] denotes the various parameters affecting the readable signal, R, such as the background signal, instrumental sensitivity and signal fading. The readable signal, R, must be related to the absorbed dose in a well-known way. The effect must also be stable and predictable over time.

The absorbed dose to the tissue, D_{tissue}, in an affected organ of the exposed person can then be expressed as (Eqn. 7.7)

$$D_{tissue} = f_{med} \cdot D_{med} = f_{tissue,air} \cdot f_{air,med} \cdot D_{med}, \tag{7.7}$$

where D_{tissue} is related to the absorbed dose in the detector medium by a conversion factor f_{med}. It is common to convert the absorbed dose observed in a detector medium to the corresponding absorbed dose in air, D_{air}, and even more common to the kerma in air, K_{air} (Eqn. 2.24). In the next step, the absorbed dose to various organs and tissues can be derived using specific conversion factors, $f_{tissue,air}$. These specific conversion factors can also be adapted to convert the value into either an equivalent dose or an effective dose.

There are several methods of measuring markers in terms of a detectable signal, categorized according to how the signal is obtained: (i) *physical dosimetry*, (ii) *biological parameters* or (iii) *computational techniques*. Retrospective dosimetry methods can also be categorized according to whether the aim is to determine the dose to a specific individual, or to obtain generic values in a given radiation environment. Figure 7.53 gives an overview of the various retrospective methods proposed by the ICRU for use in radiation protection and emergency preparedness (ICRU 1998).

	Individual D_{tissue}	Generic environmental D_{tissue}
Physical methods	• Electron paramagnetic resonance (EPR) • *In vivo* body burden measurements of radionuclide concentration	• Luminescent techniques (TLD, OSL) • Mass spectrometry
Biological methods	• Chromosome aberrations lymphocytes • Mutation frequency in somatic cells • Micronuclei	• Mapping of radionuclide ground deposition for dose modelling • Modelling

FIGURE 7.53 Overview of the various techniques used and applications in retrospective dosimetry in RN emergencies.

7.3.4 Retrospective Dosimetry: Samples Taken from Tissues

7.3.4.1 Electron Paramagnetic Resonance

Electron paramagnetic resonance (EPR) and EPR spectrometry provide a means of detecting and quantifying free radicals created by ionizing radiation in human tissue and other substances (see Section 2.3.1.3). These free radicals are markers of the absorbed dose in the exposed material. The intensity of the EPR signal is proportional to the absorbed dose over a wide dose range (0.1 to 1000 Gy) and in contrast to other physical retrospective techniques, the EPR signal is not lost during read-out, i.e. an EPR dosimeter can be used for the determination of the integrating dose, and the sample can be re-analysed. EPR analysis can therefore be considered to be *non-destructive* in the sense that the material analysed is not changed in any way when obtaining a signal related to the radiation dose.

In EPR spectrometry, the material containing the free radicals is placed in a magnetic field, which typically ranges from 0.25–0.5 T (Fig. 7.54). A variable microwave signal is transmitted through the sample and when the microwave energy (proportional to the frequency of the microwaves) is the same as the energy difference between the spin states of the electrons in the magnetic field, a resonance signal is emitted from the unpaired electrons of the free radicals in the sample (a so-called signal echo). The frequencies of the signal response correlate to various distinct spin states (so-called spin resonances) of the free radicals, and the amplitude of these resonance peaks is proportional to the population of free radicals, which in turn is proportional to the absorbed dose to the material. This technique is similar to nuclear magnetic spectrometry, but in EPR it is the change in the magnetic spin moment of electrons that is probed, instead of the nuclear spin.

Several solid crystalline materials provide an EPR response following exposure to ionizing radiation, for example, amino acids, carbohydrates, and natural materials such as keratin and tooth enamel. The radiation-induced radicals in these materials are long-lived (and are sometimes extremely stable). Of these materials, the hydroxyapatite in tooth enamel (Fig. 7.55) has been found to be a good material for retrospective dose determinations using EPR. When irradiated, the number of free radicals induced is proportional to the absorbed dose

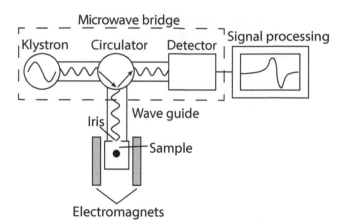

FIGURE 7.54 Principle of electron paramagnetic resonance. The sample is inserted into a magnetic field and irradiated with microwaves generated by the klystron.

over a very wide dose interval (approximately 1–1000 Gy). Efficient deconvolution of the signal allows accurate separation of the radiation-induced signal from the partly overlapping organic one. The high-sensitivity microwave bridges (a term that includes the transmission lines of microwaves from the microwave source to the sample; Fig. 7.54), now used in EPR spectrometers, have made it possible to use only small amounts of tooth enamel for accurate analysis. When using higher microwave frequencies, for example, 37 MHz Q-band spectrometers, only 2–10 mg of enamel is needed. The enamel is relatively easily obtained from a molar.

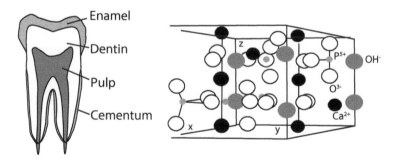

FIGURE 7.55 Cross section of a tooth, and structure of the EPR-sensitive crystal hydroxyapatite, $Ca_{10}(PO_4)_6(OH)_2$.

This technique has several advantages over other retrospective techniques: (i) a linear dose–response relation between the EPR signal and the absorbed dose for external exposures from 100 mGy up to many hundreds of Gy, (ii) stability of the markers (free radicals) ranging from decades up to centuries after the initial exposure, (iii) possibility of correlating the exposure of tooth enamel to radiation from the important fission products $^{90}Sr/^{90}Y$ and the corresponding whole-body exposure to these fission products. This has proven to be very difficult using other methods.

The drawbacks of the EPR technique are that although it is considered to be non-destructive in terms of signal read-out, the most sensitive material hitherto found in man, tooth enamel, must be physically removed from a living person for analysis. Furthermore, background signals emanating from exposure to both background radiation and UV light (for example, during dental examinations), increase the uncertainty in the assessed dose.

So-called *additive doses* are often used to extract the dose, D_{med}, from the original signal. This means that after the original signal has been recorded, the sample is irradiated with known absorbed doses and the measurement is then repeated (Fig. 7.56). This gives a more accurate dose, provided the time pattern of the gradual concentration fading of the induced radicals (in terms of the half-life of the radicals) between the exposure event and first measurement is well known. This can be compared with so-called regenerative dose assessments or protocols used with luminescent crystals (see Section 7.3.5.1).

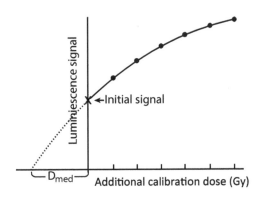 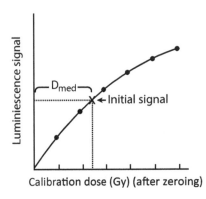

FIGURE 7.56 Left: Additive dose protocol, in which the initial luminescence, L_0, is read out, then the sample is further irradiated with known absorbed doses and the signal is accrued in a linear fashion. Right: A regenerative dose protocol in which the signal emanating from the markers in the sample is zeroed (e.g., by heating or exposure to light), and the sample is then irradiated with a known absorbed dose.

When the absorbed dose in tooth enamel, D_{med}, is used to obtain the corresponding dose to the radiation-sensitive organs in an individual, D_{tissue}, additional obstacles that contribute to the uncertainty in the analysis must be overcome. One of these is that, compared to soft tissue, hydroxyapatite has a sharp increase in its cross section for the photoelectric effect below 100 keV, which makes accurate dose predictions more difficult. Another problem is that the teeth are located close to air cavities and sharp gradients in tissue density. When this part of the body is exposed to external radiation, boundary effects are created that give rise to an inhomogeneous distribution of the absorbed dose, which in turn will make it difficult to accurately relate the absorbed dose in teeth to that in the other tissues of the body.

As a result of the problems associated with the analysis of tooth enamel, considerable efforts have recently been made to investigate the EPR response of other materials that may be found close to exposed individuals. Examples of such materials are the glass on smartphone touch screens, household sugar, and sucrose, for example, in chewing gum. The radical structure of sucrose has been thoroughly investigated using high-resolution EPR techniques, and has been used for the determination of an accidental exposure in Norway (Sagstuen et al. 1983).

7.3.4.2 Cytogenetic Dosimetry: Chromosome Aberrations in Lymphocytes and Other Somatic Cells

Four criteria are often mentioned when describing an appropriate biodosimeter, i.e., a tissue in the body that exhibits detectable changes that are traceable to a specific radiation exposure. The first criterion is that the signal (denoted R in Eqn. 7.6) must be detectable with any kind of irradiation geometry, and also for partial-body exposures. The second is that the signal is stable and remains constant over time after a given exposure. It is also preferable (third criterion) that the signal is caused only by radiation, and not by other stressors, and that it is possible to calibrate the signal response *in vitro*. This last criterion is comparable with the additive doses used in EPR to indirectly calibrate and relate the original signal to an absorbed dose, D_{med}.

In retrospective biological dosimetry the prime marker for exposure to ionizing radiation is aberrations in the chromosomes of lymphocytes. Exposure of lymphocytes in the peripheral blood, and in blood-forming organs such as bone marrow, results in radiation-induced aberrations in the chromosome structure (see Section 2.3.1) that can be visualized by various techniques. Lymphocytes are suitable for biodosimetry because they circulate in the human body, and the aberrations thus essentially represent a whole-body irradiation exposure (see Fig. 7.57). Lymphocytes in blood samples taken from an individual suspected of being exposed to ionizing radiation can be made to proliferate by *in vitro* cultivation over approximately 48 hours at body temperature (37 °C). By adding a substance called colchicine, lymphocytes can be arrested in the phase of the cell cycle in which the chromosomes are highly condensed (mitosis). Chromosomes on microscopic slides can be visualized and studied by various techniques, such as visible light or fluorescence microscopy. The following effects can then be studied.

- Dicentric chromosomes (so-called dicentric assay) hitherto considered a gold standard in biological dosimetry and a technique used since the 1960s.

- Chromosomal translocations visualized by fluorescent in situ hybridization (FISH) a method of staining individual chromosomes different colours.

- Micronuclei are chromosomal fragments that can be visualized in interphase cells (i.e., cells not undergoing mitosis).

- DNA repair foci are protein foci that form in cell nuclei around sites of DNA damage, which can be visualized by fluorescence microscopy.

- Gene expression in specific radiation-responsive genes, in lymphocytes, or in genes coding for glycophorin A (GPA) in erythrocytes.

Dicentric and translocation assays of lymphocytes have been used most often in retrospective dosimetry as they are highly sensitive to radiation (see Table 7.9). The detection limit and background signal (in terms of the frequency of non-radiation-induced aberrations) may vary due to (i) biological variation between individuals, (ii) the geographical location of the exposed individuals, (iii) atmospheric pollution giving rise to similar aberrations, (iv) exposure of individuals to other sources of aberrations, including stressors such as smoking and chemicals, and (v) exposure to pharmaceuticals. A smaller number of cells must be counted in dicentric assays (between 500 and 1000 for reliable statistics) than with the FISH technique, to predict an absorbed dose with the same precision. Theoretically, the frequency distribution of aberrations per cell follows a Poisson distribution (see Section 4.1.1). Deviation from a Poisson distribution may be the result of inhomogeneous radiation

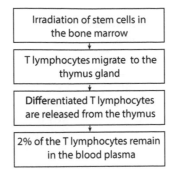

FIGURE 7.57 The event chain of irradiation, dispersion and sampling of exposed T lymphocytes in a human body.

exposure over the body of an individual, and should therefore also be assessed in order to obtain the maximum amount of information about the exposure. The dose–response relationship in these assays has been found to depend on the laboratory performing the analysis, and it is therefore recommended that each laboratory maintains a specific, well-calibrated dose–response curve.

Assays for dicentric chromosomes and chromosomal translocations (Fig. 7.58) have until recently been performed entirely manually, including the very time-consuming step of scoring under a microscope. Recently, an automated, image-based system for automatic identification and counting of mitotic cells on slides has become available (Romm et al. 2013). This allows rapid scoring, but cultivation still requires at least 48 hours. There is currently no automated system for the analysis of chromosomal translocations.

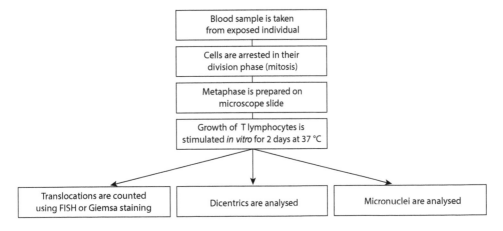

FIGURE 7.58 The event chain of irradiation, dispersion and sampling of exposed T-lymphocytes in a human body, and the subsequent options for assaying chromosome aberrations.

In addition to chromosome aberrations, so-called micronuclei can be used as a biodosimeter. Micronuclei are small nuclei that form when a chromosome fragment or a whole chromosome is not incorporated into the daughter nucleus (see Fig. 7.59). The advantage of detecting micronuclei is that it is less time- and resource–demanding than dicentric assays or FISH. Automated scoring is also possible. The detection limit of D_{med} is similar to that

TABLE 7.9 Characteristics of the two standard techniques used for lymphocyte assays to determine the absorbed dose to an exposed individual.

Dicentric chromosome assay	Translocation analysis
Counting of dicentric chromosomes in T lymphocytes. Homogeneous staining of mitotic chromosomes. Technique available since the 1960s	Conventional cell staining up until the early 1990s, after which a fluorescent agent was used to stain the aberrations (fluorescent *in vitro* hybridization, FISH). The FISH technique allows staining of single chromosomes and visualization of exchanges.
Detection limit: $D_{med} = 150$ mGy (ICRU, 2002).	Detection limit: D_{med} about 300 mGy (ICRU, 2002). Twice as high as for dicentric assays because of higher background variation depending on age and sex.
Relatively short half-time of markers (<1.5 y)	Longer half-times of markers (1–10 y). However, suffers from an inability to correctly determine doses from protracted exposures, e.g. accumulated exposures of up to 10 Gy have not demonstrated any significant signal, in terms of elevated frequency of translational aberrations.
Dose–response relationship: linear–quadratic for low-LET radiation and linear for high-LET radiation (α, neutrons).	Dose–response relationship: Sigmoid shaped (reaches saturation for high absorbed doses).
Time-consuming assessment which, until recently, was carried out manually.	

of FISH (300 mGy vs. 200 mGy), but the half-time of the marker is shorter (<1 y). The drawback of micronuclei scoring is the poor reproducibility between various laboratories performing such tests, compared with dicentric assays and FISH.

Erythrocytes can also be used as a biodosimeter by probing mutations in the glycophorin A (GPA) genes of these cells. The technique may, many times, be faster than those used for lymphocytes. However, the current detection level of this technique is about 300 mGy, which is about 2–3 times higher than for dicentric assays and FISH in lymphocytes. (Further details on this technique can be found in Ravi and Solomon 2002.)

Biological dosimetry has been very useful in determining the absorbed dose to individuals in single-accident scenarios. Despite the poor detection limit compared with luminescence techniques (at best 50 mGy (Trompier et al. 2011) vs. 1 mGy or less for thermoluminescent materials), biological methods have the advantage that D_{med} can be used to assess D_{tissue} (Eqn. 7.7) or whole-body dose with lower uncertainty than when using a detector material worn by the person. As mentioned previously, the relative frequencies of the observed aberrations provide an indication of whether the person has been exposed homogeneously or if only certain body organs are involved. The laboratory equipment needed is relatively complex and the number of staff needed to carry out biosimetric assays is relatively large compared with physical retrospective methods. Currently, only about fifty

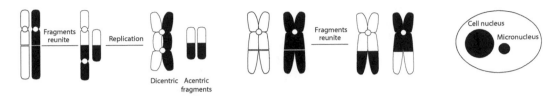

FIGURE 7.59 Left: dicentric chromosome. Middle: translocation of chromosomes. Right: micronucleus in peripheral T lymphocytes.

laboratories worldwide can carry out biodosimetric assays at the volumes that would be needed in the event of a major RN accident.

7.3.4.3 In Vivo Body Burden Measurements

Whole-body counting of humans is an important source of data in RN emergencies when accurate determinations of internal contamination need to be made. Three main techniques are used for in vivo measurements in radiation protection and emergency preparedness, which are described below.

Mobile in vivo *measurements.* These can be made using an unshielded geometry in which the subject leans over a mobile γ-ray detector with a sufficient response in the energy range 60–3000 keV (Fig. 7.60). With the advent of electrically cooled HPGe detectors, it is now, in theory, also possible to perform on-site, high-resolution *in vivo* gamma-ray spectrometry at remote locations where access to liquid nitrogen (the traditional method of cooling HPGe detectors) and the power grid may be limited. The counting efficiency of these devices is still lower than that of NaI(Tl) crystals traditionally used for mobile *in vivo* measurements. Although NaI(Tl) has a limited energy resolution, it is still valuable for screening of long-lived gamma-ray-emitting fission products such as ^{137}Cs, and has been used in screening and follow-up of exposed individuals, for example, in areas affected by Chernobyl fallout in Belarus and Russia. A faster method of screening is offered by detector portals, commonly employed by the nuclear power industry. These consist either of proportional counters, or plastic scintillators. However, such detectors have virtually no energy resolution, and are instead used qualitatively for the detection of elevated levels of gamma radiation.

Stationary whole-body counting. Measurements of the body concentration of radionuclides can be performed in so-called low-activity laboratories. To optimize the signal-to-background ratio, such laboratories require high-quality shielding materials. These shielding materials must not only be able to suppress the ambient cosmic radiation background, but also themselves be inherently free from radionuclides, both naturally occurring (such as the radium daughter ^{214}Bi in concrete) or other man-made sources (such as ^{60}Co in steel). A typical design for a low-background *in vivo* body counting laboratory is to use 10 cm of iron shielding, preferably from a source where the iron was smelted before 1945, as the metal will then contain very little pollution. The inner walls of this iron-clad chamber are covered with 2–3 mm of a high-Z material such as lead, preferably from a mine (so as to reduce the presence of ^{210}Pb with $T_{1/2} = 22$ years). This high-Z material is used to attenuate secondary γ-rays and charged particles generated by the interaction of cosmic γ-rays with the shielding. In addition to this passive shielding, background radiation can be further suppressed by employing an overpressure in the counting room, using air that has passed through absolute filters, for example, containing charcoal. This will reduce the build-up of ^{222}Rn and Rn daughters in the counting room walls and equipment.

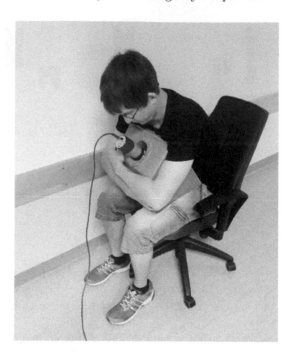

FIGURE 7.60 Mobile *in vivo* gamma-ray spectrometry using the so-called Palmer geometry.

Low- or high-resolution gamma-ray spectrometry is performed in the shielded counting room with the subject reclining in a chair or lying on a bed (see Fig. 7.61). Scanning times are typically 5–20 minutes. Longer measurement times can be stressful for individuals with claustrophobia. Typical detection levels for the whole-body burden using the definition in Eqn. 4.64 can be as low as 40 Bq for ^{137}Cs (corresponding to an annual absorbed dose to the whole body of 12.8 μGy y^{-1}) for a dual 7.54 cm ($\oslash$) by 7.54 cm NaI(Tl) detector system in a scanning bed geometry. Calibration of counting efficiency is required using phantoms of tissue-equivalent materials (such as distilled water or polystyrene) of various sizes representing body weights in the range 20–100 kg. Calibration corrects for the dependence of internal absorption of the γ-radiation on the patient's mass.

Gamma camera imaging. The systems described above are mainly used for *in vivo* identification and quantification of the total body content, Q, of radionuclides. However, in the case of internal contamination, it is preferable to use detector systems with high spatial resolution, such as gamma cameras, to determine in which organ of the body the radionuclides have accumulated. These systems, which are used in nuclear medicine and radiology departments in hospitals, are designed for imaging the organ uptake of beta- and gamma-ray-emitting radionuclides in the human body. However, gamma camera systems are not normally available for mass screening of members of the public. The clinical environment does not offer the possibility of shielding the system from background radiation. Typical detection levels are therefore a factor of ten times higher than in shielded whole-body counting systems.

In addition to determining the whole-body content of fission products such as ^{137}Cs, and neutron activation products such as ^{58}Co and ^{60}Co, low-background rooms for in vivo counting are used for the quantification of the specific uptake of ^{131}I in the thyroid, ^{241}Am in lung tissue, ^{238}Pu in the liver, and, by using techniques such as anticoincidence coupling

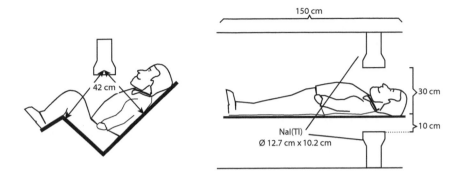

FIGURE 7.61 Left: Patient reclining in a chair with the gamma detector positioned so that the torso and legs are equidistant from the centre of the detector volume. Right: A dual detector scanning bed geometry with the patient lying between the upper and lower detectors. This system is less affected by inhomogeneous radionuclide distribution in the body.

of the NaI(Tl) detector and other gamma detectors, pure beta emitters such as ^{90}Sr/ ^{90}Y in bone and skeletal tissue. The ICRP publishes updated age-dependent dose coefficient tables [μGy per kBq] for a number of radionuclides and body organs, e.g. ICRP 119 (ICRP 2012). The absorbed dose can be estimated by assuming that a person's whole-body burden, Q [Bq], is an equilibrium value resulting from either a protracted intake, I [Bq d^{-1}] of the radionuclide (chronic intake), or from a single intake, I_0 [Bq] of the radionuclide. The exact time and duration of the intake is often not well known. Therefore, when an individual exhibits an elevated whole-body burden, it is common to err on the side of safety in occupational monitoring and assume that a single intake occurred directly after the previous measurement.

7.3.4.4 Excretion Measurements for Internal Dosimetry

The throughput of whole-body counting systems is limited (a maximum of 8–10 people a day). Mobile *in vivo* body burden measurements are also limited by the need to find a location where the background gamma radiation is relatively low compared to that in the general environment. A good alternative is thus to estimate internal contamination by sampling excretion from individuals, and then applying biokinetic models to estimate the body burden and the associated organ doses, D_{tissue}. This method employs a combination of sampling and modelling to obtain D_{tissue}.

The daily excretion in urine, E_u [Bq d^{-1}], at a given sampling time, t, after the onset of internal contamination can be expressed as

$$E_u(t) = f_1 \cdot f_u \cdot \int_0^t I(t) \cdot q(t), \qquad (7.8)$$

where f_1 is the gastrointestinal uptake fraction of the ingested or inhaled radionuclide, f_u is the fraction of the total daily excretion being excreted through urine, $I(t)$ is the ingestion rate of the radionuclide by the patient, and $q(t)$ is the so-called retention function of the radionuclide when it enters the systemic tissues of the body via the plasma (see Section 2.8.1).

The retention function can be experimentally obtained from biokinetic studies in which a well-defined single intake is traced, for example, by surveying the time evolution of the body burden by means of *in vivo* organ or whole-body counting. The retention curve is often specific for a given element, but the retention curves for various radioisotopes will be modulated by their respective physical half-life. For the element caesium the retention curve $q(t)$ in adults is given by

$$q(t) = 0.9 \cdot e^{\frac{ln(2)}{110} \cdot t} + 0.1 \cdot e^{\frac{ln(2)}{2} \cdot t}. \tag{7.9}$$

The ICRP (ICRP 1993) has published element-specific retention curves for various important elements in radiation protection and emergency preparedness, such as Cs, Co, U, Pu Sr, Am and I. Using these biokinetic models, assays of the radionuclide concentration in urine samples from an individual can be converted into a corresponding whole-body burden, $Q(t)$, at the time of sampling. In cases where the date of intake of the radionuclide is unknown, an intake must be assumed, either based on a previous measurement of the individual (in analogy with *in vivo* whole-body counting), or, if there is a lack of data, based on the suspected date of onset of internal contamination. Figure 7.62 illustrates how a sample of urine containing a certain activity of ^{210}Po can be used to determine the original intake if a single intake is assumed, shown as a function of the assumed elapsed time after intake.

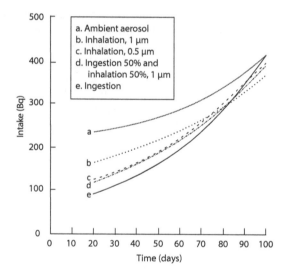

FIGURE 7.62 Intake vs. sampling time for ^{210}Po in a 24-h urine sample containing 30 mBq of the radionuclide.

In the restoration phase after an RN emergency it is likely that there will be a continuous transfer of many of the long-lived radionuclides remaining in the environment to the residents in the area. Individuals are thus subject to a protracted intake of the radionuclides, and it can then be assumed that the radionuclide has attained equilibrium via ingestion or inhalation, i.e., a resident's daily excretion is equal to their daily intake of the radionuclide. Eqn. 7.8 can then be rewritten more simply as

$$E_u = c(t_{eq}) \cdot \frac{f_u}{f_1} \cdot Q_{eq} = \left(\frac{1}{F_{i,u}} \right) \cdot Q_{eq}, \tag{7.10}$$

where $c(t_{eq})$ is a constant related to the retention of the element, and Q_{eq} is the equilibrium whole-body burden of the sampled individual. Once Q_{eq} has been estimated, published data from the ICRP (e.g. ICRP, 1993) and MIRD (Snyder et al. 1975) can be used to obtain conversion factors between the equilibrium body burden and the internal whole-body dose rate. These factors are based on models and computer simulations using mathematical body phantoms. $F_{i,u}$ is the conversion factor between the equilibrium whole-body burden and daily urine excretion of a given radionuclide, i.

However, it is both time- and labour-demanding to carry out 24-hour urinary sampling on large groups of people, and therefore small quantities of urine (spot samples), typically 100–500 ml, are used instead. The observed activity concentration in the urine samples cannot be directly related to Q_{eq}, because the radionuclide concentration in the excreted urine varies greatly during the day. This variation is related to the human body's diurnal metabolic cycle. Instead, the observed radionuclide concentration in the spot urine sample is corrected to the corresponding 24-hour urinary excretion by normalizing it to biomarkers that are assumed to vary in a similar way to the radionuclide during the day. However, these biomarkers are assumed to be released at a relatively constant rate for a given individual on a twenty-four hour level, and the concentration ratio between the biomarker and the radionuclide can thus be used to predict the diurnal excretion of the radionuclide.

One such biomarker is *creatinine*, a metabolic compound generated in the muscle tissues, which is excreted at a relatively constant rate for a given individual. In Eqn 7.11 is given the expression for the whole-body content of a radionuclide i, Q_i, as a function of $F_{i,u}$, the standard creatinine excretion, KRE_{24h}, the creatinine content found in the spot sample, Kre_u, and the radionuclide concentration in the urine, $c_{RN,u}$. KRE_{24h}, is approximately 10 mmol d^{-1} for adult females and 15 mmol d^{-1} for adult males. Kre_u denotes the creatinine concentration in the spot sample [mmol L^{-1}], and $c_{i,u}$ the radionuclide concentration in the spot sample [Bq L^{-1}].

$$Q_i = F_{i,u} \cdot \left(\frac{KRE_{24h}}{Kre_u} \right) \cdot c_{i,u} \qquad (7.11)$$

Another biomarker that can be used, especially in assays for alkali metals such as ^{134}Cs and ^{137}Cs, is the urinary K content, either in terms of ^{40}K activity or as stable K [g], (1 g of stable K corresponds to 31.0 Bq of ^{40}K). This can be used to normalize the radionuclide content in the spot sample to give the corresponding 24-hour urinary excretion. Standard values of diurnal excretion are available for both creatinine and stable K. However, stable K appears to vary between ethnic groups, and the normalization method should therefore be used with caution. For example, Russian rural inhabitants excrete, on average, three times more potassium per day than the Japanese (Rääf 2002).

7.3.5 Retrospective Dosimetry Using Environmental Radiometry

7.3.5.1 Thermostimulated and Optical Luminescence

The use of luminescent materials as radiation detectors was discussed in Section 4.2. The luminescent materials used for retrospective dose assessments are given in Table 7.10. To date, quartz and feldspar have been used most for large-scale reconstructions of the absorbed dose to affected populations (Bøtter-Jensen et al. 2000), while porcelain and ceramics have been used to estimate doses outside dwellings (Ramzaev and Göksu 2006). Household salt (mainly NaCl) and mobile phone components have recently been studied to establish how their luminescent properties compare to those of dosimeters, with promising results.

The first step in obtaining the absorbed dose to an irradiated person in the vicinity of the investigated luminescent materials is to extract the accident dose, D_{acc}. For TL and OSL in, for example, grains of quartz and feldspar extracted from bricks and tiles, D_{acc} is related to the signal from the material due to man-made radiation, and is related to the total absorbed dose and background radiation dose to the brick according to Eqns. 7.12 and 7.13:

$$D_{acc} = D_{tot} - D_{bkg} \tag{7.12}$$

where

$$D_{bkg} = a \cdot \left(b_\alpha \dot{D}_\alpha + b_\beta \dot{D}_\beta + b_\gamma \dot{D}_\gamma + b_\gamma \dot{D}_{cosmic} \right) \tag{7.13}$$

and a is a proportionality constant between the sum of four background components. D_α is the contribution to the dose rate from emitted α-particles in the brick, D_β is the corresponding contribution from emitted β-particles, D_γ is the γ-ray dose rate from γ-rays in the brick or on the surface of the brick, and D_{cosmic} is the dose rate from cosmic radiation. For bricks, the total background dose rate varies from 2 to 4 mGy y^{-1}; 60% of which is due to β-radiation, 35% γ-rays, and the remainder from cosmic radiation.

The grain size of the extracted quartz from the bricks can be chosen so that the contribution from β-radiation dominates the dose contribution, which is useful in reconstructing the age of the material. Hence, single grain sizes ranging from 90 to 150 μm in diameter are used. About 50 mg SiO$_2$ is usually required for a reliable dose assessment. The surface layer that has been exposed to visible light is first removed by etching with hydrofluoric acid, HF, as exposure to visible light bleaches the latent signal used for OSL assessments, and material exposed to ambient light does not provide reliable information.

To obtain a reliable gamma dose estimate, TL or OSL signals must be obtained from sufficiently deep inside the sampled material to ensure charged particle equilibrium, which for photon energies up to 1.25 MeV is at a depth of about 5 mm. Figure 7.63 illustrates two types of materials used for retrospective dosimetry.

TABLE 7.10 Examples of crystalline materials with known OSL properties that can be used for retrospective dosimetry.

Material	References (examples)
Quartz in building materials	Bøtter-Jensen et al. (2000); Thomsen et al. (2002)
Household salt	Bailey et al. (2000); Bernhardsson et al. (2009); Christiansson et al. (2014;) Spooner et al. (2012)
Porcelain	Ramzaev et al. (2006)
Mobile phone components	Beerten et al. (2010); Inrig et al. (2010); Woda et al. (2012); Bassinet et al. (2014)
Dental repair material	Geber-Bergstrand et al. (2012)
Dust (silicate)	Jain et al. (2006)
Fingernails and toenails	Sholom et al. (2011)
Banknotes and coins	Sholom and McKeever (2014)
Salted snacks and nuts	Christiansson et al. (2016)
Desiccants	Geber-Bergstrand et al. (2014)

Dose assessment using TL and OSL is similar to that using for EPR. The raw signal from exposure of the sample to radiation is first read out by either TL or OSL stimulation, after

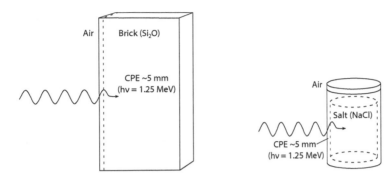

FIGURE 7.63 Left: D_{med} can be obtained from grains of quartz, extracted from bricks beneath the depth of charged particle equilibrium (CPE), by means of OSL. Right: D_{med} can be obtained from household salt contained in light-tight containers by measuring the OSL from small samples (5–10 mg).

which regenerative irradiation and read-out take place. These additional or regenerative doses (where the signal is reset between each new irradiation) are used to correlate the original signal to the accidental dose, D_{acc}. Corrections must be made for so-called fading of the latent signal between exposure and sample read-out. The luminescence signal emanating from deep traps is preferred, and the signal from shallow traps, which contributes to the uncertainty in the measurement, is reduced by pre-heating the sample to empty the shallow traps. The detection limit for OSL from quartz in bricks is about 20 mGy, but can be as low as 0.1 mGy in household salt.

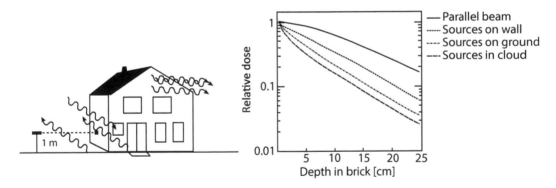

FIGURE 7.64 Left: D_{air} is obtained from computational methods relating the measured D_{acc} in detector medium (e.g. household salt or quartz in brick) and D_{air} free in air 1 m above ground for various γ fluences. Right: Relative absorbed dose as a function of depth in brick of a building for various ambient exposure geometries.

In the second step of the dose assessment, D_{acc}, obtained either by EPR or OSL/TL in a radiation-sensitive material, is converted into D_{tissue}. This is often done by first converting D_{acc} into an absorbed dose in air close to the sample, as illustrated in Figure 7.64. The translation between D_{acc} in a detector medium to absorbed dose in air can be achieved by computational methods such as MC simulation or modelling. The dose gradients in bricks or tiles can also be correlated to various gamma irradiation geometries, which improves the possibility of reconstructing the dose rate in air at the reference point, D_{air}. Once D_{air} has been estimated, the absorbed dose to the tissue of a person in the vicinity of the sample can be calculated by means of Eqn. 7.7.

7.3.5.2 Mass Spectrometry

Mass spectrometry of long-lived radionuclides or stable neutron activation products can be used to reconstruct the dose to tissue of individuals at an RN accident site long after the event has taken place. Mass spectrometry can be performed using traditional accelerator mass spectrometry (AMS), or inductively coupled plasma mass spectrometry (ICPMS). These techniques are illustrated in Figure 4.16, and are also briefly discussed in Section 4.2.6. Mass spectrometry is used to quantify and assess the concentration of neutron activation products and to reconstruct the neutron fluxes during the incident.

The flux of prompt γ-rays and thermal neutrons can be estimated from samples using the neutron activation products ^{36}Cl (formed in the reaction ^{35}Cl(n,γ)^{36}Cl) and ^{41}Ca (^{40}Ca(n,γ)^{41}Ca) as markers. The presence of ^{63}Ni in copper may be a marker for a flux of fast neutrons ($E_n > 1$ MeV) via the reaction ^{63}Cu(n,p)^{63}Ni. The radionuclide ^{63}Ni, which is a β-only emitter ($E_{\beta,\text{Max}} = 65.9$ keV), can also be determined using gas-flow proportional counters or GM counters. An example of important dose reconstructions based on mass spectrometry of neutron activation products is the assessment of ^{63}Ni in copper (electrical wires, etc.), which enabled dose reconstruction of fast neutrons 100–1000 m away from ground zero at Hiroshima and Nagasaki (ICRU 2002). Another example is the measurement of surplus ^{129}I in thyroid samples for the reconstruction of absorbed doses in the thyroid by children following the Chernobyl accident.

7.3.5.3 Retrospective Dosimetry Using in Situ Gamma Radiometry

Doses to inhabitants in an area affected by radioactive fallout can be reconstructed by determining the dose rate to air, D_{air}, at a reference point, usually 1 m above ground over an undisturbed reference surface. The absorbed dose to tissue at time, t, in a particular group of humans, i, is then calculated by assuming various conversion factors (Eqn. 7.14):

$$\dot{D}_{tissue}(i,t) = \dot{D}_{air}(t) \cdot k(i) \cdot \sum_j f_j \cdot p_j(i) \tag{7.14}$$

where $k(i)$ is the conversion factor between the absorbed dose to air and that to tissue, which is dependent on the age and gender of the exposed person. The factor $p_j(i)$ is a behaviour or *occupancy factor* that accounts for the relative time spent in location j. The site-specific factor f_j depends on the γ-ray fluence from the fallout and on how the radiation is shielded by buildings and structures in location j. The absorbed dose rate to air at time t can be expressed as

$$\dot{D}_{air}(t) = \dot{D}_{air}(t=0) \cdot r(t), \tag{7.15}$$

where $r(t)$ is a time-dependent factor that takes into account both the radioactive decay of the radionuclides in the fallout and shielding effects following the migration of these radionuclides into the ground.

The absorbed dose rate to air or the kerma rate in air, related by

$$\dot{K}_{air} = \dot{D}_{air}(1 - g(E_\gamma)) \tag{7.16}$$

can be determined relatively accurately using an ionization chamber, preferably with an air-equivalent material. However, due to the low counting efficiency of such instruments, in environmental measurements, the air kerma is usually determined using ion chambers filled with pressurized argon gas, and the kerma rate in Ar is converted to that in air, K_{air}.

Table 7.11 gives examples of site-specific factors, f_j, and occupancy factors, p_j, for various types of sites and categories of individuals. This illustrates how the absorbed tissue dose rate to various individuals in an area (D_{tissue} in Eqn. 7.14) can be reconstructed based on exactly where they live, and on their occupancy, which takes into account the time spent at work and in leisure activities.

Example 7.1 *Occupancy and site-specific factor*
As an example of using Table 7.11 and Eqn. 7.14, consider adult indoor workers living in an area with an absorbed dose rate of 1.0 $\mu Sv\ h^{-1}$, mainly due to fallout of ^{137}Cs on the ground. The sum in Eqn. 7.14 will then be $0.28 \cdot 0.55 + 0.42 \cdot 0.07 + 0.23 \cdot 0.07 + 0.01 \cdot 0.50 + 0.02 \cdot 1 = 0.22$. With $k(i) = 0.83$, the conversion factor between absorbed dose to air and whole-body for a rotational irradiation geometry of ^{137}Cs γ rays (taken from ICRP, 2010), the absorbed dose to whole-body will be $1.0 \cdot 0.83 \cdot 0.22 = 0.18$ $\mu Sv\ h^{-1}$.

TABLE 7.11 Site-specific transfer coefficients, f_j, and occupancy factors, $p_j(i)$, for groups of individuals typical of a Russian village.

Site, j	Site-specific factor, f_j	Occupancy factor, $p_j(i)$		
		Indoor workers	Outdoor workers	School children
Dwelling/garden				
Outdoors	0.55	0.28	0.20	0.39
Brick house	0.07	0.42	0.42	0.55
Workplaces				
Building	0.07	0.23	0.08	0.00
Tilled areas	0.50	0.01	0.18	0.00
Recreational areas				
Forest, meadow	1	0.02	0.01	0.04

Source: Data from ICRU (2002).

If inhabitants living in an affected area are to be surveyed using dosimeters, e.g. TL-sensitive LiF, the dosimeters must be calibrated and worn according to a well-defined geometry so that the dosemeter readings can be related to, for example, the operational dose quantity $H_p(10)$ (Section 2.5). The absorbed dose rate at a reference point (D_{air} in Eqn. 7.14) can then be used to estimate the whole-body effective dose rate at time, t, to a stationary person located at the reference point:

$$E_{ext} = D_H \cdot \left(\frac{D_{air}(\Phi)}{D_H} \right) \cdot \left(\frac{E(\Phi)}{D_{air}(\Phi)} \right) \qquad (7.17)$$

where D_H is the dose recorded by the dosimeter or a pre-calibrated dose unit such as

$H_p(10)$, and D_{air}/D_H is the ratio between the absorbed dose to air at the aforementioned reference point (1 m above ground) and the dosimeter, which depends on the radionuclide composition and the deposition geometry. Such ratios have been obtained experimentally (Wøhni 1995). The term $E(\phi)/(D_{air}(\phi)$ is an age- and gender-dependent ratio between the effective dose to a reference person and the absorbed dose to air for a given irradiation geometry. For areas affected by extensive ground deposition, the reference geometry can be described by either a surface source or a thin slab of homogeneously distributed radionuclides in the ground. The ICRU has tabulated such ratios for various fallout geometries (ICRU, 1998; ICRP, 2012). Eqn. 7.17 has been used for the reconstruction of long-term external exposures to residents in Belarus and Russia affected by the Chernobyl fallout.

The ratio (E/K_{air}) varies with (i) the deposition geometry, (ii) the photon energy and (iii) the age and physiology of the exposed person. The dependency on physiology is due to the fact that the relative positions of the organs at risk in the human body depend on the height, and thus the age, of the person. Figure 7.65 shows a plot of the E/K_{air} ratio as a function of photon energy for an ideal planar deposition. The ratio is about 20% higher for a 5-year-old child than for an adult.

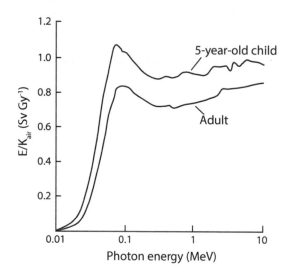

FIGURE 7.65 The ratio E/K_{air} as a function of photon energy for an infinite planar deposition of a gamma emitter for an adult and a 5-year-old child (voxel phantom). Data from Conti et al. (1999).

For dose reconstructions using personal dosimeters in a control group for the whole affected population in the area, it is important to consider the tissue equivalence of the dosimeter. As previously mentioned, LiF has an effective atomic number of 8.3, and is relatively close to that of soft tissue ($Z_{eff} = 7.4$).

7.3.5.4 Dose Reconstruction Using Radionuclide Concentration Data in Foodstuffs

Reconstruction of the internal absorbed dose to an affected population can be made by measuring the radionuclide concentrations in various components, q, of their diet. The average ingestion rate, $I_j(t, i)$ [Bq d^{-1}] of radionuclide j, for individuals among a population, i, with a certain dietary habit can be expressed as

$$I_j\left(t,i\right) = F\left(t,i\right) \cdot \sum_{q=l}^{m} C_{j,q}\left(t\right) \cdot V_q\left(i\right), \tag{7.18}$$

where $F(t,i)$ is a correction factor accounting for uncertainties and other inconsistencies between the actual food intake and those claimed by the population. Inconsistencies are caused by factors such as losses of the radionuclide during cooking and preparation. $C_{j,q}$ is the activity concentration of a given radionuclide (j) in a given food component, q, and $V_q(i)$ is the daily ingestion rate of food component q.

Once the average value of I_j has been estimated, it is also possible to estimate the average internal effective dose, $E_j(i)$ caused by the continuous intake (Eqn. 7.19), by integrating the product of the ingestion rate and the retention function of the radionuclide, $q_j(t,i)$. The retention function can vary with age and gender, and between ethnic groups. The dose coefficient, $e_{k,j}(i)$ [Sv Bq^{-1}] is also dependent on sex and age.

$$E_j\left(i\right) = e_{k,j} \cdot \int_{t_0}^{t_1} I_j\left(t,i\right) \cdot q_j\left(t_1 - t, i\right) dt \tag{7.19}$$

The dose coefficient $e_{k,j}(i)$ and retention functions can be obtained for radiation workers from ICRP data (ICRP 2012). It should be noted that in the estimation of the effective dose, the ICRP has chosen conservative values of the retention functions, i.e., these models will never risk underestimating the true value of E_j.

7.3.6 *In Situ* Radiometry

7.3.6.1 *Stationary in Situ Gamma-Ray Spectrometry*

The detection of gamma-ray-emitting radionuclides in the environment can be divided into stationary measurements at a particular point, and mobile, *in situ* surveys. Stationary measurements are suitable for on-line monitoring of ambient gamma radiation levels. Many countries have developed national networks of gamma monitoring stations. The gamma detectors used are often high-pressure ion chambers, GM counters or plastic scintillators, all of which have a gamma dose response that is relatively independent of photon energy.

The purpose of *in situ* gamma-ray spectrometry is to quantify the ground deposition of a radioactive element such as the fission product ^{137}Cs, which was widely dispersed following the Chernobyl accident in 1986. As mentioned in Section 5.1.1, atmospheric depositions exhibit high spatial variation on all geographical scales, and to minimize the inherent uncertainty associated with the inhomogeneity of radionuclide deposition, other perturbing factors such as irregular terrain, or high vegetation that shields the detector from γ-rays should be avoided. The term field of view (FOV) [m] defines the effective radius around the gamma detector setup that contributes to a specific fraction of the total primary γ-ray flux, and thus to the detector signal. The FOV can be defined as the radius of a circular area in which more than 90% of all the counted γ-ray photons originate. The FOV of an *in situ* gamma detector setup will thus directly affect the spatial resolution of the survey system. Therefore, *in situ* gamma-ray spectrometry should be performed in an open flat area, free from tall vegetation within a radius of at least 60–90 m from the measuring point (Fig. 7.66). This radius represents the average FOV of many high-resolution gamma detector systems, typically consisting of HPGe detectors with relative efficiencies ranging from 30 to 150% (note that HPGe detectors can have a relative efficiency >100%, as discussed in Section 4.6.4).

The position of the measuring point is nowadays determined using GPS. If follow-up

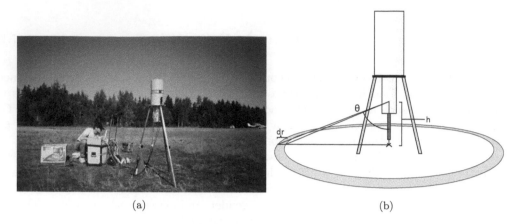

(a)   (b)

FIGURE 7.66   (a) An unshielded HPGe-detector set up in an open field for *in situ* spectrometry. (b) Illustration of the detector setup with an altitude h of the centre of the detector crystal. The vertical angle, θ, describes the angle of incidence of γ photons impinging on the detector crystal. The γ fluence contribution from a circular segment with radius r, and segment thickness dr, is integrated from $\theta = 0$ to $90°$ to obtain the γ fluence rate per unit activity for this particular field geometry.

measurements are to be performed in the future, the measuring point should be determined to a precision of better than one metre. The HPGe detector is mounted on a tripod and positioned so that the centre of the detector crystal is 1 m above the ground. The detector should be positioned vertically with the crystal facing the ground. The nitrogen Dewar or electronic cooling system must not shield the detector from radiation from the ground if the equivalent surface deposition, A_{esd}, is to be determined (see below).

Generic calibration factors are available in the literature (Finck 1992) for HPGe detectors and various deposition geometries, often for a crystal positioned 1 m above ground. The detector should thus be positioned at this height with an error of at most 10 cm, in order to make use of these calibration factors with reasonable accuracy, and to maintain compatibility with other similar measurements. For high-resolution *in situ* gamma-ray spectrometry measurements, the activity deposition, A_{insitu}, is related to the net counts in the full energy peak, N_{insitu}, according to

$$A_{in\,situ} = \frac{N_{in\,situ}}{t_m} \cdot \left(\frac{1}{\varepsilon_{in\,situ} \cdot \left(\frac{\phi}{A} \right)_{dep}} \right), \tag{7.20}$$

where ε_{insitu} [cps s^{-1}] is a calibration factor that depends on the deposition distribution and the properties of the detector crystal, and $(\phi/A)_{dep}$ is the photon fluence rate per unit activity [s^{-1} m^2 Bq^{-1}] for this geometry. The calibration factor ε_{insitu} can in turn be expressed as

$$\varepsilon_{in\,situ} = \left(\frac{N_f}{N_0} \right) \cdot \varepsilon_i, \tag{7.21}$$

where ε_i [cps s^{-1}] is the intrinsic efficiency of the detector (as defined in Eqn. 4.30), and (N_f/N_0) is a factor that takes into account the variation in the intrinsic efficiency with the angle of incidence of the photons, and the angular distribution of the photon fluence for a given deposition distribution. The intrinsic efficiency as a function of the angular incidence of photons can be experimentally determined in a laboratory using a point source at a fixed distance (Fig. 7.67).

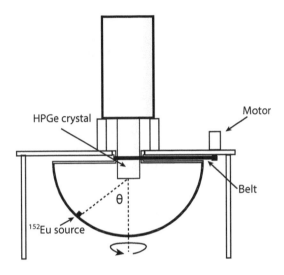

FIGURE 7.67 Calibration setup for determination of the intrinsic efficiency of the detector as a function of the vertical angle, θ, by mounting a standard source on a rotating arc.

The most basic assumption of source distribution is the homogeneous deposition of activity over an infinite plane surface with no penetration into the ground. This represents the ideal case of dry deposition of radionuclides from fresh fallout. In the absence of knowledge on the penetration of the radioactive fallout into the soil (Section 3.5.2), the net number of counts registered in the full energy peak of the spectrometer, N_{insitu}, is converted into a so-called equivalent surface deposition, A_{esd}.

The angular distribution of photons can be calculated for an ideal surface distribution, and together with the intrinsic efficiency ε_i gives a factor $(N_f/N_0)_{surface}$ for the detector crystal, for the specific source distribution. For an equivalent surface deposition, ε_{insitu} is then denoted ε_{esd}. Combining this quantity with the total fluence rate per unit activity for that deposition geometry, $(\phi/A)_{esd}$, gives an expression for the equivalent surface deposition, A_{esd}:

$$A_{esd} = \frac{N_{in\,situ}}{t_m} \cdot \left(\frac{1}{\varepsilon_{esd} \cdot \left(\frac{\phi}{A} \right)_{esd}} \right). \tag{7.22}$$

For example, if air attenuation is disregarded, the primary photon fluence 1 m above the ground per unit activity deposited, $(\phi/A)_{surface}$, is 1.8 s^{-1} Bq^{-1} m^{-2} for an infinite plane deposition of ^{137}Cs on the surface of the ground. Using this simple conversion factor enables rapid generation of fallout maps in the early phase following extensive radioactive fallout, and also improves compatibility with deposition data provided by other organizations.

An alternative to determining the quantities $(N_f/N_0)_{surface}$ and ε_i is to determine a direct conversion factor between A_{esd} and a calibration source mimicking the surface equivalent deposition. However, it is not easy to construct such a calibration source due to regulatory and practical limitations on the amount of activity, for example, of ^{137}Cs, that can be spread on an outdoor surface. In some countries, circular areas with well-known activities exist for the calibration of vehicle- and aircraft-borne gamma detectors used in geological surveys.

7.3.6.2 Mobile Gamma Radiometry

Mobile gamma radiometry involves *in situ* gamma-ray spectrometry at various geographical locations, with the aim of providing accurate determinations of the ground deposition of atmospheric fallout, A_{dep} [Bq m^{-2}] (e.g., ^{134}Cs or ^{137}Cs), or the soil activity concentration of radionuclides, A_{soil} [Bq m^{-3}] (e.g., the naturally occurring radionuclides ^{238}U, ^{232}Th and ^{40}K). It also includes vehicle-borne radiometry, such as illustrated in Figure 7.68, or radiometry on foot, where the detector is carried in a backpack. These systems are often combined with high-accuracy GPS data. Data from the detector can be recorded together with the GPS coordinates in digital geographical information systems, thereby generating radiological maps of the surveyed area, virtually in real time.

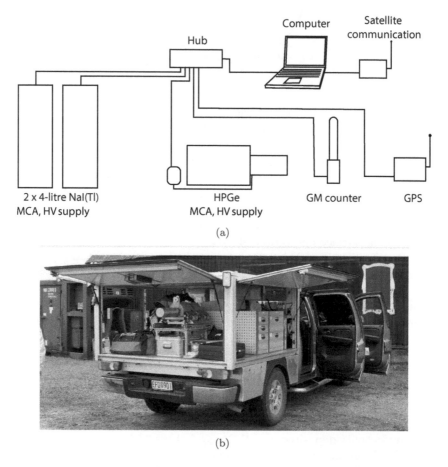

FIGURE 7.68 (a) Schematics view of the components used in a mobile gamma survey system. (b) Car-borne mobile γ survey system, with detector units mounted on a "cargo".

For small-scale events, including scenarios where an orphan source is to be located, car- or airborne measurements are carried out with gamma dose rate meters and nuclide-specific equipment. These kinds of measurements are usually denoted CGS and AGS, respectively (car-borne and airborne spectrometry). Once located, the source can be identified and quantified by on-site measurements. Even if the strength and identity of the radionuclide are known, it is still important to rule out the possibility that it has been fragmented or dispersed.

For large-scale accidents involving extensive radioactive fallout over large areas, mobile radiometry is used to map the variations and gradients in the ground deposition. A number of source–detector geometries are relevant depending on the scenario. Below are given expressions for primary γ-ray fluence rate, $\dot{\phi}$ [m^{-2} s^{-1}] at the detection point for various radionuclide distribution geometries and exposure scenarios (see also Section 2.6).

- Localized source (point geometry), e.g., when searching for orphan sources

$$\dot{\phi} = \frac{A_{source} \cdot n_\gamma}{4\pi r^2} \cdot e^{-\mu_{air} \cdot r}, \tag{7.23}$$

where A_{source} is the activity of the source [Bq], n_γ is the gamma photon probability per decay, μ_{air} is the mass attenuation coefficient of the primary photons in air and r is the detector–source distance.

- Semi-infinite volume source (homogeneous), e.g., when performing a geological survey of naturally occurring radionuclides such as ^{238}U (mainly γ-rays from ^{214}Bi), ^{232}Th (mainly γ-rays from ^{208}Tl), and ^{40}K

$$\dot{\phi} = \frac{A_{soil} \cdot n_\gamma}{2\mu_{soil}} \cdot \mu_{air} \cdot h \int_{\mu_{air} \cdot h}^{\infty} t^{-2} e^{-t} dt, \tag{7.24}$$

where A_{soil} is the radionuclide activity concentration in soil [Bq m^{-3}] and $t \equiv \mu_{air} \cdot h \cdot \sec(\theta)$.

- Infinite plane surface, for surveys of fresh NPP and nuclear weapons fallout

$$\dot{\phi} = \frac{A_{dep} \cdot n_\gamma}{2} \int_{\mu_{air} h}^{\infty} t^{-1} e^{-t} dt, \tag{7.25}$$

where A_{dep} is the radionuclide areal activity density [Bq m^{-2}]

- Slab geometry (slab layer depth z), for fresh wet-deposited fallout, and older fallout (>1 year)

$$\dot{\phi} = \frac{A_{soil} \cdot n_\gamma}{2\mu_{soil}} \cdot \mu_{air} \cdot h \cdot \left\{ \int_{\mu_{air} \cdot h}^{\infty} t^{-2} e^{-t} dt - \int_{\mu_{soil} \cdot z + \mu_{air} \cdot h}^{\infty} t^{-2} e^{-t} dt \right\}, \tag{7.26}$$

where μ_{soil} is the attenuation coefficient in the soil and A_{soil} is the radionuclide activity concentration in soil [Bq m^{-3}].

Mobile or *in situ* γ-ray detectors used for measurements of the radionuclide distributions given above must be calibrated using various reference sources or mathematical modelling. In addition to the geometry, the angular dependency of the intrinsic efficiency of the γ-ray detectors must also be taken into account.

Detectors sensitive to γ-rays (and, since the beginning of the 2000s, neutrons), such as GM counters, NaI(Tl), HPGe, and ^{3}He-proportional counters, may be mounted on a vehicle or an aircraft. Detector signals may then either be recorded at regular time intervals, dt, and the information binned and integrated over the whole time interval, or the information may be recorded in so-called *list mode*, where the time of the detected events is saved as one vector of the data, providing more flexibility in binning the pulses into other time intervals. Mobile γ-ray surveys generate data that can be described by a vector, consisting of a time coordinate and a value representing the detector output. A sequence of data points is thus measured at fixed time intervals, to form a time series, in analogy with the environmental sampling discussed in Chapter 5. In mobile γ-ray spectrometry, a spectrum is collected at uniform time intervals, t, which range from less than a second (AGS) to a few seconds (CGS). In γ-ray spectrometry data, several vectors are generated, one for each ROI in the pulse height distribution. When a GPS is connected with the system, the γ counts registered during dt in a particular ROI at time t can be paired with the geographical coordinates fed by the GPS sensor at time t. If the aim of the mobile survey is to map a complete area, the data generated can be stored as a matrix with two spatial coordinates (preferably in Cartesian coordinates).

When searching for an orphan source, the fluctuations in the detector signal during the mobile survey must be statistically tested against the hypothesis of the absence of radiation above the background (related to the null-hypothesis, H_0, discussed in Section 4.6.7). An alarm criterion is therefore set by the operator, based on a preset confidence level (criterion), such as the probability of a false alarm of less than 5%. If the pulse rate increases above the alarm level, the operator should then return to that point and investigate whether a γ-ray source is present in the vicinity. For further details, see Kock (2012).

A factor related to the alarm level for enhanced γ-ray activity is the detection limit of a single γ-ray source. The detection limit in a vehicle-borne survey will depend on the speed of the vehicle, the time required for pulse integration, the counting efficiency of the gamma detector, and the source–detector distance. A typical detection limit (MDA) for vehicle-borne surveys using a 4-litre NaI(Tl) detector travelling at a speed of 50 km h^{-1} is approximately 1 GBq for a source located 50 m away from the path of the vehicle (Hjerpe 2004). MDA was discussed for stationary γ-ray spectrometry in Section 4.6.7.

The FOV of the detector and the interval between two consecutive time coordinates, Δt, or the distance, Δr, between two neighbouring spatial coordinates, represents the area over which the gamma detector reading is expected to extend. However, the number of counts recorded during a pulse acquisition window will be based on the detection of γ-rays from a relatively large FOV (up to 100 m). This means that successive data points may contain counts from overlapping areas, which means that two successive data points are not statistically independent of each other. In other words, the data points are autocorrelated. This is a statistical term that means each value may be correlated to other values measured in the spatial vicinity, or to other values collected closely in time. In γ-ray spectrometry, autocorrelation means, in practice, that the difference in the true deposition between two adjacent areas can actually be greater than that indicated by the detector reading. This must be accounted for when analysing the data to determine whether variations and gradients observed by the mobile detector system originate from an orphan source or a local hot spot in the deposition.

Radiometric data collected over space and time in a surveyed area can be plotted in a so-called *semivariogram*, allowing the physical range over which the data are autocorrelated to be represented graphically (Fig. 7.69). The *sill* expresses the maximum range of the data, and the *nugget* is a measure of the minimum difference between two adjacent measuring points. In an area over which the deposition is assumed to vary randomly around an average value, the nugget should ideally be zero, but in practice it is not, because of the inherent stochastic fluctuation in the number of counts recorded in the detector system.

On-line visualization of γ-ray surveys is achieved by integrating multiple time series of

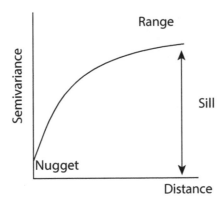

FIGURE 7.69 Typical semivariogram of gamma spectrometric data acquired in an area with a fluctuating gamma dose rate.

either γ-ray intensity or count rates in particular energy intervals during successive acquisitions. One method of incorporating both spectrometric information and the gradients in the time series is a so-called waterfall plot (Fig. 7.70). The colour coding and the time series plots in juxtaposed windows are a visual tool for the operators in the vehicle, or the backup force who receive the mobile radiometry data remotely and continuously analyse and assess the radiation environment in the surveyed area.

The strategy for aerial measurements using mobile gamma surveying in the initial phase after extensive fallout is to obtain maps over the affected areas to provide information on where remedial actions such as relocation or decontamination must be carried out. Airborne surveys are typically made at an altitude of 30–50 m, with a lateral flight path separation of 100–200 m, at velocities of 200–250 km h^{-1}. The time required to cover a zone around a NPP site with a line density of 200 m is 12–24 h. The reliability of the γ-ray spectrometry system can be checked by regularly performing standard itineraries or flying over a well-known area, to check for constancy and repeatability over time. To convert the γ-ray pulse height distributions into the activity deposition on ground, A_{dep}, calibration measurements must be carried out using standard surfaces with various γ-ray-emitting radionuclides. Car-borne measurements can provide important information on the actual γ-ray dose rate at ground level and the local distribution of ground deposition in urban areas, to detect hot spots. Car-borne surveys near a damaged facility can also provide information on the initial passage of the plumes (Fig. 7.71).

Foot-borne mobile surveys are required for mapping off-road terrain, and in dense urban areas. Foot-borne or cart-based mobile surveys can be used to locate lost sources or for detailed mapping of surfaces that are being considered for decontamination. Using cart-based systems also allows the use of collimated γ-ray detectors, which significantly improves the spatial resolution and decreases the autocorrelation between adjacent data points. Figure 7.72 gives an illustration of operators mapping an area possibly containing sealed sources. Note that if dispersed or resuspended radioactivity is suspected, protective clothing must be worn during the survey.

7.3.7 Networks of Radiometry Systems for Monitoring

A network of radiation detectors designed for outdoor long-term use is crucial for environmental monitoring and to provide an early warning of elevated radiation levels. Many

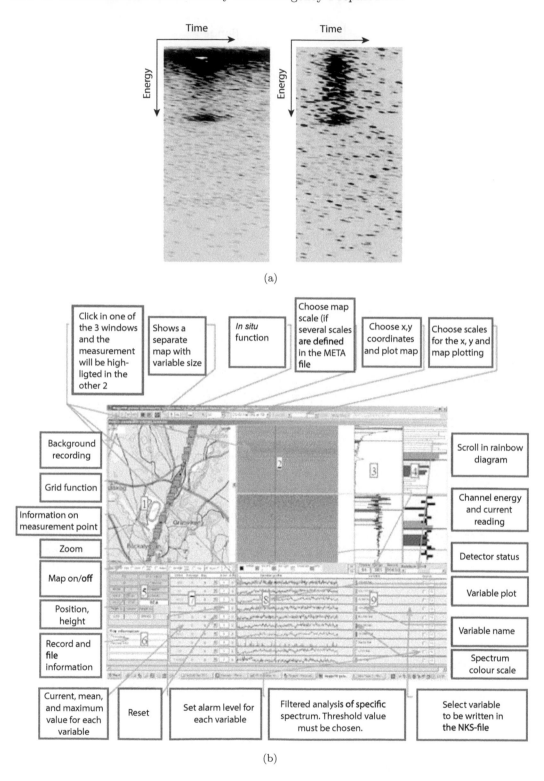

FIGURE 7.70 (a): A waterfall plot of mobile gamma spectrometric data, with photon energy on the downward y-axis, and time on the x-axis. (b): Various graphical representations of time series and spectrometric data acquired by a mobile gamma spectrometric system. This window is the working tool for the operators carrying out the mobile survey. See also colour images.

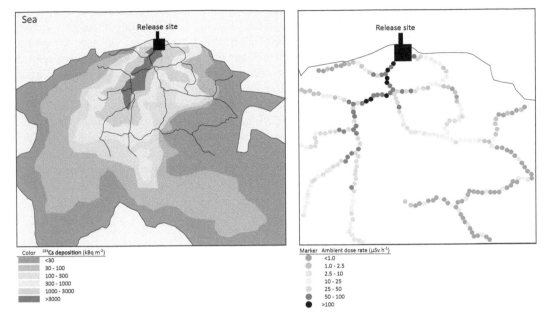

Color	^{134}Cs deposition (kBq m^{-2})
	<30
	30 - 100
	100 - 300
	300 - 1000
	1000 - 3000
	>3000

Marker	Ambient dose rate (μSv h^{-1})
	<1.0
	1.0 - 2.5
	2.5 - 10
	10 - 25
	25 - 50
	50 - 100
	>100

FIGURE 7.71 Left: Schematic activity deposition data map from airborne surveys of region close to an affected NPP releasing fission products and fission products that have been neutron activated, such as ^{134}Cs. Black lines symbolize road network close to the site. Right: Schematic doserate map from car-borne measurements close to the affected NPP site, indicating the large gradients and highly inhomogeneous nature of the ground deposition. See also colour images.

(a)   (b)

FIGURE 7.72 (a): Cart carrying mobile γ-ray spectrometry equipment. Collimation is used to improve the spatial resolution of measurements of ground deposition. (b): A mobile γ-ray spectrometry system carried as a backpack, used for mapping off-road areas.

industrialized countries developed air monitoring stations in the 1950s as a consequence of the nuclear weapons threat during the Cold War. The air activity concentrations of fission products such as ^{131}I, ^{137}Cs and ^{133}Xe were determined, and could be used to estimate the dry deposition [Bq m^{-2}] of these radionuclides. Initially, some stations were operated manually, and the data were reported and samples sent to a central laboratory affiliated to the institute responsible for the measurements. Precipitation was often collected at air sampling stations, which enabled the estimation of wet-deposited fallout of ^{90}Sr/^{90}Y and ^{137}Cs. After the Chernobyl accident in 1986 there was increased awareness of the need for an early warning system, and some of these stations were automated to reduce the delay between radiometry measurements and the reporting of data to the institute or authority. Some examples of measuring networks in Sweden and in Europe are given below.

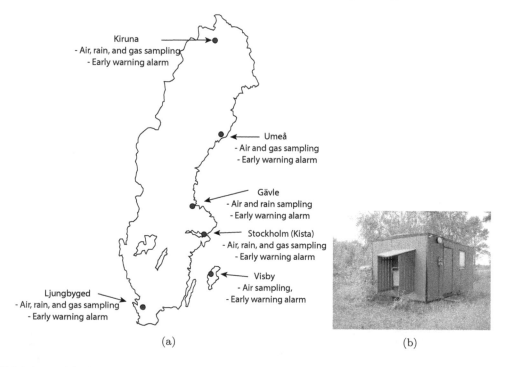

(a)   (b)

FIGURE 7.73   (a): A map of Sweden showing the location of air and precipitation sampling stations. (b): A photograph of one of the six air and precipitation sampling stations in Sweden. Air activity concentration of ^{137}Cs in μBq m^{-3} measured in Stockholm during the period 1957–2013 is shown in Figure 1.27.

Regular air and precipitation sampling are carried out at six stations in Sweden (see Fig. 7.73). The air sampling system also includes a filter for the collection of volatile radionuclides such as gaseous iodine and ^{133}Xe. The air filters are replaced twice a week, and the precipitation filter and the charcoal filter for the volatile radionuclides are changed weekly. The samples are then dispatched to a laboratory for radiometric assessment. Air filter data are aggregated to create weekly values, and the precipitation filter values are combined for a month.

Another Swedish network for radiation monitoring measures the gamma dose rate using GM counters at 28 locations throughout Sweden (Fig. 7.74). These stations are co-located with meteorological stations managed by the Swedish Meteorological and Hydrological Institute. The GM counter systems allow for a wide range of gamma dose rates to be measured:

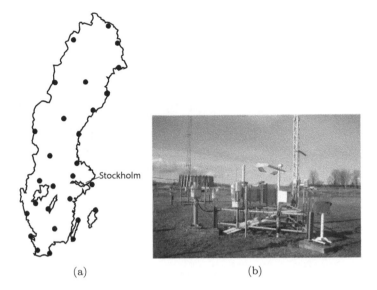

FIGURE 7.74 (a): A map of γ dose rate monitoring stations in Sweden. (b): A γ monitoring station, including the wind gauge and precipitation collector. Photograph reproduced with permission from the Swedish Radiation Safety Authority.

10 μSv h^{-1} to 10 Sv h^{-1}. The sensitivity of the GM counters makes them suitable for the rapid detection of both fast and slow changes in the ambient gamma dose rates. Data are automatically transferred to the National Radiation Safety Authority. If any of these stations reports a dose rate exceeding a pre-set alarm level, the officer on duty at the authority will be alerted automatically.

Sweden also has a (local-level) network, consisting of measuring sites for ambient gamma dose rates in each municipality (including over 280 sites). Local authorities in the municipality carry out a dose rate measurement at a pre-defined reference point every seven months (Fig. 7.75). The data are recorded and compiled in a national database. These measurements provide baseline levels for natural and existing radiation levels, which can be used in the event of an RN emergency (the non-natural radiation is mainly from the ^{137}Cs Chernobyl fallout).

There are also international networks of measurement systems for detecting enhanced radiation levels. Several European countries have joined the Radiological Data Exchange Platform, EURDEP (EURDEP, 2015), in which stationary monitoring measurements originating from national monitoring programmes are compiled into a common searchable database, enabling on-line monitoring of the radiological situation across the continent (Fig. 7.76). The IAEA coordinates a network for air sampling for non-proliferation surveillance. The aim of the International Radiation Monitoring Information System (IRMIS; IAEA 2016) is to provide radiation monitoring data routinely and reliably to the relevant authorities in countries who have signed the Convention on Early Notification of a Nuclear Accident (Early Notification Convention) and the Convention on Assistance in the Case of a Nuclear Accident or Radiological Emergency (Assistance Convention). It has previously been recognized that national radiation detection networks could not easily be merged into an international warning system, because of problems associated with data formats, etc. IRMIS attempts to gather only certified data that have been submitted by national authorities, in

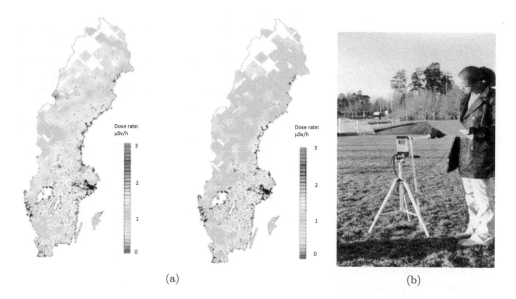

(a)   (b)

FIGURE 7.75 (a): Summer and winter maps of the ambient gamma dose rate (as defined in Section 2.5) measured with a portable dual GM counter instrument. The gamma dose background varies with season in the northern temperate and subarctic zones due to snowfall; the snow layer providing a shield from ground shine. See also colour images. (b): A photograph of a portable dual GM counter mounted on a tripod at a reference point 1 m above the ground.

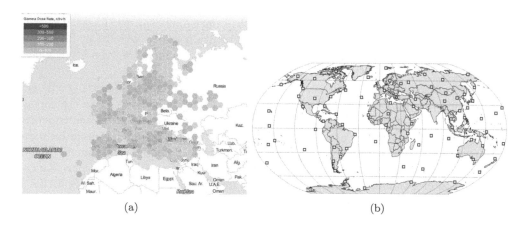

(a)   (b)

FIGURE 7.76 (a): The European network of gamma radiation monitoring stations included in EURDEP. (b): Global map of air sampling stations for non-proliferation control included in the CTBTO network. Data reproduced with permission from EURDEP and the CTBTO.

order to avoid the need for further data authentication and validation. The data format is based on a similar format to that used in EURDEP.

The Comprehensive Nuclear Test Ban Treaty Organization (CTBTO 2015) oversees an international monitoring system of more than 200 certified measurement stations, including seismic detection and the detection of airborne radionuclides that are associated with nuclear detonations, such as the noble gas ^{133}Xe (Fig. 7.76). The CTBTO has recently received global attention from detecting seismic activity due to detonations localized to North Korea (for example in 2013 and 2016), and the analysis regarding the potential involvement of fissile materials in these events.

7.4 TREATMENT, MONITORING, AND TRIAGE IN PATIENT CARE

7.4.1 The Medical Response to an RN Emergency

According to the IAEA (IAEA 2000), the aim of the medical response to an RN emergency must be to reduce the radiological consequences of the accident. The most important aim is to prevent radiation-induced tissue effects (such as acute radiation syndrome), but also to reduce the stochastic effects of radiation exposure. Even when only medical countermeasures are considered (in contrast to physical measures such as evacuation and relocation, described in Section 7.2), several organizations are involved, and it is therefore vital that all those involved in emergency response have a well-defined role in the management of the emergency. It is also necessary that this role is well understood throughout the organization, from the national coordinator to the members of the response team at the accident site. The main tasks of the medical emergency response team are therefore (i) to treat life-threatening injuries, (ii) to provide other medical care at the site (and associated duties, see below), and (iii) to take part in continuous training in the management of RN emergencies.

Scenarios involving RN emergencies, such as NPP accidents and accidents related to other industrial and medical radiation sources, will give rise to exposures that may cause acute radiation effects to the whole body, including the bone marrow and skin. Damage or accidents related to nuclear arms manufacturing may cause exposure to alpha emitters such as ^{238}Pu, ^{239}Pu, ^{240}Pu, and ^{210}Po, the latter of which can be used as a modulated neutron initiator (^{210}Po+^{9}Be) in some types of nuclear weapons.

The most common radiological accidents leading to medical complications are associated with medical radiation and orphan radiation sources that have either been lost or stolen (recall Fig. 7.2). These types of incidents are revealed by the onset of radiation skin damage to those exposed to the source. There is often a considerable time delay between the onset of exposure and the final diagnosis (up to several weeks). The exposures are often combined with external body contamination. These kinds of accidents are the result of poor knowledge and experience of radiation and its health consequences among the people involved, such as operators of radiographic equipment and scrap metal dealers. Moreover, the fact that the accidents have been discovered after a relatively long time is also due to poor knowledge of the effects of radiation among general practitioners who first encounter the symptoms of the exposed individuals.

For a country or a region to achieve adequate medical preparedness for RN emergencies, an infrastructure must exist that enables *the functional requirements* of the response to be carried out, as well as responders who can meet the objectives of this response. The infrastructure must include an authority or a formally appointed organization responsible for operational measures, for designating the responsibilities of the organizations involved, for plans and procedures related to the responders, for coordinating the various responding organizations, and for providing logistic support such as the coordination of medical care centres. The functional requirements include (i) initial assessment and classification of the accident, (ii) a notification and alert system, from the first worker on the scene (response initiator) to other levels such as medical centres and national expert assistance, (iii) accident condition mitigation, (iv) emergency worker protection, (v) first responder assistance, and (vi) media relations.

The IAEA also suggests that the functional requirements of the medical assistance in an RN emergency include the following: (i) the capability to provide immediate on-site first aid, (ii) the development of guidelines for decontamination, (iii) the provision of initial treatment and transportation for injured and exposed victims, (iv) predefined medical centres that can carry out specialized treatment of radiation injuries, and (v) planning for triage and treatment of patients using the medical resources available.

The information required for local authorities to be able to manage a medical RN emergency must at least include the locations of the radiation sources used locally. This information may sometimes be treated as classified by national radiation safety authorities. The information must include the type of source, the source strength and categorization (see Section 7.1), as well as the types of radiation-generating devices, such as x-ray units for industrial use. Furthermore, local authorities must also have an overview of the regular transport of radioactive cargo through the area. The authorities must consult various experts in order to identify possible accident scenarios related to these sources, and to estimate the numbers of people that might be affected in these scenarios. The accident scenarios form a basis for conceiving appropriate emergency planning and for dimensioning the resources involved.

The local authorities providing a base for off-site emergency management should prepare and have available the following lists:

- Names and contact information of medical support staff in the municipality/county.

- Important medical facilities on the local, regional and national levels having some auxiliary capacity in a crisis situation.

- Specialized care for the treatment of radiation injuries at national level.

- Specialized radiation injury care abroad.

- Equipment needed for monitoring (such as decontamination facilities), treatment and triage.

- Ambulance and pre-hospital care management.

The basic principles of medical care of exposed individuals includes general as well as specific methods of mitigating the possible effects of radiation exposure and contamination. Medical care is also divided into care at the accident site (on-site care) and care at medical centres (off-site care). These will be briefly discussed in the following sections.

7.4.2 Monitoring at the Accident Site

7.4.2.1 Initial Emergency Response at an RN Accident Site

The IAEA (IAEA 2000) defines seven general steps in the initial medical RN emergency response:

1. The first rescue worker to arrive at the accident site must assume the role of officer in command until relieved or a replacement arrives. In an RN emergency this person is defined as the response initiator (Fig. 7.77).

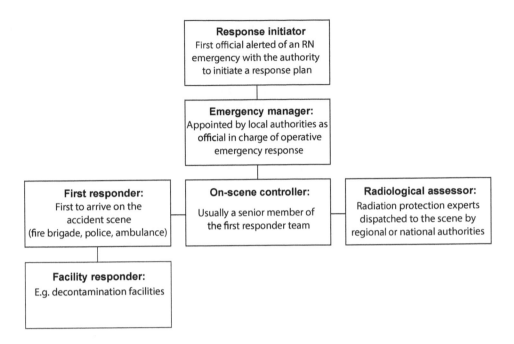

FIGURE 7.77 On-site medical emergency response as proposed by the IAEA (IAEA 2000).

2. Personal dosimeters, as well as protective clothing should be worn (if available) by the rescue workers. No rescue workers should enter the accident site without these items, unless life-saving actions are to be carried out.

3. Several medical actions are now implemented: (i) localizing injured individuals and rescuing them as soon as possible, (ii) medical triage (see below) of the affected victims to identify people who must be treated at once for life-threatening injuries, (iii) removal of casualties from the hot zone (hazard area) to stop radiation exposure as soon as possible, and finally (iv) calling for additional medical assistance if necessary. It is especially emphasized that in this step life-saving actions should not be delayed because of the presence of radiation.

4. Decontamination procedures are now initiated. Once specialized decontamination teams have arrived they must perform the following: (i) triage and isolation of contaminated persons, (ii) removal and packaging of all contaminated clothing and personal belongings, (iii) preparation of casualties for transport to a medical centre, for example, by covering wounds with sterile dressings, (iv) wrapping affected patients in blankets before transportation to the medical centre by ambulance to prevent contamination of ambulance personnel and equipment.

5. The police make lists of names and addresses of the patients and make records of any personal affects that have been collected during decontamination. Some of the victims may be questioned regarding the event, especially if a malevolent act is suspected to be the cause of the RN accident.

6. Notification of the medical centre of the arrival of the casualties (especially the A&E unit), including the nature of the accident (which hazardous substances have been involved) and extent of injuries.

7. Contamination monitoring of rescue workers, material and equipment for external contamination, and decontamination are done by the radiological assessor or by the facility responder. If there is no medical urgency regarding the personnel involved, they should be instructed not to leave the accident site before being checked for external contamination. It is important that no one involved at the scene of the accident removes any equipment or other objects from the site without them being checked for contamination.

7.4.2.2   Monitoring of Casualties and Personnel

In another IAEA document, (IAEA 2006), further operative instructions are given on the role of various actors, such as the radiological assessor at the RN accident site. The potential presence of antagonists is included in these instructions.

- Do not delay treatment and transportation of severely injured persons by performing decontamination procedures. It is stressed that life-saving operations must be prioritized. As in the previous steps, it is recommended to simply remove the victims' clothes and wrap the victims in blankets if they are in need of urgent medical treatment.

- Search for weapons: If malevolent acts or the presence of terrorists are suspected, body searches for weapons must be conducted of all persons leaving the accident site. It is important that the rescue workers can operate safely and not be in danger from armed perpetrators.

- Before using portable radiometry equipment such as the γ intensity survey meter described in Section 7.2.2.3, a test is to be carried out at a point distant from the hazard zone of the accident site to ensure that the equipment is functioning correctly. A spare portable instrument should be kept in reserve near the accident site, but in a non-contaminated zone.

- A contamination control area should be established near the hazard zone, where the gamma dose rate should not exceed 0.5 μSv h^{-1}. It should be noted that the set-up of full-scale decontamination facilities may take some time (up to 1 d) depending on the availability of such in the near geographical region.

- Personnel and members of the public at the accident site are instructed to not eat, drink, smoke or touch their mouths before washing their hands. They should refer to the media for further instructions from the emergency management.

A location with as low a gamma background as possible must be identified inside the cordoned area for monitoring individuals using portable gamma meters, for example, mobile *in vivo* measurements (Section 7.3.3). The personnel engaged in the screening of individuals for external contamination must wear gloves and change them regularly after contact with contaminated individuals. The radiometry instruments should be protected, for example, by wrapping in thin plastic. (Examples of appropriate instruments are given in Section 7.2.2.3.) Regular measurements and contamination control of the operators themselves, as well as of the equipment, must be carried out to check for gradual build-up of contamination. For contamination monitoring, for example, using beta probes, attention should be focused on the hands, feet and hair, as well as pockets and any "dirty" patches on clothing. The probe should be kept at a distance of 10 cm away from all these surfaces. However, this measuring geometry will only detect the presence of β-particle- and γ-ray-emitting radionuclides. If it is suspected that α-particle emitters are present, the probe must be held only a few centimetres away from the surface. Note that for α-particle monitoring, the detector entrance window must *not* be covered with protective plastic, as this would prevent almost all the α-particles from reaching the detector. All these measurements must be carefully documented. The IAEA (IAEA 2006) and *The TMT Handbook*, 2009, provide examples of suitable forms for the documentation of contamination monitoring of individuals.

The following instructions must be given to the monitored individual: If the external γ-ray dose rate is less than 1 μSv h^{-1}, or the total β/γ contamination rate is less than 10 kBq cm^{-2}, the person is instructed to change clothes and follow the instructions given by the media. If either of the aforementioned levels is exceeded, the person must be sent to the decontamination facility, if available at the site, otherwise the same instructions as those for lower levels are given.

Contamination monitoring should also be performed on materials, equipment and vehicles leaving the hot and warm zones. The monitoring area should be close to the outer cordon of the cold zone, and the gamma dose rate should be <1 μSv h^{-1}. As with casualties, equipment and materials are to be monitored using β/γ probes at a distance of 10 cm from the surface. On vehicles, scanning should focus on tyres and wheels, air filters and inner compartments. A designated decontamination unit should be set up for equipment and materials. Waste water should be collected if possible.

7.4.2.3 Decontamination of Casualties and Personnel

As previously mentioned, the IAEA stresses that the transport of severely injured casualties from the accident site should not be delayed by performing decontamination. Removing the victims clothes will remove roughly 90% of the external contamination. At the site of a large incident or accident, a decontamination facility (see Fig. 7.78) should be set up close to the boundary between the warm and cold zones, where full-scale decontamination (including wet showering) can be carried out. Decontamination of severely injured casualties must be performed at medical centres and not at the accident site.

FIGURE 7.78 A decontamination unit providing complete decontamination of individuals exposed to hazardous substances such as chemicals and radioactive materials (photograph by R. Finck).

The radiological assessor, often a radiation protection expert from a nearby university or research establishment, will probably have to communicate to other first responders that externally irradiated individuals do not need to be decontaminated. Medical staff can also misinterpret the term *radioactive contamination*. It must be stressed that cross contamination from one person or object to another is unlikely to cause high radiation exposures of the newly contaminated individual. Another misconception among medical staff, also mentioned in Section 7.2, is that radioactive contamination mimics the process of biological contamination in which a newly contaminated person can actually be harmed more severely than the person transmitting the contamination. Moreover, radiological decontamination is not as dependent on time as chemical decontamination. The radiological assessor must be aware that many first responders are more comprehensively trained in handling chemicals than radioactive material, and that emergency medical staff are probably more familiar with biologically hazardous substances than radioactive materials.

While waiting for decontamination, wet wipes should be distributed for cleaning the hands and face, as well as plastic bags for the collection of belongings such as keys, credit cards, coins and medication. Forms should be filled in for each individual, including personal data and information on decontamination monitoring. Information should also be included so that it is possible to identify the plastic bag containing the individual's personal belongings. When the form is completed it should thus contain data concerning (i) personal identity, e.g. name, address, date of birth, social security number, (ii) the activity measurements made on the patient and the results, and (iv) the name of the person carrying out the measurements. A copy of this form is also handed to the individual.

The effectiveness of decontamination must be evaluated by repeating gamma and β/γ probe measurements on the decontaminated individual. The decontamination of seriously injured casualties should be carried out at a medical centre. However, if such procedures are deemed necessary on-site, they must be carried out by trained staff in order not to cause further injury to the casualty. Teams of at least two people are required for decontamination of each person. *The TMT Handbook* recommends that the procedures that are to be carried out are explained to the individual.

In the first step of decontamination, the individual is *disrobed*, if necessary, by cutting and removing their clothes. It is not recommended that clothing be pulled over the head. Clothing is then placed in bags that are labelled with a tag identifying the owner. Spectacles are washed and dried before being returned to the individual.

The second step is washing, in which areas of the body are gently washed with warm water. About 5 ml of soap per litre of water is recommended. Affected areas are wiped gently with a sponge, after which they are rinsed with warm water. Do not use hair conditioners as these may fix the contamination to the hair. Washing should start with the hair and face and then move downwards, with the individual standing. After this, all open wounds are flushed with saline and covered with dressings. Special attention should be paid to skin folds, hair, nails, ears, and legs if they have not been covered by clothing. During washing and rinsing, avoid scrubbing which may cause irritation and ablation of the skin.

Washing and rinsing should be continued until there is no noticeable difference in the external gamma dose rate between two repeated measurements, following a full washing procedure. *The TMT Handbook* suggests that the ambition should be to reduce the external dose rate to below twice the background at the decontamination facility, but if no difference is seen between repeated cycles, decontamination should be stopped. Finally, the individual is dried and given clean clothing.

As mentioned in the previous section, the IAEA recommends that individuals who exhibit external γ dose rates less than 1 μSv h^{-1} be instructed to go home and decontaminate themselves by thorough washing. They are to be instructed to undress at the doorway to their house or garage, removing their clothing and placing it in plastic bags. The nose must be blown gently, and eyes and ears must be washed. A shower is then taken using warm water and soap, but avoiding scrubbing or mechanical damage to skin. Baths and showers should be thoroughly rinsed, as well as the vehicles in which the person drove home. More detailed instructions can be found in *The TMT Handbook,* 2009.

For contaminated objects, if the γ dose rate exceeds 10 μSv h^{-1}, or if β/γ contamination exceeds 10 kBq cm^{-2}, the object is to be cleaned with solvents, scrubbing with brushes, and flushing with water, for example, from a fire extinguisher. If the decontaminated object still exhibits a γ dose rate >1 μSv h^{-1} the following alternatives are suggested. In the case of noncritical equipment, the object should be isolated at the accident site. A receipt is given to the owner. If the object consists of equipment that could be useful for the rescue operations, it should only be used for critical response activities. Generally, contaminated objects should not be released from the accident site unless the radiological assessor deems that their use compensates for the potential radiological hazard resulting from the contamination.

7.4.2.4 Management and Triage of Patients in Cases of Mass Casualties

The term triage is used in RN emergencies, as in other crisis situations including war. Triage is used to manage mass casualties rationally, by grouping victims based on their medical condition. The name is derived from its original meaning of division into three categories by the French army in the First World War: (i) those who are dead or who will die no matter what treatment they receive (focus on pain relief), (ii) those who will live regardless of treatment (the "walking wounded"), and (iii) those who will benefit from medical treatment of various kinds (Fig. 7.79). Modern-day triage typically has four or five categories.

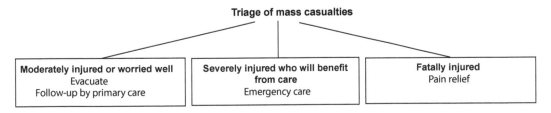

FIGURE 7.79 Principle of triage as defined in the original use of the term.

The IAEA suggests that a designated area should be set up at the accident site for triage of patients. The area should be identifiable by a flashing blue light. Victims in the triage category "dead and the severely injured" must be given pain relief and provided with identity tags. First aid must be given to those who will still benefit from emergency medical care. If there is no significant external contamination, or if it is not possible to remove the clothes of severely injured victims, wrapping blankets around them may limit the contamination of equipment such as stretchers. An on-site estimate must be made of the number of casualties, as well as of the number of casualties the rescue teams can attend to at a time. Family members should not be separated from each other.

Further categorization of casualties must be carried out on-site to identify those who require specialist care that is not available at regional or even national medical facilities. If it is suspected that antagonists are present, the police must aid in the surveillance of the triage site to increase the safety of other casualties and personnel. A special information centre or hub should be established for receiving concerned individuals at the accident site who need medical advice. To alleviate the work load of the medical staff, those classified as worried well should be referred to such information hubs. It is very likely that in many small countries, national expertise and resources must be coordinated to sustain the functions of emergency management at the accident site.

In summary, the radiological assessor, together with the pre-hospital team must carry out the following tasks at the accident site:

- Organize a measuring site and a system for managing the measuring equipment.

- Set-up a decontamination facility.

- Measure external contamination before and after decontamination of casualties or personnel.

- Determine whether the radiation remaining on the induvial is due to internal or external contamination.

- Check that that the radiation level in the triage area is safe for patients and personnel.

- Manage waste and potentially contaminated personal belongings, and managing decontaminated personal belongings and equipment.

- Direct and guide the worried well to information hubs.

- Document dosimetry data for the casualties and others who have been monitored for contamination.

7.4.3   Monitoring at Hospitals

Large regional hospitals will usually have a medical radiation physicist responsible for radiation protection issues in clinical practice, such as x-ray diagnostics, nuclear medicine and radiation therapy. Such people are crucial in drawing up a local RN emergency plan for the A&E unit at the hospital. Plans for the reception of casualties who have been exposed to radiation, internally or externally, can be made by the radiation physicist and a qualified nurse or physician in charge of the A&E unit (Fig. 7.80). In the case of mass casualties involving hazardous substances such as radioactive materials, plans must be made together with the directors of the hospital and with local and regional authorities (see Section 7.2).

Other resources needed include radiation monitoring instruments which must be regularly tested and calibrated. If instruments in storage are obsolete, faulty or not properly calibrated, this may result in a false assessment of capacity and erroneous activity measurements, which may affect the initial emergency response. Moreover, protective clothing and equipment must be available, and it should also be fit for use in a wet decontamination unit (see Fig. 7.81a). Local medical services, such as standard clinical chemistry laboratories that can carry out lymphocyte counting, for example, are required to treat patients exhibiting ARS. At least one hospital covering a region with approximately one million inhabitants should have a decontamination unit with showers where waste water can be collected, not only for RN emergencies, but also for incidents involving exposure to chemicals. A regional hospital should also have designated staff in their A&E unit who have been trained in dealing with patients exposed to hazardous substances.

The operative role of the radiation protection expert when medically assisting in an RN emergency is to provide the treating medical team with accident reconstruction and dose estimates for the patients. The tools used for this purpose (see Section 7.3.4) are (i) physical and biological dosimetry, enabling assessment of external exposures, (ii) an environmental survey at the accident site, enabling further reconstruction of the exposure events, and (iii) environmental and metabolic models that facilitate predictions of the internal exposure.

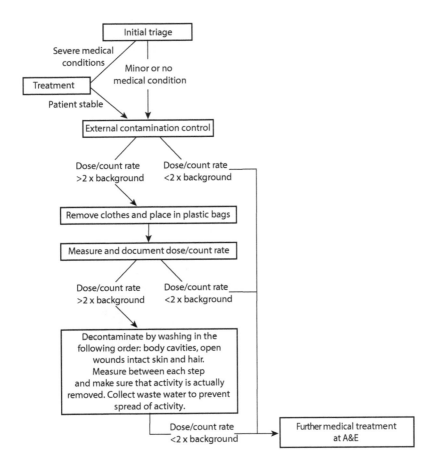

FIGURE 7.80 Flow chart describing the reception, management and treatment of a potentially contaminated patient from an RN accident site, as employed at a Swedish regional hospital.

The role of the treatment unit at the hospital is to make a clinical evaluation of the patient based on input from triage and information received from the on-site pre-hospital team. Together with supplementary information regarding radiological aspects, staff at the hospital can then make a prognosis and adapt therapies according to the patient's response to treatment.

When patients arrive at the A&E unit from the accident site, an area must be designated to receive them where it is possible to perform a rapid second contamination measurement and, if necessary, carry out further decontamination. If the person is still considered to be contaminated, but has a life-threatening injury, they must be taken to a treatment room dedicated to acute treatment of contaminated casualties. This room should preferably be close to the decontamination room. It is important that a radiation protection expert, such as the hospital physicist, is present, together with the emergency team. The role of the hospital physicist is to continuously assist with radiation measurements, and to interpret the health hazard, not only to the patient, but also to the medical staff. In most cases, the exposure of to hospital staff to radiation is negligible, even if the patient has been severely externally or internally contaminated.

Radiation exposure of healthcare providers is often negligible when treating patients with a surface contamination on the order of 1 kBq to 1 MBq m^{-2} of γ-ray emitters. For example, the dose rate to the members of a medical team standing approximately 20 cm away from a patient is 1.13 μGy h^{-1} per MBq m^{-2} surface contamination by ^{60}Co (Bridges 2006). For internal contamination, the ambient dose rate on the surface of an adult male patient is estimated to be at most 0.33 μGy h^{-1} per MBq of inhaled ^{137}Cs (Eastburg 2010). As a comparison, a deadly amount of ingested ^{137}Cs of 1 GBq would result in a maximum surface γ-ray dose rate of 33 mGy h^{-1}, and an internal dose rate of 13.7 mGy h^{-1} (based on a homogeneous internal distribution of the radionuclide).

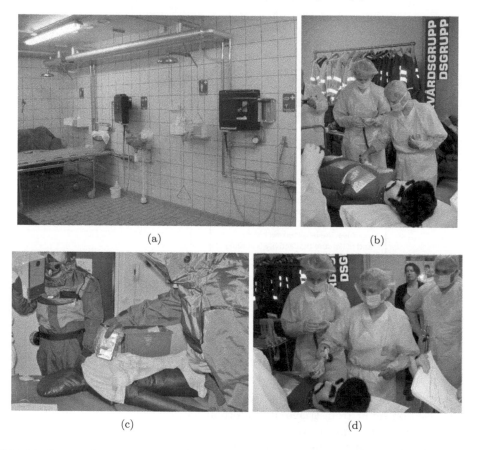

(a)   (b)

(c)   (d)

FIGURE 7.81 Images from the decontamination facility at Malmö University Hospital during training in 2010. (a): The reception area with decontamination facilities. (b) Arrival of the patient. (c) Contamination monitoring of clothing and personal effects. (d) Contamination measurements are made on the patient using a handheld α/β sensitive probe. Note that for enabling particle detection the probe must be held closer to the skin and without a protective plastic film at the entrance window of the probe.

Once the trauma or injury has been temporarily stabilized, the patient can be decontaminated using similar procedures to on-site decontamination. Based on external contamination surveys, the team can decide whether further decontamination is useful, or if treatment of the patient can continue immediately. The patient and the members of the team leave the decontamination area through a second exit, but only if it is confirmed that they are free from external contamination. Figure 7.81 gives an illustration of some of the aforementioned procedures with a contaminated patient.

Once a patient has entered the A&E unit, normal treatment begins. The task is now to establish the patient's accumulated radiation exposure, both from internal and external contamination. An initial estimate of the internal absorbed dose can be made using an exposure history based on information received from the accident site, which can indicate intake routes, such as inhalation or uptake through open wounds. Other useful information includes the physical state of the radioactive material (e.g. aerosol or fragments), together with useful dose conversion factors (for example, ICRP 1993; ICRP 1995; ICRP 2012). However, many modern hospitals have diagnostic equipment such as gamma cameras and SPECT systems that can be used to visualize and quantify uptake in the patient. Low-background whole-body measurements can be made, and the presence of gamma emitters in the body can be determined with a relatively high accuracy (Section 7.3.4). These measurements can then be used to obtain a more accurate estimate of the internal absorbed dose, and can be used as the basis for decisions on treatment, such as the use of Prussian blue to hasten the excretion of ingested radioactive caesium (as was used in the Goiania accident in 1987: Melo et al. 1998).

The external exposure of the patient is assessed by the on-site radiological assessor or the radiation protection expert at the A&E unit, based on dose rate data from the accident site, or on readings from personal dosimeters worn by the first responders. However, in an RN emergency it is likely that there will initially be no, or only very uncertain, data on the external exposure of the patient. The emergency medical team must therefore focus on symptoms indicating acute radiation-induced tissue effects and estimating the patient's total radiation exposure, using biological and physical retrospective dosimetry, based on samples taken from the patient (blood, hair, teeth, nails, etc.), or from their personal belongings. In retrospective dosimetry assessments, there is a time delay of at least two days when using biological dosimetry (apart from gross lymphocyte counts), and at least one day of delay when using physical dosimetry, assuming that materials and equipment are available for techniques such as EPR or OSL (see Section 7.3.4). During this time, the biological symptoms guide treatment and the prognosis of the patient. If compiled systematically, these can be used to obtain a relatively accurate prognosis concerning the radiological injury to the patient.

One approach to grading the severity of the acute radiation symptoms is the METREPOL system discussed in *The TMT Handbook* (and in Section 2.3.1.5). The METREPOL system relies on the qualitative judgement of radiation-induced damage in key regions of the body: (i) the neurovascular system, (ii) the blood, (iii) the skin, and (iv) the gastrointestinal tract. The injuries are scored on a four-level scale according to the severity of damage. Neurovascular (N) symptoms include nausea, vomiting, fatigue, fever, headache, depressed blood pressure, and cognitive deficits. Haematopoietic (H) symptoms include effects on the blood, such as depressed lymphocyte concentration (lymphocyte counts), and infections that may become manifest. Cutaneous (C) effects include effects on the skin such as erythema, swelling and oedema, blistering, and ultimately, necrosis. Hair loss is also an effect of acute doses to the skin of the head. Gastrointestinal (G) effects include diarrhoea, gastrointestinal bleeding, and abdominal cramps and pain. *The TMT Handbook* gives a brief description of the various degrees of severity of these symptoms. Section 2.3.1.5 also gives an example of how a METREPOL score is calculated.

Another example of a primary scoring system for predicting the prognosis is illustrated in Table 7.12. It is intended for use in the simplified screening of ARS in the case of mass casualties, and is based on the time delay in the debut of some key ARS-related symptoms, such as diarrhoea, nausea, and depressed lymphocyte concentration in the blood. *The TMT Handbook* discusses how the scoring is interpreted if the scores fall in different columns of Table 7.12. Caution must be employed in the scoring of these symptoms as several of them can also be related to psychosomatic stress, or other stressors at the accident site. Experience has shown that many of the worried well also exhibit very similar symptoms. Because of the time delay between the end of irradiation and the onset of symptoms, there is a period during which radiological experts cannot provide any concrete information to many affected patients: that is, whether or not they have been severely exposed.

TABLE 7.12   Scoring system for acute radiation syndrome for the rapid screening of mass casualties.

	Score I	Score II	Score III
Time after assumed exposure	<12 hours	<5 hours	<30 minutes
Erythema	0	+/-	+++ <3 h
Asthenia	+	++	+++
Nausea	+	+++	(-)
Vomiting (per day)	1	1–10	>10, intractable
Diarrhoea (no. of excretions per day)	2–3, bulky	2–9, soft	>10, watery
Abdominal pain	Minimal	Intense	Excruciating
Headache	0	++	Excruciating + intracranial hypertension
Fever	<38 °C	38–40 °C	>40 °C
Blood pressure	Normal	Temporary decrease	Systolic <80 mm Hg
Loss of consciousness	0	0	Temporary/Coma
Lymphocytes at 24 h	>1500 per μL	<1500 per μL	<500 per μL
Lymphocytes at 48 h	>1500 per μL	<1500 per μL	<100 per μL
Strategy	Outpatient monitoring	Hospitalization for curative treatment	Hospitalization (multiple organ failure and death predicted

Source: Data from Rojas-Palma et al. (2009).

The magnitude of the worried well dilemma was evident in the Goiania accident in 1987, when an abandoned ^{137}Cs therapy source was stolen by scrap metal thieves (see e.g., IAEA 1988). Unaware of the radiation hazard, the thieves started to open the casing of the source to retrieve the valuable metal parts of the shield. This led eventually to the dispersion of the ^{137}Cs source, in the form of CsCl, appearing as a bluish phosphorescent powder, seen by many hundreds of individuals. It was unfortunate that the "glowing salt" was regarded as an amusement for several families, including children, leading to a high number of exposed individuals. The time delay between the onset of exposure of the first person, to the discovery of the radiation exposures by the local authorities, was about fourteen days. The debut of ARS in one of the most exposed individuals occurred as early as two days after the first exposure. This person sought medical help but the practitioner misinterpreted his symptoms as an allergic reaction. The practitioner first began to suspect radiological exposure when several patients from the area sought treatment for similar ailments, and when personal testimony of the adverse effects of the particular substance was given by one of the family members affected. Within days, full-scale monitoring had begun, including assistance from international experts. A mass triage centre was set up at a sports arena, and in all about 112 000 people were monitored, of which approximately 250 were deemed to be significantly internally contaminated by ^{137}Cs. However, only 20 individuals were found to be exposed internally and externally to such an extent that acute effects could be observed. Of these 20 individuals, four suffered fatal injuries from the irradiation.

Other experience has also shown that γ-ray and x-ray irradiation at levels far below the exposures required for acutely observable radiation-induced effects (<1000 mSv) cause great alarm, and apparently warrant hospitalization. This happened in the Aitik incident in Sweden in 2010, where maintenance work was performed on an x-ray apparatus. Due to a failure in the alert system, the technicians were unaware that the x-ray unit was still turned on and emitting radiation. This incident is described by the Swedish Radiation Emergency Medicine Centre (2012). Of a total of fifteen people suspected to have been irradiated, eight were hospitalized, albeit for only a few days. Symptoms such as headaches, fatigue, nausea and abdominal pain were reported. In the final audit conducted by the Swedish Radiation Safety Authority, it was concluded that the most exposed individual probably received an effective dose in the range 160–650 mSv (the wide range indicating the large uncertainties associated with reconstructing radiation exposures retrospectively). None of the radiation exposures could therefore have caused the initially experienced symptoms.

Another lesson learnt from experience is that there is an unwarranted fear of radioactivity among medical staff, both physicians and nurses, who in some cases have refused to treat patients who were exposed to radiation (Vano et al. 2011). It is thus important that radiation protection experts and on-site radiological assessors affirm that external exposure from α, β, and γ radiation does not cause any delayed radiation of bystanders or medical personnel. The difference between internal contamination and external exposure is often not understood.

The medical response is the most essential part of an emergency response to a radiological or nuclear crisis involving the exposure of individuals. The local and regional resources available, as well as the functional requirements mentioned in Section 7.4.1, must be assessed in the RN emergency plan. Given the high turnover of medical staff at A&E units, the local hospital physics staff must regularly inform and train nurses and physicians in the routines for managing and treating individuals suspected of having been acutely exposed to radiation. In national planning, it is important to have a plan for consolidating hospital and medical resources in several regions, as well as at a national level. It is also important to plan for access to biodosimetry and specialized radiation syndrome treatment facilities. This must be considered in view of the global scarcity of laboratories with the required competence and capacity to handle the large numbers of samples generated by an event leading to mass casualties.

Although the main effort in RN emergency preparedness should be to mitigate the risk of accidents, the reason for which is briefly described, for example, by Ortiz et al. (2000), it is important to learn from previous experience and avoid past mistakes in dealing with RN emergencies when they do occur. Coeytaux et al. (2015) provide a systematic review of reported radiation overexposures between 1980 and 2013. The IAEA (IAEA 2015c) describes several RN incidents, for which the emergency management and the medical response are also described. These publications provide an important understanding of how authorities, experts, the public, and the individuals affected interact.

7.5 REFERENCES

- Aoyama, M., Hirose, K. and Igarashi, Y. 2006. Re-construction and updating our understanding on the global weapons tests ^{137}Cs fallout. *J. Environ. Monit.*, 8, 431–438.

- Bailey, R. M., Adamiec, G. and Rhodes, E.J. 2000. OSL properties of NaCl relative to dating and dosimetry. *Radiation Measurements.* 32, 717–723.

- Bassinet, C., Woda, C., Bortolin, E., Della Monaca, S., Fattibene, P., Quattrini, M.C., Bulanek, B., Ekendahl, D., Burbidge, C.I., Cauwels, V., Kouroukla, E., Geber-Bergstrand, T., Mrozik, A., Marczewska, B., Bilski, P., Sholom, S., McKeever, S.W.S., Smith, R.W., Veronese, I., Galli, A., Panzeri, L. and Martini, M., 2014. Retrospective radiation dosimetry using OSL of electronic components: Results of an interlaboratory comparison. *Radiation Measurements.* doi: 10.1016/j.radmeas.2014.03.016.

- Beerten, K., Reekmans, F., Schroeyers, W., Lievens, L. and Vanhavere, F. 2010. Dose reconstruction using mobile phones. *Radiation Protection Dosimetry.* 144(1–4), 580–583.

- Bernhardsson, C., Christiansson, M., Mattsson, S. and Rääf, C. L. 2009. Household salt as a retrospective dosimeter using optically stimulated luminescence. *Radiation and Environmental Biophysics.* 48(1), 21–28.

- Bridges, A. 2006. *Estimating the Radiation Dose to Emergency Room Personnel in an Event of a Radiological Dispersal Device Explosion.* Thesis. Georgia Institute of Technology.

- Bøtter-Jensen, L., Solongo, S., Murray, A. S. and Jungner, H. 2000. Using the OSL single-aliquot regenerative-dose protocol with quartz extracted from building materials in retrospective dosimetry. *Radiation Measurements.* 32(5–6), 841–845.

- Christansson, M., Bernhardsson, C., Geber-Bergstrand, T., Mattsson, S. and Rääf, C. 2014. Household salt for retrospective dose assessments using OSL: Signal integrity and its dependence on containment, sample collection, and signal read-out. *Radiation and Environmental Biophysics.* 53(3), 559–569.

- Christiansson, M., Geber-Bergstrand, T., Bernhardsson, C., Mattsson, S. and Rf, C. 2016. Performing retrospective dosimetry using salted snacks and nuts: A feasability study. *Radiat Prot Dosimetry*, DOI: 10.1093/rpd/ncw044.

- Coeytaux, K., Bey, E., Christensen, D., Glassman, E. S., Murdock, B. and Doucet, C. 2015. Reported radiation overexposures accidents worldwide, 1980–2013: A systematic review. Open Access Journal Website: http://www.ncbi.nlm.nih.gov/pmc/articles/PMC4366065/. Accessed 2016-03-13.

- Collison, R. 2005. Personal photon dosimeter trial—Devonport Royal Dockyard. *J. Radiol. Prot.* 25(1), 1–32.

- Conti, C. C., Bertelli, L. and Lopes, R. T. 1999. Age-dependent dose in organs per unit air kerma free-in-air: conversion coefficients for environmental exposure. *Rad Prot Dosimetry*. 86, 39–44.

- CTBTO. Preparatory Commission for the Comprehensive Nuclear Non-Test-Ban Treaty Organization. 2016. https://www.ctbto.org/tiles/pdf/CTBTO-Map-IMS-2015-10-23-All_Stations-Overview.pdf. Accessed 2016-03-13.

- Degteva, M. O., Kozheurov, V. P. and Vorobiova, M. I. 1994. General approach to dose reconstruction in the population exposed as a result of the release of radioactive wastes into the Techa River. *Science of the Total Environment*. 142(1–2), 49–61.

- DHH, Department of health and human services, Assistant Secretary for Legislation. 2016. Website: http://www.hhs.gov/asl/testify/t971001a.html. Accessed 2016-03-13.

- Eastburg, A. 2010. *Assessing the Dose after a Radiological Dispersal Device (RDD) Attack Using a Military Radiac Instrument*. Thesis. Georgia Institute of Technology.

- Eckerman K. F. and Leggett R. W. 1996. *DCFPAK: Dose Coefficient Data File Package for Sandia National Laboratory*, ORNL/TM-13347. Oak Ridge National Laboratory, Oak Ridge, TN, USA.

- EU, European Union. 2016. Eur-Lex. Website: http://eur-lex.europa.eu/legal-content/EN/TXT/PDF/?uri=CELEX:32016R0052&from=EN. Accessed 2016-03-20.

- EURDEP, European Radiological Data Exchange Platform. 2016. Webpage: https://eurdep.jrc.ec.europa.eu/Basic/Pages/Public/Home/Default.aspx. Accessed 2016-03-13.

- EUROANOS. 2009. *Generic Handbook for Assisting in the Management of Contaminated Inhabited Areas in Europe Following a Radiological Emergency*, EURANOS(CAT1)-TN(06)-09-02.

- Euratom. 2013. *Council Directive 2013/59/Euratom*. http://eur-lex.europa.eu/LexUriServ/LexUriServ.do?uri=OJ:L:2014:013:0001:0073:EN:PDF. Accessed 2016-03-20.

- Finck, R. 1992. *High Resolution Field Gamma Spectrometry and Its Application to Problems in Environmental Radiology*. Thesis, Lund University, Sweden.

- Fogh, C. L., Andersson, K. G., Barkovsky, A. N., Mishine, A. S., Ponamarjov, A. V., Ramzaev, V. P. and Roed, J. 1999. Decontamination in a Russian settlement. *Health Phys*. 76(4), 421–430.

- Froidevaux, P., Baechler, S., Bailat, C. J., Castella, V., Augsburger, M., Michaud, K., Mangin, P. and Bochud, F.O. 2013. Improving forensic investigation for polonium poisoning. *Lancet*. 382(9900), 1308.

- Geber-Bergstrand, T., Bernhardsson, C., Mattsson, S. and Rääf, C. 2012. OSL from tooth enamel and dental repair materials irradiated under wet and dry conditions. *Radiation and Environmental Biophysics*. 51, 443–449.

- Geber-Bergstrand, T., Bernhardsson, C., Christiansson, M., Mattsson, S. and Rääf C. L. 2014. Desiccants for retrospective dosimetry using optically stimulated luminescence (OSL). *Radiation Measurements*. 78, 17–22.

- Glasstone, S. and Dolan, P. J. 1977. *The Effects of Nuclear Weapons*. United States department of Defence and United States department of Energy. USA.

- Hjerpe, T. 2004. *On-line Mobile in Situ Gamma Spectrometry*. Thesis. Radiation Physics Lund. Faculty of Science, Lund University, Sweden. ISBN 91-85313-00-9.

- IAEA, International Atomic Energy Agency. 1988. *The Radiological Accident in Goiânia*. International Atomic Energy Agency. Vienna. http://www-pub.iaea.org/mtcd/publications/pdf/pub815_web.pdf. Accessed 2016-03-13.

- IAEA 2000. *Generic Procedures for Assessment and Response during a Radiological Emergency*. IAEA-TECDOC-1162. International Atomic Energy Agency. Vienna.

- IAEA, International Atomic Energy Agency. 2002a. *General Safety Standard—Report 2 Preparedness and Response for a Nuclear or Radiological Emergency*. International Atomic Energy Agency. Vienna.

- IAEA, International Atomic Energy Agency. 2002b. *Preparedness and Response for a Nuclear or Radiological Emergency*. IAEA Safety Standards Series No. GS-R-2. International Atomic Energy Agency. Vienna.

- IAEA, International Atomic Energy Agency. 2002c. *The Radiological Accident in Gilan*. International Atomic Energy Agency. Vienna. http://www-pub.iaea.org/MTCD/publications/PDF/Pub1123_scr.pdf. Accessed 2016-03-13.

- IAEA, International Atomic Energy Agency. 2003. *Method for Developing Arrangements for Response to a Nuclear or Radiological Emergency—EPRMETHOD (2003)*. International Atomic Energy Agency. Vienna.

- IAEA, International Atomic Energy Agency. 2005a. *Categorization of Radioactive Sources*. IAEA Safety Standards Series No. RS-G-1.9. International Atomic Energy Agency. Vienna.

- IAEA, International Atomic Energy Agency. 2005b. *Reassessment of the Atomic Bomb Radiation Dosimetry for Hiroshima and Nagasaki. Dosimetry System 2002. DS02. Volume 2*. International Atomic Energy Agency. Vienna.

- IAEA, International Atomic Energy Agency. 2006. *Manual for First Responders to a Radiological Emergency*. EPR–2006. International Atomic Energy Agency. Vienna.

- IAEA, International Atomic Energy Agency. 2007a. *Arrangements for Preparedness for a Nuclear or Radiological Emergency*. IAEA Safety Standards Series No. GS-G-2.1. International Atomic Energy Agency. Vienna.

- IAEA, International Atomic Energy Agency. 2007b. *Disposal of Radioactive Waste*. Specific Safety Requirements No. SSR-5. International Atomic Energy Agency. Vienna.

- IAEA, International Atomic Energy Agency. 2008. *Nuclear Safety Infrastructure for a National Nuclear Power Programme Supported by the IAEA Fundamental Safety Principles*. INSAG-22. International Atomic Energy Agency. Vienna. http://www-pub.iaea.org/MTCD/publications/PDF/Pub1350_web.pdf. Accessed 2016-03-13.

- IAEA, International Atomic Energy Agency. 2009. *Security of Radioactive Sources*. IAEA Nuclear Security Series No. 11. International Atomic Energy Agency. Vienna.

- IAEA, International Atomic Energy Agency. 2012. *Lessons Learned from the Response to Radiation Emergencies (1945–2010)*. EPR-Lessons Learned 2012. International Atomic Energy Agency. Vienna.

- IAEA, International Atomic Energy Agency. 2013. *Preparedness and Response for a Nuclear or Radiological Emergency in the Light of the Accident at the Fukushima Daiichi Nuclear Power Plant*. International Atomic Energy Agency. Vienna.

- IAEA, International Atomic Energy Agency. 2015a. IAEA, website: http://www-ns.iaea.org/tech-areas/emergency/ines.asp. Accessed 2016-03-23.

- IAEA, International Atomic Energy Agency. 2015b. *Nuclear Power Reactors in the World*. Reference Data Series No. 2. International Atomic Energy Agency. Vienna.

- IAEA, International Atomic Energy Agency. 2015c. IAEA Publications on Accident response. Website: http://www-pub.iaea.org/books/IAEABooks/Publications_-on_Accident_Response. Accessed 2015-12-01.

- IAEA, International Atomic Energy Agency. 2016. http://www-ns.iaea.org/downloads/iec/info-brochures/13-28111-irmis.pdf. Accessed 2016-03-13.

- ICRP 1991. *1990 Recommendations of the International Commission on Radiological Protection*. ICRP Publication 60, Ann. ICRP 21 (1–3).

- ICRP 1993. *Age-Dependent Doses to Members of the Public from Intake of Radionuclides—Part 2 Ingestion Dose Coefficients*. ICRP Publication 67. Ann. ICRP 23 (3–4)

- ICRP 1995. *Age-Dependent Doses to Members of the Public from Intake of Radionuclides— Part 4 Inhalation Dose Coefficients*. ICRP Publication 71. Ann. ICRP 25 (3–4).

- ICRP 1996. *Age-Dependent Doses to the Members of the Public from Intake of Radionuclides—Part 5 Compilation of Ingestion and Inhalation Coefficients*. ICRP Publication 72. Ann. ICRP 26 (1).

- ICRP 2007. *The 2007 Recommendations of the International Commission on Radiological Protection*. ICRP Publication 103. Ann. ICRP 37 (2–4).

- ICRP 2009a. *Application of the Commission's Recommendations for the Protection of People in Emergency Exposure Situations*. ICRP Publication 109. Ann. ICRP 39 (1).

- ICRP 2009b. *Application of the Commission's Recommendations to the Protection of People Living in Long-term Contaminated Areas after a Nuclear Accident or a Radiation Emergency*. ICRP Publication 111. Ann. ICRP 39 (3).

- ICRP, 2010. *Conversion Coefficients for Radiological Protection Quantities for External Radiation Exposures*. ICRP Publication 116, Ann. ICRP 40(2–5).

- ICRP 2012. *Compendium of Dose Coefficients Based on ICRP Publication 60*. ICRP Publication 119. Ann. ICRP 41 (Suppl.).

- ICRU 1998. *Conversion Coefficients for use in Radiological Protection against External Radiation*. ICRU Publication 57. Nuclear Technology Publishing, Ashford, UK. ISBN 1 870965 94

- ICRU 2002. *Retrospective Assessment of Exposures to Ionising Radiation.* ICRU Publication 68. ICRU Publications, Bethesda, USA.

- Inrig, E.L., Godfrey-Smith, D. I. and Larsson, C. L. 2010. Fading corrections to electronic component substrates. *Radiation Measurements.* 45(36), 608–610.

- Jain, M., Andersen, C. E., Bøtter-Jensen, L., Murray, A. S., Haack, H. and Bridges, J. C. 2006. Luminescence dating on Mars: OSL characteristics of Martian analogue materials and GCR dosimetry. *Radiation Measurements.* 41(7-8), 755–761.

- Katata, G., Chino, M., Kobayashi, T., Terada, H., Ota, M., Nagai, H., Kajino, M., Draxler, R., Hort, M. C., Malo, A., Torii, T. and Sanada, Y. 2015. Detailed source term estimation of the atmospheric release for the Fukushima Daiichi nuclear power plant station accident by coupling simulations of an atmospheric dispersion model with an improved deposition scheme and oceanic dispersion model. *Atmos. Chem. Phys.* 15, 1029–1070.

- KIT. 2016. Karslruhe Institute of Technology. Website: https://www.iket.kit.edu/english/294.php. Accessed 2016-03-20.

- Kock, P. 2012. *Orphan Source Detection in Mobile Gamma-Ray Spectrometry: Improved Techniques for Background Assessment.* Thesis. Faculty of Science, Lund University, Sweden. ISBN 978-91-7473-385-3.

- McColl, N. P. and Prosser, S. L. 2002. *Emergency Data Handbook.* NRPB-W19. National Radiation Protection Board, UK. ISBN 0 85951 490 0.

- Melo, D. R, Lipsztein, J. L., Oliveira, C. A. N., Lundgren, D. L., Muggenburg, B. A and Guilmette, R. A. 1998. Prussian blue decorporation of ^{137}Cs in humans and beagle dogs. *Radiation Protection Dosimetry.* 79(1–4), 473–476.

- Moseley, T. 2010. *Transport of Radioactive Materials by Road.* AURPO. Guidance Note No. 6. Association of University Radiation Protection Officers. UK.

- Nordqvist, M. 2013. *Modeling Protection Coefficents of Buildings during a Release of Radioactive Materials.* MSc thesis. Uppsala University. Website: https://www.diva-portal.org/smash/get/diva2:608098/FULLTEXT01.pdf. Accessed: 2016-03-25.

- Ortiz, P., Oresegun, M. and Wheatley, J. 2000. Lessons from major radiation accidents. *Proceedings of the 10th International Congress of the IRPA* (Hiroshima, May 2000). T-21-1, P-11-230. Japan Health Physics Society, Tokyo (Japan).

- Rääf, C. L. 2002. Multiregression analysis on predictors determining the urinary potassium excretion and radiocaesium body burdens in two different ethnic groups, (ed. F. Bréchignac) In: *The Radioecology-Ecotoxicology of Continental and Estuarine Environments ECORAD 2001*, Proceedings Vol II of the international congress, 3–7 September 2001, Aix-en-Provence, France, Radioprotection – colloques 37, C1–1335–1340.

- Ramzaev, V. and Göksu, H. Y. 2006. Cumulative dose assessment using thermoluminescence properties of porcelain isolators as evidence of a severe radiation accident in the Republic of Sakha (Yakutia), Russia, 1978. *Health Physics.* 91(3), 263–269.

- Ramzaev, V., Barkovsky, A,, Mishine, A. and Andersson, K. G. 2013. Decontamination tests in the recreational areas affected by the Chernobyl accident: Efficiency of decontamination and long-term stability of the effects. *Journal of the Society for Remediation of Radioactive Contamination in the Environment.* 1(2), 93–108.

- Ravi, M. and Solomon, F. D. P. 2002. A rapid biodosimetric technique at the human glycophorin-A locus. *Int J Hum Genet.* 2(4), 251–254.

- Roed, J., Andersson, K. G. and Sandalls, J. 1990. Reclamation of nuclear contaminated urban areas. In *Proceedings of the BIOMOVS Symposium on the Validity of Environmental Transfer Models in Stockholm*, 157–167, SSI , ISBN 91-630-0437-2.

- Rojas-Palma, C., Liland, A., Naess Jerstad, A., Etherington, G., del Rosario Pérez, M., Rahola, T. & Smith, K. (Eds.). 2009. *TMT handbook, Triage, Monitoring and Treatment of people exposed to ionising radiation following a malevolent act.* SCK-CEN, NRPA, HPA, STUK, WHO. http://www.tmthandbook.org/. Accessed 2016-03-14.

- Romm, H., Ainsbury, E., Barnard, S., Barrios, L., Barquinero, J. F., Beinke, C., Deperas, M., Gregoire, E., Koivistoinen, A., Lindholm, C., Moquet, J., Oestreicher, U., Puig, R., Rothkamm, K., Sommer, S., Thierens, H., Vandersickel, V., Vral, A. and Wojcik, A. 2013. Automatic scoring of dicentric chromosomes as a tool in large scale radiation accidents. *Mutat Res.* 756, 174–183.

- Roscoe, R. J., Deddens, J. A,, Salvan, A. and Schnorr, T. M. 1995. Mortality among Navajo uranium miners. *Am J Public Health.* 85(4), 535–540.

- Sagstuen, E., Theisen, H. and Henriksen, T. 1983. Dosimetry by ESR spectroscopy following a radiation accident. *Health Phys.* 45, 961–968.

- Sholom, S., DeWitta, R., Simon, S.L., Bouville, A. and McKeever, S. W. S. 2011. Emergency optically stimulated luminescence dosimetry using different materials. *Radiation Measurements.* 46(12), 1866–1869.

- Sholom, S. and McKeever, S. W. S. 2014. Emergency OSL dosimetry with commonplace materials. *Radiation Measurements.* 61, 33–51.

- Snyder, W. S., Ford, M. R., Warner, G. G. and Watson, S. B. 1975. *S, Absorbed Dose Per Unit Cumulated Activity for Selected Radionuclides and Organs.* MIRD Pamphlet No. 11. Society of Nuclear Medicine, New York, USA.

- Spooner, N. A., Smith, B. W., Creighton, D. F., Questiaux, D. and Hunter, P. G. 2012. Luminescence from NaCl for application to retrospective dosimetry. *Radiation Measurements.* 47(9), 883–889.

- Swedish Radiation Safety Authority, Danish Emergency Management Agency, Danish Health Authority, Norwegian Radiation Protection Authority, Icelandic Radiation Safety Authority, Finnish Radiation and Nuclear Safety Authority. 2014. *Protective Measures in Early and Intermediate Phases of a Nuclear or Radiological Emergency. Nordic Guidelines and Recommendations.* http://www.stralsakerhetsmyndigheten.se/ Global/Pressmeddelanden/2014/Nordic%20Flagbook%20February%202014.pdf. Accessed 2016-03-14.

- The Swedish Radiation Emergency Medicine Centre. 2012. Website: https://sremc.files.wordpress.com/2012/12/rapport-kcrn-r-olycka-gc3a4llivare-nov2010-ls.pdf. [In Swedish]. Accessed 2016-03-14.

- Thomsen, K. J, Bøtter-Jensen, L., Murray, A. S. and Solongo, S. 2002. Retrospective dosimetry using unheated quartz: A feasibility study. *Radiation Protection Dosimetry.* 101(1–4), 345–348.

- Trompier, F., Bassinet, C. and Della Monaca, S. 2011. Overview of physical and biophysical techniques for accident dosimetry. *Radiation Protection Dosimetry.* 144(1–4), 571–574.

- UNSCEAR, United Nations Scientific Committee on the Effects of Atomic Radiation. 2000. *Sources and Effects of Ionizing Radiation, Vol.II: Effects, Annex J.* New York: United Nations.

- UNSCEAR, United Nations Scientific Committee on the Effects of Atomic Radiation. 2008. *Sources and Effects of Ionizing Radiation, Vol.I: Sources of Ionizing Radiation, Annex B.* New York: United Nations.

- Vano, E., Ohno, K., Cousins, C., Niwa, O. Boice, J. 2011. Radiation risks and radiation protection training for healthcare professionals: ICRP and the Fukushima experience. *J. Radiological Protection.* 31(3), 285–287.

- Woda, C., Fiedler, I. and Spöttl, T. 2012. On the use of OSL of chip card modules with molding for retrospective and accident dosimetry. *Radiation Measurements.* 47(11–12), 1068–1073.

- Wöhni, T. 1995 External Doses from Radioactive Fallout. Dosimetry and Levels. PhD Thesis, Trondheim University, Norway. http://www.iaea.org/inis/collection/NCLCollectionStore/_Public/28/008/28008834.pdf. Accessed 2016-03-14.

7.6   EXERCISES

7.1   A man in his sixties seeks help on a Monday morning at the A&E unit of a hospital where you hold the position of medical physicist. The man claims he is experiencing persistent irritation on his right thigh, and some additional vague symptoms, such as slight nausea and slight fatigue. He says that on Saturday he was walking his dog in the vicinity of an accident where two trucks collided. The reception staff at the A&E unit refer the patient to the temporary triage area adjacent to the ambulance hall, where several cases of potential contamination have been referred during Saturday and Sunday as the result of a major transport accident that occurred at noon on Saturday. A truck transporting radioactive substances, Class 7, Grade III, skidded on spilled oil at a roundabout and collided with a truck, with the result that the transport vehicle overturned. The impact caused a small explosion due to a leaking fuel line. The freight contained a number of packages of radioactive material and some packages were broken. There is conflicting information on the sources in the cargo, and it is still not completely clear whether some additional radioactive materials had been unofficially included in the load which, according to the driver, had not been properly documented. The maximum dose rate at the site of the accident was measured and found to be about 1 mSv h^{-1}, before the rescue team together with the radiological assessor managed to locate and take care of the sources. The drivers of the two vehicles involved were decontaminated at the emergency unit on Saturday evening. There were also numerous witnesses and other passers-by who tried to help the injured, who have also been monitored. The man now presenting at the emergency unit exhibits no external contamination. All you know is that for a short time he was present at the accident site, and that it is possible that not all the radioactive sources being carried by the affected vehicle have been retrieved yet. Draw up a questionnaire together with the medical team, based on plausible scenarios that could explain the man's symptoms. Suggest further investigations of the man in question if you think he may have been externally exposed.

7.2 At 9 a.m. a man in his 30s and his girlfriend entered the A&E unit at the hospital where you work as a medical physicist. He is complaining of nausea, pain in his airways and general fatigue. He is instructed to take a queue ticket and wait his turn. After about 30 minutes he vomits and is admitted to the emergency unit where he is cared for in one of the booths by a nurse. The emergency unit is busy and it takes another 30 minutes before a physician has time to attend to him. The discussion between the patient and the doctor reveal nothing that can explain his condition. The emergency doctor in charge of the emergency unit decides that the patient should have a CT scan of the thorax and abdomen at the hospital's x-ray department. Blood samples are collected and sent to the hospital's laboratory for analysis, including a blood count, CRP level, liver status and blood gas/electrolytes. His blood pressure, heart rate and ECG are checked. The patient vomits again, and a nurse is called to help the patient and clean up the vomit on the patient's clothes and bed.

The CT examination of the man is performed at 11 a.m. Preliminary examination shows nothing pathological. Initial results from the blood tests indicate relatively normal blood values, possibly with elevated CRP. The doctor at the emergency unit informs the patient that they have not found anything wrong, but that he should remain in the hospital until 3 p.m. for observation. If his condition does not deteriorate, he can return home. His girlfriend is not satisfied with this, and asks her boyfriend if he might have ingested something when he was working with chemicals at home in the basement. The man becomes visibly agitated and does not want to discuss this. The girlfriend becomes very upset, and a counsellor is called in. During the discussion between the counsellor and the girlfriend, it eventually emerges that the boyfriend has for some time talked about experimenting with radiation. He has ordered "gadgets" on the web and has told his girlfriend that he would "cook up something to show those local bureaucrats". The counsellor, who, incidentally, is the in charge of CBRNE incidents for the region, contacts the medical physicist (you) and tells you about his suspicions. You bring a gamma survey meter and measure the blanket and paper sheets from the gurney on which the patient was treated. The instrument sounds and displays dose levels of at least 100 mSv h^{-1}. What roles will you and your colleagues play in both the short and long term (days), and which radiation protection resources will be required?

7.3 An explosion occurred at 6.30 a.m. in the restricted zone of a nuclear power plant, causing the rupture of a water pipe. A worker, an adult male in his thirties, who was wearing protective clothing, was injured and became unconscious. The injured man was removed from the restricted zone, was wrapped in a blanket, and transported urgently by ambulance to the regional A&E unit at the hospital. A manager at the accident site informed the ambulance staff that the injured man had recently undergone a security audit and a complete health check. The man is estimated to have been unconscious at the scene until ten minutes before he was found by colleagues. His protective suit and the respirator were partially torn by the explosion.

The gamma dose rate at the scene, measured by the staff at the facility, was approximately 5 mSv h^{-1}. There was no time to confirm that no external contamination was present on the worker. The main facts of the above information were communicated to the national Emergency Services Operator (999 or 112) about 25 minutes before the ambulance arrived. The Emergency Services Operator has had time to call you as a radiation protection expert, and you are now present at the reception of the A&E unit. The patient arrives in the ambulance at the designated area. Your task as the radiation protection expert is now to use your expertise to provide assistance and recommendations for the continued management of the patient based on the recommendations prevailing at your own hospital and in *The TMT Handbook*.

7.7 FURTHER READING

Fedchenko, V. (Ed.). 2015. *The New Nuclear Forensics. Analysis of Nuclear Materials for Security Purposes*. SIPRI, Stockholm International Peace Research Institute. Oxford University Press, Oxford, UK.

IAEA, International Atomic Energy Agency. 2014 *Advances in Nuclear Forensics: Countering the Evolving Threat of Nuclear and Other Radioactive Material out of Regulatory Control*. Summary of an International Conference Held in Vienna, Austria, 7–10 July 2014. IAEA Proceedings Series. International Atomic Energy Agency. Vienna.

IAEA, International Atomic Energy Agency. 2015 *Preparedness and Response for a Nuclear or Radiological Emergency*. IAEA Safety Standards Series No. GSR Part 7. International Atomic Energy Agency. Vienna.

Index